ISBN 978-1-5281-1259-8
PIBN 10909169

TWELFTH ANNUAL REPORT

OF THE

COMMISSIONERS

OF THE

State Reservation at Niagara

FOR THE FISCAL YEAR,

From October 1, 1894, to September 30, 1895.

TRANSMITTED TO THE LEGISLATURE JANUARY 28, 1896.

WYNKOOP HALLENBECK CRAWFORD CO.,
STATE PRINTERS,
ALBANY AND NEW YORK.
1896.

COMMISSIONERS.

ANDREW H. GREEN, *President.*

JOHN M. BOWERS.
WILLIAM HAMILTON.
ROBERT L. FRYER.
GEORGE RAINES.

TREASURER AND SECRETARY,
HENRY E. GREGORY.

SUPERINTENDENT,
THOMAS V. WELCH.

STATE OF NEW YORK.

No. 40.

IN ASSEMBLY,

JANUARY 28, 1896.

TWELFTH ANNUAL REPORT

OF THE

Commissioners of the State Reservation at Niagara for the Year 1895.

To the Honorable the Speaker of the Assembly:

SIR.—I herewith transmit for presentation to the Legislature the twelfth annual report of the Commissioners of the State Reservation at Niagara, for the fiscal year ended September 30, 1895.

Respectfully yours,

ANDW. H. GREEN,
President.

REPORT.

To the Honorable the Legislature of the State of New York:

The Commissioners of the State Reservation at Niagara, as required by law, submit their report for the fiscal year begun October 1, 1894, and ended September 30, 1895, being their twelfth annnal report.

The tenth year of the existence of the Reservation ended July 15, 1895. The changes that have taken place during this decade of years have been indicated in the annual reports. A recital of them conveys but an inadequate impression of what has been accomplished. One can only realize what has been done towards restoring the scenery of the Falls to a state resembling that in which it was before commerce and civilization began to utilize the waters of the Niagara river, by a visit to the Falls. No one who remembers the condition of the shore and islands of the Niagara in the vicinity of the Falls during the summer of 1885 could fail to be gratified by their appearance in the summer of 1895.

The commissioners are not inclined to overestimate what has been done by themselves or their predecessors. Indeed, they are quite willing to admit that much more might have been done and that the work of restoration might have been much further advanced had the appropriations been adequate. When it is recollected that only since 1887 have they had an annual appropriation for maintenance, and that not until 1889 was an appropriation for special improvements granted to them, and when the limited amount of these allowances is considered, the results are perhaps not unsatisfactory.

It probably did not occur to those who interested themselves in the work preliminary to the establishment of the Reservation, the work of arousing public sentiment and securing legislative approval of the project, that the State, which in 1885 was willing to expend a large sum of money to make Niagara free, would so soon tolerate, authorize and legalize schemes that could not but be injurious to the Reservation and in antagonism to the purpose of the State in establishing it.

By the act of 1885 the lands of the Reservation were made free to all mankind forever. Or, in the language of the statute, "They shall forever be reserved by the State for the purpose of restoring the scenery of the Falls of Niagara to and preserving it in its natural condition; they shall forever be kept open and free of access to all mankind without fee, charge or expense to any person for entering upon or passing to or over any part thereof." The State assumed a trusteeship over them. Henceforth they were to be regarded as in a measure sacred. No new structures, save those necessary for the comfort and convenience of the people, no money-making schemes, no speculative enterprises, no tramways, no advertisements were to be permitted within the Reservation. The lands of the Reservation were to be restored to their natural condition And although corporations and individuals have sought to secure a foothold within the Reservation for private pecuniary profit, they have thus far always been defeated.

The Legislature might, by enacting laws allowing private interests to obtain rights within the Reservation, nullify the act of 1885. The commissioners are convinced, however, that should such a course be pursued, or attempted, the opposition would be so strenuous as to prevent its eventual success. It is not unreasonable to expect that the Legislature will co-operate with the commissioners, and sustain them in their efforts to protect the Reservation from encroachment and injury.

A bill was presented to the Legislature last winter, the purpose of which was to authorize a company to erect a tower on the American side of the river and from this tower to stretch cables across the gorge to the Queen Victoria park, and along these cables to run cars to enable passengers to get views of the Falls and the river. Such a proposition seems too preposterous to merit serious consideration; nevertheless, it may be brought forward again, and unless stoutly opposed, may yet become an accomplished fact.

Another scheme last winter was to permit a tramway company to run its tracks into the Reservation. A bill was introduced in the Senate entitled, "An act to confirm the right of way of the Niagara Falls and Suspension Bridge Railway Company."

Section 2 of the bill reads as follows: "The rights of said railway company were not cut off or extinguished by the proceedings of the Commissioners of the State Reservation at Niagara Falls under and by virtue of chapter three hundred and thirty-six of the laws of eighteen hundred and eighty-three and any act or acts amendatory thereof, as no compensation was awarded to said company in said proceedings."

The position of the commissioners, with reference to this contention, is, that by the proceedings taken by the commissioners to acquire the property now included within the Reservation, the rights of every person and corporation to or in land of the Reservation, were completely extinguished. The proceedings were entirely legal and regular. Due notice was given to all persons interested. If any wrong was done to any property-owner or person, or corporation claiming rights to or in the property, there was ample opportunity for appeal to the courts, and for substantial redress.

It is hardly necessary to repeat that the commissioners are strenuously opposed to all schemes that are in any way intended to permit private persons or corporations to obtain rights within, or to encroach upon the Reservation. That such measures should be presented to the

Legislature with any prospect of favorable consideration is unaccountable, especially in view of the expressed opinions of the authorities distinctly charged with the protection and preservation of this property of the State.

It would seem to be the duty of the commissioners to take, if possible, a more determined attitude than ever in opposition to all measures of this sort. With reference to diversions of the water of the river, they maintain that, as the boundary of the Reservation extends to the line that divides the United States from the Dominion of Canada, it is their duty to protest against the drawing off of the water into canals and tunnels, as directly tending to work injury to the property of the State and its sublimest spectacle.

The method by which it is possible to inflict the greatest injury upon the Reservation and to diminish the value of the State's property at Niagara has been referred to in previous reports; and that is, to grant to private corporations the right to divert the waters of the Niagara river above the Falls, by drawing off sufficient quantities of which, it will not be very long before the volume going over the American fall will be noticeably diminished. If the Falls cease to be attractive by reason of the diminution of the volume of water; if, instead of being the most stupendous of all the world's cataracts, they become reduced and disappointing, it can readily be seen that the usefulness and attractiveness of the Reservation will become seriously impaired.

It need hardly be repeated that the American fall will suffer much more from diversions on the American side than the Horseshoe or Canadian fall.

Eight corporations claim to have already secured the right to divert the waters of the upper Niagara, and without contributing a dollar by way of compensation.

That these grants are indefensible and antagonistic to the public interest, would seem to require no demonstration. And it is not

The Whirlpool

unreasonable to question the existence of the moral or legal right to grant to private corporations, without compensation, what belongs to the people. But what has already been done at the urgency of private speculators may be repeated, the more especially as the efforts made to secure in the new Constitution a clause prohibiting such action, were unsuccessful.

The Commissioners of the Queen Victoria Niagara Falls Park have granted to one corporation, at least, the right to divert the water on the Canadian side of the river, as well as to erect structures and dig a tunnel. But for this grant or concession they have exacted what would seem to be a somewhat insufficient and inadequate compensation. The commissioners are not informed whether such action on the part of the Ontario authorities has met with public approval on the Canadian side. But whether or not the Canadian people approve of the eventual injury to the Queen Victoria Park and the diminution of the flow at the Falls, they can at least commend their commissioners for having the sagacity to secure for the rights granted at least some remuneration.

But whether compensation is exacted or not, such diversions are likely to work injury to the Falls, and the time may come when the great spectacle, unequalled as it is and so famous not in this country only, but throughout the civilized world, will cease to be an attraction. If the State of New York and the Province of Ontario fail to interpose for the protection of the great cataract from threatened injury, the question arises whether protection can not be insured by the intervention of the governments of the two nations whose territories are bounded by these waters.

There is no doubt that the Niagara, though a tideless river, and one in which navigation is interrupted by rapids and falls, is a public and navigable stream or watercourse, a watercourse that carries the discharges of the four great lakes into the fifth. Again, the Niagara is an international boundary stream. Through it runs the imaginary boundary line that separates the United States from the Dominion of Canada. This fact and the fact that it is a public navigable stream, secure to the federal government jurisdiction over its waters, either to prevent excessive diversions on the Canadian side, or to enter into relations with the government of Great Britain, in order to effect, by treaty or convention, the preservation of the Falls of Niagara. The Niagara, on the American side of the boundary line, is subject to the control of Congress and to the admiralty jurisdiction of the United States courts under the Constitution.

The right to control the waters of a river which is a boundary between two countries, belongs to both countries. This right is incidental to sovereignty. The withdrawal of excessive quantities of water by one country would naturally give rise to complaint from the other. Should Canada or Ontario authorize or permit the diversion of a sufficient quantity of the water of the upper Niagara to inflict injury upon the opposite bank, or the riparian owners thereof, it would be incumbent upon the federal government to remonstrate and prevent it. The State of New York, it need hardly be said, can not enter into negotiations with a foreign power.

While it is true that the State of New York is the owner of the river-bed to the boundary line, it can hardly be questioned that the United States, through the executive, in the exercise of the treaty-making power appertaining to sovereignty, can enter into an agreement with Great Britain to prevent the further diversion of the water of the upper Niagara.

The cataract being so unique a spectacle, and belonging, one may say, to the human race, the nations within whose boundaries it lies may be regarded as trustees for its protection and preservation, and therefore it becomes obligatory upon them to see to it that private corporations shall not be permitted to diminish its grandeur.

If the United States and Great Britain refuse to interpose, there is nothing to prevent the State of New York and the Province of Ontario from drawing off so much of the water of the upper Niagara as to make the cataract practically disappear.

But inasmuch as natural objects of great fame and dignity are now regarded by enlightened people as entitled to governmental protection, and States as under an implied trust or obligation to preserve them for the benefit, instruction and pleasure of the people, it would seem to be the most natural and proper course for the two governments to agree that hereafter the protection required and expected shall be accorded.

Whether or not the diversions authorized, or claimed to be authorized, will have a serious effect upon the cataract, need not be here discussed. So long as the upper river remains beyond the pale of international protection, there is constant danger to the cataract.

If, to cultivated men both in this country and in England, it seemed to be reprehensible for the State of New York and the Province of Ontario to permit the banks and islands of the Niagara, adjacent to the Falls, to remain under private control, with all the concomitant and inevitable defacements, unsightlinesses and injuries to the scenery, what shall be said when they grant away to private corporations the right to divert large volumes of the water of the upper Niagara, and thus lay profane hands upon the great cataract itself?

Surely, here is an opportunity for the two governments, in the exercise of the power which appertains to them as sovereigns, to interpose, and as trustees for the peoples of both countries and for mankind, to decree, that hereafter such diversions shall cease; that the Falls of Niagara shall be preserved in all their natural grandeur, in order that men of all nations may resort thither for edification and delight henceforth and forever.

The monthly receipts from the Reservation were as follows:

DATE.	From the Inclined Railway.	From rentals.	Dividends from Cataract Bank.	Interest on balances in bank.
1894.				
October	$214 20	....	$56 69	....
November	55 65	....	....	....
December	40 35	....	....	$42 28
1895.				
January	25 65	....	....	....
February	9 05	....	....	....
March	....	....	28 34	44 38
April	32 80	....	....	....
May	203 70	....	....	....
June	467 15	....	113 38	52 01
July	1,016 30	$550 00	....	....
August	1,969 75	570 00	....	....
September	906 25	970 00	56 69	23 39
	$4,940 85	$2,090 00	$255 10	$162 06
				255 10
				2,090 00
				4,940 85
Total				$7,448 01

Monthly pay-rolls have been as follows:

DATE.	Amounts.	Supplementary.	Total.
1894.			
October	$1,477 41	....	....
November	2,222 05	....	....
December	1,145 29	....	....
1895.			
January	1,198 05	....	....
February	1,199 42	$52 50	....
March	1,179 66	....	....
April	1,438 42	....	....
May	1,498 41	1,012 37	....
June	1,863 17	437 50	....
July	1,475 16	....	....
August	1,478 05	....	....
September	1,470 16	....	....
	$17,645 25	$1,502 37	$19,147 62

Expenditures, as per abstract (maintenance), were as follows:

Abstract LXXXVII	$2,176 92
Abstract LXXXVIII	4,508 43
Abstract LXXXIX	2,055 50
Abstract XC	2,864 52
Abstract XCI	7,485 62
Abstract XCII	1,507 18
Abstract XCIII	25 00
Abstract XCIV	5,037 06
Total	$25,660 23

Improvement abstracts as follows:

Abstract X. Series D	$1,167 97	
Abstract XI. Series D	9 19	
		$1,177 16
Abstract II. Series E	$3,012 62	
Abstract III. Series E	572 54	
Abstract IV. Series E	28 44	
		3,613 60
Abstract I. Series F	$2,541 78	2,541 78
Total		$7,332 54

The receipts and earnings of the Reservation have been sent to the State Treasurer monthly, and interest on balances in the Manufacturers and Traders' Bank, Buffalo, has been remitted to the same official quarterly.

The Comptroller has advanced to the commissioners quarterly a fourth part of the $25,000 appropriated by chapter 654, Laws of 1894.

Of the treasurer's report herewith submitted, exhibiting in detail all receipts and disbursements for the fiscal year ended September 30, 1895, the following is a summary:

Balance on hand October 1, 1894		$3,990 79
RECEIPTS.		
Inclined railway	$4,940 85	
Rentals	2,090 00	
Interest	162 06	
Dividends	255 10	
		7,448 01
From the State treasury, chapter 654, Laws of 1894		25,000 00
From the State treasury, chapter 726, Laws of 1893		1,177 16
From the State treasury, chapter 358, Laws of 1894		3,613 60
From the State treasury, chapter 932, Laws of 1895		2,541 78
Total		$43,771 34

PAYMENTS.

Pay-rolls at Niagara (maintenance)	$19,147 62	
Repairs, materials, superintendent's expenses, etc.	5,166 62	
Commissioners, treasurer, traveling expenses, etc	1,345 99	
		$25,660 23
Remitted to State Treasurer		7,448 01
Improvements		1,177 16
Improvements		3,613 60
Improvements		2,541 78
Balance		3,330 56
Total		$43,771 34

Total receipts since organization of commissioners, 1883	$391,765 69
Total disbursements	388,435 13
Total	$3,330 56

The Legislature has made appropriations for maintenance, current expenses and salaries, as follows:

By chapter 336, Laws of 1883	$10,000 00
By chapter 656, Laws of 1887	20,000 00
By chapter 270, Laws of 1888	25,000 00
By chapter 569, Laws of 1889	25,000 00
By chapter 84, Laws of 1890	20,000 00
By chapter 144, Laws of 1891	20,000 00
By chapter 324, Laws of 1892	20,000 00
By chapter 414, Laws of 1893	25,000 00
By chapter 654, Laws of 1894.	25,000 00
By chapter 807, Laws of 1895	25,000 00
Total	$210,000 00

For special improvements appropriations have been made as follows:

By chapter 570, Laws of 1889	$15,000 00
By chapter 302, Laws of 1891	15,000 00
By chapter 356, Laws of 1892	15,000 00
By chapter 726, Laws of 1893	25,000 00
By chapter 358, Laws of 1894	20,000 00
By chapter 932, Laws of 1895	20,000 00
Total	$110,000 00

In compliance with statutory directions, the commissioners have remitted to the State treasury the receipts from the Reservation, as follows:

From October 1, 1887, to September 30, 1888	$9,331 55
From October 1, 1888, to September 30, 1889	7,393 77
From October 1, 1889, to September 30, 1890	7,670 29
From October 1, 1890, to September 30, 1891	9,327 67
From October 1, 1891, to September 30, 1892	9,823 03
From October 1, 1892, to September 30, 1893	10,923 85
From October 1, 1893, to September 30, 1894	9,251 43
From October 1, 1894, to September 30, 1895	7,448 01
Total	$71,169 60

The following is "an estimate of the work necessary to be done and the expenses of maintaining said Reservation for the ensuing fiscal year," ending September 30, 1896:

CONSTRUCTION.

Grading, planting, etc	$25,000 00

MAINTENANCE.

Salaries, office and traveling expenses	$4,750 00
Reservation police	5,400 00
Mechanics and laborers	7,500 00
Materials, tools, etc	6,000 00
Miscellaneous	1,350 00
Total	$25,000 00

Estimated receipts from October 1, 1895, to September 30, 1896:

Inclined railway	$6,000 00
Cave of the Winds	1,200 00
Ferry and steamboat landing	750 00
Carriage service	100 00
Baggage-room	50 00
Interest	150 00
Total	$8,250 00

During the year the shores of the Reservation have been improved by grading, and by planting so as to fringe the banks in a natural and picturesque manner. Trees and shrubs have also been planted along the walks of Prospect Park. Native specimens have been chiefly used, and the aim has been to produce a wild, natural effect, in harmony with the surrounding scenery.

The work of grading and planting has been carried on in accordance with the advice of Samuel Parsons, Jr., landscape architect to the board.

A new shelter building on Goat Island was begun in August, but not having been completed during the fiscal year, a further account of it will be deferred until the next annual report.

A new terminal station at the foot of the Inclined Railway is in course of construction, and will be more particularly referred to in the next report.

For a detailed account of the work of the year, the annual report of the superintendent, which is herewith submitted, should be consulted.

The commissioners have not abandoned hope of some day having an elevator on Goat Island in place of the old Biddle stairs.

The bridges that connect Bath Island and Goat Island with the main land have required numerous repairs and considerable expenditures. They are at present safe for the ordinary travel. But they need frequent inspection, and may in the near future demand increased expenditures

New bridges of stone or steel, in place of the old ones, would be an improvement of permanent value. The commissioners have not yet determined upon plans on such bridges, but they are quite sure that money appropriated for such a purpose would be wisely expended, and they respectfully request a special appropriation of $25,000 towards these proposed structures.

Respectfully submitted.

ANDW. H. GREEN,

President.

JOHN M. BOWERS,

ROBERT L. FRYER,

WILLIAM HAMILTON,

GEORGE RAINES,

Commissioners of the State Reservation at Niagara.

Bowlder Bridge. Willow Island (West Side).

Report of the Superintendent.

To the Board of Commissioners of the State Reservation at Niagara:

GENTLEMEN. — I respectfully submit the annual report of the superintendent.

The work done during the year includes the construction of a shelter building on Goat Island, the commencement of the work of constructing a terminal station at the foot of the Inclined Railway, the completion of the work of renovating the two Whipple arch iron truss bridges, the construction of a boulder stone arched bridge at the head of Willow Island, the filling, grading and planting of the lawn below First street, the planting of the river bank between Prospect Park and Mill slip, the grading of the river bank between the bridge to the Islands and Mill slip, planting at the approach to Goat Island and the approaches to the Luna Island bridge, the planting of American elms along the Riverway and at the approach to the bridge to the Islands, the construction of a gravel walk on the Riverway between the northern entrance and Bridge street, and on the slope on Goat Island leading to the new Luna Island bridge. The road of Goat Island has been Macadamized, and the roads graveled on Bridge street, First street and Mill slip. The river has been deepened above the mouth of the inlet which supplies the Inclined Railway with water. Retaining walls and paved gutters have been constructed at the approaches to the Whipple arch bridges, stone walks and paved gutters constructed at Mill slip and at the intersection of Falls street and the Riverway, the buildings repaired and painted, and other incidental work performed.

Notices for sealed proposals for the erection of a shelter building on Goat Island and a terminal station at the foot of the Inclined Railway were published as directed. August seventh the contracts were awarded to the lowest bidders, to wit:

Shelter building.....	Wm. Shepard & Son.....	$5,677 00
Terminal station.....	R. D. Young.............	8,366 47

Shelter Building on Goat Island.

Work on the shelter building was commenced August fifteenth. The building has been completed, and the approaches filled and graded.

The building will afford the public a much needed and convenient shelter from rain.

It also contains lavatories, toilet rooms and drinking fountains.

A photograph of the shelter building is herewith submitted.

Suitable shelter buildings are also needed at the American Falls and at the Horsehoe Falls on Goat Island, at which points visitors congregate in large numbers At times of sudden rain storms the present shelters, at the entrance to Goat Island and at the Three Sisters Islands, are too far distant for the public convenience.

The Terminal Station.

Work on the terminal station was commenced September fifteenth. It has been much retarded by adverse winds which drench the workmen with spray from the Falls. The stone and brick work have been completed, and the rafters are in place. The building will soon be enclosed, but its completion before the coming spring is very doubtful.

Bridges.

The coping of the rustic stone arched bridge at the foot of Willow Island has been completed, and the structure suitably planted with vines and other plants.

A photograph of the rustic stone bridge is herewith submitted.

Boulder Bridge.

The wooden bridge at the head of Willow Island has been removed, and a single arch bridge constructed, composed of boulders taken from fields in the outskirts of the city. The approaches to the bridge have been filled and graded.

A photograph of the boulder bridge is herewith submitted.

Bridges to the Islands.

The work of substituting iron for wooden needle beams and stringers, in the two Whipple arch iron truss bridges, has been completed. The bridges have been painted, and retaining walls and paved gutters constructed at the ends of the bridges.

The bridge to Luna Island has been painted and the bridges to the Three Sisters Islands repaired and painted.

The bridge from Goat Island to the First Sister Island is in need of extensive repairs, and should be replaced by a new structure.

Plans for a rustic stone arched bridge in that locality have already been submitted.

Grading and Planting.

The lawn bounded by Buffalo avenue, First street, Mill slip and the Riverway, has been filled, graded and planted. Three thousand loads of filling, mainly from excavations made for paving city streets, were delivered without expense to the State.

April twenty-ninth, Samuel Parsons, Jr., made an examination of the grounds to be graded and planted. The river shore between Prospect Park and Mill slip has been restored, graded and planted. Loam for filling was carted to the locality from various parts of the city. Rustic stones for restoring the shore to a natural appearance were brought from Goat Island and from fields in the suburbs of the city, from which sod for covering was also carted. Considerable labor was required to conceal the outlet of the Cataract House conduit.

American elms have been planted along the Riverway and at the approach to the bridge to the islands.

Native shrubs have been planted in Prospect Park and in the territory lately graded between Prospect Park and Mill slip. The shrubs were selected by Theodore Wirth, landscape gardener, after consultation with Samuel Parsons, Jr., and were planted by Mr. Wirth under the directions of Mr. Parsons.

The rustic stone bridge and the boulder bridge have been suitably planted. The approaches to the Luna Island bridge have been planted and the banks at the approach to Goat Island.

Roads.

The road around Goat Island has been Macadamized with 1,275 yards of hand-broken stone and 750 yards of gravel. A covering of gravel has been placed on Bridge street.

The paving of Main street and the filling of the lawn necessitated the raising of the roadways about ten inches at that point, in Mill slip and in the Riverway. This was done with gravel obtained from excavations made during the winter, without expense to the State.

A carriage turn-out has been constructed in the road on Goat Island, adjacent to the new shelter building.

The road in Prospect Park has been widened and repaired, and materials obtained for further repairs of the road in Prospect Park, on the Riverway above First street, and on Bath Island, in the coming spring as soon as the weather will permit.

Walks.

The plank walks which remained on the west side of the Riverway, below Bridge street, have been removed and gravel walks ten feet wide constructed. A gravel walk, according to the plan adopted, has been constructed between Bridge street and Mill slip in the territory lately graded. A stone walk, curb and gutter have been constructed in Mill slip; stone gutters constructed at the intersection of Falls street and the Riverway, and the stone crosswalks at that point relaid. In the Riverway, along the new gravel walks, a margin eight feet wide has been filled and sodded, and the old picket fence at Prospect Park removed.

The removal of the old picket fence at Prospect Park has greatly improved the appearance of the Reservation. The grounds in the locality are very attractive and are more largely frequented than any other portion. The road at that point, leading from the islands to the upper Suspension bridge, is the most frequented driveway in the city, and should be kept free from street railroads, which would destroy its beauty and excellence as a driveway.

Inclined Railway.

The low water in the river threatened to cut off the supply for the operation of the Inclined Railway. To obviate this, the inlet pier would have to be again extended, or several ledges of rock in the bed of the river, above the mouth of the inlet, removed. As the inlet pier is already a disfigurement in the locality, the latter course has been adopted. The ledges of rock have been blasted out and a sufficient supply of water obtained, although all of the rock which has been blasted has not yet been removed.

The machinery and water gate, in the Inclined Railway building, have been repaired, and a new cable purchased for the railway.

Pending the construction of the new terminal station, temporary waiting rooms were constructed at the bottom of the Inclined Railway.

During the late severe winter the accumulations of the ice at the foot of the Inclined Railway and on the railway structure were so great that at times it was not considered prudent to allow visitors to descend. The operation of the Inclined Railway was entirely suspended from February twenty-sixth to April twelfth.

The lack of proper accommodations at the foot of the railway and the building operations in progress have also tended to lessen the receipts from the Inclined Railway for the current year.

The seemingly dangerous condition of the cliff near the Inclined Railway structure has been brought to the notice of the President of the board.

Electric Lighting.

The removal of the fence on the Riverway has caused Prospect Park to be frequented in the evening by visitors and citizens much more than formerly. Difficulty is experienced in walking or driving in the grounds after sundown, and the necessity for lighting the grounds is constantly increasing. A proposal was received from Charles N. Owen and S. M. Brooks to light Prospect Park free of charge, and a similar proposal from the Buffalo and Niagara Falls Electric Light and Power Company. August ninth the proposal of the Buffalo and Niagara Falls Electric Light and Power Company was accepted by the executive committee. A form of license which had been submitted to the President, and amended by the Attorney-General of the State, was handed to the manager of the electric light company. A number of poles and a quantity of wire have been placed in Prospect Park, but the license has never been executed and nothing further has been done by the electric light company.

Legislative Investigation.

The sub-committee of the Legislative investigating committee visited the Reservation March second, and made an investigation into the management of the Reservation affairs. The chairman of the executive committee and the superintendent were in attendance. The books and papers of the Reservation and all possible information were placed at the disposal of the committee. A tour of the grounds and an inspection of the buildings and bridges were made, and the committee informed of the work done since the establishment of the Reservation, and the needs and requirements for its future maintenance and improvement.

The Shelter Building. Goat Island.

The Street Railroad Bill.

April twenty-fourth, by direction of the President, the superintendent went to Albany and appeared before the railroad committees of the Senate and Assembly in opposition to the passage of a bill, entitled "An act to confirm the right of way of the Niagara Falls and Suspension Bridge Railway Company," and intended to permit the operation of a street railroad in a portion of the Riverway which is more largely used as a driveway than any other part of the Reservation. The superintendent also brought the matter to the attention of the Governor and the Attorney-General of the State.

The bill was not reported from the committee in either branch of the Legislature.

A copy of the proposed law is herewith submitted.

Water Supply.

Correspondence between the superintendent and the Niagara Falls Water Works Company, concerning the account for water furnished for the use of the Reservation, is herewith submitted. The account has been paid to June 1, 1893. Since that date no account has been rendered, although it has been requested by the superintendent each year.

The mains for water service in Prospect Park and on Goat Island are surface pipes, and the supply is shut off during the winter season. The need of a general plan for an adequate water service for the Reservation grounds and buildings, is now apparent.

An Elevator Wanted at the Cave of the Winds.

The waiting rooms and winding staircase at the Cave of the Winds have received the usual repairs during the year. The dressing rooms are inadequate and unsuitable. Better accommodations should be provided for the traveling public.

The Biddle staircase has been examined and pronounced safe, but it is old and dilapidated in appearance, and affords but a fatiguing method of obtaining the unrivalled view of the Falls from below. It should be replaced by a commodious elevator, with a free stairway attached. Such an elevator, operated for a fee of five cents up or down, would furnish an estimated revenue of $5,000 a year to the State, which with the present income, mainly derived from the Inclined Railway, would in time render the Reservation self-sustaining.

Employes.

The regular force employed consists of ten men, to wit: One superintendent, one clerk, six police gatemen and caretakers and two Inclined Railway men.

The following is a statement of the number of foremen, teamsters and laborers employed on maintenance and on works of improvement during the months of the year ending September 30, 1895:

PAY-ROLLS.	Maintenance.			Chap. 358, Laws 1894.			Supplementary.			Chap. 938, Laws 1895.			
	Foreman.	Laborers.	Teamsters.	Foreman.	Laborers.	Teamsters.	Gardener.	Laborers.	Teamsters.	Gardener.	Foreman.	Laborers.	Teamsters.
1894.													
October	..	16	..	1	6	3	..	..	..	..	..	..	....
November	1	24	8	..	..	..	..	..	..	..	..	..	....
December	1	5	1	..	..	..	..	..	..	..	..	..	....
1895.													
January	1	6	1	..	..	..	..	..	..	..	..	..	....
February	1	7	1	..	..	..	..	..	..	..	..	..	....
March	1	7	1	..	..	..	..	..	..	..	..	..	....
April	1	18	2	..	..	..	..	..	..	..	..	..	....
May	1	14	1	..	..	..	1	22	2	..	..	..	....
June	1	22	2	..	..	..	1	14	1	..	..	..	....
July	..	12	1	..	..	..	..	..	..	..	1	14	5
August	1	13	..	..	..	..	..	..	..	..	1	12	5
September	1	14	..	..	..	..	..	..	..	1	1	11	3

Exclusive Livery Privileges.

The legality of the exclusive livery privileges granted by the railroad companies has not yet been decided by the Court of Appeals. The manager of the livery business has continued to also manage the elevator bazaar and photograph business at Table Rock on the Canadian side, to which passengers, obtained by solicitation on the railroad trains on the American side, are systematically and expeditiously conducted. Visitors so intercepted on the railroad trains, are not aware that the elevator, bazaar and photograph business are managed in combination with the livery privileges. The result of this system is to deceive the traveling public, to destroy fair competition in business and to create and uphold a monopoly of the most vicious character.

Reservation Carriage Service.

The Reservation Carriage Service has been operated very successfully during the year. The traveling public are now generally aware of the existence of such a service, and are glad to give it their patronage. Eight carriages have been in operation during the year. In view of the privileges enjoyed by the carriage service, it is important that it be kept free from combination with liveries, hotels, stores, toll places, railways and so-called excursion companies.

Some of the carriages in use are heavy and cumbersome, and can with advantage be replaced by lighter vehicles, which would render the service more expeditious and efficient.

Licensed Carriage Drivers.

Ten complaints have been made by visitors against licensed carriage drivers. Two drivers have been excluded from the Reservation for violation of the ordinances of the commissioners. Public carriage stands have been permitted on the Riverway, and two carriages at one time have been allowed to stand on Bridge street, and two on Falls street; the privilege is subject to revocation upon cause of complaint.

The carriage stand on the Riverway, between Falls street and Niagara street, have been the subject of complaint on the part of property owners on the opposite side of the Riverway. A carriage stand at that point is in many ways objectionable, and it may be well to abolish it, or at least to limit the number of carriages which shall occupy it at one time.

Number of Visitors.

The number of visitors during the year was below the average, and is estimated at 500,000. The excursion travel was slightly below that of last year, there being 4,571 railroad cars bringing an estimated number of 274,260 persons.

The construction of the electric railroad on the Canadian side for a time materially reduced the excursion travel to the State Reservation. The Gorge Electric Railroad, just opened on the American side, may largely restore the former condition of affairs. It is also expected that the electric railroad from Buffalo, which brings its passengers to the entrance of the Reservation, will largely increase the number of visitors.

The Cave of the Winds.

During the year visitors have been conducted through the Cave of the Winds by George W. Wright, lessee. In the cave, additional safe guards for visitors have been provided. About the average number of tourists visited the cave. As the locality is between the American and the Horseshoe Falls, the view is unparalleled; but the winding stairway leading to it is so unsuitable and fatiguing, that very few visitors are able to enjoy the beauty of the scenery below the high bank. A walk can easily be constructed along the edge of the water between the American and the Horseshoe Falls, which, if a suitable elevator was provided, would undoubtedly be a point of great resort and of surpassing interest to visitors.

Steamboat Landing.

The steamboat landing has been leased and kept in repair by the "Maid of the Mist" Steamboat Company during the year. Only one boat has been in operation. The second boat has not been used, because of the moderate volume of travel and the low water prevailing at the dock.

Receipts.

The receipts of the superintendent from all sources during the fiscal year have been $7,285.95, to wit: Inclined Railway, $4,940.85; rentals, $2,090; dividends from the receiver of the Cataract bank, $255.10.

The stoppage of the Inclined Railway from February twenty-sixth to April twelfth, because of the dangerous accumulation of ice and the work of building the new terminal station, have tended to lessen the receipts from the Inclined Railway during the past year. The new terminal station, when completed, will give better accommodations than were ever before afforded. The electric railroad from Buffalo and Lewiston will increase the volume of travel, and it is probable that the receipts for the coming year will be greater than those of any year since the establishment of the Reservation.

Tabular Statement.

Detailed statements of the receipts and expenditures by the superintendent, the amount of the pay-rolls for each month and the classification of the pay-rolls and accounts are herewith appended.

Respectfully submitted,

THOMAS V. WELCH,

Superintendent.

Excursions for 1894-1895.

Date.	WHERE FROM.	Number of cars.	Estimated number of visitors.
1894.			
Oct. 2	Allentown, Pa., via Lehigh Valley R. R.	8	480
	Newton's excursion, via W. N. Y. & Pa. Ry.	8	480
7	Rochester, N. Y., via West Shore R. R.	7	420
	Buffalo, Lockport and vicinity, via N. Y. C. R. R.	13	780
	Rochester, via Erie Ry.	7	420
8	Buffalo, N. Y., Orpheus singing society	6	360
9	Reading, Pa., Hose company No. 1	2	120
19	New York State Superintendents of Instruction.	3	180
21	Buffalo, Lockport and vicinity	15	900
28	Boston, Mass., special party	2	120
	Buffalo, Rochester and Lockport	12	720
			4,980
1895.			
May 3	Lockport, N. Y., Union schools, Arbor Day	30	1,800
	Boston, Mass., Raymond party	9	540
5	Buffalo, N. Y., via Erie Ry.	3	180
8	Boston, Mass., Raymond party	8	480
12	Jamestown, N. Y., via W. N. Y. & Pa. Ry.	4	240
19	Prescott's excursion, via Erie Ry.	4	240
	Buffalo, N. Y., via N. Y. C. R. R.	6	360
	Rochester and Batavia, via Erie Ry.	2	120
20	New York city, excursion retail coal dealers	3	180
22	Buffalo, N. Y., B. P. O. of Elks	4	240
	Reunion Twenty-eighth regiment	2	120
24	Excursion U. S. army surgeons	4	240
	Queenston, Ont., Canadian encampment	4	240
25	Toronto, Ont.	3	180
	Buffalo, N. Y., Railroad passenger agents	2	120
26	Buffalo, N. Y., via N. Y. C., Erie & West Shore R. R.	10	600
30	Decoration Day excursion, via N. Y. C. & M. C. and Lehigh Valley R. R.	40	2,400
	Rochester, N. Y., via Erie R. R.	6	360
	Jamestown, via Erie R. R.	8	480
		152	9,120
June 2	Buffalo, N. Y., via Erie Ry.	2	120
	Lehigh Valley, Writer's club, Lehigh Valley R. R.	5	300
	Rochester, N. Y., via West Shore R. R.	6	360
	Buffalo, N. Y., via N. Y. C. R. R.	20	1,200
	Lockport, via N. Y. C. & Erie	8	480
	Iroquois Co., Ill.	7	420
5	Buffalo, N. Y., company soldiers, Fort Niagara.	1	60
7	Jamestown, N. Y., Swedish-Lutheran society	4	240
8	Rochester and Monroe Co. Union schools	10	600
9	Binghamton and Carbondale, via Erie Ry.	11	660
	Buffalo, Rochester and Lockport	25	1,500
11	York Co., Ont., Council	2	120
13	Lindsay, Ont., Odd Fellows	4	240
14	Boston, Mass.	4	240
	New York city bowling club	1	60

EXCURSIONS — (*Continued*).

Date.	WHERE FROM.	Number of cars.	Estimated number of visitors.
1895.			
June 14	Toronto, Ont., Bookbinders	4	240
16	Buffalo, N. Y., via Erie Ry	2	120
	Buffalo, Rochester and Lockport, via N. Y. C. R. R	30	1,800
17	Orangeville, Ont., I. O. of Foresters	4	240
	Orangeville, Ont., I. O. of Foresters (by boat)	3	180
19	Buffalo, N. Y., via steamer	1	60
	Toronto, Ont., via G. W. Ry	6	360
	Money-Penny, Hammond Co., Wholesale grocers	5	300
	Toronto, Ont., Board of Trade	2	120
20	Acton and Georgetown, Ont., Union society	8	480
	Toronto, Ont., Agricultural society	7	420
22	Toronto, Ont., Typemakers	5	300
	Buffalo, N. Y., Zither club	1	60
	Buffalo, N. Y., Union schools	10	600
	Cleveland, O., Republican national league	4	240
	Canada, employes G. T. Ry	12	720
23	Rochester, N. Y., via Erie Ry	5	300
	Buffalo, N. Y., via Steamer Columbia	5	300
	Buffalo, N. Y., Italian Bersaglen society	4	240
	Buffalo, Rochester and Lockport, via N. Y. C. R. R	40	2,400
24	Toronto, Ont., Sons of England	6	360
	Buffalo, N. Y., Public schools	2	120
	California, Phillip's excursion	8	480
	Buffalo, N. Y., via Steamer Columbia	3	180
25	Lockport, N. Y., Y. M. C. association	4	240
	Cleveland, O., via boat and rail	6	360
	American institute of electrical engineers	6	360
	Buffalo, N. Y., via Steamer Columbia	5	300
	Toronto, Ont., via Steamer Chippewa	7	420
	Elmira, N. Y., via Erie Ry	7	420
26	Buffalo, N. Y., via Steamer Pilgrim	4	240
	Rochester, N. Y., Royal Arcanum	6	360
	Carbondale, Pa., via Erie Ry	4	240
	Buffalo, N. Y., via Steamer Pilgrim	1	60
27	Clyde and Lyons, N. Y., via West Shore Ry	9	540
	Western New York Women's Auxiliary of Diocese	5	300
	Carbondale, Pa., via Erie Ry	6	360
28	Stratford, Ont., Civic holiday	10	600
	Hamburg, Ont., Civic holiday	4	240
29	Suspension Bridge, Episcopal S. S	2	120
30	Buffalo, N. Y., via Erie Ry	2	120
	North College, Glee club	1	60
	New Castile and Titusville, via B. N. Y. & Pa. Ry	9	540
	Buffalo, Rochester and Lockport	30	1,800
		415	24,900
July 1	Dominion day, Canada excursion	30	1,800
	Woodstock, Ont., via G. T. Ry	8	480
	Hamilton, Ont., via G. T. Ry	15	900
	Toronto, Ont., via Steamers Chippewa and Chicora	12	720

EXCURSIONS — (*Continued*).

Date.	WHERE FROM.	Number of cars.	Estimated number of visitors.
1895.			
July 3	Buffalo, N. Y., via Steamer Pilgrim	6	360
	Grimsly, Ont., via G. T. Ry	5	300
	Buffalo, N. Y., via Steamer Pilgrim	1	60
	Buffalo, N. Y., Carpenters' art sketch party	1	60
4	Boston, Mass., Raymond excursion	2	120
	Buffalo, Rochester and Lockport	200	12,000
	Carbondale, Pa	10	600
	Jamestown, N. Y	9	540
	Hornellsville, N. Y	8	480
6	Buffalo, N. Y., via Steamer Pilgrim	10	600
	Boston, Mass., Raymond party	14	840
7	Boston, Mass., Christian Endeavor society	22	1,320
	Lehigh Valley, via Lehigh Valley Ry	18	1,080
	Boston, Mass., Christian Endeavor society	20	1,200
	Buffalo, N. Y., via Steamer Pilgrim	8	480
8	Buffalo, N. Y., via Erie Ry	3	180
	Toronto, Ont., Saint Paul de Vincent society	8	480
9	Boston, Mass., Christian Endeavor society	10	600
	Buffalo, N. Y., via Steamer Pilgrim	3	180
	Buffalo, N. Y., via Erie Ry	2	120
	Boston, Mass., Christian Endeavor society	45	2,700
	Hornellsville, N. Y., Ladies' Unity lodge, locomotive firemen	2	120
10	Buffalo, N. Y., Paper bag manufacturers	4	240
	Buffalo, N. Y., via Steamer Pilgrim	8	480
	Toronto, Ont., Elm M. E. church	4	240
	Buffalo, N. Y., M. E. church S. S	6	360
	Rochester, N. Y., via West Shore R. R	7	420
	Toronto, Ont., via Steamer Cibola	2	120
	Akron, N. Y., via N. Y. C R. R	1	60
11	Erie, Pa., First M. E. church society	9	540
	Buffalo, N. Y., Seneca Street M. E. church society	4	240
13	Toronto, Ont., Colin M F. Co	4	240
	Buffalo, N. Y., via Steamer Pilgrim	1	60
14	Jamestown and Ridgeway	6	360
	Buffalo, N. Y., via Steamer Pilgrim	9	540
	Jamestown, N. Y., Gracius singing society	10	600
	Erie Ry., along the line	15	900
	Western N. Y. & Pa. Ry., along the line	10	600
	Harrington party (special)	1	60
	Buffalo, Rochester and Lockport	65	3,900
16	Buffalo, N. Y., via Steamer Pilgrim	4	240
	Toronto, Ont., via Steamer Chippewa	5	300
17	Buffalo, N. Y., via Steamer Pilgrim	19	1,140
	Toronto, Ont., via Steamer Chippewa	10	600
	Hamilton, Ont., G. T. Ry	6	360
	Toronto, Ont., Commercial travelers	5	300
	Dunkirk, N, Y., via Allegheny Ry	6	360
18	Buffalo, N. Y., opening Gorge Ry	4	240
	Pennsylvania State editorial association	6	360
	Detroit, Mich., Press club	15	900
19	Buffalo. N. Y., National association railroad baggage agents	5	300
	C. H. & D. Ry., along the line (five secs.)	45	2,700
	Buffalo, N. Y., via Steamer Pilgrim	6	360
	Toronto, Ont., King Street Trinity church	8	480
20	Chautauqua, Newton's excursion	10	600

EXCURSIONS — (Continued).

Date.		WHERE FROM.	Number of cars.	Estimated number of visitors.
1890				
July	20	Erie, Pa., C. M. B. A. society	8	480
		Buffalo, N. Y., Half Holiday excursion	10	600
		Buffalo, N. Y., Half Holiday excursion, via Steamer Pilgrim	14	840
		Toronto, Ont., via Steamer Chippewa	14	840
	21	Buffalo, Rochester and Lockport	65	3,900
		Buffalo, via Erie Ry	6	360
		Buffalo, via Steamer Pilgrim	11	660
	22	Toronto, Ont., A. O. U. W. society	4	240
	23	Buffalo, N. Y., via B. R. & Pa. Ry	7	420
		Buffalo, N. Y., Baptist church society	8	480
	24	Chautauqua	6	360
		Buffalo, N. Y., Spring Street M. E. church	4	240
	25	Baltimore, Md., via Lehigh Valley Ry	8	480
	27	Chautauqua	4	240
	28	Buffalo, N. Y., via Steamer Pilgrim	6	360
		Rochester	4	240
		Buffalo, Rochester and Lockport	20	1,200
		Buffalo, via B. R. & Pitts. Ry	10	600
		Lehigh Valley, special party	10	600
		Cleveland, O., via steamer	10	600
	29	Buffalo, N. Y., via Steamer Pilgrim	10	600
	30	Buffalo, N. Y., St. Vincent's Orphan Asylum	3	180
		Buffalo, N. Y., via Steamer Pilgrim	9	540
		Toronto, Ont., via Steamer Chippewa	2	120
	31	Batavia, N. Y., via old road	6	360
		New Bethlehem, Pa	6	360
		Lockport, N. Y., German Lutheran church	6	360
		Erie Ry., from C., H., V. & T. Ry	8	480
		Elmira, N. Y., and Carbondale, Pa	9	540
		Buffalo, N. Y., St. Paul's Epis. church society	8	480
			1,088	65,280
Aug.	1	Toronto, Ont., All Saints' church, via Steamer Chippewa	9	540
		Tonawanda, N. Y., German church society	10	600
	2	C., C., C. & St. L. Ry., along the line, five [illegible]	35	2,100
		Buffalo, N. Y., via Steamer Pilgrim	7	420
	3	Erie, Pa., employes [illegible] works	7	420
		N. Y. [illegible] & O. Ry. [illegible] excursion	22	1,320
		[illegible], Ohio, employes wholesale [illegible] firms	5	300
		Buffalo, N. Y., via Steamer Pilgrim	5	300
	4	Buffalo, N. Y., via Steamer Pilgrim	7	420
	5	[illegible], Ont., via Steamer [illegible]	10	600
	6	Buffalo, Rochester, Syracuse and Lockport, via West Shore Ry	45	2,700
	7	[illegible], N. Y., via W. N. Y. & Penn. Ry	7	420
		Erie, Pa., [illegible]	5	300
		[illegible]	5	300
		Buffalo, N. Y., via Steamer Pilgrim	[illegible]	[illegible]
	[illegible]	[illegible]	10	600
		[illegible], via W. N. Y. & Penn. Ry	10	600
		[illegible]	7	420
		[illegible] & [illegible]	70	4,200
		Buffalo, N. Y., via Steamer Pilgrim	2	120
		[illegible], via Erie Ry	10	600

Bowlder Bridge. Willow Island (West Side)

EXCURSIONS—(*Continued*).

Date.	WHERE FROM.	Number of cars	Estimated number of visitors.
1895.			
Aug. 8	Akron, O	18	1,080
	Youngstown, O	10	600
	Toronto, Ont., via Steamer Garden City	13	780
	Oswego, N. Y., via R. W. & O. Ry	10	600
	Pittsburg, Chenango & L. E. Ry	10	600
	Western N. Y. & Penn. Ry	10	600
	Youngstown, Pa	10	600
	Brooklyn, N. Y., Hendrickson excursion	2	120
	Nickel Plate Ry	20	1,200
9	Nickel Plate railway (ten secs.)	100	6,000
	Washington, D. C., special party	10	600
	Philadelphia, Pa., special party	10	600
	Galena & Greenville, Lebigh Valley Ry	22	1,320
10	Elmira, Ont , via G. T. Ry	6	360
	Toronto, Ont., via Steamer Empress of India	6	360
	R. W. & O. Ry., harvest excursion	15	900
	Toronto, Ont., Uniformed Catholic Knights	6	360
11	Pittsburg, Pa	13	780
	Buffalo, Rochester & Lockport, via rail and water	45	2,700
	Buffalo, via Erie Ry	7	420
	Buffalo, via Steamer Pilgrim	4	240
12	Simcoe, Ont., via Steamer Garden City	6	360
13	W. N. Y. & Penn. Ry	7	420
	Chautauqua, N. Y	5	300
	Lockport, N. Y., English Lutheran church	8	480
	Buffalo, N. Y., via Steamer Pilgrim	10	600
14	Buffalo, Rochester, Lockport and Tonawanda, N. Y., and Toronto, Ont., Ancient Order Hibernians	25	1,500
	Titusville, Pa., along L. S. & M. S. Ry	9	540
	Jamestown, N. Y	4	240
	Toronto, Ont., via G. T. Ry	9	540
	Buffalo, N. Y., via Steamer Pilgrim	9	540
15	Chicago, Ills., G. T. railway	35	2,100
	Manchester, N. Y., via Lebigh Valley Ry	12	720
	St. John's, Ont	1	60
	Buffalo, N. Y., via Steamer Pilgrim	20	1,200
	Erie railway, via along the line	20	1,200
16	Toronto, Ont., Barber, Ellis Co. employes	10	600
	Port Jervis, via G. T. Ry	8	480
	Buffalo, N. Y., via Steamer Pilgrim	11	660
	Cinn., Dayton & Hamilton Ry	65	3,900
	Berlin, Ont., 29th Batt. band	7	420
17	Hamilton, Ont., foundry empl'yes (three secs.)	35	2,100
	Bay City, Wis., via G. T. Ry	10	600
	Geneva, N. Y., via Lehigh Valley Ry	16	960
	Buffalo, N. Y., via Steamer Pilgrim	11	660
18	Buffalo, Rochester, Lockport and Syracuse	45	2,700
	Akron & Youngstown, O., via Erie Ry	12	720
	West Shore railway, along line of	9	540
19	R. A. & O. railway, along line of	12	720
	Chautauqua, N. Y	9	540
	Buffalo, N. Y., half-holiday excursion	12	720
18	Janesville, O	16	960
	Akron and Kent, O	17	1,020
	Buffalo, N. Y., telegraph operators	2	120
	Buffalo, N. Y., via Steamer Pilgrim	15	900

EXCURSIONS — (*Continued*).

Date.	WHERE FROM.	Number of cars.	Estimated number of visitors.
1895.			
Aug. 19	London, Ont., civic holiday	5	300
	St. Catherines, Ont., via G. T. Ry	4	240
	Toronto, Ont., via Steamer Chippewa	14	840
	Toronto, Ont., via Steamer Empress of India	10	600
	Seaforth, Ont., via G. T. Ry	6	360
	St. Thomas, Ont., via M. C. Ry	8	480
	Buffalo, N. Y., via Steamer Pilgrim	4	240
	London, Ont., via G. T. Ry	6	360
20	Meriden, O., & Detroit, Mich., via G. T. Ry	12	720
	South Bend, Ind., via Erie Ry	9	540
	Buffalo, N. Y., via Steamer Pilgrim	4	240
	Nickel Plate railway (two sections)	22	1,320
	New York Central railroad, over old road	30	1,800
21	Boston, Mass., Knight Templars' conv	32	1,920
	N. Y., P. & O. Ry	9	540
	Streator, Ills. (two sections)	19	1,140
	Hinsdale, O	10	600
22	Reunion 8th Heavy Artillery	6	360
	Boston, Mass., Knight Templars' conv	33	1,980
	Ottowa, Ont., Knight Templars' conv	10	600
	Springfield, O., Knight Templars' conv	10	600
	Portsmouth, O., Knight Templars' conv	7	420
23	Toronto, Ont., Queen St. M. E. church	10	600
	Buffalo, N. Y., via Steamer Pilgrim	5	300
	Columbus, O	10	600
	Chautauqua, N. Y	10	600
	Chautauqua, N. Y., via W. N. Y. & Pa. Ry	9	540
	Boston, Mass., Knight Templars' conv	20	1,200
24	Stratford, Ont., employes G. T. Ry	45	2,700
	Buffalo, N. Y., via Steamer Pilgrim	5	300
	Jackson, W. Va., Knight Templars' conv	6	360
25	Port Jervis, via Erie Ry	12	720
	Carbondale, Pa., via Erie Ry	14	840
	Corning and Wellsville, via Erie Ry	12	720
	Dayton, O, Knight Templars	5	300
	Kansas City, Mo., Knight Templars	9	540
	Batavia, N. Y., A. O. of Hibernians	12	720
	Buffalo, N. Y., via Steamer Pilgrim	5	300
	Stratford, Ont., via G. T. Ry	5	300
	Boston, Mass. (en route to), Knight Templars.	28	1,680
	Boston, Mass. (en route to), Knight Templars,	41	2,460
	Terre Haute, Ind., via Wabash Ry	19	1,140
	Erie railway, via main line	20	1,200
	Pittsburg, Pa., via P. R. & P. Ry	14	840
	N. Y. C., Mich C. & West Shore Ry	70	4,200
26	Erie, Pa., Christian Endeavor	10	600
27	B., N. Y. & Penn. R. R	10	600
	B., R. & Pitts. R. R	10	600
	B., L. & Western R. R	10	600
28	Buffalo, N. Y., via Steamer Pilgrim	9	540
	Toronto, Ont., via Steamer Chippewa	5	300
	St. Catharines, Ont., via G. T. Ry	1	60
	Buffalo, N. Y., freight agents' conv	4	240
	Cleveland, O., special party	6	360
29	Brantford, Ont., via M. C. Ry	5	300
	Carsonburg, Ont., via M. C. Ry	4	240
	Buffalo, N. Y., via Steamer Pilgrim	1	60
	Youngstown, N. Y., Pres. Ch. S. S	3	180

EXCURSIONS — (*Continued*).

Date.	WHERE FROM.	Number of cars.	Estimated number of visitors.
1895.			
Aug. 30	Buffalo, N. Y., Knight Templars	8	480
	Buffalo, N. Y., via Steamer Pilgrim	3	180
	Buffalo, N. Y., Detroit Commandery	6	360
	Brooklyn, N. Y., Eagle newspaper	2	120
31	Toronto and Waterloo, Ont	24	1,440
	Middleport, N. Y	1	60
		1,921	115,260
Sept. 1	Wilkesbarre and Scranton, Pa., via Lehigh Valley Ry	10	600
	Buffalo, Rochester, Lockport, Syracuse and Dunkirk, N. Y., and Cleveland, O	60	3,600
	Detroit, Mich., via G. T. Ry	11	660
	Rochester, N. Y., via Erie Ry	12	720
	Buffalo, N. Y., via Steamer Pilgrim	7	420
2	Labor Day excursion, via Lehigh Valley Ry	11	660
	Labor Day excursion, N. Y. C. Ry	20	1,200
	Rochester & Syracuse, via West Shore Ry	9	540
	Oswego, via R. W. & O. Ry	10	600
	Paterson, N. J., via Erie Ry	12	720
	Hamilton, Ont., via G. T. Ry	15	900
	Buffalo, N. Y., via Steamer Pilgrim	11	660
3	Cleveland, O., via L. S. and M. S. Ry	9	540
	Jamestown, N. Y., via Erie Ry	7	420
	Buffalo, N. Y., via Steamer Pilgrim	5	300
4	Brooklyn, N. Y., German Scheutzen Corps	10	600
5	Galion, O., via Erie Ry	12	720
	Sharpsville, O., via Erie Ry	11	660
	Warren, O., via Erie Ry	10	600
	Buffalo, N. Y., via Steamer Pilgrim	5	300
	Williard, Ont	1	60
	Youngstown, N. Y	1	60
6	Philadelphia, Pa., via Lehigh Valley Ry	10	600
	Kent, O., via L. S. & M. S. Ry	10	600
	Canton and Akron, O., via Erie Ry	12	720
	Buffalo, N. Y., via Steamer Pilgrim	2	120
7	Medina, N. Y., Union Ch. & S. S	7	420
	Toronto, via G. T. Ry	7	420
	Buffalo, N. Y., via Steamer Pilgrim	1	60
8	Erie Ry., along main line	12	720
	Louisville, Ky., (en route) G. A. R	18	1,080
	Boston, Mass., (en route) Am. Ry. ticket agents	21	1,260
	Syracuse and Rochester, via West Shore	12	720
	Buffalo, Lockport, Tonawanda and Dunkirk, N. Y. and Cleveland, O., via N. Y. C. & L. S. & M. S. Ry	14	840
	Buffalo, N. Y., via Erie Ry	5	300
	Buffalo, N. Y., via Steamer Pilgrim	6	360
9	Cortland, N. Y., via Erie Ry., Knight Templars	7	420
	Meadville, Pa., via Erie Ry, Knight Templars	5	300
	Hornellsville, N. Y., via Erie Railway, Knight Templars	3	180
	Chautauqua, Newton's excursion	10	600
10	New York State conclave Knight Templars, via Lackawanna Ry	12	720
	Knight Templars, via Lehigh Valley Ry	10	600
	Knight Templars, via N. Y. C. Ry	44	2,640

EXCURSIONS – (*Concluded*).

Date.	WHERE FROM.	Number of cars	Estimated number of visitors.
1895.			
Sept. 10	Knight Templars, via West Shore Ry	20	1,200
	Knight Templars, via R. W. & O. Ry	8	480
	Knight Templars, via Western N. Y. & Pa. Ry.	12	720
	Port Jervis and Carbondale, via Erie Ry	11	660
	Hornellsville, N. Y., via Erie Ry	10	600
	Batavia, N. Y., via Erie Ry	3	180
	Almond, N. Y., via Erie Ry	7	420
	Tioga, N. Y., via Erie Ry	2	120
	Jamestown, N. Y., via Erie Ry	11	660
12	Buffalo, N. Y., retail butchers' convention	7	420
	Seneca Falls, N. Y., I. O. O. F	9	540
	Toronto, Ont., via Steamer Chippewa	8	480
	Buffalo, N. Y., via Steamer Pilgrim	2	120
13	Buffalo, N. Y., via Steamer Pilgrim	2	120
	Toronto, Ont., via Steamers Chippewa and Chicora	21	1,260
14	Toronto, Ont., via Steamer Chicora	15	900
	Toronto, Ont., via Steamer Chippewa	6	360
	Buffalo, N. Y., via Steamer Pilgrim	2	120
15	Buffalo, N. Y., via Steamer Pilgrim	2	120
	Buffalo, N. Y., via Erie Ry	6	360
	Jackson, Mich., via M C. Ry	18	1,080
	Chicago, Ill., and Niles, Mich	10	600
	Rochester and Syracuse, N. Y., via West Shore Ry	12	720
	Buffalo, Lockport and Dunkirk, N. Y., via N. Y. C. Ry	22	1,320
18	Masonic Commandery, Ancient and Accepted Masons	6	360
21	Buffalo, N. Y., via Steamer Pilgrim	3	180
	Findlay, O., via Erie Ry	9	540
22	Arnot, Pa., via Erie Ry	7	420
	Elmira, N. Y., via Erie Ry	5	300
	Buffalo, N. Y., via Erie Ry	2	120
	Buffalo, N. Y., via Steamer Pilgrim	2	120
	Buffalo, N. Y., opening electric railway	20	1,200
	Buffalo, N. Y., Press Association	25	1,500
	Nickel Plate Ry, along the line	15	900
	Cleveland, O., via steamer	14	840
	Buffalo, N. Y., via N. Y. C. Ry	22	1,320
	Buffalo, N. Y., via B. N. F. Electric Ry	50	3,000
26	Lehigh Valley R. R	6	360
	Boston, Mass., Raymond excursion	1	60
27	Lehigh Valley R. R	6	360
	Erie R. R., Prescott excursion	5	300
	Pennsylvania University, foot ball team	1	60
		912	54,720

RECAPITULATION.

Year.	MONTH.	Number of cars.	Estimated number of visitors.
1894	October	83	4,980
1895	May	152	9,120
1895	June	415	24,900
1895	July	1,088	65,280
1895	August	1,921	115,260
1895	September	912	54,720
		4,571	274,260

AN ACT to confirm the right of way of the Niagara Falls and Suspension Bridge Railway Company.

The People of the State of New York, represented in Senate and Assembly, do enact as follows:

Section 1. The right of way of the Niagara Falls and Suspension Bridge Railway Company, as required by said company by order of the supreme court dated January twenty-sixth, eighteen hundred and eighty-three, in Canal street, in the former village of Niagara Falls (now the city of Niagara Falls), is hereby ratified and confirmed.

§ 2. The rights of said railway company were not cut off or extinguished by the proceedings of the commissioners of the State reservation at Niagara Falls, under and by virtue of chapter three hundred and thirty-six of the laws of eighteen hundred and eighty-three, and any act or acts amendatory thereof, as no compensation was awarded to said company in said proceedings.

§ 3. All acts or parts of acts inconsistent herewith, and especially section one hundred and nine of chapter five hundred and sixty-five of the laws of eighteen hundred and ninety-two, are hereby repealed in so far only as the same apply to Canal street in the city of Niagara Falls.

§ 4. This act shall take effect immediately.

REPORT OF THE TREASURER

FOR THE

Fiscal Year Begun October 1, 1894, and Ended September 30, 1895.

THE COMMISSIONERS OF THE STATE RESERVATION AT NIAGARA *in account with* HENRY E. GREGORY, *Treasurer.*

1894.				
Oct.	1. Balance on hand this date			$3,990 79
	RECEIPTS.			
	29. Quarterly advance from the State Comptroller on account of $25,000, etc	$6,250 00		
1895.				
Jan.	18. Quarterly advance from the State Comptroller on account of $25,000, etc	6,250 00		
Apr.	18. Quarterly advance from the State Comptroller on account of $25,000, etc	6,250 00		
Aug.	4. Quarterly advance from the State Comptroller on account of $25,000, etc	6,250 00		
				25,000 00
	Special appropriation as per chapter 726, *Laws of* 1893.			
1894.				
Nov.	16. Payment by State Comptroller on account	$473 00		
	22. Payment by State Comptroller on account	694 33		
1895.				
March	6. Payment by State Comptroller on account	9 19		
				1,177 16
	Special appropriation as per chapter 358, *Laws of* 1894.			
1894.				
Oct.	10. Payment by State Comptroller on account	$1,537 64		
	16. Payment by State Comptroller on account	1,349 98		
	22 Payment by State Comptroller on account	125 00		
Dec.	12. Payment by State Comptroller on account	572 54		

1895.				
March	6.	Payment by State Comptroller on account....	$28 44	
				$3,613 60
		Special appropriation as per chapter 932, *Laws of* 1895.		
Aug.	16.	Payment by State Comptroller on account	$876 63	
Sept.	12.	Payment by State Comptroller on accennt....	1,365 15	
	26.	Payment by State Comptroller on account....	300 00	
				2,541 78
1894.				
Nov.	1.	Draft on Bank of Niagara for October receipts,	$214 20	
Dec.	1.	Draft on Bank of Niagara for November receipts	55 65	
	31.	Draft on Bank of Niagara for December receipts	40 35	
1895.				
Feb.	1.	Draft on Bank of Niagara for January receipts	25 65	
March	1.	Draft on Bank of Niagara for February receipts	9 05	
		(No receipts in March).		
May	1.	Draft on Bank of Niagara for April receipts...	32 80	
June	1	Draft on Bank of Niagara for May receipts....	203 70	
	29.	Draft on Bank of Niagara for June receipts...	467 15	
Aug.	1.	Draft on Bank of Niagara for July receipts....	1,566 30	
Sept.	3.	Draft on Bank of Niagara for August receipts.	2,539 75	
	30.	Draft on Bank of Niagara for September receipts	1,876 25	
				7,030 85
1894.				
Oct.	18.	Dividend on deposits in Cataract Bank	$56 69	
1895.				
March	28.	Dividend on deposits in Cataract Bank.......	28 34	
June	29.	Dividend on deposits in Cataract Bank.......	113 38	
Sept.	17.	Dividend on deposits in Cataract Bank.......	56 69	
				255 10
1894.				
Dec.	31.	Interest on balances in Manufacturers and Traders' Bank..............................	$42 28	
1895.				
March	30.	Interest on balances in Manufacturers and Traders' Bank..............................	44 38	
June	29.	Interest on balances in Manufactuiers and Traders' Bank..............................	52 01	
Sept.	30.	Interest on balances in Manufacturers and Traders' Bank..............................	23 39	
				162 06
				$43,771 34

EXPENDITURES.

1894.	No. of abstract.	No. of voucher.			
Nov. 1.	LXXXVII	1,138.	Pay-roll at Niagara for October	$1,477 41	
		1,139.	Thomas V. Welch, supt., office expenses	20 94	
12.	LXXXVII	1,140.	Wm. Hamilton, comr., traveling expenses	53 70	
15.	LXXXVII	1,141.	Geo. E. Wright, repairs	116 74	
	LXXXVII	1,142.	John Irwin, repairs	8 25	
	LXXXVII	1,143.	P. C. Flynn & Son, painting	64 05	
	LXXXVII	1,144.	Wm. Shepard & Son, mason work	396 98	
	LXXXVII	1,145.	Henry E. Gregory, treas. and sec., office and other expenses	38 85	
					$2,176 92
Dec. 1.	LXXXVIII	1,146.	Pay-roll at Niagara for November	$2,222 05	
	LXXXVIII	1,147.	Thomas V. Welch, supt., office expenses, etc	47 98	
8.	LXXXVIII	1,148.	Geo. E. Wright, repairs	52 56	
	LXXXVIII	1,149.	Estate of John Johnson, repairs	60 66	
	LXXXVIII	1,150.	James O'Brien, stone	54 00	
	LXXXVIII	1,151.	Coleman Nee, stone	167 63	
	LXXXVIII	1,152.	Timothy Horan, stone	159 75	
	LXXXVIII	1,153.	P. J. Davy, plumbing work, etc.	139 77	
12.	LXXXVIII	1,154.	Niagara Sand Co	167 64	
31.	LXXXVIII	1,155.	Henry E. Gregory, treas. and sec., salary October, November and December	275 00	
	LXXXVIII	1,156.	Pay-roll at Niagara for December	1,145 29	
	LXXXVIII	1,157.	Thomas V. Welch, supt., office expenses, etc	16 10	
					4,508 43
1895.					
Feb. 1.	LXXXIX	1,158.	Pay-roll at Niagara for January	$1,198 05	
	LXXXIX	1,159.	Thomas V. Welch, supt., office expenses	13 28	
21.	LXXXIX	1,160.	Henry E. Gregory, treas. and sec., office expenses	24 72	
	LXXXIX	1,161.	The Courier Lithographing Co., map and guide	569 43	
	LXXXIX	1,162.	Maloney & McCoy, ice	64 54	

1895.		No. of abstract.	No. of voucher.			
Feb.	21.	LXXXIX	1,163.	Hardwicke & Co., hardware, etc	$17 19	
		LXXXIX	1,164.	Hardwicke & Co., hardware, etc	21 78	
		LXXXIX	1,165.	F. W. Oliver Co., iron posts.	14 40	
		LXXXIX	1,166.	F. W. Oliver Co., hardware, etc	21 31	
		LXXXIX	1,167.	D. Phillips, repairs	20 60	
		LXXXIX	1,168.	McDonald & Welch, coal	90 20	
						$2,055 50
		XC	1,169.	Jas. McCarthy, teamster	$52 50	
March	1.	XC	1,170.	Wm. Hamilton, comr., traveling expenses	67 87	
		XC	1,171.	Pay-roll at Niagara for February	1,199 42	
		XC	1,172.	Thomas V. Welch, supt., office expenses, etc	47 70	
	30.	XC	1,173.	Henry E. Gregory, treas. and sec., salary for January, February and December	275 00	
		XC	1,174.	Pay-roll at Niagara for March	1,179 66	
		XC	1,175.	Thomas V. Welch, supt., office expenses, etc	42 37	
						2,864 52
May	1.	XCI	1,176.	Pay-roll at Niagara for April.	$1,438 42	
		XCI	1,177.	Thomas V. Welch, supt., office expenses, etc	49 77	
June	1.	XCI	1,178.	Pay-roll at Niagara for May	1,498 41	
		XCI	1,179.	Thomas V. Welch, supt., office expenses, etc	49 88	
	6.	XCI	1,180.	Supplemental pay roll	1,012 37	
	8.	XCI	1,181.	G. Chormann, repairing cushions	6 50	
		XCI	1,182.	Hardwicke & Co., tools, etc	8 56	
		XCI	1,183.	Niagara Falls Printing House, stationery	3 50	
		XCI	1,184.	P. C. Flynn & Son, painting and varnishing	71 65	
		XCI	1,185.	G. H. Burdick, tools	14 90	
		XCI	1,186.	Globe Ticket Co., Inclined railway tickets	9 00	
		XCI	1,187.	F. Batchellor, grass seed	18 00	
		XCI	1,188.	F. W. Oliver Co., sundries	7 26	
		XCI	1,189.	Ellwanger & Barry, trees and shrubs	146 95	

1895.		No of abstract.	No. of voucher.			
June	8.	XCI	1,190.	Niagara Sand Co., gravel....	$29 77	
		XCI	1,191.	P. C. Flynn & Son, painting bridges..................	385 46	
	13.	XCI	1,192.	T. T. Southwick, grease.....	5 00	
		XCI	1,193.	Denther & Peck, painting materials.................	130 33	
	24.	XCI	1,194.	Pay-roll (supplemental).....	437 50	
	29.	XCI	1,195.	Pay-roll at Niagara for June.	1,863 17	
		XCI	1,196.	Thomas V. Welch, supt., office expenses..............	24 22	
		XCI	1,197.	Henry E. Gregory, treas. and sec., salary for April, May and June.................	275 00	
						$7,485 62
July	17.	XCII	1,198.	McDonald & Welch, coal....	$35 00	
		XCII	1,199.	P. W. Oliver Co., tools	12 49	
		XCII	1,200.	Jos. Mackenna & Son, office.	9 53	
		XCII	1,201.	The Howard Iron Works, park seats................	200 00	
		XCII	1,202.	Wm. Shepherd & Son, mason work	130 72	
		XCII	1,203.	P. J. Davy, plumbing	161 44	
		XCII	1,204.	Hardwicke & Co., tools	75 54	
		XCII	1,205.	The Barber Asphalt Paving Co., gravel	15 75	
		XCII	1,206.	The Barber Asphalt Paving Co., gravel	24 00	
		XCII	1,207.	Niagara Sand Co	14 93	
		XCII	1,208.	Jas. E. Rock, office	27 48	
		XCII	1,209.	Estate of John Johnson, blacksmithing	23 45	
		XCII	1,210.	Globe Ticket Co., tickets....	18 00	
		XCII	1,211.	Niagara Sand Co., gravel....	18 36	
		XCII	1,212.	Geo. E. Wright, carpenter...	226 73	
		XCII	1,213	F. W. Oliver Co., hardware ..	2 64	
		XCII	1,214	P. C. Flynn & Son, painting .	203 39	
	18.	XCII	1,215	Wm. Shepard & Son, mason work.....................	140 96	
	17.	XCII	1,216.	Hardwicke & Co., hardware.	105 87	
		XCII	1,217.	Howard H. Baker & Co., national flag...............	10 00	
		XCII	1,218.	Wm. Thurecht, gravel	19 50	
		XCII	1,219.	Deuther & Peck, painting...	18 00	
		XCII	1,220.	Niagara Falls Metal Sign Co., signs	13 40	
						1,507 18

1895.		No. of abstract.	No. of voucher.			
July	17.	XCIII	1,221.	Sam'l Parsons, Jr., traveling expenses	$25 00	
						$25 00
	18.	XCIV	1,222.	Andrew H. Green, pres., traveling expenses............	$35 85	
Aug.	1.	XCIV	1,223.	Pay-roll at Niagara for July..	1,475 16	
		XCIV	1,124.	Thomas V. Welch, supt, office expenses	41 70	
Sept.	2.	XCIV	1,225.	Pay-roll at Niagara for August	1,478 05	
		XCIV	1,226.	Thomas V. Welch, supt, office expenses...............	40 30	
	17.	XCIV	1,227.	J.W. Spencer, illustrations for report	60 00	
		XCIV	1,228.	Globe Ticket Co., tickets for Inclined railway...........	42 00	
		XCIV	1,229.	Hardwicke & Co., repairs, etc.	70 08	
	30.	XCIV	1,230.	Pay-roll at Niagara for September	1,470 16	
		XCIV	1,231	Thomas V. Welch, supt., office expenses	48 76	
		XCIV	1,232.	Henry E. Gregory, treas. and sec., salary for July, August and September	275 00	
						5,037 06

Payments out of $20,000 appropriated by chapter 726, Laws of 1893.

1894.		No. of abstract. Series D.	No. of voucher.			
Nov.	17.	X	69.	F. W. Oliver Co., work on bridge......................	$473 64	
	22.	X	70.	D. Phillips, work on bridge..	694 33	
						1,167 97
1895.		XI	71.	Jas. McCarthy, grading....	$9 19	
						9 19

Payments out of $20,000 appropriated by chapter 358, Laws of 1894.

1894.		Series E.			
Oct.	11.	II	24.	Pay-roll at Niagara	$115 25
		II	25.	Dobbie, Stuart & Co., grates,	40 15
		II	26.	W. Shepard & Son, mason work	452 03
		II	27.	T. E. McGarigle, blacksmithing	17 55
		II	28.	Fred Batchelor, seed	18 00
		II	29.	David Phillips, repairing bridge	576 66

1894.		Series E.			
Oct.	11.	II	30. Fred Batchelor, seed........	$18 00	
Nov.	17.	II	31. F. W. Oliver Co	199 40	
		II	32. D. Phillips, repairs to bridge,	605 08	
		II	33. Pay-roll at Niagara	545 50	
Nov.	22.	II	34. Vaux & Emery, plans and specifications............	125 00	
					$3,012 62
Dec.	13.	III	35. F. W. Oliver Co., bridge repairs...	$16 73	
		III	36. Wm. Shepard & Son, bridge repairs	240 45	
		III	37. David Phillips, bridge repairs	315 36	
					572 54
1895.					
March	6.	IV	38. Jas. McCarthy, grading	$28 44	
					28 44

Payments out of $20,000 appropriated by chapter 932, Laws of 1895.

1895.		Series F.			
Aug.	16.	I	1. Pay-roll at Niagara...........	$697 75	
		I	2. Rochester Printing Co., printing...................	20 00	
			3. Geo. E. Matthews & Co., printing	19 32	
			4. The Courier Company, printing	19 72	
			5. The Daily Cataract, printing,	13 00	
			6. The Niagara Falls Gazette, printing	6 50	
			7. Union and Advertiser Co., printing..................	18 75	
			8. Niagara Sand Co., gravel...	62 51	
		I	9. Niagara Sand Co., gravel...	19 08	
Sept.	12.	I	10. Pay-roll at Niagara.........	834 00	
		I	11. Wm. Shepard & Son, building bridge, etc............	507 15	
		I	12. Dennis McDonald, inspector,	24 00	
	26.	I	13. Vaux & Emery, plans, etc..	300 00	
					2,541 78

Remittances to the State Treasurer.

1894.			
Nov.	1.	Draft for October receipts.....................	$214 20
Dec.	1.	Draft for November receipts..................	55 65
	31.	Draft for December receipts..................	40 35

1895.				
Feb.	1.	Draft for January receipts..................	$25 65	
March	1.	Draft for February receipts..................	9 05	
May	1.	Draft for April receipts......................	32 80	
June	1.	Draft for May receipts......................	203 70	
	29.	Draft for June receipts......................	467 15	
Aug.	1.	Draft for July receipts......................	1,566 30	
Sept.	3.	Draft for August receipts..................	2,539 75	
	30.	Draft for September receipts................	1,876 25	
				$7,030 85
1894:				
Nov.	1.	Dividend on deposits in Cataract bank.......	$56 69	
1895.				
March	30.	Dividend on deposits in Cataract bank.......	28 34	
June	29.	Dividend on deposits in Cataract bank.......	113 38	
Sept.	30.	Dividend on deposits in Cataract bank.......	59 69	
				255 10
1894.				
Dec.	31.	Interest on balances in Manufacturers and Traders' Bank............................	$42 28	
1895.				
March	30.	Interest on balances in Manufacturers and Traders' Bank............................	44 38	
June	29.	Interest on balances in Manufacturers and Traders' Bank............................	52 01	
Sept.	30.	Interest on balances in Manufacturers and Traders' Bank............................	23 39	
				162 06
				$40,440 78
1895.				
Sept.	30.	Cash balance in treasurer's hands............		3,330 56
				$43,771 34

HENRY E. GREGORY, *Treasurer.*

We, the undersigned, hereby certify that we have examined the foregoing report of the Treasurer, and the vouchers and other papers, and we find the report and accompanying documents correct, and that the treasurer has properly accounted for all moneys received and disbursed by him during the year ended September 30, 1895.

ROBERT L. FRYER,

WM. HAMILTON,

Commissioners.

CLASSIFICATION OF ACCOUNTS.

Fences	$8 52
Railings	78 50
Stairways	33 89
Trees	146 95
Stationery	3 50
Water pipes	232 96
Ice	64 54
Treasurer and secretary	1,100 00
Police	5,400 00
Salaries (superintendents and clerk)	2,899 98
Goat island	1,311 75
Seed	18 00
Settees	200 00
Coal	125 20
Signs	18 40
Prospect park	1,762 00
Roads	3,252 33
Walks	2,431 18
Bridges	974 37
Buildings	730 09
Commissioners' expenses	157 42
Inclined railway	2,459 67
Treasurer and secretary (office expenses)	38 57
Niagara office	463 81
Tools	261 37
Guard railing	50 30
Map and guide	569 43
Teaming	52 50
Grading	468 00
Planting	232 00
National flag	10 00
Expense	110 00
	$25,660 23

SPECIAL IMPROVEMENTS UNDER CHAPTER 726 OF THE LAWS OF 1893.

Bridge from Bath Island to Goat Island	$1,167 97
Grading	9 19
	$1,177 16

SPECIAL IMPROVEMENTS UNDER CHAPTER 358 OF THE LAWS OF 1894.

Roads	$733 43
Walks	431 50
Grading	263 19
Bridges	252 65
Seed	36 00
Shelter on Goat Island	125 00
Bridges to Goat Island	1,771 83
	$3,613 60

SPECIAL IMPROVEMENTS UNDER CHAPTER 932 OF THE LAWS OF 1895.

Boulder bridge	$597 36
Grading	478 76
Roads	572 46
Walks	374 51
Printing	97 29
Retaining wall (Goat Island)	97 40
Inclined railway station	300 00
Shelter building	24 00
	$2,541 78

Rustic Bridge. Willow Island.

Letters.

NEW YORK, *October* 17, 1894.

HON. WALTER Q. GRESHAM,

Secretary of State, Washington:

DEAR SIR.—In the New York State Convention, to revise the Constitution, whose labors have recently been completed, the protection and preservation of the Falls of Niagara from injury and diminution by private corporations was the subject of extended and somewhat heated discussions.

Certain corporations claim to have obtained the right to divert large quantities of the water of the Niagara above the Falls for manufacturing purposes, both in this State and Canada.

As President of the Niagara commissioners, I have opposed the granting of these rights or licenses by the Legislature, on the ground that large diversions of the upper river will certainly be followed by a diminished flow at the Falls, and that, eventually, the grandeur of the great cataract will be impaired, the value of the State's property depreciated, and one of the sublimest objects of nature become less and less attractive.

The apologists and representatives of the corporations before the convention argued that a constitutional amendment prohibiting or limiting the diversion of the Niagara would be useless, since the corporations would be able to secure the right or license to use the water from the Ontario or Canadian authorities, and transfer their plants and bases of operations to that side of the river, with the consequent pecuniary and other material advantages.

4

The Falls of Niagara belong partly to Canada and partly to the United States. It would naturally be supposed, therefore, that each country would be jealous of any encroachment by the other, and that both would be equally zealous in guarding the Falls from injury or partial destruction, and that some international agreement could be speedily arrived at that would guarantee the desired exemption and protection for all time.

It is hardly necessary for me to advance arguments in favor of international protection of the Falls of Niagara against the designs ot wealthy corporations. To permit the latter to inflict permanent injury upon the most famous natural object in the western hemisphere for purposes of private advantage, would be to demonstrate an indifference that would justify the ridicule and excite the contempt of foreign nations and of cultured people everywhere.

The possession of such a natural wonder must be regarded as a trust — faithlessness to which should be followed by universal condemnation.

Something, I hope, may forthwith be done to save the great cataract from the greedy grasp and destructive clutch of money-mongering corporations.

The subject is one of so great importance that it seems to me a correspondence between the United States Department of State, and the Canadian Minister of State, should be opened, with a view of learning the attitude of the Canadian government, and that proper procedure may be determined upon to effect the permanent protection of the Falls by international agreement.

Very respectfully yours,

ANDREW H. GREEN,

President.

NEW YORK, *October* 19, 1894.

J. W LANGMUIR, Esq., *Chairman Commissioners Queen Victoria Niagara Falls Park, Toronto, Canada:*

DEAR SIR.—You are no doubt aware of the recent discussions in the New York State constitutional convention on the proposed amendment to prevent further diversion of the waters of the upper Niagara.

To prohibit the New York Legislature from granting to corporations the right or license to divert the water of the river would not, it was argued, be effectual in preventing the taking of water and the consequent diminution of the Falls, inasmuch as the Ontario authorities could still grant such rights or licenses.

In order, therefore, to exempt the river and the Falls from the attacks and depredations of the corporations, it would seem that the only course to pursue is to secure an international agreement, to the effect that hereafter no right or license to divert the water of the river shall be granted to any corporation on either side of the river.

It is quite unnecessary for me to enlarge upon the great importance of preserving the cataracts at Niagara in undiminished volume and dignity, and the equally great importance of checking the destructive designs of corporations organized for mere money-making. I trust that you will agree with me that every reasonable effort to accomplish the desired result will be cordially furthered by your board. I should be glad to hear from you on this subject at your earliest convenience, and oblige,

Very respectfully,

ANDREW H. GREEN,

President.

No. 214 BROADWAY,
NEW YORK, *July* 17, 1895.

Hon. THEODORE E. HANCOCK,

Attorney-General, Albany, N. Y.:

MY DEAR SIR.—It appears that the Niagara Falls Hydraulic Power and Manufacturing Company are now actually engaged in widening a water-way from the Niagara river just above and around the Niagara Falls.

The Commissioners of the State Reservation are not aware of any lawful authority to increase the take of water from the river by this company, nor indeed to take any water from the river.

The commissioners therefore respectfully request the Attorney-General to enquire as promptly as possible in this matter, and to advise them whether this company has any statutory authority to take water from the Niagara river and, if none exists, to take the necessary measures to protect the interests of the State in the premises.

Will you please let me hear from you on this matter as early as practicable, and oblige,

Yours very respectfully,

ANDW. H. GREEN,

President.

July 18, 1895.

Hon. THEODORE E. HANCOCK,

Attorney-General, Albany, N. Y.:

DEAR SIR.—Supplementing my letter of yesterday respecting the taking of water from the Niagara river, as the work of widening the channel of the canal of the company mentioned in that letter is now actually going on, I will submit it to your judgment whether immediate proceedings should not be taken to stop this work.

Would you favor me with a line indicating the proceedings that you propose to take in the premises, and much oblige,

Yours very respectfully,

ANDW. H. GREEN,

President.

STATE OF NEW YORK:

ATTORNEY-GENERAL'S OFFICE,
ALBANY, *November* 16, 1895.

Hon. ANDREW H. GREEN,

President Niagara Reservation Commission, New York City:

DEAR SIR.—Some time ago the question of the right of the Niagara Falls Hydraulic Power and Manufacturing Company to enlarge the capacity of their canal, by which a portion of the water of the Niagara river is diverted for manufacturing purposes, was submitted to me for examination by you. The question is one involving great interests, not only to the corporation referred to, but to the State itself, and I have therefore considered it with a great deal of care, before venturing to express an opinion.

The facts in the case may be briefly stated. The canal in question was originally constructed in the year 1859. Its dimensions were seventy feet wide by fourteen feet deep. The inlet is at Port Day, about one mile above the Falls, and it runs through a strip of land 100 feet wide, to the mills on the bank of the river below the Falls, where the waters, after supplying power to various industries, are discharged into the river.

About the year 1878, the title to the land in the 100 foot strip, as I am informed by Mr. Schoelkopf, of Niagara Falls, was acquired by the present owners, since which time the canal has been in active operation, and has supplied power to mills of a sufficient capacity to employ a large number of hands, residents of the City of Niagara Falls, and whose continued prosperity, to a very large degree, is dependent upon the operation of the mills in which they are employed. Some time after the acquisition of title to the strip of land by the present owners, they made application to the Land Commissioners of the State of New York for a grant of land under water adjoining the inlet to the canal.

In the papers submitted on that application it was stated to be the intention of the owners to increase the capacity of the canal, and thereby increase its production of horse power. The grant was made by the commissioners, with the condition that no structures were to be built upon the granted land without the consent of the Niagara Reservation Commission. Thereafter, application was made to the Reservation Commission for leave to erect cribs on the land under water, the purpose of which was to prevent the flow of ice and other refuse into the canal, to the detriment of the interests of the Niagara Falls Hydraulic Power and Manufacturing Company.

The capacity of the canal at that time, if I am correctly informed was 200,000 cubic feet per minute. No objection was made (at least publicly) to this diversion of the waters of the river at that time. Since then, however, various grants of privileges by the Legislature of the State have been given to several corporations, to divert the waters of the Niagara river for power purposes. In consequence of these grants apprehension has been created as to the probable effect upon the flow of water over the Falls, and your commission, actuated by commendable zeal to protect the great natural beauty of the Reservation, have determined that further encroachments upon the stream shall be prevented, if possible

The law under which your board was created (chapter 336, Laws of 1883) states that the object of the creation of the commission was to preserve the scenery of the Falls of Niagara. It provides for the condemnation of the lands to be selected by the commission, and for the compensation to be paid to the owners of the property condemned. In carrying out the provisions of the law several million dollars have been expended by the State of New York, which will be converted into a mere waste of public moneys, if the flow of water over these Falls is to be seriously diminished.

While this is, of course, a very serious consideration, I have not permitted myself to lose sight of the importance to the industries dependent upon the maintenance of the canal for their power, which action on the part of the State authorities will have.

It is a very grave duty to be compelled to pass upon public questions wherein such great private interests are concerned. Nevertheless, it is one which I see no way to escape, and, while from certain considerations, I would be pleased to arrive at a different conclusion, I am compelled to hold, from my examination of the law on the subject, that the Niagara Falls Hydraulic Power and Manufacturing Company may be restrained from increasing the capacity of the canal. It is only fair, however, that my reasons for this conclusion should be stated. They are as follows:

The Niagara river is a public, navigable stream, to the bed of which and the water flowing over it, the State and not the riparian owner, has title.

It would be a waste of time to attempt to show why this proposition is correct. It is sufficient to say that it has been amply supported by judicial decisions and is now the established law.

Ill. C. R. R. Co. *v.* Ill., 146 U. S., 387.
Smith *v.* Rochester, 92 N. Y., 479.
Matter of St. Reservation, 16 Abb. N. C., 395.

The sole question, therefore, for determination is, "Can an owner of the soil adjoining a navigable stream divert the water for private manufacturing purposes without the consent of the State?" Let us examine it.

By the term "navigable," it must be remembered, is not meant "capable of being navigated." As used in this discussion, "navigable stream" means one which is navigable in the legal sense. Rivers may be navigable in fact but not in law, or they may be navigable in law but only in part navigable in fact. A mere local interruption of actual navigability, therefore, will not change the character of a stream in its legal aspect.

The river being navigable, in the legal sense, the title to the bed of the stream and to the water flowing over it is in the State, at least to the boundary line between the State and Canada.

People v. Appraisers, 33 N. Y., 464.
Crill v. Rome, 47 How. Pr., 398.
Morgan v. King, 35 N. Y., 451.
People v. Tibbetts, 19 N. Y., 523.
Ex parte Jennings, 6 Cow., 518.

Therefore, leaving out of view for the present the grant of land under water to the Hydraulic Power and Manufacturing Company, the State could unquestionably deprive the corporation of all use of the waters of the river for power purposes by devoting the stream to other public use.

Smith v. Rochester, 92 N. Y.

Whether or not that has been done by the laws establishing the Niagara Reservation, I will discuss hereafter. I prefer at this point to consider the abstract question of the right of an owner of land adjoining a navigable stream to divert a considerable portion of the waters for manufacturing purposes, without a grant or prescriptive right.

Nuisances may always be abated by action in the name of the aggrieved party. Public nuisances include any encroachment upon highways or navigable streams, and it is not an essential characteristic of the encroachment upon the stream that it should be an actual hindrance to navigation.

Wood on Nuisances, 2d Ed., Secs. 478-480, and cases cited.

The diversion of water from a public stream for any other than domestic purposes is a nuisance, and therefore may be abated at the suit of the Attorney-General.

Philadelphia v. Gelmartin, 71 Penn. St., 140.

The Niagara Falls Hydraulic Power and Manufacturing Company is organized under the act of 1875, chapter 611. Its objects are declared to be the development of the hydraulic canal in Niagara Falls, and the establishment and conducting of various manufacturing interests. Under its charter it is not only supplying its own mills but is furnishing other industries with power for a consideration. So far as the latter fact is concerned, certainly no question can be raised as to the rights of a riparian owner to the use of water for his own benefit I assume the fact that the capacity of the canal at the outset was sufficient for all the purposes of the power company, and that the increased capacity is desired for the purpose of enabling the corporation to derive a revenue from its sale of power to others. I have no hesitation in declaring this to be unlawful. A non-riparian owner is not entitled to any benefits of a stream other than those enjoyed in common by the public, and a riparian owner at the most is entitled only to personal benefits derivable from use devoted to personal purposes solely. They do not include the transmission of power to property located upon premises may be far removed from the lands of the riparian owner.

The case last cited was an action brought by the owner of a boat which had been prevented from navigating the Schuylkill, by reason of the diversion of the waters of that stream by the city of Philadelphia for domestic or other purposes. The court in its opinion says

"In deciding upon the question of illegality in drawing off the water from the navigation, we are carried beyond its use for power, to inquire into the character of the consumption claimed as an overruling necessity. We have already seen that the city is a large vendor of water from which she is deriving *revenue*, for all the purposes of the arts, manufacturing, business and pleasure. These uses are not domestic, that is, such as are for the preservation of the life and health of the population and their creatures, but are simply utilitarian or business uses, and far exceed those needed for domestic purposes. And even as to those termed domestic, a distinction must be noted between the use proper and that which is lavishly expended in pavement washing, baths, etc. It is perfectly obvious, therefore, that the city drew off water not only for driving and lifting power, but for a consumption far beyond any imperious necessity, and for purposes wholly subordinate to the right of navigation. She chose to prefer the pecuniary interests of her citizens and doing an injury thereby, she must make compensation to the injured parties. I mean not by these remarks to draw any comparison between the importance of the use of the water for the great purposes of industry, wealth and cleanliness, of a city so populous as Philadelphia, and the use of it for navigation during a few days of drought. The question for us is that of legal right, not comparative weight. Such important interests as those of the city are not likely to lead to the substitution of might for right; yet, they are not of that imperious necessity which justifies might, and changes wrong into right. Administrators of the law, we can not bend or break the law before a large interest, more than we can before one that is small. The doctrine of imperious necessity is not in this case."

It is historical that the Niagara river at Port Day has been navigated by vessels of large burden, and, indeed, to a point some distance below. The erection of cribs to divert ice and other refuse from the canal inlet is, therefore, an actual obstruction to navigation, and it is not necessary to show present use of the river at this point for navigation purposes. Once a highway, always a highway, is true of navigable streams. (See [illegible] v [illegible], 36 Cal., 193; Wood on Nuisances, 478, 485).

Ex parte Jenkins (6 Cowen, 518), is also of interest on this point. That was a proceeding brought in mandamus to compel commissioners appointed to appraise damages occasioned by the diversion of the stream of the Chittenango, for the purposes of the Erie canal, which diversion prevented the use of the water of the stream by riparian owners for power purposes in operating mills. The court, in awarding mandamus, says:

"The objection is contained in the affidavits of Mr. Seymour, that in point of fact, the State had not parted with the land upon which the Chittenango passes, at the places claimed, but had bounded purchases of land on the margin of the stream; so that, as he believes, (and he believes the other appraisers were satisfied of the fact being so) the State was still the owner of the land covered by the waters of the stream, and had not parted with it or contracted to part with it, to any person whatever; or authorized the use of the water for hydraulic purposes at the places in question. If the construction set up by the commissioners be the true one, if the State owns the land covered by the water, it is clear that, though the relators may be entitled to the use of the water flowing by and touching upon them, for all ordinary purposes, yet they can not build mills upon and raise the water of the stream. They are trespassers, and the State may claim not only the waters, but the mills themselves, so far as they encroach upon the stream."

I will not consider the effect of the grant by the Land Commissioners, of lands under water to the corporation operating the canal.

The powers of the Land Commissioners, at the time the grant was made, were conferred by section 67, page 633, volume one, eighth edition, Revised Statutes. It reads:

"The Commissioners of the Land Office shall have power to grant, in perpetuity or otherwise, so much of the lands under the waters of navigable rivers or lakes as they shall deem necessary to promote the commerce of this State, or proper for the purpose of beneficial enjoyment of the same by the adjacent owner."

The Court of Appeals, in passing upon the character of such grant, says: "In every such grant there was an implied reservation of the public right, and so far as it assumed to interfere with it, or to confer the right to impede or obstruct navigation, or to make an exclusive appropriation of the use of navigable waters, the grant was void."

Again: "Public grants to individuals under which rights are claimed in impairment of public interests, are construed strictly against the grantee, for it is reasonable to suppose that if they were intended to have this operation, the intention would have been expressed in plain and explicit language."

People v. N. Y. & Staten Island Ferry Co., 68 N. Y., 71.

I have been unable to find any language in the grant to the Niagara Falls Hydraulic Power and Manufacturing Company which can be construed as authorizing them to divert the waters of the Niagara river. Applying the principles in the case last cited, it is certain that that grant can afford no defense to an action brought to restrain the unlawful taking of the waters.

It now remains to determine whether or not the waters of the Niagara river have been devoted by the Legislature to a public use to an extent that will prevent the diversion of the water above the Falls for power purposes. The objects and purposes of the statutes creating the Niagara Reservation were to preserve a great natural waterfall and its environments for the enjoyment of the people of this State. In fact, the statutes themselves declare that the commissioners shall take all proper steps to restore and afterwards to preserve the scenery as nearly in its natural state as possible.

The flow of water over the Falls is an essential element in the preservation of the scenery, and if it can be shown (as I am informed it can) to be the fact that the diversion of the large quantities of water through the canal of the Niagara Falls Hydraulic Power and Manufacturing Company has a diminishing effect upon the flow of the water over the Falls, the diversion is a nuisance and can be restrained

All of which is respectfully submitted.

T. E. HANCOCK,
Attorney-Genera.

Cave of the Winds—February 14, 1896.

American Falls—February, 1896.

THIRTEENTH ANNUAL REPORT

OF THE

Commissioners of the State Reservation

AT NIAGARA.

TRANSMITTED TO THE LEGISLATURE FEBRUARY 1, 1897.

WYNKOOP HALLENBECK CRAWFORD CO.
STATE PRINTERS,
ALBANY AND NEW YORK.
1897.

COMMISSIONERS.

ANDREW H. GREEN, *President.*

JOHN M. BOWERS, ROBERT L. FRYER,

WILLIAM HAMILTON, GEORGE RAINES.

Treasurer and Secretary,

HENRY E. GREGORY.

Superintendent,

THOMAS V. WELCH.

STATE OF NEW YORK.

No. 42.

IN ASSEMBLY,

FEBRUARY 1, 1897.

THIRTEENTH ANNUAL REPORT

OF THE

Commissioners of the State Reservation at Niagara.

To the Honorable the Legislature of the State of New York:

The Commissioners of the State Reservation at Niagara, as required by law, submit their report for the fiscal year begun October 1, 1895, and ended September 30, 1896, being their thirteenth annual report.

The work incidental to maintenance has been carried on with the usual regularity; and the principal improvements begun during the previous year have been brought to completion. The new Shelter building, on Goat Island, and the new terminal station at the foot of the Inclined railway, are substantial structures that may be expected to be gratifying to the taste as well as suitable for the use of tourists and visitors.

The Shelter building is situated on the river bank among the trees, on Goat Island, near the bridge leading from Bath Island. It is an inconspicuous structure; and the graceful lines and warm color of the roof, the red-brown tone of the stonework, harmonize with the natural surroundings. It is very readily accessible, and the accommodations for visitors are ample. The walls are of Medina stone, with arched openings; the posts, of Georgia pine, have curved brackets, supporting the overhanging roof. The floor is asphalt, and the ceiling is of narrow Georgia pine and semi-circular in shape. The lavatories are lined with brick, with slate fittings, and are thoroughly ventilated.

Permanent wooden seats are attached to the walls, both inside and outside of the building.

The total cost of the Shelter building was $6,572.94.

The old terminal station at the foot of the Inclined Railway was destroyed by a mass of ice and snow descending upon it in March, 1893. The new building is a much more commodious structure. It includes ticket offices for the Inclined railway and "Maid of the Mist," and a large waiting-room. A hanging balcony encircles the river side of the building, and provides protection from the spray. The copper tile roof is supported on chamfered eight-inch Georgia pine posts by an open timbered roof construction of sufficient strength. The floor is of hard wood. The walls are battered and built of native stone laid in cement mortar with square window openings.

The total cost of this new terminal station was $9,749.85.

Persons desiring to acquaint themselves with the details of the year's work are referred to the annual report of the superintendent, which is appended to this report.

The Commissioners desire to impress upon the Legislature the great importance of protecting the Reservation from the insidious designs of corporations and individuals who seek to secure from the Legislature and from the Commissioners rights and privileges at Niagara in direct contravention of the purpose of the State in establishing the Reservation, and in flagrant opposition to the public interests.

While it is true that much has been said on this subject in previous reports, the Commissioners nevertheless feel that, as custodians of the State's property at Niagara, they should do what they can to prevent further attempts on the part of corporate or private interests to get a foothold within the Reservation itself, or to divert for money-making purposes the waters of the Niagara river. Every winter there is the probability that some bill may be introduced at Albany having for its object the granting of a franchise to some corporation proposing to undertake operations at Niagara in opposition to the interests of the people.

The success of one or more corporations in utilizing the water power of the upper Niagara in manufacturing enterprises, or the development of electrical power, has naturally directed the eyes of shrewd and speculative men towards a similar use of the river. It has been pointed out that no less than eight corporations have secured alleged rights to divert the water of the river. In the opinion of the Commissioners the only safe course for the Legislature to pursue is that of invariably refusing to enact all bills that directly or remotely have for their object the diversion of the waters of the upper river, or the entrance into the Reservation of any corporation for private, money-making purposes.

In spite of the opposition of the Commissioners, the Legislature during the last session passed a bill, entitled "An act confirming and defining certain riparian rights of the Niagara Falls Hydraulic Power and Manufacturing Company."

By this bill, "the right of the Niagara Falls Hydraulic Power and Manufacturing Company, to take, draw, use, and lease and sell to others to use the waters of the Niagara River for domestic, municipal, manufacturing, fire and sanitary purposes and to develop power therefrom for its own use and to lease and sell to others to use for manufacturing, heating, lighting and other business purposes is hereby recognized, declared and confirmed."

The quantity of water to be taken by the company is limited by a canal 100 feet in width, and deep enough to carry a maximum uniform depth of 14 feet of water.

With reference to this and other similar measures, it may be said that, simply because certain speculative and manufacturing concerns in Niagara county desire to secure enormous profits from the use of the waters of the Niagara river, there is no valid reason why the State of New York should grant away rights and privileges of immense value without compensation. The Commissioners believe that, in opposing this and other measures having the same object, they represent the best sentiment of the people of the State, and they repeat that in their opinion the Legislature should co-operate with them, and refuse to approve bills of this character. The Commissioners are of the opinion that the Legislature should not grant to a private corporation, without compensation, that which belongs to the people.

It is hoped that an international agreement may eventually be entered into between Great Britain and the United States, by which the waters of the upper Niagara shall be permanently exempted from diversion.

Frequent reference has been made to the brevity of the visits of most travelers to Niagara. A locality so interesting on account of its variety of striking and impressive scenery, or by reason of its relation to scientific speculations, should, one might think, invite visits measured by weeks rather than by days or hours. It is not to be expected that the larger number of visitors will spend much time at Niagara. But many persons of leisure, it seems, are not wont to tarry much longer. They can make the circuit of the Reservation and of the Queen Victoria Park within, perhaps, three or four hours. They can take the trip to Lewiston, down the gorge by the new trolley line, and return within an hour and a half. They can thus see the Falls, the Rapids, the Whirlpool and all the principal features of the Niagara scenery within half a day. One may leave New York at 6 p. m., be at Niagara the next morning at 7 o'clock, see all the points of interest, and be at home the next morning, the whole trip being made in 36 hours.

Three or four decades ago the number of summer and autumn resorts was very small. People bent upon taking a vacation had a limited range of selection. Niagara was the principal natural object in the country, a visit to which was by no means to be omitted. The tourist was less driven by business cares and exactions. Rest and recreation were more common. Life was slower and more deliberate. But Niagara as a place of resort was less interesting then than now.

The shores of the river and island were defaced by wretched or hideous structures, tolls and fees were exacted at every point, the hack drivers were aggressive, insolent and rapacious. Under State control a great change has taken place. The hideous buildings have been removed, the shores are freed from unsightly objects, and no fees are charged, except at the Inclined railway and the

Cave of the Winds. One may walk or drive all over the Reservation without being requested to pay a cent.

In bright cool autumn weather Niagara may be said to be at its best. There is no oppressive heat to make exercise objectionable; no crowds of visitors in excursion parties. One may wander at will along the Goat Island paths, linger at the points of finest and grandest prospect, saunter along the river road to Port Day, delighting his senses with the different aspects of the wonderful scenery. He may cross the Suspension bridge, and from that giddy height have an extensive view of the Falls and the Niagara gorge. Then landing upon Canadian soil he has the impressive sight of both the cataracts before him. From different points in the Queen Victoria Park the views of the Canadian Falls are unsurpassable, as well as of the rapids and cascades above. One may walk for hours in the Dufferin Islands and have enchanting views of the dashing river.

From July 15, 1885, the day the Reservation was opened to the public, to September 30, 1896, the State granted the Commissioners in the annual appropriation and supply bills:

For maintenance	$205,000 00
For special improvements	120,000 00
Total	$325,000 00
During the same period the Commissioners have remitted to the State Treasury as receipts from the Reservation	77,348 69
Leaving	$247,651 31

As the amount actually appropriated by the State for the Reservation during a period of a little more than 11 years, or about $22,513.75 a year; or, if we leave out of consideration the amount appropriated for special improvements and deduct the receipts of the Reservation remitted to the State Treasury from the aggregate amount appropriated by the State for maintenance, we have	$205,000 00
	77,348 69
	$127,651 31

Or about $11,604.66 a year for 11 years and a fraction of a year.

The monthly receipts from the Reservation were as follows:

1895.	Inclined railway.	Rentals.	Interest on balances in bank.	Dividend
October	$197 40	......	......	......
November	34 90	......	......	......
December	20 15	......	$30 05	......
1896.				
January	222 35	......	......	$28 34
February	70 70	......	......	......
March	71 60	......	29 12	......
April	57 20	......	......	......
May	247 95	......	......	......
June	431 25	......	34 34	......
July	986 40	$310 00	......	......
August	1,212 70	560 00	......	......
September	689 80	920 00	24 84	......
	$4,242 40	$1,790 00	$118 35	$28 34
				118 35
				1,790 00
				4,242 40
Total				$6,179 09

Monthly pay-rolls have been as follows:

1895.		Supplementary.	Total.
October	$1,966 79	......	......
November	1,410 41	$614 76	......
December	1,155 29	334 50	......
1896.			
January	1,249 54	......	......
February	1,140 17	......	......
March	1,176 79	......	......
April	1,997 91	......	......
May	1,497 42	428 50	......
June	1,404 91	......	......
July	1,475 67	......	......
August	1,798 41	......	......
September	1,790 29	......	......
	$18,063 60	$1,377 76	$19,441 36

Expenditures as per abstract (maintenance) were as follows:

Abstract XCV	$6,868 93
Abstract XCVI	1,528 20
Abstract XCVII	2,856 82
Abstract XCVIII	1,729 23
Abstract XCIX	7,474 85
Abstract C	7,293 25
Total	$27,751 28

Improvement abstracts as follows:

Abstract II, Series F	$8,131 86
Abstract III, Series F	6,433 39
Abstract IV, Series F	2,864 82
Total	$17,430 07
Abstract I, Series G	$2,941 22

The receipts and earnings of the Reservation have been sent to the State Treasurer monthly, and interest on balances in the Manufacturers and Traders' Bank, Buffalo, has been remitted to the same official quarterly.

The Comptroller has advanced to the Commissioners quarterly a fourth part of the $25,000 appropriated by chapter 948, Laws of 1895.

Of the treasurer's report herewith submitted, exhibiting in detail all receipts and disbursements for the fiscal year ended September 30, 1896, the following is a summary:

Balance on hand October 1, 1895		$3,330 56

RECEIPTS.

Inclined railway	$4,242 40	
Rentals	1,790 00	
Interest	118 35	
Dividend	28 34	
		$6,179 09
From the State Treasury, chapter 807, Laws of 1895.		25,000 00
From the State Treasury, chapter 932, Laws of 1895.		17,430 07
From the State Treasury, chapter 950, Laws of 1896.		2,941 22
		$54,880 94

PAYMENTS.

Pay-rolls at Niagara (maintenance)	$18,063 60	
Repairs, materials, superintendent's expenses, etc.	8,354 97	
Commissioners, treasurer, traveling expenses, etc.	1,332 71	
		$27,751 28
Remitted to State Treasurer		6,179 09

Improvements	$17,430 07
Improvements	2,941 22
Balance	579 28
Total	$54,880 94
Total receipts since organization of the Commissioners, 1883	$443,316 07
Total disbursements	442,736 79
Balance	$579 28

The Legislature has made appropriations for maintenance, current expenses and salaries, as follows:

By chapter 336, Laws of 1883	$10,000 00
By chapter 656, Laws of 1887	20,000 00
By chapter 270, Laws of 1888	25,000 00
By chapter 569, Laws of 1889	25,000 00
By chapter 84, Laws of 1890	20,000 00
By chapter 144, Laws of 1891	20,000 00
By chapter 324, Laws of 1892	20,000 00
By chapter 414, Laws of 1893	25,000 00
By chapter 654, Laws of 1894	25,000 00
By chapter 807, Laws of 1895	25,000 00
By chapter 948, Laws of 1896	25,000 00
Total	$235,000 00

For special improvements appropriations have been made as follows:

By chapter 570, Laws of 1889	$15,000 00
By chapter 302, Laws of 1891	15,000 00
By chapter 356, Laws of 1892	15,000 00

By chapter 726, Laws of 1893....................	$25,000 00
By chapter 358, Laws of 1894....................	20,000 00
By chapter 932, Laws of 1895....................	20,000 00
By chapter 950, Laws of 1896....................	10,000 00
Total....................................	$120,000 00

In compliance with statutory directions, the Commissioners have remitted to the State Treasury the receipts from the Reservation, as follows:

From October 1, 1887, to September 30, 1888......	$9,331 55
From October 1, 1888, to September 30, 1889......	7,393 77
From October 1, 1889, to September 30, 1890......	7,670 29
From October 1, 1890, to September 30, 1891......	9,327 67
From October 1, 1891, to September 30, 1892......	9,823 03
From October 1, 1892, to September 30, 1893......	10,923 85
From October 1, 1893, to September 30, 1894......	9,251 43
From October 1, 1894, to September 30, 1895......	7,448 01
From October 1, 1895, to September 30, 1896......	6,179 09
Total	$77,348 69

The following is "an estimate of the work necessary to be done and the expenses of maintaining said Reservation for the ensuing fiscal year," ending September 30, 1897:

CONSTRUCTION.

Grading, planting and other improvements.......	$25,000 00

MAINTENANCE.

Salaries, office and traveling expenses...........	$4,750 00
Reservation police............................	5,400 00

Mechanics and laborers	$7,500 00
Materials, tools, etc.	6,000 00
Miscellaneous	1,350 00
	$25,000 00

Estimated receipts from October 1, 1896, to September 30, 1897:

Inclined railway	$6,000 00
Cave of the Winds	1,200 00
Ferry and steamboat landing	750 00
Carriage service	100 00
Baggage-room	50 00
Interest	150 00
Total	$8,250 00

Respectfully submitted,

ANDREW H. GREEN,

President.

JOHN M. BOWERS,

ROBERT L. FRYER,

WILLIAM HAMILTON,

GEORGE RAINES,

Commissioners of the State Reservation at Niagara.

Boulder Bridge—Willow Island

REPORT

OF THE

Superintendent of the State Reservation at Niagara

For the Fiscal Year Ending September 30, 1896.

2

REPORT.

To the Board of Commissioners of the State Reservation at Niagara:

Gentlemen.— I respectfully submit the report of the superintendent.

The work done during the year includes the completion of the Shelter building on Goat Island, and the construction of the terminal station at the foot of the Inclined Railway; the erection of six hundred and twenty-five feet of iron guard railing on Goat Island; the grading and planting of the river shore between Mill Slip and First street; the grading and planting of the shore of Bath Island; the planting out of the cribwork on the southern shore of Goat Island; the construction of a rustic stone-arched bridge at the overflow below First street; the widening of the Midway Road on Goat Island; the construction of a gravel walk from Mill Slip to First street, and the construction of gravel walks adjacent to the Midway Road, from the Spring to the Three Sister Islands, and from the American Falls to the Parting Waters on Goat Island.

The observatory at Hennepin's View has been rebuilt and new walks constructed at Hennepin's View, at the new Shelter Building on Goat Island and at the new Terminal Station at the foot of the Inclined Railway.

SHELTER BUILDING ON GOAT ISLAND.

The Shelter Building was completed December 21, 1895. The approaches to the Shelter have been filled, graded and covered with gravel, and the adjacent gravel walks rearranged. The Shelter is built of red Medina sandstone with spacious arched openings and wide verandas. The Georgia pine ceiling of the main building is semi-circular; the floors are of concrete and the roofs of red tile. The main building is supplied with a drinking fountain of Medina stone and Tennessee marble and the plumbing in the lavatories is of the most approved quality. The building and verandas contain stationary seats of Georgia pine. The Shelter is a great accommodation to the traveling public.

A photograph of the Shelter is herewith submitted.

THE TERMINAL STATION.

The work of constructing the lower Terminal Station for the Inclined Railway was continued during the winter and spring months, and was seriously retarded by the spray and ice. The completed building was accepted by the architects June 9, 1896. It is a substantial stone structure with Gothic roof covered with "Spanish" roofing tiles, cresting and finials of copper. The outside posts are fitted with glazed portable partitions, so that the view on three sides is unobstructed. The Terminal Station is furnished with lavatories, ticket offices for railroad and steamboat, and a covered veranda extends around the three sides of the building. A photograph of the Terminal Station is herewith submitted.

The stone walks leading to the Terminal Station have been rearranged; the structure inclosing the Inclined Railway has been repaired and painted; a new manilla cable and a new steel safety cable have been attached to the cars of the Inclined Railroad; the cars have been thoroughly renovated and the machinery repaired; two drinking fountains have been provided in the Terminal Station, and a telephone service provided between the top and bottom of the Inclined Railroad. The buildings and machinery are now in good condition. Some uncertainty exists as to the effect of the accumulations of ice on the copper roofing of the Terminal Station.

RESTORATION OF THE RIVER SHORE.

The artificial stone walls and cribwork along the river shore between Mill Slip and First street have been removed, and the shore riprapped with very large stones obtained from the excavation on the lands of the Niagara Falls Hydraulic Power and Manufacturing Company. Many of the stones weigh five tons and upwards and are calculated to withstand the great force of the current in that locality. The Inlet Pier for the conduit for the Inclined Railway has also been riprapped so as to conceal the timberwork. The artificial cribwork on the southern shore of Bath Island has been removed and the greater portion of the shore of the island riprapped with large stones in a similar manner. The opportunity of obtaining the stone in sufficient size and quantity is very exceptional and the work done is of a most important and permanent character in protecting the banks from erosion and giving the shores a natural and rustic appearance. The shores have been graded, planted with shrubs and covered with turf, and the formal character of the territory obliterated.

A photograph of the river shore, showing in detail the nature of the work done during the past season, is herewith submitted.

Pockets have also been made in the cribwork along the southern shore of Goat Island and willows and vines planted so as to cover the cribwork and give the shore a more natural appearance. The slope below Bridge street has been top-dressed and planted with native forest trees.

RUSTIC WEIR BRIDGE.

The wooden bridge and iron railing at the overflow of the raceway at First street have been removed and a rustic stone-arched bridge and forest coping constructed. The rustic water-worn stones were collected in the fields in the outskirts of the city. The weir bridge and the rustic bridge and boulder bridge at Willow Island are attractive features along the riverway.

A photograph of the rustic weir bridge is herewith submitted.

THE OLD FRENCH LANDING PLACE.

The pier at "Port Day" has been filled and extended, with material excavated from the hydraulic canal, and a roadway constructed, forming a loop, affording a continuous driveway at the eastern terminus of the Reservation and commanding fine views of the river and the islands. The pond, which the roadway encircles, has been cleansed of all grass and weeds, and will form an attractive feature of the Reservation. A stone-arched bridge with a bay, commanding a fine view of the river, has been constructed over the outlet of the pond. A temporary wooden bridge spans the inlet which admits the stream to produce a current through the pond. The road around the pond has not yet been graded. When it is completed, and the embankment covered with loam and planted, it will add greatly to the attractions of the riverway.

ROADS.

The Mid-way road on Goat Island which was too narrow to allow vehicles to pass conveniently, has been increased in width to 20 feet and macadamized and graded so as to correspond with the road around the island.

The Riverway between First street and Fourth street has been macadamized, and 1,800 yards of gravel obtained from the excavations for the State Park Hotel, at a cost of 10 cents per yard, delivered, have been spread upon the roads on the Riverway, in Prospect Park and on Bath Island.

Stone gutters have been constructed near the Rapids Gate House in Prospect Park, and on the Riverway between Bridge street and Niagara street.

WALKS.

A gravel walk for the convenience of visitors has been constructed from the Three Sister Islands to the Spring on Goat Island.

A gravel walk according to the plan proposed, has also been constructed from the American Falls to the head of Goat Island.

A gravel walk has been constructed adjacent to the Mid-way Road on Goat Island, from Mill Slip to First street on the mainland, and at Hennepin's View in Prospect Park.

IRON GUARD RAILINGS.

The large number of wheelmen frequenting Goat Island, and the consequent danger of accident along the high bank, made it necessary to provide additional safeguards.

Six hundred and twenty-five feet of iron railing of the pattern designed by the late Calvert Vaux, have been erected between the Cave of the Winds and the Horseshoe Falls. The standards are set on posts of Medina sandstone, four feet long, embedded in the ground. The iron railing now extends from the American to the Horseshoe Falls.

HENNEPIN'S VIEW.

The observatory at Hennepin's View, which was in bad condition, has been entirely rebuilt. Landslides from the cliff in that locality, rendered the site of the old structure unsafe. The new structure has been built further back from the face of the cliff so as to be safe and convenient for visitors.

BUILDINGS.

The office of the Commissioners on Bath Island, and the summer houses in Prospect Park have been repaired and repainted.

The Biddle Stairway at the Cave of the Winds has been repaired.

A band stand has been erected in Prospect Park, and the iron guard railings in Prospect Park, on Goat Island and on Luna Island repainted, and other incidental work performed.

ELECTRIC LIGHTING.

Since June 1, 1896, Prospect Park and the Riverway have been lighted by 20, 2,000 candle-power arc lamps, at a cost of $50 per month. The poles, arms and wires are the property of the State, and were provided at an additional cost of $361.70.

Since the electric lights have been provided, visitors can walk or drive with safety in the evening, and the grounds are frequented much more than formerly.

Fifty additional park seats have been provided for the accommodation of visitors.

LIBERTY POLE.

After consultation with the executive committee, consent was given for placing a city liberty pole at the intersection of Falls street and the Riverway, adjacent to the soldiers' monument. A large American flag, visible the entire length of Falls street is hoisted every day, and serves to guide visitors to the entrance to the Reservation.

THE CLIFF NEAR THE INCLINED RAILWAY.

The apparently dangerous condition of the cliff adjacent to the Inclined railway, has been mentioned by the superintendent in a previous report. An examination has been made under the direction of the State Engineer and Surveyor, who reports that he is of the opinion that there is no cause for apprehended danger at this time.

The cliff on Goat Island appears to be dangerous in some places, and such portions of the loose rock as are accessible are removed from time to time.

NUMBER OF VISITORS.

The number of visitors during the year was below the average and probably did not exceed half a million. The excursion travel fell below that of last year, there being 4,363 railroad cars, carrying an estimated number of 261,780 persons.

A statement of the excursion parties to the reservation during the year is herewith submitted.

LICENSED CARRIAGE DRIVERS.

Thirty complaints have been made against licensed carriage drivers during the year. Six drivers have been excluded from the Reservation for violation of the ordinances of the commissioners. The Buffalo and Niagara Falls Electric Railway which brings its passengers to the entrance of the Reservation renders the police supervision of the licensed carriage drivers at that point more difficult.

THE STEAMBOAT LANDING.

The ice bridge of the last winter almost entirely destroyed the dock at the steamboat landing. It has been rebuilt in a very substantial manner by the steamboat company. During the greater part of the season only one boat was used because of the moderate volume of travel. Not as much trouble has been had with low water at the dock as was experienced last year. As a rule the water in the river has been higher than last year, although on several occasions it was very low.

Photographs are herewith submitted of the ledge above the Three Sister Islands at low water, and of the Cave of the Winds as it appeared February 14, 1896, when no water was falling over the central fall between Luna Island and Goat Island, the first instance of the kind ever recorded.

RESERVATION CARRIAGE SERVICE.

The Reservation carriage service has been in successful operation during the year. The manager of the service is considering the advisability of placing one or more electric carriages in the service.

THE CAVE OF THE WINDS.

About the average number of tourists visited the Cave of the Winds. As the locality is between the American and the Horseshoe Falls, the view is unparalleled, but the winding stairway leading to it is so unsuitable and fatiguing that very few persons are enabled to enjoy the beauty of the scenery below the high bank, where a walk can be easily made along the edge of the water from the American to the Horseshoe Falls.

The waiting rooms and winding stairway at the Cave of the Winds have received the usual repairs during the year. The dressing rooms are inadequate and unsuitable. Better accommodations should be provided for the traveling public.

The Biddle staircase has been examined and pronounced safe,

but it is old and dilapidated in appearance and affords but a fatiguing method of obtaining the view of the falls from below, it should be replaced by a commodious elevator with a free stairway attached. Such an elevator operated for a nominal fee of 5 cents up or down would furnish an estimated revenue of $5,000 a year, which, with the present revenue mainly derived from the Inclined railway, would almost render the Reservation self-sustaining.

EMPLOYES.

The regular force employed exclusive of laborers consists of 10 men, to wit: One superintendent, 1 clerk, 6 police gatemen and constables, and 2 Inclined railway men; 1 of whom is employed during the summer season only.

The following is a statement of labor employed during the year:

STATEMENT *of number of employes on maintenance and improvement pay-rolls, for the year ending September* 30, 1896.

PAY-ROLLS.	MAINTENANCE.			CHAPTER 932, LAWS 1895.			CHAPTER 950, LAWS 1896.		
	Foreman.	Laborers.	Teamsters.	Gardener.	Laborers.	Teamsters.	Foreman.	Laborers.	Teamsters.
1895.									
October	1	27	1	1	1	8	....	....	....
November	1	22	11	1	11	....	....	....	....
December	1	31	1	....	....	....	....	....	....
1896.									
January	1	7	1	....	....	....	....	....	....
February	1	8	1	....	....	....	....	....	....
March	1	7	1	....	....	....	....	....	....
April	1	46	3	....	....	....	....	....	....
May	1	24	4	....	....	....	....	....	....
June	1	12	1	....	....	....	1	11	1
July	1	13	1	....	....	....	1	12	1
August	1	23	1	....	....	....	1	13	3
September	1	25	1	....	....	....	1	14	4

WATER SUPPLY.

The mains for water service in Prospect Park and on Goat Island are surface pipes, and the supply is shut off during the winter season. A general plan is needed for an adequate water service for the Reservation grounds and buildings.

IMPROVEMENTS NEEDED.

The grounds are now graded and planted from Prospect Park to First street. Above First street, the grounds are in practically a natural condition, requiring very little to be done.

The road along the river should be widened, and, in the neighborhood of the Whitney homestead, the road might well be moved to the water's edge, so as to avoid the grade at that point. The present retaining wall might then be removed, and the bank graded down.

Gravel walks are also needed from First street to Port Day, where the loop of the new driveway should be completed.

Shelter buildings are needed at the American Fall and at the Horseshoe Fall, on Goat Island, at which points visitors congregate in large numbers. At times of sudden rainstorms, the present shelters are too far apart for public convenience.

The wooden bridges to the Three Sister Islands, and at Terrapin Point, are old and unsuitable. They should be replaced by structures capable of accommodating the increased number of visitors. If this is not done, in order to guard against the possibility of accident, the present bridges should practically be rebuilt during the coming season.

RECEIPTS.

The receipts by the superintendent during the year have been $4,242.40 from the Inclined railway, and $1,790 from rentals, making a total of $6,032.40.

TABULAR STATEMENTS.

Detailed statements of the receipts and expenditures of the superintendent, the amount of the pay-rolls for each month and the classification of the pay-rolls and accounts are hereto appended.

Respectfully submitted,

THOMAS V. WELCH,

Superintendent.

EXCURSIONS 1895-1896.

Date		Excursion	No. of cars.	No. visito
1895.				
Oct.	2.	Buffalo, N. Y., via steamer Pilgrim....	2	1
	4.	Pennsylvania county commissioners ..	9	5
	6.	Philadelphia, Pa., via Lehigh Valley Ry.	9	5
		Buffalo, Rochester and Lockport, via West Shore and New York Central..	20	1,2
		Buffalo, N. Y., via Electric Railway....	34	2,0
	7.	American Bottlers' Association convention	6	3
	9.	Commercial Travelers' convention......	4	2
	13.	Boston, Mass., N. E. Ticket Agents' Association...........................	3	1
		Buffalo, N. Y., via trolley.............	16	9
		Buffalo, N. Y., via New York Central Railway........................	10	6
		Rochester, N. Y., via Erie Railway.....	5	3
	16.	Washington and Philadelphia, via Lehigh Valley Railway...............	8	4
	16.	Via Nickel Plate Railway.............	8	4
	17.	Beer Brewers' Association convention..	3	1
		Union Veteran Legion................	6	3
	18.	Union Veteran Legion................	6	3
1896.				
May	7.	Buffalo, N. Y., via N. Y. C., W. S. Ry. and trolley.......................	30	1,8

1896.			No. of cars.	No. of visitors.
May	8.	Lockport, N. Y., Arbor Day Union school	42	2,520
		Buffalo, N. Y., via boat	4	240
		Jamestown, N. Y., via boat	8	480
		Rochester and Lockport, Erie and N. Y. C. Ry	10	600
	15.	Ticket Brokers' Association convention	1	60
	16.	Buffalo Field Club	4	240
	22.	Buffalo Horseshoers' convention	2	120
		Cleveland, O., M. E. conference	3	180
	24.	Buffalo, N. Y., via trolley	15	900
		Buffalo, N. Y., via N. Y. C. Ry	10	600
	26.	Hamilton and St. Catherines, via G. T. Ry	14	840
	30.	Via W. N. Y. and Pa. Ry	10	600
		Brooklyn, N. Y., Hendrickson party	2	120
		Rochester, via West Shore Ry	8	480
		Buffalo, N. Y., via trolley and rail	20	1,200
		Hornellsville, N. Y., via Erie Ry	7	420
		Rochester, N. Y., via Erie Ry	1	60
		Carbonsdale and Jamestown, via Erie Ry	8	480
	31.	Buffalo, N. Y., via steamer and rail	20	1,200
		Buffalo, N. Y., via Erie Ry	10	600
June	6.	Toronto, Ont., Masonic Society	4	240
	7.	Buffalo, N. Y., via Erie Ry	2	120
	9.	Snidersville, N. Y., public school	1	60
	13.	Delegates to St. Louis convention	10	600
		Toronto, Ont., employes Kean's shoe *manufactory*	8	480

1896.			No. of cars.	No. of visitors.
June	13.	Orangeville, Ont., Order of Foresters...	10	600
		Rochester and Buffalo, half-holiday....	6	360
	14.	Rochester and Buffalo, via Erie Ry.....	8	480
		Rochester, Buffalo, and Lockport, via N. Y. C. and West Shore............	25	1,500
		Buffalo, via trolley...................	20	1,200
		Dunkirk, via C. & B. line and trolley....	10	600
	19.	Hornellsville A. O. U. W..............	5	300
		Jamestown A. O. U. W...............	7	420
		Via W. N. Y. & P. R. R...............	10	600
		Via Erie Ry., Prescott................	8	480
		Buffalo, N. Y., Bryant and Stratton College...........................	1	60
	20.	Buffalo, N. Y., Lovejoy Street M. E. Church..........................	8	480
		Toronto, Ont., O'Keefe's brewery employes........................	4	240
		Buffalo, N. Y., via Erie Railway.......	6	360
	21.	Syracuse, N. Y., Liederkranz Society...	4	240
		Buffalo, N. Y., Trolley, N. Y. C. and Erie Railroad.........................	33	1,980
		Lockport, N. Y., Erie and N. Y. C. Ry...	9	540
	24.	Brooklyn, N. Y., Hendrickson party....	2	120
		Buffalo, N. Y., Married Men's Association............................	5	300
	25.	International League Press Association.	8	480
		Equitable Aid Union.................	12	720
		Epworth League.....................	10	600
		White Mountain tourists.............	10	600
	26.	*Dundas*, Ontario, Knox Church Society.	4	240

LOW WATER ABOVE THREE SISTER ISLANDS—NOVEMBER, 1896.

WEIR BRIDGE AT WILLOW ISLAND.

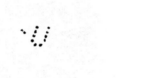

1896.			No. of cars.	No. of visitors.
June	26.	New York State Pharmaceutical Association	5	300
	27.	American Association Photographers..	7	420
		Buffalo, N. Y., Half-holiday	8	480
		Jamestown, N. Y., via Erie Ry	6	360
		Toronto, Ont., via steamer Chippewa...	5	300
	28.	Buffalo, N. Y., via Erie Ry	3	180
		Erie Ry., via Main line	8	480
		Buffalo, N. Y., via trolley	20	1,200
		Buffalo, N. Y., via N. Y. C. Ry	10	600
		Buffalo, N. Y., via Erie Ry	4	240
	29.	Lockport, N. Y., via Erie and Central...	5	300
		Buffalo, N. Y., via steamer Shrewsbury.	10	600
		Toronto, Ontario, via steamer Empress of India	5	300
	30.	Lockport, N. Y., Congregational Church.	3	180
		Toronto, Ontario, via steamer Chippewa.	20	1,200
		Chilsonburg, Ontario, Baptist Society..	10	600
uly	1.	Canada, Dominion Day excursions	30	1,800
		Buffalo and Cleveland, via steamer and trolley	20	1,200
	4.	Buffalo, N. Y., via various railroad, trolley and steamboats	150	9,000
		Buffalo, N. Y., via steamer Shrewsbury.	15	900
		Hornellsville, N. Y., via Erie Ry	7	420
		Jamestown, N. Y., via Erie Ry	5	300
	5.	Brooklyn, N. Y., Hendrickson party....	5	300
		New York city Weinachts "Schillerbund"	10	600
		New York *city "Frohbichen"* Society..	8	480

1896.		No. of cars.	No. of visitors.
July	5. New York city "Kreutzner Quartette".	9	540
	Buffalo, N. Y., via N. Y. C. Ry., and trolley	90	5,400
	Rochester, Lockport, Syracuse and Jamestown, via N. Y. C. Ry	50	3,000
	Buffalo, N. Y., via steamer Harrison	12	720
	Buffalo, N. Y., via Erie Ry	6	360
	Jamestown, N. Y., via Erie Railroad	4	240
	7. Cleveland, Ohio, via C & B Line	8	480
	Tonawanda, N. Y., M. E. Church	3	180
	National Teachers' Association	15	900
	8. Davisville, Ontario, via steamer Chippewa	6	360
	Brantford, Ontario, via C. H. & D Ry	6	360
	Buffalo, N. Y., via steamer Harrison	2	120
	Buffalo, N. Y., via Erie Ry	3	180
	9. Buffalo, N. Y., National Teachers' Association	25	1,500
	Buffalo, N. Y., reunion One Hundred and Sixteenth Regiment New York State Volunteers	8	480
	Erie, Pa., First Presbyterian Church	8	480
	Buffalo, N. Y., Holy Trinity Church	7	420
	Buffalo, N. Y., Covenant Church	7	420
	8. Buffalo, N. Y., Nat'l Educational Conv	60	3,600
	Buffalo, N. Y., via steamer Harrison	15	900
	9. Buffalo, N. Y., school teachers' excursion	6	360
	Buffalo, N. Y., school teachers' excursion	10	600
	10. Toronto, Ont., M. E. Church	8	480
	Buffalo, N. Y., per steamer "Idle Hour".	6	360

1896.			No. of cars.	No. of visitors.
July	10.	Buffalo, N. Y., N. E. A. teachers......	30	1,800
	11.	Buffalo, N. Y., Saturday half holiday...	10	600
		Buffalo, N. Y., N. E. A. convention.....	20	1,200
	12.	Buffalo, N. Y., via steamer Shrewsbury.	4	240
		Buffalo, N. Y., via Erie Ry............	5	300
		Cleveland, O., via C. & B. line..........	10	600
		Rochester and Lockport via Erie and New York Central.................	15	900
		Buffalo via N. Y. C., trolley and steamer.	100	6,000
	13.	Buffalo via steamer Harrison..........	2	120
	14.	Buffalo via steamer Shrewsbury.......	3	180
		Brantford, Ont., via G. T. Ry.........	6	360
		New York City Baptist Y. P. Union....	2	120
		Brantford, Brant Avenue M. E. Church.	5	300
		Buffalo, N. Y., Trained Nurses........	2	120
		Buffalo, N. Y., via steamer Idle Hour...	3	180
	15.	Buffalo, N. Y., Del. Nat. Hardware Merch..........................	5	300
		Hamilton, Ont., via C. H. & D. Ry......	15	900
		Hamilton, Mich., via M. C. Ry.........	16	960
		Buffalo, N. Y., via steamer Shrewsbury.	5	300
		Toronto, Ont., via steamer Chippewa...	5	300
	17.	Toronto, Ont., Broadway Tabernacle Church.........................	7	420
		Buffalo, N. Y., New York State Bankers.	5	300
		Cleveland, O., Buckeye Press Ass'n.....	4	240
		Employes, Michigan Central..........	20	1,200
		St. Thomas, Ont....................	16	960
		Detroit, Mich., Michigan Central employes.........................	4	240

1896.			No. of cars.	No. of visitors.
July	17.	Buffalo, N. Y., via steamer Harrison...	2	120
		Toronto, Ont., via steamer Chippewa...	5	300
	18.	Paris, Ont., employes woolen works.....	8	480
		Hamilton, Ont., employes wire works...	8	480
		Erie, Pa., C. M. B. A.................	6	360
		Western New York and Pa. Ry........	8	480
		Buffalo, N. Y., Orpheus Singing Society.	6	360
		Buffalo, N. Y., half holiday...........	10	600
		Buffalo, N. Y., via steamer Harrison...	3	180
		Toronto, Ont., via Empress of India....	15	900
	19.	Buffalo, N. Y., via Erie Ry............	6	360
		Buffalo, N. Y., via steamer Shrewsbury.	5	300
		Cleveland, O., via C. & B. line.........	8	480
		Toronto, Ont., Canadian Catholic Foresters........................	2	120
		Buffalo, Rochester and Lockport, via N. Y. C. Ry., trolley and steamer.....	90	5,400
	20.	Toronto, Ont., First Baptist Church....	3	180
	21.	Toronto, Ont., Queen St. M. E. Church..	3	180
		W. N. Y. & Pa. Ry., Newton's excursion.	8	480
		Cleveland, O., special party...........	6	360
		National Union Convention...........	2	120
		Rochester, N. Y., via Erie Ry..........	5	300
		Buffalo, N. Y., via steamer Harrison....	2	120
	22.	Dundas, Ont., via C. H. & B...........	12	720
		Toronto, Ont., via steamer Chippewa..	5	300
		Staford, LeRoy & Batavia, via Erie.....	8	480
		Buffalo, N. Y., Paper Makers' conv....	5	300
	23.	Toronto, Ont., wholesale and retail grocers........................	10	600

1896.			No. of cars.	No. of visitors.
July	23.	Dundas, Ont., First Baptist Church....	8	480
		Springwater, N. Y., via Erie Ry........	8	480
		LeRoy, N. Y., Baptist Church	8	480
		Toronto, Ont., wholesale and retail confectioners	8	480
	24.	Owen Sound, Ont., Ind. Order Foresters.	6	360
		Canadian Order Railway Cond........	8	480
	25.	Boston, Mass., special party...........	8	480
		Chautauqua	16	960
		Buffalo, N. Y., special trolley party....	6	360
		Buffalo, N. Y., half holiday...........	32	1,920
		Buffalo, N. Y., via steamer Harrison....	2	120
		Toronto, Ont., Watson's Candy Manufactory	3	180
	26.	Buffalo, N. Y., via steamer Shrewsbury.	11	660
		Carbondale, Pa., via Erie Ry. (two sections)	23	1,380
		Buffalo, N. Y., via Erie Ry............	5	300
		Buffalo, N. Y., via trolley............	60	3,600
		L. S. & M. S. Ry....................	10	600
		Buffalo, N. Y., via steamer Idle Hour..	8	480
		Buffalo, Rochester and Lockport, via N. Y. C. and Erie Ry.................	20	1,200
	27.	Toronto, Ont., Christie Brown Bisquit Mfg. Co	16	960
		Brantford, Ont., Union Churches......	10	600
		Buffalo, N. Y., via steamer Harrison..	3	180
	28.	Ottawa, Ont., via G. T. Ry...........	12	720
		Buffalo, N. Y., via steamer Shrewsbury.	5	300
	29.	*West Shore Ry. excursion* along line...	8	480

1896.			No. of cars.	No. of visitors.
July	29.	Oskawa, Ont., Odd Fellows and Foresters	10	600
	30.	Cleveland and Columbus, O., via C., C., C. and St. L. Ry	16	960
		Philadelphia, Pa., via Lehigh Val. Ry..	14	840
		Western N. Y. and Pa. Ry	8	480
	31.	Toronto, Ont	6	360
		Toronto, Ont., via G. T. Ry	6	360
Aug.	1.	Toronto, Ont., "Order True Blues," Orange Lodge	8	480
		Buffalo, N. Y., half holiday	20	1,200
		Toronto, Ont., Machinists and Molders' Union	7	420
		Stratford, Ont., Foresters, via G. T. Ry.	12	720
		Buffalo, N. Y., via steamer Shrewsbury.	8	480
	2.	Buffalo, N. Y., via Erie Ry	5	300
		Buffalo, N. Y., via steamer Harrison...	10	600
		Pittsburg, Pa., via B., R. and P. Ry....	10	600
		Bradford, Pa., via West. N. Y. and Pa.	6	360
		Buffalo, N. Y., via trolley	20	1,200
		Buffalo, Rochester and Lockport, via N. Y. C. R. R.	20	1,200
		Buffalo, N. Y., via steamer Idle Hour..	5	300
	3.	Toronto, Ont., wholesale and retail stationers	3	180
		Hamilton, Ont., Sons of Scotland	8	480
	3.	Hamilton Ontario, Masonic Society....	9	540
		Buffalo, N. Y., via steamer Shrewsbury.	2	120
		Barry, Ontario, via Portsmouth	2	120
	5.	*Pike, N. Y.*, Odd Fellows, via Erie Ry..	6	360

1896.		No. of cars.	No. of visitors.
Aug.	5. Leroy, N. Y., Union Churches.........	7	420
	Meadville, Pa., via W. N. Y. and Penn. Ry.............................	8	480
	Port Perry, Ontario, Fire Brigade......	9	540
	Port Perry, Ontario, Methodist and Baptist Churches.....................	7	420
	Buffalo, N. Y., via steamer Harrison...	10	600
	Hornellsville, N. Y., via Erie Ry.......	5	300
	Rochester, N. Y., via Erie Ry..........	4	240
	Jamestown, N. Y., via Erie Ry........	4	240
	Gowanda, N. Y., Presbyterian Church..	8	480
	Batavia and Leroy, N. Y., via N. Y. C. R. R.............................	10	600
	6. Buffalo, N. Y., via steamer Shrewsbury.	4	240
	Mahora, Ontario, via steamer Chicora..	10	600
	Toronto, Ontario, St. Mary's Catholic Church..........................	7	420
	East Aurora, N. Y., Union Churches...	8	480
	Big Four Ry.........................	10	600
	Buffalo, N. Y., Odd Fellows...........	9	540
	7. Buffalo, N. Y., Odd Fellows...........	15	900
	Toronto, Ontario, liquor distillers......	6	360
	Lake Erie & Western Ry., along line (three sections).....................	32	1,920
	Boston, Mass., Raymond party.........	2	120
	Buffalo, N. Y., via steamer Shrewsbury.	3	180
	8. Charlotte, N. Y., via R., W. & O. Ry....	14	840
	Rochester, N. Y., via West Shore Ry...	14	840
	Kent, Ohio, via steamboat and trolley..	10	600
	R., W. & O. Ry., along line, harvest ex.		

1896.		No. of cars.	No. of visitors.
Aug.	8. Buffalo, N. Y., half holiday...........	20	1,200
	B., N. Y. & Penn. Ry., along line.......	9	540
	Cincinnati, Ohio, Germania Singing Club...........................	4	240
	Buffalo, N. Y., Odd Fellows Convention.	6	360
	Buffalo, N. Y., Christian Endeavor Convention........................	10	600
	9. Buffalo, N. Y., via steamer Shrewsbury.	10	600
	Buffalo, N. Y., via Erie Ry............	6	360
	Buffalo, N. Y., via Central Ry.........	15	900
	Buffalo, N. Y., via trolley............	30	1,800
	Buffalo, N. Y., via Erie Ry............	4	240
	Buffalo, N. Y., via Steamer Idle Hour..	10	600
	Rochester, N. Y., via West Shore......	8	480
	Syracuse, N. Y., via N. Y. C. Ry........	8	480
	10. Lake Simcoe, Ontario, "Arc Line Home Circle".........................	7	420
	Toronto, Canada, civic holiday.........	12	720
	Buffalo, N. Y., special trolley party....	6	360
	St. Thomas, Toronto and Woodstock civic holiday excursion.............	35	2,100
	11. Dunkirk, N. Y., St. Mary's Seminary..	6	360
	Chautauqua.......................	10	600
	Fort Niagara celebration.............	25	1,500
	Canandaigua National Science Camp..	2	120
	Auburn, N. Y., Hop Growers Picnic....	10	600
	Milton, Ontario, Union Sunday School..	8	480
	Western N. Y. and Penn. Ry., along line............................	9	540
	New York State Hotel Men, Convention.	4	240

1896.			No. of cars.	No. of visitors.
Aug.	11.	Buffalo, N. Y., via steamer Shrewsbury.	3	180
	12.	Coburg and Port Hope, Ont., Foresters.	12	720
		Plon, Ont., via G. T. Ry.	1	60
		Buffalo, N. Y., via Steamer Shrewsbury.	2	120
	13.	Buffalo, N. Y., Sisters of Charity	4	240
		Bradford Ont., Fire Department and friends	6	360
		Lockport, N. Y., German Lutheran Church	6	360
		St. Thomas, Ont., Board of Trade and friends	6	360
		Cincinnati, Ft. Wayne and Peoria, L. E. & W. Ry	12	720
		Branton, Ont., via Steamer Chippewa	7	420
		Toronto, Ont., Knight Street Methodist Church	6	360
		Buffalo, N. Y., via Steamer Shrewsbury.	3	180
	14.	Galt, Ont., Civic holiday	6	360
		Washington and Philadelphia, special party	17	1,020
		Cincinnati, O., via L. E. and Western Ry.	10	600
	15.	Buffalo, N. Y., via Steamer Shrewsbury.	5	300
		Toronto, Ont., via Steamer Chippewa	3	180
		Chautauqua, N. Y	9	540
		Buffalo, N. Y., via N. Y. C. & West Shore Ry.	20	1,200
	16.	Jamestown, N. Y., Pancius Singing Club	8	480
		Lake Shore and M. S. Ry., along line	8	480
		Pittsburg, Pa., via B. R. & P. Ry	9	540

1896.			No. of cars.
Aug.	16.	Rochester and Syracuse, via N. Y. C....	10
		Bradford, Pa., Western N. Y. & Pa.....	8
		Cleveland, O., via steamer and trolley..	8
		Buffalo, N. Y., via trolley..............	20
		Buffalo, N. Y., via N. Y. C. Ry.........	15
		Cleveland, O., via Erie Ry.............	8
		Buffalo, N. Y., via Erie Ry............	3
		Rochester, N. Y., Poland Singing Club..	4
		Buffalo, N. Y., via Steamer Shrewsbury.	6
	18.	Buffalo, N. Y., via Steamer Shrewsbury.	1
		Burlington, Ont., via G. T. Ry.........	4
		Binghamton, N. Y.,.via Erie Ry........	9
		Nickel Plate Ry., along line...........	12
		Western N. Y. and Pa., along line......	10
		Erie Ry., along line..................	10
	17.	London, Ont., Civic holiday...........	10
		Canada, G. T. Ry., employes...........	8
	19.	Clarkson, Ont., A. O. U. W............	6
		Lockport, N. Y., firemen en route......	6
		Buffalo, N. Y., W. C. T. U.............	5
		Bradford, Pa., via Western N. Y. and Pa. Ry.............................	5
		L. S. and M. S. Ry., Y. M. C. A.........	6
	21.	St. Catherines, Ont., via G. T. Ry......	1
		Buffalo, N. Y., via Steamer Shrewsbury.	3
		Yonkers, N. Y., Fire Department and friends...........................	6
		East Aurora, N. Y., Union Churches...	6
		Cattaraugus., Co., Farmers Picnic......	10
		Toronto, Ont., A. O. U. W. lodges......	8

1896.		No. of cars.	No. of visitors.
Aug. 21.	Lockport, N. Y., State Firemen's Convention	12	720
	St. Louis, Mo., via Wabash Ry., and C. & B. line	10	600
	Western N. Y. and Penn. Ry., along line.	9	540
	Georgetown, Ontario	9	540
22.	Stratford, Ontario, via Grand Trunk Ry. (three sections)	20	1,200
	Toronto, Canada, via Grand Trunk Ry., Coleman Safe Works	8	480
	Buffalo, N. Y., via steamer Shrewsbury.	11	660
	Reunion Col. P. A. Porter's Regiment, Eighth New York Heavy Artillery	6	360
	Society of Engineers' Convention	•2	120
17.	Buffalo, via steamer Shrewsbury	2	120
18.	Buffalo, via steamer Shrewsbury	2	120
	Lockport, Pine Street Lutheran Church.	10	600
23.	New York city Schlichlenzer Club	2	120
	Rochester and Syracuse, via N. Y. C. Ry.	20	1,200
	Buffalo, via N. Y. C. Ry	15	900
	Buffalo, via trolley	20	1,200
	Buffalo, via steamboat	5	300
	Cleveland, Ohio, via C. & B. line	10	600
24.	Pittsburg, Pa., Comm. Ry. Agency	2	120
	Buffalo, N. Y., via trolley, special	4	240
26.	Buffalo, N. Y., via trolley, special	6	360
	Cleveland, Ohio, via C. & B. line	8	480
	Toronto, Ontario	7	420
	Buffalo, N. Y., via steamer Shrewsbury.	1	60
27.	*Buffalo, N. Y., via steamer* Shrewsbury.	2	120

1896.			No. of cars.	No. visit
Aug.	27.	Cleveland, Ohio, Knights Pythias......	6	
		Del., Lack. & West. Ry., along line.....	14	
	28.	Washington, D. C., special party......	8	
		Philadelphia, Pa., special party.......	8	
		Buffalo, N. Y., via trolley, special party.	5	
	29.	Reunion Twenty-third Independent Battery New York Volunteers..........	2	
		American Association Advancement Science Convention................	6	
		Lehigh Valley Ry., along line.........	12	
		Buffalo, N. Y., half-holiday excursion...	20	1
		Ohio and Indiana, via Nickel Plate Ry..	20	1
		Erie Ry., along line..................	10	
		Buffalo, N. Y., Knights Pythias........	8	
		Cleveland, Ohio, via C. & B. line.......	8	
		Cleveland, Ohio, via L. S. & M. S. Ry...	10	
		St. Paul, Minn., G. A. R., en route......	10	
		Rochester and Syracuse, via N. Y. C. and W. S. Ry........................	15	
		Buffalo, N. Y., via trolley.............	25	1
		West. N. Y. and Penn. Ry., along line..	9	
		Buffalo, N. Y., via steamer Shrewsbury.	2	
	30.	Buffalo, N. Y., via steamer Shrewsbury.	7	
		Jamestown, N. Y., via Erie Ry........	2	
	31.	Buffalo, N. Y., via steamer Shrewsbury.	2	
		Baltimore, Md., Iron Hall Underwriters.	1	
Sept.	2.	West Seneca, N. Y., St. Patrick's S. S..	2	
		Buffalo, N. Y., National Association State Engineers	6	
		Buffalo, N. Y., U. S. Veterinary Medical Association	5	

1896.			No. of cars.	No. of visitors.
Sept.	2.	Buffalo, N. Y., via steamer Shrewsbury.	3	180
		Jamestown, N. Y., via Erie Ry........	5	300
	3.	Buffalo, N. Y., via steamer Shrewsbury.	5	300
	5.	Buffalo, N. Y., via steamer Shrewsbury.	5	300
		Brooklyn, N. Y., via Erie Ry..........	9	540
	6.	New York, N. Y., via Erie Ry.........	9	540
		Buffalo, N. Y., via steamer Shrewsbury.	7	420
		New York, N. Y., Co. B, 71st Regiment..	1	60
		Cleveland, O., via C. and B. line......	8	480
		Pittsburg, Pa., via Allegany Valley Ry.	10	600
		Buffalo, N. Y., via trolley	20	1,200
		Buffalo, N. Y., via N. Y. C. and West Shore Ry	10	600
		Boston, Mass., Cook's tourist party....	3	180
		Brooklyn, N. Y., Hendrickson party...	3	180
		Li Hung Chang party................	1	60
	7.	Labor day excursion, via N. Y. C. R. R..	25	1,500
		Labor day excursion, via West Shore R. R	20	1,200
		Labor day excursion, via Erie R. R....	10	600
		Labor day excursion, via R., W. and O. R. R	10	600
		Labor day excursion, via W. N. Y. and Penn. R. R.......................	10	600
		Buffalo, N. Y., via steamer Idle Hour..	10	600
		Cleveland, O., via. C. and B. line......	10	600
		Buffalo, N. Y., via trolley............	50	3,000
	8.	Erie Ry., via main line...............	24	1,440
		Western N. Y. and Penn. Ry. along line.	12	720
		Buffalo, N. Y., Deaf and Dumb Institute.	1	60

1896.			No. of cars.	No. of visitors.
Sept.	8.	American Road Masters Ass'n Conv...	4	240
		Ellicottville, N. Y., Union Churches...	5	300
		Cleveland, O., via steamer Shrewsbury.	5	300
		Erie Ry., via along line..............	41	2,460
	11.	American Society Ry. Supts. Conv....	5	300
		Michigan Press Ass'n Convention......	4	300
		Toronto, Ont., via Erie Ry............	2	120
		Toronto, Ont., via steamer Chippewa..	9	540
		Toronto Road Masters, via steamer Chippewa	5	300
		Buffalo, N. Y., via steamer Shrewsbury.	1	60
		Detroit, Mich., Press Association......	4	240
	13.	Pittsburgh, Pa., via B. R. and Pitts. Ry.	10	600
		Watertown, N. Y., via R., W. and O., 39th Separate Co	4	240
		Buffalo, N. Y., via C. and B. line......	6	360
		Buffalo, N. Y., via N. Y. C. R. R.......	15	900
		Buffalo, N. Y., via trolley	24	1,440
		Rochester, N. Y., via West Shore......	9	540
		L. S. and M. S. Ry., via along line.....	10	600
	15.	Pittsburgh, Pa., via B. R. and Pitts. Ry.	8	480
		Insurance Union Convention..........	4	240
	16.	Democratic Delegates to State Conv..	6	360
		Buffalo, N. Y., via steamer Shrewsbury.	12	720
	17.	Buffalo, N. Y., via National Builders' Association	12	720
	19.	Buffalo, N. Y., American Public Health Association	4	240
	20.	Johnsonburg, via Erie Ry.............	12	720
		Philadelphia, Pa., via Lehigh Valley...	8	480

1896.		No. of cars.	No. of visitors.
Sept. 20.	Buffalo, N. Y., via trolley	10	600
	Buffalo and Rochester, via West Shore and N. Y. C	12	720
	Carbondale, Pa., via Erie Ry	10	600
	Buffalo, N. Y., via Erie Ry	2	120
	Buffalo, N. Y., via steamer Shrewsbury.	2	120
	Buffalo, N. Y., Board of Health Commissioners	7	420
22.	Binghamton, N. Y., via Del. & Lack. Ry.	5	300
	Niagara on the Lake, from encampment Canada Militia	6	360
28.	Agents Metropolitan Life Insurance Co., Convention	6	360
		577	34,620

RECAPITULATION.

		No. of cars.	Est. visitors.
1895.	October	149	8,940
1896.	May	229	13,740
	June	379	22,740
	July	1,558	93,480
	August	1,471	88,260
	Setember	577	34,620
		4,363	261,780

THE COMMISSIONERS OF THE STATE RESERVATIO
NIAGARA, IN ACCOUNT WITH HENRY E. GREG
TREASURER.

1895.				
Oct.	1.	Balance on hand this date............		$3,8
		Receipts.		
Oct.	24.	Quarterly advance from the State Comptroller for maintenance.................	$6,250 00	
1896.				
Jan.	15.	Quarterly advance from the State Comptroller for maintenance.................	6,250 00	
April	22.	Quarterly advance from the State Comptroller for maintenance.................	6,250 00	
July	25.	Quarterly advance from the State Comptroller for maintenance.................	6,250 00	
				25,0

Special appropriation as per chapter 932, Laws of 1895:

1895.			
Oct.	11.	Payment by State Comptroller on account........	$903 45
Nov.	2.	Payment by State Comptroller on account........	2,838 50
	14.	Payment by State Comptroller on account........	674 75

Shore Below Willow Island.

SHELTER—GOAT ISLAND.

1895.			
Nov. 21.	Payment by State Comptroller on account........	$2,788 83	
Dec. 11.	Payment by State Comptroller on account........	926 33	
1896.			
Jan. 10.	Payment by State Comptroller on account........	73 50	
23.	Payment by State Comptroller on account........	5,997 53	
29.	Payment by State Comptroller on account........	302 36	
March 10.	Payment by State Comptroller on account........	60 00	
May 8.	Payment by State Comptroller on account........	56 00	
June 9.	Payment by State Comptroller on account........	20 00	
20.	Payment by State Comptroller on account........	2,788 82	
			$17,430 07

Special appropriation as per chapter 950, Laws of 1896:

1896.			
July 8.	Payment by State Comptroller on account........	$496 75	
18.	Payment by State Comptroller on account........	709 75	
Aug. 12.	Payment by State Comptroller on account........	996 34	
Sept. 9.	Payment by State Comptroller on account........	738 38	
			2,941 22

Nov.	2.	Draft on Bank of Niagara for October receipts..........	$197 40	
Dec.	2.	Draft on Bank of Niagara for November receipts	34 90	
	31.	Draft on Bank of Niagara for December receipts	20 15	
1896.				
Feb.	1.	Draft on Bank of Niagara for January receipts.........	222 35	
March	2.	Draft on Bank of Niagara for February receipts	70 70	
	31.	Draft on Bank of Niagara for March receipts	71 60	
May	2.	Draft on Bank of Niagara for April receipts	57 20	
June	1.	Draft on Bank of Niagara for May receipts	247 95	
	30.	Draft on Bank of Niagara for June receipts	431 25	
Aug.	1.	Draft on Bank of Niagara for July receipts	1,296 40	
Sept.	1.	Draft on Bank of Niagara for August receipts	1,772 70	
	30.	Draft on Bank of Niagara for September receipts	1,609 80	
				$6,
1896.				
Jan.	21.	Dividend on deposits in Cataract Bank.		
1895.				
Dec.	31.	Interest on balances in Manufacturers and Traders' Bank..................	$30 05	

1896.			
March 31.	Interest on balances in Manufacturers and Traders' Bank..................	$29 12	
June 30.	Interest on balances in Manufacturers and Traders' Bank..................	34 34	
Sept. 30.	Interest on balances in Manufacturers and Traders' Bank..................	24 84	
			$118 35
	Total....................................		$54,880 94

EXPENDITURES.

Date.		No. of Abstract.	No. of Voucher.		Amount.	Total.
1895.						
Oct.	15..	XCV	1233	Hardwicke & Co., hardware, etc.	$35 74	
			1234	Hardwicke & Co., hardware, etc.	10 48	
			1235	George E. Wright, carpenter's work	369 38	
			1236	John Irwin & Co., work on electric bells	11 26	
			1237	P. J. Davy, plumbing	9 89	
			1238	F. W. Oliver Co., hardware	1 40	
			1239	William F. Wall Rope Co., cable	142 49	
			1240	George Haeberle, repairs	35 29	
			1241	E. O. Babcock, stationery	20 88	
			1242	P. C. Flynn & Son, painting	29 25	
	22..	....	1243	Milton C. Johnson & Co., stationery	27 13	
			1244	Henry E. Gregory, treasurer and secretary, office and other expenses	105 65	
	24..	....	1245	Alexander Henschel, office expenses	12 30	
Nov.	2..	....	1246	Pay-roll for October	1,966 79	
			1247	Thomas V. Welch, superintendent, office expenses	35 74	
	8..	....	1248	E. H. Cannon, stove	35 05	
Dec.	2..	....	1249	Pay-roll for November	1,410 41	
			1250	Thomas V. Welch, superintendent, office expenses	29 76	
	6..	....	1251	Supplementary pay roll	614 76	
			1252	Colman Nee, stone for road	127 00	
			1253	John Sullivan, stone for road	144 00	
			1254	Timothy Horan, stone for road	237 00	
			1255	D. Phillips, excavating, etc.	718 35	
Dec.	13..	....	1256	Pay-roll	334 50	
			1257	Colman Nee, repairs to road on Goat Island	10 00	
			1258	James O'Brien, repairs to road on Goat Island	45 00	
			1259	John Beagan, repairs to road on Goat Island	63 00	
			1260	Cornelius Burns, repairs to road on Goat Island	99 00	
	17..	....	1261	W. A. Philpott, carpentering	12 50	
			1262	George E. Wright, repairs, etc.	17 18	
			1263	William Shepard & Son, mason work on bridges	28 00	

			1264	Mackworth Bros. Co., repairing roofs	0 30	
			1265	Thomas E. McGarigle, repairing tools	23 88	
			1266	Thomas E. McGarigle, repairing tools	11 91	
			1267	Hardwicke & Co., hardware, etc	50 62	
			1268	Hardwicke & Co., hardware, etc	34 04	$6,868 93
Dec.	31..	XCVI	1269	Henry E. Gregory, treasurer and secretary, salary three months	275 00	
			1270	Alexander Henschel, clerk to president	25 00	
			1271	Pay-roll for December	1,155 29	
			1272	Thomas V. Welch, superintendent, office expenses	47 55	
			1273	Henry E. Gregory, treasurer and secretary, office expenses	25 36	1,528 20
1896.						
Jan.	3..	XCVII	1274	William Hamilton, commissioner, traveling expenses	82 03	
	17..	....	1275	[illegible]ge Haeberle, repairs	95 97	
			1276	George Haeberle, repairs	43 81	
Feb.	3..	....	1277	Pay roll for January	1,249 54	
			1278	[illegible]V. Welch, superinten[illegible]dent, office expenses	34 46	
March	2..	....	1279	Pay-roll for February	1,140 17	
			1280	Thomas V. Welch, superintendent, office [illegible]	49 76	
	6..	....	1281	Maloney & McCoy, ice	59 40	
			1282	Howard Iron Works, cast steel rope	57 75	
			1283	F. W. Oliver Co., repairs	6 91	
			1284	Hardwicke & Co., tools, etc	8 12	
			1285	Hardwicke & Co., [illegible]ols etc	25 37	
			1286	Hardwicke & Co., repairing inclined railway car	3 53	2,856 82
March	31..	XCVIII	1287	Henry E. Gregory, treasurer and secretary, salary three months	275 00	
			1288	Pay-roll for March	1,176 79	
			1289	[illegible]as V. [illegible]h, superintendent, [illegible]ffice expenses	47 99	
			1290	Henry E. Gregory, [illegible]urer and secretary, [illegible]ffice expenses	19 90	
			1291	William Shepard & Son, inclined railway repairs	18 45	
			1292	McDonald & Welch, coal	102 50	
			1293	F. W. Oliver Co., repairs	14 10	
			294	P. C. Flynn & Son, painting	74 50	1,729 23
May	1..	XCIX	1295	Pay roll for April	1,997 91	
			1296	Thomas V. Welch, superintendent, office expenses	49 85	
May	18..	....	1297	George Haeberle, repairs, etc	37 27	
			1298	William Hamilton, commissioner, traveling expenses	37 47	

EXPENDITURES—(*Continued*).

Date.	No. of Abstract.	No. of Voucher.		Amount.	Total.
1896.					
May 18..	....	1299	Hardwicke & Co., hardware, etc	$7 27	
		1300	Power City Lumber Co., lumber	7 50	
		1301	Power City Lumber Co., lumber	22 01	
		1302	Frederick Batchelor, seed	27 00	
		1303	Hardwicke & Co., repairs, etc	138 39	
		1304	R. D. Young, gravel	180 00	
June 1..	....	1305	Pay-roll for May	1,497 42	
		1306	Thomas V. Welch, superintendent, office expenses, etc	28 18	
5..	....	1307	Supplementary pay-roll	428 50	
		1308	Niagara Sand Co., gravel	24 41	
		1309	Braas Bros., repairs	113 94	
		1310	P. C. Flynn & Son, painting	103 98	
		1311	George E. Wright, repairs	30 75	
		1312	P. J. Davy, plumbing	35 53	
		1313	J. L. Shepard & Co., mason work, etc	90 22	
		1314	Hardwicke & Co., hardware	45 03	
30..	....	1315	Pay-roll for June	1,404 91	
		1316	Thomas V. Welch, superintendent, office expenses, etc	40 69	
		1317	Henry E. Gregory, treasurer and secretary, salary three months	275 00	
		1318	Thomas V. Welch, superintendent office expenses	99 67	
		1319	Peter Henderson & Co., tools	9 00	
		1320	J. L. Shepard & Co., roads and walks	61 55	
		1321	Hardwicke & Co., hardware, etc	129 11	
		1322	P. C. Flynn & Son, painting	455 80	
		1323	Braas Brothers, repairs	50 47	
		1324	Braas Brothers, carpenter's work	46 02	
					$7,474 85
Aug. 1..	C	1325	Pay-roll for July	1,475 67	
		1326	Thomas V. Welch, superintendent, office expenses, etc	49 95	
8..	....	1327	Howard Iron Works, park settees	200 00	
		1328	Thomas E. McGarigle, tools, etc	9 86	

			No.		Amount
			1329	Estate of John Johnson, tools	25 06
			1330	P. C. Flynn & Son, painting	179 79
			1331	Niagara Falls Hydraulic P. & M. Co., electric lighting	100 00
			1332	Niagara Falls Hydraulic P. & M. Co., electric lighting	361 00
			1333	Brass Brothers, carpenter's work	208 13
Sept.	1..	....	1334	Pay-roll for August	1,798 41
			1335	Thomas V. Welch, superintendent, office expenses	47 74
	5..	....	1336	Niagara Falls Hydraulic P. & M. Co., electric lighting	50 00
			1337	E. O. Babcock, stationery	18 85
			1338	P. C. Flynn & Son, painting	64 90
			1339	H. H. Baker & Co., national flag	10 00
			1340	Brass Bros., carpenter's work	139 72
			1341	Hardwicke & Co., hardware, etc	4 79
			1342	Hardwicke & Co., hardware, etc	44 10
			1343	Hardwicke & Co., hardware, etc	33 70
	30..		1344	Pay-roll for September	1,790 29
			1345	Thomas V. Welch, superintendent, office expenses, etc	30 46
			1346	Henry E. Gregory, treasurer and secretary, salary three months	275 00
			1347	F. W. Oliver Co., tools, etc	7 81
			1348	F. W. Oliver Co., tools, etc	1 66
			1349	Hardwicke & Co., hardware, etc	43 59
			1350	John Irwin & Co., telephone	40 00
			1351	Thomas E. McGarigle, machinist's work	85 79
			1352	P. C. Flynn & Son, painting	30 85
			1353	Niagara Falls Hydraulic P. & M. Co., electric lighting	50 00
			1354	Globe Ticket Co., tickets for inclined railway	12 50
			1355	P. J. Davy, plumbing	86 68
			1356	Frederick Batchelor, seed	11 00
			1357	Frederick Batchelor, seed	6 00

7,

PAYMENTS OUT OF $20,000, AS PER CHAPTER 932, LAWS OF 1895.

Date.	Series F.			Amount.	Total.
1895.					
Oct. 14..	II.......	14	Pay-roll	$691 75	
		15	Dennis McDonald, [illegible]	100 00	
		16	Estate of John Johnson, repairs of tools	35 20	
		17	William S[illegible]d & Son, [illegible] work	31 50	
		18	Peter Hen[illegible]n & Co., grass [illegible]ed	45 00	
Nov. 2..		19	[illegible] Shepard & Son, [illegible] building	2,[illegible]88 50	
14..		20	Pay-roll	[illegible]86 75	
		21	[illegible] McDonald, [illegible]	108 00	
21..		22	R. D. Young, terminal station	2,[illegible]88 83	
Dec. 11..		23	Pay-roll	487 50	
		24	Ellwanger & Barry, shrubs	30 18	
		25	Ellwanger & Barry, shrubs	58 15	
		26	Ellwanger & Barry, shrubs	26 50	
		27	Dennis Mc[illegible]ald, inspector	104 00	$8,131 86
1896.					
Jan. 10..	III	28	Samuel Parsons, Jr., [illegible] [illegible], traveling [illegible]	26 50	
		29	[illegible] Wirth, [illegible] [illegible], traveling [illegible]	47 00	
23..		30	William [illegible] & [illegible], shelter [illegible]	2,838 50	
		31	William [illegible] & [illegible], shelter [illegible], extra work	30 21	
		32	R. D. Young, [illegible] station	2,788 82	
		33	D. McDonald, inspector	20 00	
Jan. 29..	....	34	Vaux & Emery, [illegible] building	176 36	
		35	[illegible] L. Emery, traveling [illegible]	74 25	
		36	[illegible] Vaux, traveling [illegible]	51 75	
Mar. 12..	....	37	Dennis McDonald, [illegible]	60 00	6,433 39
May 8..	IV	38	Dennis McDonald, inspector	56 00	
June 10..	....	39	Dennis McDonald, inspector	20 00	
25..	....	40	Robert D. Young, terminal station	2,788 82	2,864 82

PAYMENTS OUT OF $10,000, AS PER CHAPTER 950, LAWS OF 1896.

		Series G.				
July	8..	I........	1	Pay-roll	496 75	
	23..	....	2	R. D. Young, terminal station	563 27	
			3	Vaux & Emery, terminal station	146 48	
Aug.	13..	....	4	Pay-roll	543 13	
			5	Kearney & Barrett, iron railing	284 40	
			6	James Reynolds, iron railing	58 56	
			7	J. L. Shepard & Co., mason work	110 25	
Sept.	12..	....	8	Pay-roll	540 88	
			9	Jackson Architectural Iron Works, iron railing	35 00	
			10	W. A. Shepard, mason work	52 50	
			11	P. J. Davy, iron railing	110 00	
						2,941 22

REMITTANCES TO THE STATE TREASURER.

1895.				
Nov.	2..	Draft for October receipts	$197 40	
Dec.	2..	Draft for November receipts	34 90	
	31..	Draft for December receipts	20 15	
1896.				
Feb.	1..	Draft for January receipts	222 35	
March	2..	Draft for February receipts	70 70	
	31..	Draft for March receipts	71 60	
May	2..	Draft for April receipts	57 20	
June	1..	Draft for May receipts	247 95	
	30..	Draft for June receipts	431 25	
Aug.	1..	Draft for July receipts	1,296 40	
Sept.	1..	Draft for August receipts	1,772 70	
	30..	Draft for September receipts	1,609 80	
				6,032 40
Feb.	1..	Dividend on deposits in Cataract Bank	28 34	
				28 34
1895.				
Dec.	31..	Interest on balances in M. & T. Bank	30 05	

REMITTANCES TO THE STATE TREASURER — (*Concluded*).

Date.		Amount.	Total.
1896.			
March 31..	Interest on balances in M. & T. Bank	$29 12	
June 30..	Interest on balances in M. & T. Bank	34 34	
Sept. 30..	Interest on balances in M. & T. Bank	24 84	
			$118 35
	Cash balance in treasurer's hands		579 28
	Total		$54,880 94

HENRY E. GREGORY,
Treasurer.

We, the undersigned, hereby certify that we have examined the foregoing report of the treasurer, the vouchers and other papers, and we find the report and accompanying documents correct, and that the treasurer has properly accounted for all moneys received and disbursed by him during the year ended September 30, 1896.

ROB'T L. FRYER,
WM. HAMILTON,
Commissioners.

ACCOUNTS.

CLASSIFICATION OF ACCOUNTS.

Fences	$13 05
Iron railing	173 90
Stairways	9 45
Furniture	13 25
Freight	20 35
Stationery (Niagara office)	66 86
Water-pipes	346 71
Ice	61 40
Treasurer and secretary	1,100 00
Seed	44 00
Settees	203 24
Coal	102 50
Signs	38 19
Sidewalks	90 30
Commissioners' expenses	119 50
Treasurer and secretary (traveling expenses)	8 80
Treasurer and secretary (office expenses)	104 41
Tools	242 95
National flag	10 00
Expense	75 00
Inclined railway	2,033 62
Walks	1,311 73
Buildings	1,570 38
Salaries (superintendent and clerk)	2,899 98
Police	5,325 00

Prospect park	$2,367 64
Goat island	1,599 25
Roads	6,346 97
Niagara office	571 67
Bridges	129 29
Parapet wall	5 25
Gutters	37 27
Electric lighting	561 00
Observatory	148 37
	$27,751 28

Special Improvements under Chapter 932, Laws of 1895.

Seed	$45 00
Grading	1,318 75
Walks	88 50
Shelter building	6,423 94
Terminal station	8,740 10
Tools	35 20
Planting	433 25
Services	10 50
Shrubs	334 83
	$17,430 07

Special Improvements under Chapter 950, Laws of 1896.

Roads	$163 00
Walks	959 76
Terminal station	709 75
Iron railing	933 71
Bridges	175 00
	$2,941 22

PROSPECT PARK. FROM THE TOWER.

FOURTEENTH ANNUAL REPORT

OF THE

COMMISSIONERS

OF THE

ATE RESERVATION

AT NIAGARA,

·the Fiscal Year from October 1, 1896, to September 30, 1897.

Commissioners:

ANDREW H. GREEN, President,

I. BOWERS, ROBERT L. FRYER,

M HAMILTON, GEORGE RAINES.

Treasurer and Secretary:

HENRY E. GREGORY.

Superintendent:

THOMAS V. WELCH.

Landscape Architect:

SAMUEL PARSONS, Jr.

IITTED TO THE LEGISLATURE JANUARY 25, 1898.

WYNKOOP HALLENBECK CRAWFORD CO.
STATE PRINTERS,
NEW YORK AND ALBANY.
1898.

STATE OF NEW YORK.

No. 31.

IN ASSEMBLY,

JANUARY 25, 1898.

FOURTEENTH ANNUAL REPORT

OF THE

Commissioners of the State Reservation at Niagara.

NEW YORK, *January* 24, 1898.

To the Honorable, the Speaker of the Assembly, Albany:

Sir.— I transmit herewith for presentation to the Legislature the Fourteenth Annual Report of the Commissioners of the State Reservation at Niagara, for the fiscal year ended September 30, 1897.

Respectfully,

ANDW. H. GREEN,

President.

REPORT.

To the Honorable, the Legislature of the State of New York:

The Commissioners of the State Reservation at Niagara, as required by law, submit their report for the fiscal year begun October 1, 1896, and ended September 30, 1897, being their fourteenth annual report.

The work of the year has been mainly in continuance of the system of improvement and restoration which the commissioners have had in view, and in accordance with their previously adopted plan.

The river shore at different points has required grading, filling and planting, and this work has been carried on notably between Port Day and First street. The crib work on the southern shore of Goat Island has been covered with soil and planted so as to make it invisible from the island The shore of Bath Island has been graded, planted and rip-rapped. A loop driveway at Port Day has been constructed, graded and gravelled. A stone arched bridge has been erected at the overflow below First street, and another stone arched bridge has been erected at Port Day. The river road between Falls street and Niagara street has been macadamized. Seventh street from Buffalo avenue to the river road has been widened. This is not intended as a complete statement of the year's work, details of which are contained in the report of the superintendent, appended to this report.

The Reservation has been public property for more than 12 years (July 15, 1885 to September 30, 1897). In each annual report of the commissioners, the work of the preceding year has been narrated. First, in 1885-86, soon after the opening of the Reservation, they undertook the work of demolition. They caused the structures that defaced the river shore and the islands to be removed. This having been accomplished, the work of construction and restoration was proceeded with; gradually, it must be admitted, since the amount of money placed at their disposal by the Legislature was at no time large enough to allow of their entering into contracts for extensive improvements. The work has necessarily been done piecemeal, year by year, as the money was granted. With a sufficient sum of money at their disposal, the commissioners might ten years ago have undertaken to carry out, on a large scale and in a comprehensive way, the general plan of improvement they had adopted. It is profitless, however, to refer to what might have been done. Under the circumstances, and with annual appropriations for improvements since 1889 only, the commissioners are inclined to think that what they have effected in the way of improvements at Niagara is not inconsiderable.

The following especially may be mentioned: A substantial crib-work or solid timber frame filled with stone, has been con structed along the southern shore of Goat Island to prevent erosion; the bridges to Bath and Goat Islands have been strengthened; a new macadamized road has been laid on Goat Island, and the damaged bank of the island restored; Bath Island has been enlarged and planted; the large area between the Grove and Bridge street has been graded and planted, and the high retaining wall at its base partly taken down—a great improvement; a very substantial bridge to Luna Island has been erected in place of the old wooden structure that had been there so long;

a suitable iron railing has been placed at dangerous points; a shelter building has been erected on Goat Island; and a new terminal station has been constructed at the foot of the inclined railway.

The following publications have been printed with the annual reports of the commissioners:

(Fourth Report)—Catalogue of Niagara Flora, by Hon. David F. Day, 65 p.

(Sixth Report)—The History of the Niagara River, by Prof. G. K. Gilbert, 24 p. with illustrations.

(Seventh Report)—Report on the Outline and Crest of the Falls, with a survey to determine the crest line of the falls, with tables, maps and illustrations, by Hon. J. Bogart and A. S. Kibbe, 27 p.

(Eighth Report)—Niagara Falls, Its Past, Present and Prospective Condition, by Jas. Hall, LL. D., 24 p.

(Ninth Report)—Bibliography of Hennepin, by C. K. Remington, Esq., 19 p.

Bibliography of La Salle, by C. K. Remington, Esq., 5 p.

Map of Historic Niagara, by W. C. Johnson, Esq.

(Tenth Report)—Historic Niagara, by Hon. P. A. Porter, 15 p.

Niagara Bibliography, by C. K. Remington, Esq., 35 p.

(Eleventh Report)—Report to the Constitutional Convention relating to the Diversions of Waters of the Niagara River, by Hon. J. A. Barhite, 13 p.

Niagara Bibliography (continued), by C. K. Remington, Esq., 7 p.

The Duration of Niagara Falls and the History of the Great Lakes, by J. W. Spencer, Ph. D., 126 p. with illus. and index.

In view of the recent development of power at Niagara by corporations for manufacturing and other purposes, and of the apparent impossibility of preventing further grants to corporations, the future of the falls and river becomes a matter of speculation and conjecture. With reference to this, the remarks of an eminent man of science have been circulating in the public prints as follows:

"The originators of the work so far carried out and now in progress hold concessions for the development of 450.000 horse-power from the Niagara River. I do not myself believe and such limit will bind the use of this great natural gift, and I look forward to the time when the whole water from Lake Erie will find its way to the lower level of Lake Ontario through machinery doing more good for the world, than that great benefit which we now possess in contemplation of the splendid scene which we have presented before us at the present time by the waterfall of Niagara. I wish I could think it possible that I could live to see this grand development."

While beauty and grandeur in natural scenery make some impression upon the modern man of science, it is vague and easily erased, in comparison with that produced by utility. The Falls of Niagara are no doubt a sublime spectacle, but with the sensations of beauty and awe in the scientific mind there is commingled some more or less conscious disturbance and disapproval in seeing so mighty a mass of water rolling, dashing, foaming along and tumbling over a precipice; and the scientific beholder perhaps involuntarily ejaculates: "To what purpose is this waste? How much better would it be if the river and falls could be utilized in turning immense turbine wheels and generating

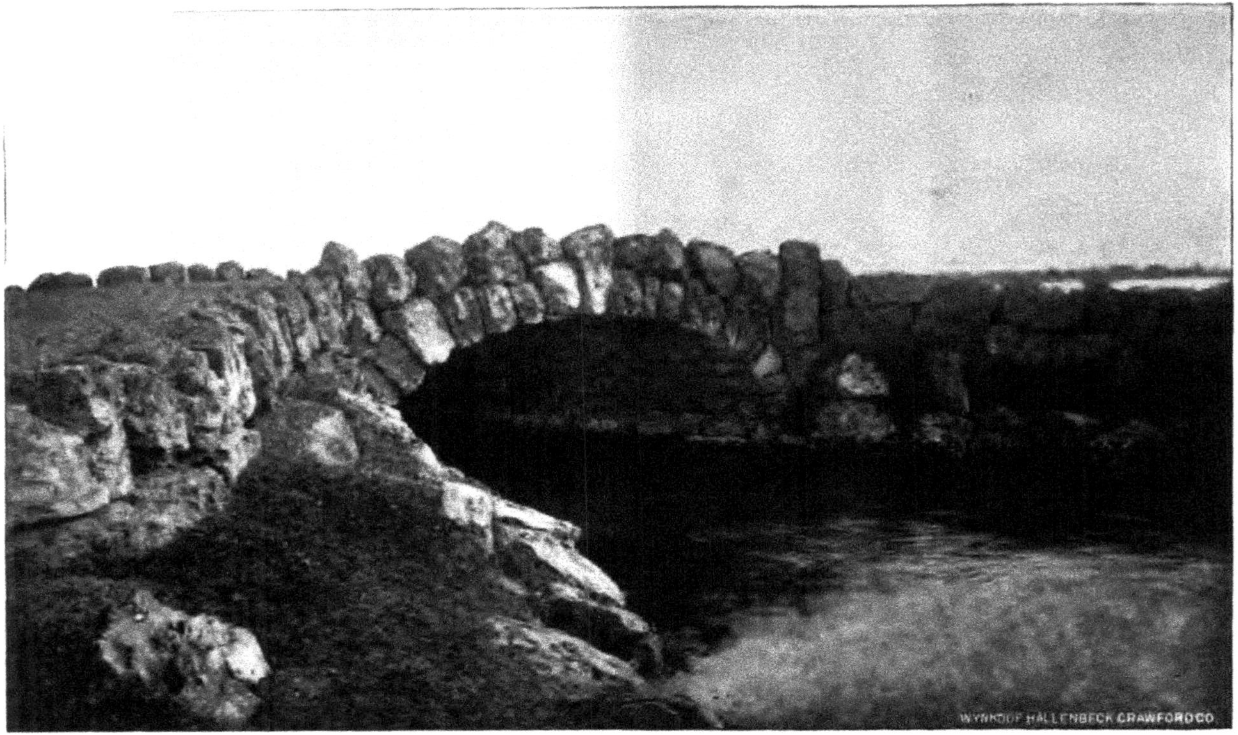

Port Day. Bridge.

PORT DAY. BRIDGE. LOOKING UP STREAM.

The falling water accomplishes nothing. It should be operating mills, giving employment to men and women, producing something tangible, useful, marketable! Beauty and sublimity are there, it is true; but beauty and sublimity have no commercial value; they put money into nobody's pocket, unless we except the owners of railroads, hotels and livery stables.

Such would seem to be the attitude of not a few scientific and practical men towards Niagara.

The conflict between utility and beauty, between the practical and the ideal, is perpetual.

Contractors need stone for roads, and the Palisades of the Hudson are shattered by dynamite to supply the material. Lumber men must have logs, and Adirondack forests go down before the woodman's axe.

While the attitude of its Legislature towards Niagara has not been consistent, the State of New York has nevertheless done much that is creditable towards protecting and preserving natural objects of great beauty and celebrity. In addition to rescuing the lands adjacent to the falls from private ownership and causing them to be restored to some semblance of their original attractiveness, she has appointed a commission to see to it that steps be taken to prevent further destruction of the Palisades. She has decreed that the devastation of the Adirondack forest lands shall cease, and that vast tracts of those lands shall be placed beneath the aegis of State protection.

She has incorporated The Trustees of Scenic and Historic Places and Objects.

But we repeat, the State has not maintained a consistent policy with reference to Niagara. After putting an end to private ownership and taking the cataract, the rapids and lands adjacent under her care and protection, she incontinently grants to corporations the right to withdraw large quantities of water from the upper river, and so to diminish the magnificence of the world-famous spectacle.

The owners of riparian lands adjacent to the falls in 1885 claimed compensation for their right of property in the unused water-power. But their claims were disallowed. Only a few years afterward the State by legislative acts grants to corporations, without compensation, the privilege of taking enormous quantities of water from the river.

In their last annual report the commissioners said: "It is hoped that an international agreement may eventually be entered into between Great Britain and the United States, by which the waters of the Upper Niagara shall be permanently exempted from diversion."

In view of the probability that an attempt may soon be made to divert large quantities of the upper river into Canadian territory, it seems to the commissioners especially important that measures should promptly be taken to effect the permanent exemption that is so desirable, and by which alone the falls can be surely protected. The withdrawal of sufficiently large quantities of water into Canadian territory and under Canadian authority, would undoubtedly give rise to international complications, that would require diplomatic treatment.

Interference with navigation and obstruction to commerce are matters that would probably not present themselves for adjustment. The question is, whether the most famous natural phenomenon on the continent is to be diminished in volume, and made less and less attractive to the people of this and other countries, if not eventually destroyed, by private corporations taking and diverting the waters of the upper river for their own use and emolument, with the sanction of Provincial, or Colonial and State governments.

While it would seem to be a well settled rule of international law that no right of property can be claimed in the tide waters of the sea or the running waters of rivers, yet if, to use the language of the late Dr. Travers Twiss, "the free and common use of a thing of this nature * * be prejudicial, or dangerous to a nation, the care of its own safety will entitle it so far, and so far only, to control the use of it by others, as to secure that no prejudice or danger result to itself from their use of it. A nation may accordingly have a right of empire over things which are nevertheless by nature *communis usus*, and over which it cannot acquire an absolute right of property; as, for instance, over portions of the high seas, or over *rivers which form the boundary of its territory*." ("Law of Nations," by Travers Twiss, LL. D.) (See 12th Annual Report Commissioners of the State Reservation at Niagara, pp. 9-11.)

It was the same spirit which moved the State of New York in 1885, and the Province of Ontario in 1888, to rescue the picturesque and attractive scenery of the banks and islands of the Niagara from the injury and desecration inflicted upon it in the days of private ownership—scenery, which including, as it does, so grand an exhibition of the forces of nature, so extraordinary a display of natural beauty and sublimity, could no longer, by peoples calling themselves civilized, be left exposed to the degradation and defacement that inevitably attend private ownership.

The question may well be asked: Has the interest in natural scenery of exceptional grandeur that prompted and approved the establishment of the State Reservation at Niagara and the Queen Victoria Niagara Falls Park, wholly disappeared; or has the interest in the development of electrical power and manufacturing enterprises become so general and so dominant as to produce indifference to the future of Niagara Falls, and even tacit approval of schemes that cannot but be effective in diminishing the volume of the cataract, and defacing it as a world-famous spectacle? At any rate, it must be admitted that the river and falls are at present not safe-guarded from diversion and destruction. And the only power that can provide absolutely sure protection is the combined power of the government of the United States and that of Great Britain, known as Her Majesty's Government. In the opinion of the commissioners this protection should be sought; and in their efforts to secure it, they invite the support and co-operation of all people interested in the preservation of the great Cataract.

From July 15, 1885, the day the Reservation was opened to the public, to September 30, 1897, the State granted the commissioners in the annual appropriation and supply bills the following amounts:

For maintenance	$235,000 00
For special improvements	135,000 00
Total	$370,000 00
During the same period the commissioners have remitted to the State treasury, as receipts from the Reservation	84,900 52
Leaving	$285,099 48

As the amount actually appropriated by the State for the Reservation during a period of twelve years two months and a half, or about $23,500 a year; or, if we leave out of consideration the amounts appropriated for special improvements, and deduct the receipts of the Reservation remitted to the State treasury from the aggregate amount appropriated by the State for maintenance, we have.............................. $235,000 00

	84,900 52
	$150,099 48

or about $12,500 a year.

THE MONTHLY RECEIPTS FROM THE RESERVATION WERE AS FOLLOWS:

1896.	Inclined railway.	Rentals.	Interest on balances in bank	Dividend.
October	$182 40	$300	......	....
November	51 30	......	......	....
December	29 45	......	$11 79	$28 34
1897.				
January	63 45	......	......	....
February	83 05	......	......	....
March	38 95	......	33 99	....
April	53 85	......	......	28 34
May	163 45	......	......	....
June	352 50	......	37 36	....

1897.	Inclined railway.	Rentals.	Interest on balances in bank.	Dividend.
July	$822 85	$310	......	....
August	2,229 65	810	......	....
September	925 50	970	$25 61	....
	$4996 40	$2,390	$108 75	$56 68
				108 75
				2,390 00
				4,996 40
Total				$7,551 83

MONTHLY PAY ROLLS HAVE BEEN AS FOLLOWS:

1896.	
October	$1,490 03
November	1,344 92
December	1,283 91
1897.	
January	1,182 17
February	1,237 16
March	1,274 67
April	1,769 29
May	1,791 68
June	1,703 79
July	1,777 30
August	1,795 02
September	1,722 78
	$18,372 72

EXPENDITURES AS PER ABSTRACT (MAINTENANCE) WERE AS FOLLOWS:

Abstract CI	$4,860 34
Abstract CII	4,883 85
Abstract CIII	7,608 64
Abstract CIV	7,467 92
Total	$24,820 75

IMPROVEMENT ABSTRACTS WERE AS FOLLOWS:

Abstract II Series G	$935 43
Abstract III Series G	4,278 08
Abstract IV Series G	1,433 35
Abstract V Series G	440 86
	$7,087 72
Abstract I Series H	8,014 96

The receipts and earnings of the Reservation have been sent to the State Treasurer monthly, and interest on balances in the Manufacturers & Traders Bank, Buffalo, has been remitted to the same official quarterly.

The Comptroller has advanced to the commissioners quarterly a fourth part of the $25,000 appropriated by chapter 948 Laws of 1896.

Of the treasurer's report herewith submitted, exhibiting in detail all receipts and disbursements for the fiscal year ended September 30, 1897, the following is a summary:

Balance on hand October 1, 1896		$579 28

RECEIPTS

Inclined Railway	$5,196 40	
Rentals	2,190 00	
Interest	108 75	
Dividends	56 68	
		7,551 83
From the State treasury, chapter 948 Laws of 1896.		25,000 00
From the State treasury, chapter 950, Laws of 1896.		7,087 72
From the State treasury, chapter 790, Laws of 1897.		8,014 96
		$48,233 79

PAYMENTS.

Pay-rolls at Niagara (maintenance)	$18,372 72	
Repairs, materials, superintendent's expenses, etc.	5,088 93	
Commissioners, treasurer, traveling expenses, etc.	1,359 10	
		$24,820 75
Remitted to State Treasurer		7,551 83
Improvements		7,087 72
Improvements		8,014 96
Balance		758 53
		$48,233 79

All the expenses of the commissioners in attending meetings during the year, including railway fares and hotel bills, amounted to $97.74.

Fort Day. Flood. Looking down Stream.

Total receipts since organization of the Commission, 1883	$490,970 58
Total disbursements	490,212 05
Balance	$758 53

The Legislature has made appropriations for maintenance, current expenses and salaries, as follows:

By chapter 336, Laws of 1883	$10,000 00
By chapter 656, Laws of 1887	20,000 00
By chapter 270, Laws of 1888	20,000 00
By chapter 569, Laws of 1889	25,000 00
By chapter 84, Laws of 1890	20,000 00
By chapter 144, Laws of 1891	20,000 00
By chapter 324, Laws of 1892	20,000 00
By chapter 414, Laws of 1893	25,000 00
By chapter 654, Laws of 1894	25,000 00
By chapter 807, Laws of 1895	25,000 00
By chapter 948, Laws of 1896	25,000 00
By chapter 306, Laws of 1897	25,000 00
Total	$260,000 00

For special improvements appropriations have been made as follows:

By chapter 570, Laws of 1889	$15,000 00
By chapter 302, Laws of 1891	15,000 00
By chapter 356, Laws of 1892	15,000 00
By chapter 726, Laws of 1893	25,000 00
By chapter 358, Laws of 1894	20,000 00

By chapter 932, Laws of 1895	$20,000 00
By chapter 950, Laws of 1896	10,000 00
By chapter 790, Laws of 1897	15,000 00
Total	$135,000 00

In compliance with statutory directions, the commissioners have remitted to the State treasury the receipts from the Reservation, as follows:

From October 1, 1887, to September 30, 1888	$9,331 55
From October 1, 1888, to September 30, 1889	7,393 77
From October 1, 1889, to September 30, 1890	7,670 29
From October 1, 1890, to September 30, 1891	9,327 67
From October 1, 1891, to September 30, 1892	9,823 03
From October 1, 1892, to September 30, 1893	10,923 85
From October 1, 1893, to September 30, 1894	9,251 43
From October 1, 1894, to September 30, 1895	7,448 01
From October 1, 1895, to September 30, 1896	6,179 09
From October 1, 1896, to September 30, 1897	7,551 83
Total	$84,900 52

The following is "an estimate of the work necessary to be done and the expenses of maintaining said Reservation for the ensuing fiscal year," ending September 30, 1898:

CONSTRUCTION.

New bridges	$100,000 00
Elevator on Goat Island	35,000 00
Grading, planting and other improvements	25,000 00

MAINTENANCE.

Salaries, office and traveling expenses	$4,750 00
Reservation police	5,400 00
Labor	7,500 00
Materials, tools, etc	6,000 00
Miscellaneous	1,350 00
Total	$25,000 00

Estimated receipts from October 1, 1897, to September 30, 1898.

Inclined railway	$5,500 00
Cave of the Winds	1,200 00
Ferry and steamboat landing	750 00
Carriage service	100 00
Baggage room	50 00
Interest	100 00
	$7,700 00

The commissioners have already adopted a plan for a new bridge to the First Sister Island, for which they desire an adequate amount. They also have in view the construction of a new bridge or bridges from the mainland to Goat Island, substantial structures of stone and steel, in place of the old bridges now in use. For this improvement they estimate that a large sum, perhaps $100,000, would be necessary.

They would also repeat their request for an appropriation of $35,000 for an hydraulic elevator at the Cave of the Winds on

Goat Island. This would prove to be a source of profit to the State, and in a few years would realize the amount required for its construction.

Respectfully submitted,

ANDW. H. GREEN,

President.

JOHN M. BOWERS,

ROBERT L. FRYER,

WILLIAM HAMILTON,

GEORGE RAINES,

Commissioners of the State Reservation at Niagara.

REPORT OF THE SUPERINTENDENT

OF THE

State Reservation at Niagara

FOR THE

Fiscal Year Ending September 30, 1897.

Report of the Superintendent.

To the Board of Commissioners of the State Reservation at Niagara:

Gentlemen.—I respectfully submit the report of the Superintendent.

The greater part of the work of improvement during the year, has been between First street and the upper terminus of the Reservation at Port Day. The work was undertaken after the inspection made by the executive committee, Samuel Parsons, jr., and the Superintendent, on June 12th last. The fact that materials for grading and filling, as well as for roads, walks and bridges, could be obtained near at hand without cost, except the expense of hauling, led to the prosecution of the work of improvement in that locality. Near the upper end of the Reservation, a great pile of stones had been collected, some weighing several tons, suitable for riprapping the river shore and for building retaining walls and bridges. Mixed with the larger stones were smaller stones, suitable for foundations for roads and walks.

IMPROVEMENTS.

The work undertaken and concluded, includes a large amount of filling and the construction of a stone arch bridge and retaining walls at Port Day; the construction of a loop driveway commencing at Seventh street and traversing the pier at Port Day; where a spacious "lay by" for carriages has been constructed, and returning to the mainland below Seventh street by the bridges and the outer part of the loopway drive constructed last year.

ROADS.

A driveway has also been constructed connecting the loopway with Buffalo avenue adjacent to the Reservation, and Seventh street has been macadamized and graded so as to be in keeping with the loop driveway. The lake or pond within the loop has been cleansed and the shores graded, riprapped and prepared for planting. The margins of the new roads and walks have been filled and sodded. A portion of the pier at Port Day and the shore of the pond has been top dressed and sown with grass seed, and the mounds and elevations prepared for planting, adjacent to the pond, have been graded and covered with sod. A retaining wall below Fourth street has been removed and the slope graded and covered with sod, and a large amount of filling done in that locality. The western end of Bath Island and the southern shore below the bridge to Goat Island have been riprapped with large stones, and the shores filled, graded and sown with grass seed. The timber crib work on either side of the bridge, at the approach to Goat Island, has been removed and the shore riprapped and graded.

WALKS.

Gravel walks have been constructed on First street between Buffalo avenue and the Riverway, on Buffalo avenue between Seventh street and Quay street, and on the Riverway between First street and the terminus of the Reservation at Port Day, the latter more than half a mile in extent. The plank walks in Prospect Park leading from the bridge entrance have been removed and substantial gravel walks have been constructed.

MAINTENANCE OF BRIDGES.

The work of maintenance done, included the thorough repairs of the bridges to the Three Sister Islands, and the bridge leading to Terrapin Point at the Horseshoe Falls. The cribwork supporting the second pier of the bridge to Bath Island, has been largely rebuilt and the crib refilled with stone.

BUILDINGS.

The gate house at the foot of Falls avenue which obstructed the roadway at that point, has been removed to a suitable and convenient location and repaired and repainted. Repairs have been made upon the Cave of the Winds building and the summer houses and out-houses in Prospect Park. The cellar in the Inclined Railway building, has been cemented, the Inclined Railway conduit repaired, and the inlet riprapped with large stones, the cars renovated and new cables and wheels attached. Two additional drinking fountains have been placed in Prospect Park, the park seats repaired and repainted, the signs and finger-boards repainted, the parapet wall in Prospect Park repaired and the iron guard railing in Prospect Park and on the Islands repaired and repainted. The walk around the head of Goat Island from the Three Sister Islands to the spring has been graveled and a new walk constructed on the Third Sister Island. A stone cross walk has been constructed at Bridge street, and repairs made upon the stone gutters in Prospect Park.

ROADS.

The Riverway between First street and Seventh street, has been macadamized and graveled, also the hills at Falls avenue and at the approach to the bridge at Bath Island. Repairs have been made upon the road on Goat Island, and the trees and shrubs planted have been cultivated, and the lawns top-dressed and other incidental work preformed.

INSPECTION BY GOVERNOR BLACK.

On June 11th an official inspection of the Reservation was made by the Governor and Lieutenant-Governor of the State, the Speaker of the Assembly, Senator Ellsworth and Assemblyman Dudley.

G. A. R. ENCAMPMENT.

During the week of the Grand Army Encampment at Buffalo, August 23d to 30th, the Reservation was visited each day by great multitudes of people without accident or disorder of any kind.

PORT DAY BRIDGE.

The temporary wooden bridge at the inlet to the pond at Port Day, has been removed, and a rustic stone arched bridge constructed, connecting the pier with the new loop driveway. The bridge is an elliptic arch with rustic coping and a bay, commanding fine views up and down the river. A spacious turnout for carriages has been constructed adjacent to the bridge.

A photograph of the Port Day bridge is herewith submitted.

THE LOOP DRIVEWAY.

The loop driveway at Port Day, has been completed. The roadway has been macadamized and graveled, and the river side of the roadway, graded and covered with sod. The driveway conforms with the line of the shore and removes the former artificial appearance of the pier at Port Day. On the pier, retaining walls have been built and walks constructed leading to the driveway. The loop driveway had already become a favorite route for visitors in carriages, wheelmen and pedestrians, because of the river breeze and the fine views which it affords, and also because it provides a continuous thoroughfare within the Reservation which did not formerly exist.

A photograph of the Loop Driveway is herewith submitted.

FRENCH LANDING BRIDGE.

The stone arched bridge constructed last year, at the outlet of the pond at the point known as the "Old French Landing," has been macadamized and graveled, and the approaches planted. Rustic stones have been placed in the margin of the loop driveway adjacent to the bridge so as to point out the line of the roadway. The new gravel walk along the Riverway, has been arranged with reference to the bridges.

A photograph of the French Landing Bridge is herewith submitted.

PORT DAY POND.

The dumping of refuse, and the draining of sewers into the pond at Port Day, have been prohibited and discontinued. The angle of the pond at the pier has been filled in, and the shore of the pond brought to flowing lines. The inlet has been deepened and a strong current brought through the pond to keep the water in motion. The debris and eel grass have been largely removed. Ice cutting in winter, has been discontinued, and the pond has become a favorite skating ground for the public.

A photograph of the Port Day Pond with the Port Day bridge, is herewith submitted.

PORT DAY PARK.

The work done during the past year at Port Day, has changed the appearance of the locality. What was formerly a dumping ground for refuse of every kind, has now assumed a park-like appearance. The plank walk on Buffalo avenue has been removed, and a gravel walk substituted. A road has been constructed from Buffalo avenue, branching toward the Port Day bridge, and toward the French Landing Bridge at the outlet of the pond. A gravel walk has been constructed around the margin of the pond which has been graded and prepared for planting. There are numbers of walnut trees and good sized maples in the locality; and the young trees planted four years ago, are making rapid growth. Additional walks, and the shrub-planting and surface-grading yet to be done, will make that part of the Reservation exceedingly attractive.

NUMBER OF VISITORS.

The number of visitors during the year, was far larger than usual. The "excursionists" nearly doubled in number, there being an estimated number of 511,420 this year as against 261,780 last year. This increase was largely due to the National Encampment of the Grand Army of the Republic held in Buffalo. Nearly all of those in attendance, visited the Falls.

A statement of the excursion parties to the Reservation during the year, is herewith submitted.

LICENSED CARRIAGE DRIVERS.

Nineteen complaints have been made against licensed carriage drivers during the year. No carriage drivers have been excluded from the Reservation for violation of the ordinances of the Commissioners. The complaints are almost entirely confined to half a dozen persons who are unfit to act as drivers of public carriages, and who should not be licensed as such.

THE RESERVATION CARRIAGE SERVICE.

The Reservation Carriage service has been in operation during the year. Now that the loop driveway has been completed, regular trips should be made hereafter along the riverway to Port Day and return.

THE STEAMBOAT LANDING.

The landing has been used by the Maid of the Mist Steamboat Company. Only one boat has been in operation. The level of the water in the river has been a trifle higher than last year.

EMPLOYES.

The regular force employed exclusive of laborers, consists of ten men, to wit: One superintendent, one clerk, six police gatemen, and care takers and two Inclined Railway men, one of whom is employed during the summer season only.

The following is a statement of the laborers employed during the year:

STATEMENT

Of number of employes on maintenance and improvement payrolls for the year ending September 30, 1897:

PAY ROLLS.	MAINTENANCE.				CHAP. 950, LAWS 1896.				CHAP. 790, LAWS 1897.			
	Foreman.	Asst. foreman.	Laborers.	Teamsters.	Foreman.	Asst. foreman.	Laborers.	Teamsters.	Foreman.	Asst. foreman.	Laborers.	Teamsters.
1896.												
October	1	...	16	....	1	....	30	3	....	....	....	...
November ..	1	..	16	....	1	....	37	7	....	....	....	...
December ...	1	...	13	....	1	....	15	7	....	....	....	...
1897.												
January	1	...	13	1	1	....	6	2	....	....	....	...
February ...	1	...	13	....	1	....	2	1	....	....	....	...
March	1	...	12	....	1	....	2	1	....	....	....	...
April.......		...	25	1	....	....	4	1	....	....	....	...
May	1	...	32	1	....	....	3	1	1	....	20	2
June		1	20	1	....	....	....	....	1	....	35	5
July		1	26	1	....	...	....	....	1	2	53	8
August.....		1	26	1	....	....	....	....	1	2	54	7
September ..		1	21	1	....	....	....	....	1	2	57	5

PAN-AMERICAN EXPOSITION, 1899.

It is proposed to hold an exposition on Cayuga Island, situated in the Niagara river, a short distance above the Falls, in the year 1899. The exposition will be especially devoted to the various countries of North and South America, and it is intended to illustrate the progress made in the New World during the present century. The directors of the enterprise, estimate the expenditure to be made in preparing for the exposition, at $5,000,000. It will be the only large exposition held within the borders of the State since the Crystal Palace Exposition in New York City in 1853. On August 26, 1897, President McKinley visited the locality, and inaugurated the work. It is proposed to keep the exposition open from May 1st to November 1st, 1899, and it is estimated that the number of visitors will exceed five millions, as all who come to the exposition, will probably visit the Falls and other points within the Reservation, preparation should be made in advance for the accommodation of a greatly increased number of visitors during the year of the exposition, probably ten times as many as come in an ordinary year. The bridges, stairways and guard railings will require to be strengthened to guard against accident, and additional accommodations provided.

IMPROVEMENTS NEEDED.

SHELTERS.

Shelter buildings are needed at the American Falls and at the Horseshoe Falls on Goat Island, at which points visitors congregate in large numbers. At times of sudden rain storms, the present shelters are too far apart for public convenience.

WATER PIPES.

The mains for water service in Prospect Park and on Goat Island are surface pipes, and the supply is shut off during the winter season. A general plan is needed for an adequate water service for the Reservation grounds and buildings.

ROADS.

On the riverway in the neighborhood of the Whitney homestead, the road might well be moved toward the water's edge so as to avoid the grade at that point. The present retaining wall might then be removed and the bank graded down.

ELEVATOR AT THE CAVE OF THE WINDS.

As this locality is between the American and the Horseshoe Falls, the view is unparalleled, but the winding stairway leading to it is so unsuitable and fatiguing, that very few persons are enabled to enjoy the beauty of the scenery below the high bank, where a walk can be easily made along the edge of the water from the American to the Horseshoe Falls.

The waiting rooms and winding stairway at the Cave of the Winds, have received the usual repairs during the year. The dressing rooms are inadequate and unsuitable. Better accommodations should be provided for the traveling public.

The Biddle staircase has been examined and found safe, but it is old and dilapidated in appearance, and affords but a fatiguing method of obtaining the view of the Falls from below. It should be replaced by a commodious elevator with a free stairway attached. Such an elevator operated for a nominal fee of five cents up or down, would furnish an estimated revenue of

$5,000 a year, which, with the present revenue mainly derived from the Inclined Railway, would almost render the Reservation self-sustaining.

GRADING AND PLANTING.

Planting, surface grading and additional gravel walks are needed at Port Day and on Bath Island. A portion of the territory at Port Day is now ready for planting whenever the Board may direct.

PROSPECT POINT.

The river shore within the Reservation is almost entirely cleared of artificial constructions down as far as Prospect Point. At Prospect Point, a stone wall juts out into the stream. To obtain a near view of the Falls, visitors descend a series of steps into an enclosed excavation behind the wall. The original shore at Prospect Point, sloped down to the water's edge with nothing to intercept the view. At present the stone wall stands in the foreground. The enclosure behind the wall is a receptacle for ice, snow and slush during the winter months. This artificial excavation might now be filled so as to restore the natural slope of the shore, the stone wall taken down to grade, the outside rip-rapped with large stones of the locality now at hand, and the public safeguarded by a substantial iron railing of the pattern designed by Mr. Vaux, which would not obstruct the view. It seems advisable that this work be undertaken at this time, as the constant falling off of the cliff at that point, will in any event soon necessitate the removal of the whole or a part of the present parapet wall. On April 9th, last, a large mass of rock fell from the cliff below the parapet wall. With a careful study of the locality, provisions may be made for the public safety, and at the same time, for the removal of the present artificial constructions at the principal point of view within the Reservation.

Prospect Point. Former Natural Condition.

PROSPECT POINT. PRESENT ARTIFICIAL CONDITION.

Photographs showing the former natural state of Prospect Point, and its present artificial condition, are herewith submitted.

TABULAR STATEMENTS.

Detailed statements of the receipts and expenditures of the superintendent, the amount of the pay rolls for each month and the classification of the pay rolls and accounts, are hereto appended.

Respectfully submitted,

THOMAS V. WELCH,

Superintendent.

EXCURSIONS, 1896-97.

1897		No. of Cars.
April	27. Royal Arcanum, Jersey City, N. J.	3
May	4. International Union Journeymen Bakers, Brooklyn	1
	7. Arbor Day Excursion, Lockport, N. Y.	30
	9. Buffalo via. N. Y. C. and Trolleys.	7
	9. Brooklyn, Journeymen Bakers...	1
	15. Buffalo, N. Y., High School	10
	24. Queens Birthday and Convention Railroad Trainmen	40
	27. Mt. St. Stephen's School, Buffalo.	2
	29. Bakers Buffalo Bowling Club . . .	1
	30. Henderson Party, Brooklyn, N. Y.	2
	30. Main Line Lehigh Valley	8
	30. New York City via West Shore Railroad	5
	30. Binghamton via Erie	7
	30. Buffalo via Trolley	20
	31. Memorial Day, Buffalo, N. Y.	50
	31. Return of railroad men from Los Angeles, Cal.	2
	31. American Union Swedish Singers to New York	4
	31. Syracuse	12
	31. W. Y. N. & P.	6

1897.		No. of Cars.	No. of Visitors.
[Ma]y	31. Inaugural Excursion	4	240
[Ju]ne	4. Order Mystic Circle, Buffalo, N.Y.	3	180
	6. Buffalo via Central, West Shore and Trolley	55	3,300
	8. Members of York County Council, Toronto, Ont.	2	120
	8. Convention National Electrical Ass'n	20	1,200
	8. Universal Postal Congress of 63 nations	4	240
	11. Order Sons of Israel, New York City	9	540
	11. Rochester Chamber of Commerce with Governor Black	8	480
	11. United States Ass'n of Brewers..	9	540
	13. Hotel Men's Mutual Benefit Ass'n, Chicago	3	180
	13. Church of Ascension	50	3,000
	13. Buffalo	8	480
	13. Druggists' Convention, Cleveland, Ohio	2	120
	13. Farmers' Picnic, Eden Center, N. Y.	2	120
	13. Order of Y. P. S. C. E., St. Catherine, Ont.	[illegible]	120
	13. Local Half Holiday Excursion....	10	600
	14. Canadian Militia, Niagara-on-the-lake, Ont.		[illegible]

1897.		No. of Cars.	No. of Visitors.
June	16. Wholesale and Retail Coal Ass'n, Ohio	6	360
	16. American Railway Telegraph Superintendents	4	240
	17. American Railway Telegraph Superintendents	2	120
	17. Main Line, Erie Railroad	6	360
	18. Main Line, W. N. Y. & P.	8	480
	18. Western New York & Ohio	8	480
	18. Sons of Scotland, Markham, Toronto, etc.	8	480
	19. Old Settlers and Descendants of Benj. Long, Erie & Niagara Cos.	2	120
	19. Employes Bicycle Mfg. Co., Toronto	3	180
	19. Y. M. C. A. of Buffalo	8	480
	19. General Half Holiday	10	600
	20. W. N. Y. & P.	9	540
	20. Nickel Plate	8	480
	20. Buffalo via N. Y. C. and Trolley	40	2400
	22. Queen Victoria Diamond Jubilee Excursion	12	720
	23. Fast Freight Agents	1	60
	23. Newton Party, Chautauqua	5	300
	23. Union School, Clarence Center, Erie Co.	4	240
	25. Pennsylvania Ass'n	6	360
	25. Employees of Perry Morey, Columbus, Ohio	6	360

1897.		No. of Cars.	No. of Visitors.
[J]une	26. Connecticut via Nickel Plate......	6	360
	26. Order of Foresters, Lindsay, Ont..	3	180
	27. Buffalo, Lockport, Rochester and Cleveland	63	3,780
	28. American Society Civil Engineers	1	60
	29. American Institute Eye and Ear	6	360
	Doctors	6	360
	29. Retail Wall Paper Dealers.......	5	300
	30. Niagara River Excursion	7	420
	30. Union Churches, Pen Yan	8	480
	30. Special Buffalo Party	14	840
	30. Delegates to Christian Endeavor Convention, California	15	900
[J]uly	1. Dominion Day Excursion, Buffalo and Toronto	59	3,540
	2. Hahneman Convention	2	120
	2. Toronto Plumbers	7	420
	3. Master Plumbers, Canada	10	600
	3. B. P. O. Elks, Camden, N. J......	1	60
	3. Saturday Half Holiday..........	15	900
	4. Henderson Party, Brooklyn, N. Y.	1	60
	4. Pittsburgh, Cleveland, Buffalo...	94	5,640
	5. Fourth of July Excursion	150	9,000
	6. Convention Straw Board Makers.	1	60
	6. Doctors, Buffalo	5	300
	8. City Officials, Chattanooga, Tenn.	3	180
	8. Lutheran Church, Buffalo	11	660
	8. Hornellsville	9	540
	9. Cincinnati	20	1,200

1897.		No. of Cars.	No. of Visito
July	9. Fishing Club, Buffalo...........	8	48
	10. Half Holiday, Buffalo and Port Jervis......................	16	90
	11. General Excursion	25	1,50
	14. Epworth League	130	7,80
	15. Epworth League	140	8,40
	15. Oil City, Hornellsville, Allegany Valley, Rochester	29	1,74
	16. Epworth League	10	60
	16. Beta Thita Pi Society...........	1	[illegible]
	16. Special Party, Cleveland	10	60
	16. Employees Michigan Central, Windsor & Welland, Ont.	12	72
	17. Louisiana Rifles	3	18
	17. Special Party, Cleveland, Ohio...	9	54
	17. Employees W. J. Gates, Stationer, Toronto	8	48
	17. Chautauqua Party	6	36
	18. Photographers' Convention, Jimtown	3	18
	18. General Excursion	35	2,10
	19. Epworth League	12	72
	19. Epworth League	83	4,98
	19. St. Vincent de Paul Society, Toronto	10	60
	20. Epworth League	20	1,20
	20. Big Four Railroad Employees and Caledonian Society, Toronto . .	14	84

1897.		No. of Cars.	No. of Visitors.
July	21. Employees Wholesale and Retail Grocers, Hamilton	8	480
	21. Galion and points on C. C. C. & St. Louis Line	10	600
	21. Ladies' Auxiliary Locomotive Engineers, Erie	2	120
	22. Union Methodist Colored Picnic..	6	360
	22. Order of Foresters, St. Mary's Tavestock, etc., Canada	8	480
	22. Colored Methodists via Erie	1	60
	23. Special Party, Cleveland, Ohio...	8	480
	23. Henderson Party, Brooklyn......	1	60
	23. Western points	13	780
	24. Half Holiday Excursion via N. Y. C. & Lehigh.................	30	1,800
	24. Odd Fellows Picnic, Jamestown, N. Y.	10	600
	24. Mechanical Engineers, Rochester, N. Y.	2	120
	24. Newton Party, Chautauqua	13	780
	25. Fort Wayne & Nickel Plate	12	720
	25. Buffalo, Rochester and Lockport .	50	3,000
	25. Cleveland	8	480
	27. Christian Endeavorers	18	1,080
	28. Christian Endeavorers	20	1,200
	28. Yonondio Lodge, F. and A. M., Rochester	6	360
	28. Hoppickers Picnic, Auburn and Canandaigua	9	540

1901.			No. of Cars.	No. of Visitors.
July	28.	National Science Camp, Canandaigua	1	60
	28.	Order of Chosen Friends, Hamilton, Ont.	2	120
	29.	Christian Endeavorers	9	540
	30.	Christian Endeavorers	9	540
	30.	From the West via Wabash Railroad	21	1,260
	30.	From Lansing, Mich	10	600
	30.	Biscuit and Confectionery Makers, Toronto	7	420
	30.	Lawrence Hose Co., Port Credit	......	680
	30.	Iron Moulders, Toronto by boat	......	680
	31.	Catholic Church Choir, Middleport	1	60
	31.	Sherman Greys, Erie	7	420
	31.	Pittsburgh Party	8	480
	31.	Chautauqua and Mayville and Half Holiday	24	1,440
Aug.	1.	Carbondale, Pa., Jersey City, N. J., Rochester, N. Y.	18	1,080
	1.	Buffalo via boat	......	400
	1.	Lake Shore & M. C.	20	1,200
	1.	Buffalo via Trolley	......	9,000
	1.	Buffalo via N. Y. C.	20	1,200
	1.	29th Separate Co., Washington, D. C.	4	240
	1.	McKeesport, Pa.	20	1,200
	1.	Main Line, Erie Railroad	6	360
	1.	Main Line, R. W. & O.	12	720

1897.		No. of Cars.	No. of Visitors.
Aug.	2. Dominion Civic Holiday, Toronto.	20	1,200
	2. Emancipation Day	5	300
	2. Special Party, Buffalo	5	300
	3. Chicago	12	720
	3. Christian Endeavorers en route home	10	600
	4. German Lutheran Church, Lockport	6	360
	5. Parkdale Presbyterian Church, Toronto	8	480
	5. Special Party, Buffalo	10	600
	5. A. O. W., Coburgh, Ont.	10	600
	6. Via Nickel Plate and Lehigh Valley	116	7,000
	7. German Picnic, Oswego, N. Y. . . .	15	900
	7. German American Pioneers	20	1200
	7. Chautauqua	9	540
	7. Half Holiday, Buffalo	20	1,200
	8. Wisconsin Press Ass'n	10	600
	8. Buffalo	70	4,200
	9. Hamilton, Dundees, London, Chatham, Ont	33	2,000
	9. Buffalo.	15	900
	10. Susquehanna	10	600
	10. German Ulk Dramatic Ass'n, New York	6	360
	10. Chautauqua Party	8	480
	11. Foresters and City Fire Department, Aurora, Ont.	6	360

1897.		No. of Cars.	No. of Visitors.
Aug.	11. Buffalo	10	600
	12. Western States	50	3,000
	12. Independent Order of Red Men	2	120
	13. Western States	15	900
	13. Raymond & Whitcomb Excursion	2	120
	13. Sons of Scotland, Ontario via boat	......	500
	13. Toronto	7	420
	13. Buffalo via boat	......	120
	13. Jamestown, N. Y.	8	480
	14. Elmira, Ont.	4	240
	14. Belleville, Toronto, Port Dalhousie	12	720
	14. Buffalo via boat	......	180
	14. Chautauqua	9	540
	14. General Half Holiday	25	1,500
	15. General Excursions	75	4,500
	16. London, Ont.	6	360
	16. Buffalo	6	360
	17. English Lutheran Church, Lockport	6	360
	17. Chautauqua	8	480
	17. Western points	20	1,200
	17. Queen St. East Methodist Church, Toronto	7	420
	17. Cleveland, Ohio	10	600
	18. Odd Fellows via D. L. & W.	10	600
	18. Cleveland, Ohio	8	480
	18. Y. M. C. A., Rochester	8	480
	18. Firemen, Bowmansville, N. Y.	7	420

1897.		No. of Cars.	No. of Visitors.
Aug.	18. Buffalo via boat	......	240
	18. Toronto via boat	......	480
	18. Toronto via G. T. R.	6	360
	19. Youngstown, Ohio	12	720
	19. Forresters, Peterboro, Ont.	6	360
	19. American Ass'n of Obstetricians and Gynaecologists	2	120
	20. Washington and Baltimore	15	900
	20. Indian Orphans, Versailles, Erie County, N. Y.	1	60
	20. Y. M. C. A., Rochester	12	720
	21. Chautauqua	10	600
	21. Arion Society, New York	6	360
	21. Stratford, Ont.	9	540
	21. Half Holiday	25	1,500
	22. All sources	333	20,000
	23. G. A. R. 31st Annual Encampment	667	40,000
	24. G. A. R. 31st Annual Encampment	1,250	75,000
	25. G. A. R. 31st Annual Encampment	667	40,000
	26. G. A. R. 31st Annual Encampment	417	25,000
	26. Main Line R. W. & O.	15	900
	27. G. A. R.	416	25,000
	28. G. A. R.	167	10,000
	29. G. A. R. and others	333	20,000
	30. G. A. R. and others	167	10,000
	31. G. A. R. 31st Annual Encampment	100	6,000
Sept.	1. Batavia, N. Y.	12	720
	1 Cleveland, Ohio	10	600
	1. Via Erie Railroad	10	600

1897.		No. of Cars.	No. of V
Sept.	2. Buffalo via boat	......	
	2. Toronto via boat	......	
	2. Erie Railroad	10	
	2. W. N. Y. & P.	12	
	3. Detroit	10	
	2. Granger Picnic via W. N. Y. & P.	10	
	4. General Half Holiday	20	
	5. Baltimore and Washington	10	
	5. Main Line Erie	10	
	5. Cleveland	10	
	5. Buffalo	6	
	5. Jamestown	9	
	6. Oil City	12	
	6. Buffalo, Rochester, etc.	12	
	6. Main Line Erie	10	
	6. Main Line N. Y. C.	20	1
	6. Trolleys, etc.	50	8
	7. Buffalo, Rochester and Pittsburgh	10	
	7. Jamestown	10	
	7. Elmira, N. Y.	9	
	7. Buffalo Trolley	6	
	8. Buffalo and Cleveland	12	
	8. Toronto	10	
	9. Gainesville and Youngstown, Ohio	18	1
	9. Buffalo	11	
	9. Toronto via boat	......	
	11. Rochester and Syracuse	12	
	11. Batavia	10	

1897.		No. of Cars.	No. of Visitors.
Sept.	11. General Half Holiday	30	1,800
	12. Raymond & Witcomb Excursion	2	120
	12. Cleveland, Ohio	10	600
	12. Buffalo, Rochester and Syracuse	60	3,600
	13. Columbia Commandery K. T., N. Y.	2	120
	13. Norwich Commandery K. T., N. Y.	3	180
	13. Improved Order Red Men, Buffalo	11	660
	14. Electric Illuminating Companies, Annual Meeting	3	180
	14. State Convention Street Railway Ass'ns	6	360
	14. State Convention Select Knights	6	360
	14. Buffalo	10	600
	15. Electric Illuminating Companies Annual Meeting	4	240
	15. State Convention Street Railway Ass'ns	4	240
	15. State Convention Select Knights	6	360
	15. Buffalo	10	600
	17. Washington, D. C.	10	600
	18. Union Churches, Royalton Center	5	300
	18. Union Churches, Medina	6	360
	19. General Half Holiday	10	600
	19. Pittsburgh, Pa.	10	600
	19. Cleveland, Ohio	8	480
	19. Buffalo	40	2,400
	22. Convention Fraternal American Insurance Union	2	120

1897.	No. of Cars.	No. of Visitors.
Sept. 23. Celebration Opening Steel Arch Bridge	100	6,000
24. Celebration Opening Steel Arch Bridge	100	6,000
25. Celebration Opening Steel Arch Bridge	100	6,000
26. General Excursions	25	1,500
30. Western Points	5	300
Total	8,434	511,420

RECAPITULATION.

April	3	180
May	221	12,720
June	483	22,620
July	1,313	79,460
August	5,475	339,460
September	939	56,980
Total	8,434	511,420

REPORT OF THE TREASURER

For the Fiscal Year Begun October 1, 1896, and Ended September 30, 1897.

French Landing Bridge. West View.

Mrol

The Commissioners of the State Reservation at Niagara, in Account with Henry E. Gregory, Treasurer.

1896.				
Oct.	1.	Balance on hand this date....................................		$579 28
		RECEIPTS.		
Nov.	9.	Quarterly advance from the State Comptroller.	$6,250 00	
1897.				
Jan.	18.	Quarterly advance from the State Comptroller.	6,250 00	
April	15.	Quarterly advance from the State Comptroller.	6,250 00	
July	29.	Quarterly advance from the State Comptroller.	6,250 00	
				25,000 00
		Special appropriation as per chaper 950, Laws of 1896:		
1896.				
Oct.	7.	Payment by State Comptroller on account....	$935 43	
Nov.	9.	Payment by State Comptroller on account....	1,730 80	
Dec.	11.	Payment by State Comptroller on account....	2.547 28	
1897.				
Jan.	12.	Payment by State Comptroller on account....	1,060 60	
Feb.	9.	Payment by State Comptroller on account....	215 75	
March	9.	Payment by State Comptroller on account....	157 00	
April	3.	Payment by State Comptroller on account....	188 50	
May	10.	Payment by State Comptroller on account....	149 13	
June	8.	Payment by State Comptroller on account....	103 23	
				7,087 72
		Special appropriation as per chapter 790, Laws of 1897:		
1897.				
June	8.	Payment by State Comptroller on account....	$237 00	
July	9.	Payment by State Comptroller on account....	1,406 48	
Aug.	9.	Payment by State Comptroller on account....	2,970 86	
Sept.	9.	Payment by State Comptroller on account....	2,714 54	
Sept.	23.	Payment by State Comptroller on account....	686 08	
				8,014 96

1896.				
Nov.	2.	Draft on Bank of Niagara for October receipts	$482 40	
Dec.	2.	Draft on Bank of Niagara for November receipts	51 30	
Dec.	31.	Draft on Bank of Niagara for December receipts	29 45	
1897.				
Feb.	1.	Draft on Bank of Niagara for January receipts	63 45	
March	1.	Draft on Bank of Niagara for February receipts	83 05	
March	31.	Draft on Bank of Niagara for March receipts	38 95	
May	1.	Draft on Bank of Niagara for April receipts	53 85	
June	1.	Draft on Bank of Niagara for May receipts	163 45	
June	30.	Draft on Bank of Niagara for June receipts	352 50	
Aug.	2.	Draft on Bank of Niagara for July receipts	1,132 85	
Sept.	1.	Draft on Bank of Niagara for August receipts	3,039 65	
Sept.	30.	Draft on Bank of Niagara for September receipts	1,895 50	
				$7,386
1896.				
Dec.	24.	Dividend on deposits in Cataract Bank	$28 34	
1897.				
May	6.	Dividend on deposits in Cataract Bank	28 34	
				56
1896.				
Dec.	31.	Interest on balances in Manufacturers and Traders' Bank	$11 79	
1897.				
March	31.	Interest on balances in Manufacturers and Traders' Bank	33 99	
June	30.	Interest on balances in Manufacturers and Traders' Bank	37 36	
Sept.	30.	Interest on balances in Manufacturers and Traders' Bank	25 61	
				10[illegible]
				$48,23[illegible]

EXPENDITURES.

No. of abstract.		No. of voucher.			
5.	CI.	1358	Wm. Hamilton, Com'r, traveling expenses	$40 39	
		1359	Henry E. Gregory, Treasurer and Secretary, office expenses	40 40	
9.		1360	Pay-roll for October	1,490 03	
		1361	Thomas V. Welch, Supt., office expenses	28 89	
		1362	N. F. Hydraulic Power and Mfg. Co., electric lighting.........	50 00	
		1363	Pay-roll for November	1,344 92	
		1364	Thomas V. Welch, Supt., office expenses	28 43	
8.		1365	N. F. Hydraulic Power and Mfg. Co., electric lighting	50 00	
		1366	P. J. Davy, water supply......	41 35	
		1367	Perkins & Mathews, plumbing repairs.....................	11 42	
		1368	E. H. Cannon, stove...........	20 70	
		1369	Hardwicke & Co., hardware...	56 23	
		1370	Hardwicke & Co., hardware ...	44 83	
		1371	Power City Lumber Co., lumber,	5 81	
		1372	Geo. Haeberle, repairs.........	10 28	
31.		1373	Henry E. Gregory, Treasurer and Secretary; salary, Oct., Nov. and Dec.	275 00	
		1374	Pay-roll for December...........	1,283 91	
		1375	Thomas V. Welch, Supt., office expenses	37 75	
					$4,860 34
97.					
8.	CII.	1376	Power City Lumber Co., repairing walks	$27 59	
		1377	McDonald & Welch, coal	52 25	
		1378	Timothy Horan, stone.........	34 00	
		1379	C. Burns, stone................	20 00	
		1380	T. Dunlevy, stone.............	64 00	

1897.	No. of abstract.	No. of voucher.			
Jan. 8.	CII.	1381	C. Nee, stone	$47 00	
		1382	F. W. Oliver Co., tools	28 16	
		1383	N. F. Hydraulic Power and Mfg. Co., electric lighting	50 00	
Feb. 1.		1384	Pay-roll for January	1,182 17	
		1385	Thomas V. Welch, Supt., office expenses	17 80	
Feb. 5.		1386	Estate of J. Johnson, blacksmithing	50 51	
Feb. 16.		1387	N. F. Hydraulic Power and Mfg. Co., electric lighting	50 00	
March 1.		1388	Pay-roll for February	1,237 16	
		1389	Thomas V. Welch., Supt., office expenses	36 55	
March 5.		1390	Hardwicke & Co., tools, etc.	29 01	
		1391	N. F. Hydraulic Power and Mfg. Co., electric lighting	50 00	
		1392	Maloney & McCoy, ice	59 00	
		1393	Braas Bros., repairs	16 75	
March 31.		1394	W. A. Shepard, mason	10 50	
		1395	McDonald & Welch, coal	57 75	
		1396	N. F. Hydraulic Power and Mfg. Co., electric lighting	50 00	
		1397	N. F. Hydraulic Power and Mfg. Co., changing wires	56 52	
		1398	Henry E. Gregory, Treas. and Secy., salary, January, February and March	275 00	
		1399	Henry E. Gregory, Treas. and Secy., office and traveling expenses	70 96	
		1400	Pay-roll for March	1,274 67	
		1401	Thomas V. Welch, Supt., office expenses	20 56	
		1402	Hardwicke & Co., tools, etc.	15 94	
					$4,

No. of abstract.	No. of voucher.		
CIII.	1403	Alex. Henschel, clerk to President	$50 00
	1404	Wynkoop Hallenbeck Crawford Co., paper	5 00
	1405	J. Warren Mead, tools	5 55
	1406	Pay-roll for April	1,769 20
	1407	Thomas V. Welch, Supt., office expenses	49 77
	1408	P. J. Davy, plumbing, etc	82 41
	1409	Fred Batchelor, seed	18 00
	1410	F. W. Oliver Co., tools	3 80
	1411	Harry ap' Rees, painting	73 00
	1412	N. F. Hydraulic Power and Mfg. Co., electric lighting	50 00
	1413	Platt & Washburn Refining Co., oil	7 50
	1414	W. A. Shepard, cementing	88 73
	1415	Hardwicke & Co., tools., etc	116 37
	1416	Braas Bros., repairs, etc	38 64
	1417	Power City Lumber Co., repairing	5 14
	1418	Pay-roll for May	1,791 68
	1419	Thomas V. Welch, Supt., office expenses	43 23
	1420	Perkins & Mathews, repairing fountains	58 87
	1421	W. A. Shepard, mason work	21 85
	1422	Charlotte Haeberle, rollers	3 00
	1423	P. J. Davy, plumbing	84 36
	1424	N. F. Hydraulic Power and Mfg. Co., electric lighting	50 00
	1425	Fred Batchelor, seed	6 15
	1426	P. C. Flynn & Son, painting	109 65
	1427	Power City Lumber Co	19 95
	1428	Braas Bros. repairing fountains.	51 10

1897.	No. of abstract.	No. of voucher.			
June 4.	CIII.	1429	Courier Lithographing Co., maps and guides	$452 34	
		1430	Harry ap' Rees, painting	88 75	
June 30.		1431	Pay-roll for June	1,703 79	
		1432	Thomas V. Welch, Supt., office expenses	42 03	
		1433	Henry E. Gregory, Treas. and Secy., salary, April, May and June	275 00	
		1434	Braas Bros., repairs	35 94	
		1435	Omar V. Sage, stationery	14 30	
		1436	Perkins & Mathews, plumbing.	47 02	
		1437	N. F. Metal Sign Co., signs	12 37	
		1438	N. F. Hydraulic Power and Mfg. Co., electric lighting	50 00	
		1439	Harry ap' Rees, painting	74 50	
		1440	McDonald & Welch, coal	21 00	
		1441	F. W. Oliver Co., tools	47 80	
		1442	F. W. Oliver & Co., tools	9 36	
		1443	Hardwicke & Co., hardware, etc.	83 65	
		1444	Hardwicke & Co	47 75	
					$7,
July 13.	CIV.	1445	William Hamilton, Comr., traveling expenses	57 35	
		1446	Wm. F. Wall, agt. cable	147 75	
Aug. 2.		1447	Pay-roll for July	1,777 30	
		1448	Thomas V. Welch, Supt., office expenses	39 08	
Aug. 5.		1449	Hardwicke & Co., tools, etc	17 77	
		1450	Power City Lumber Co	10 44	
		1451	Power City Lumber Co	58 96	
		1452	Robert Hamilton, repairs	35 00	
		1453	Thos. E. McGarigle, repairs	132 32	
		1454	N. F. Hydraulic Power and Mfg. Co., electric lighting	50 00	
		1455	F. W. Oliver Co., tools	5 60	

1897.	No. of abstract.	No. of voucher.			
Aug. 5.		1456	J. E. Perkins, plumbing	$14 54	
		1457	P. C. Flynn & Son, painting	228 14	
		1458	Braas Bros., building, etc	14 33	
		1459	Braas Bros., repairs	3 62	
		1460	Braas Bros., repairs	398 59	
		1461	Braas Bros., repairs	29 65	
Sept. 1		1462	Pay-roll for August	1,795 02	
		1463	Thomas V. Welch, Supt., office expenses	36 47	
Sept. 4.		1464	P. C. Flynn & Son, painting	38 10	
Sept. 4		1465	Globe Ticket Co., Inc. railway tickets	25 00	
		1466	Braas Bros., repairs	209 65	
		1467	O. V. Sage, letter heads	4 00	
		1468	N. F. Hydraulic Power and Mfg. Co., electric lighting	50 00	
		1469	Hardwicke & Co., tools and supplies	38 54	
		1470	E. O. Babcock, stationery	42 50	
Sept. 30.		1471	Pay-roll for September	1,722 78	
		1472	Thomas V. Welch, Supt., office expenses	26 21	
		1473	Henry E. Gregory, Treas. and Sec'y salary, July, August and September	275 00	
		1474	Estate of John Johnson, repairing tools	18 60	
		1475	F. W. Oliver Co., tools, etc	5 71	
		1476	Hardwicke & Co., tools	16 95	
		1477	Coleman Nee, stone	34 00	
		1478	Cataract Ice Co., ice	42 95	
		1479	C. Burns, stone	16 00	
		1480	N. F. Hydraulic Power and Mfg. Co., electric lighting	50 00	$7,467 92

PAYMENTS out of $10,000 as per Chap 950, Laws of 1896:

1896		Series G.				
Oct.	8.	II.	12	Pay-roll, grading, etc	$661 63	
			13	Dobbie, Stuart & Co., iron railing	167 20	
			14	W. A. Shepard, mason work	106 60	$935 43
Nov.	10.	III.	15	Pay-roll	$1,245 01	
			16	Geo. Cook, landscape gardener	212 29	
			17	W. A. Shepard, bridge	248 50	
			18	Samuel Parsons, Jr., landscape architect	25 00	
Dec.	11.		19	Pay-roll	1,522 88	
			20	R. D. Young, cement walks	78 96	
			21	Wm. Young, drawing stone	110 63	
			22	N. F. Hydraulic Power and Mfg. Co., material for roads	250 00	
			23	W. A. Shepard, mason work	334 88	
			24	Ellwanger & Barry, trees	81 68	
			25	Geo. E. Wright, extra work	25 00	
			26	Geo. Cook, landscape gardener	143 25	4,278 08
1897						
Jan.	12.	IV.	27	Pay-roll	$829 50	
			28	W. A. Shepard, bridge work	183 60	
			29	Wm. Young, drawing material	47 50	
Feb.	9.		30	Pay-roll	215 75	
March	9.		31	Pay-roll	157 00	1,433 35
April	3.	V.	32	Pay-roll	$188 50	
May	10.		33	Pay-roll	149 13	
June	8.		34	Pay-roll	103 23	440 86

PAYMENTS out of $15,000, as per chapter 790, Laws of 1897:

1897.		Series H.				
June	8.	I.	1	Pay-roll	$237 00	
July	9.		2	Pay-roll	1,128 50	
			3	Niagara Sand Co	277 98	
Aug.	12.		4	Pay-roll	2,297 48	
			5	W. A. Shepard, work on bridges	305 62	
			6	Niagara Sand Co	275 96	
			7	Braas Bros., bridges	91 80	

Sept. 10.	I.	8	Pay-roll	$2,383 47	
		9	W. A. Shepard, retaining wall	156 10	
		10	Niagara Sand Co.	165 97	
		11	Fred Batchelor, seed	9 00	
Sept. 25.		12	Pay-roll	686 08	
					$8,014 96

1896.	**REMITTANCES TO THE STATE TREASURER.**		
Nov. 2.	Draft for October receipts	$482 40	
Dec. 2.	Draft for November receipts	51 30	
Dec. 31.	Draft for December receipts	29 45	
1897.			
Feb. 1.	Draft for January receipts	63 45	
March 1.	Draft for February receipts	83 05	
March 31.	Draft for March receipts	38 95	
May 1.	Draft for April receipts	53 85	
June 1.	Draft for May receipts	163 45	
June 30.	Draft for June receipts	352 50	
Aug. 2.	Draft for July receipts	1,132 85	
Sept. 1.	Draft for August receipts	3,039 65	
Sept. 30.	Draft for September receipts	1,695 50	
			7,886 40
1896.			
Dec. 31.	Dividends on deposits in Cataract Bank	$28 34	
1897.			
June 1.	Dividend on deposits in Cataract Bank	28 34	
			56 68
1896			
Dec. 31.	Interest on balance in Manufacturers and Traders' Bank	11 79	
1897.			
March 31.	Interest on balances in Manufacturers and Traders' Bank	33 99	
June 30.	Interest on balances in Manufacturers and Traders' Bank	37 36	
Sept. 30.	Interest on balances in Manufacturers and Traders' Bank	25 61	
			108 75
Sept. 30.	Cash balance in bank		758 53
	Total		$48,233 79

HENRY E. GREGORY,
Treasurer.

We the undersigned hereby certify that we have examined the foregoing report of the treasurer for the fiscal year ended September 30, 1897, the vouchers and other papers, and we find the report and accompanying documents correct and that the treasurer has properly accounted for all moneys received and disbursed by him during the year ended September 30, 1897.

ROBERT L. FRYER,
WM. HAMILTON,
Commissioners of State Reservation at Niagara.

CLASSIFICATION OF ACCOUNTS.

Maintenance.

Commissioners' expense	$97 74
Treasurer and secretary, office expenses	64 56
Niagara office expenses	446 12
Salaries (superintendent and clerk)	2,674 98
Police	5,400 00
Inclined railway	1,676 02
Prospect Park	2,605 56
Goat Island	2,297 87
Roads	2,260 48
Walks	1,693 08
Coal	131 00
Electric lighting	656 52
Ice	101 95
Seed	25 25
Buildings	1,239 93
Water supply	41 35
Tools	362 91
Bridges	968 29

Water pipes	102 29
Expense	55 00
Signs	70 11
Seats	99 72
Stairways	27 69
Folders	4 82
Maps and guides	452 34
Drinking fountains	109 97
Total	$24,820 75

Improvements Under Chap. 950 *Laws of* 1896.

Grading	$2,405 01
Walks	1,059 82
Iron railing	167 20
Roads	1,305 23
Planting	489 17
Bridges	1,554 61
Trees	81 68
Terminal station	25 00
Total	$7,087 72

Improvements Under Chap. 790 *Laws of* 1897.

Roads	$2,910 18
Grading	1,599 32

Walks	$1,753 86
Inclined railway	13 20
Bridges	1,288 27
Retaining walls	441 13
Seed	9 00
Total	$8,014 96

APPENDIX.

A CATALOGUE

OF THE

FLOWERING AND FERN-LIKE PLANTS

GROWING WITHOUT CULTIVATION IN THE VICINITY OF THE FALLS OF NIAGARA.

Prepared at the request of the Commissioners of the State Reservation at Niagara, by DAVID F. DAY.

INTRODUCTION.

In December, 1886, the writer, in answer to the request of the Commissioners of the State Reservation at Niagara, promised to prepare for their use a catalogue of the plants growing upon the reservation and its vicinity. He had already on hand the record of his observations made in the neighborhood of the Falls, during a period of more than twenty years. But he well knew that in order to give to the promised catalogue, such a degree of accuracy and completeness as would make it of value to botanists and the public, it would be necessary to revise and renew his observations in the field. To this task, he devoted such leisure as was at his command during the year 1887. The results are presented in this Catalogue. Still he does not doubt that further investigations, made in the vicinity of the Falls, will considerably increase the number of species here recorded. In the more difficult genera of the *Cyperaceae and Graminae*, demanding always in a large degree the skill of the specialist, there must be omissions, more or less numerous and important. Yet it is probable that no species, really characteristic of the flora of Niagara, has been overlooked.

To aid him in making the list complete, the writer has regarded it as his duty to consult, so far as was within his power, the observations made in the neighborhood of the Falls, by all other botanists. It is, however, a matter for great regret that references to the botany of the Falls, especially in the reports of the earlier explorers, have proved so few in number. It seems probable that Peter Kalm, the friend and correspondent of the great Linne, left some record of the botanical observations, which he made during his visit at Niagara, in the year 1750. But, the author failed to find any mention of its publication, either in the Swedish tongue or in an English translation. If his journal still exists, its publication, at the present day, could not but be welcomed as an important contribution to the literature of American botany. It seems not unlikely that the species of *Hypericum* and *Lobelia*, which bear his name, were discovered by him near Table Rock. It is to be doubted whether either the elder or the younger Michaux visited the neighborhood of the great cataract, and it is certain that the enterprising spirit of Pursh brought him no nearer than the site of the present city of Elmira. Nuttall, who botanized near the Falls sometime previous to the year 1818, mentions but one plant, *Utricularia cornuta*, as found by him in their vicinity. Torrey doubtless visited the region — possibly was familiar with it — yet, in his Flora of the State of New York, published in the year 1843, of the 1,511 species of plants, which he described, only fifteen are attributed to Niagara, and none of these, upon his own authority. In the *Flora of North America*, of Torrey and Gray, published in 1838-1842, Niagara is mentioned as a station only five times.

The labors of later botanists have been far more useful in the preparation of the list. The MS. journals of the Hon. George W. Clinton, while engaged in his arduous labors upon the botany of Buffalo and its vicinity, have proved of the greatest value; and the "*Flore Canadienne*" of Abbe Provancher and the *Catalogue of Canadian Plants*" of Professor Macoun, the learned and indefatigable botanist of the Canadian Geological Survey, and the "*Canadian Filicineæ*," the joint work of Professor Macoun and Dr. Burgess, of London, Ontario, have been of important service.

FRENCH LANDING BRIDGE. EAST VIEW.

Wherever use has been made of these or of other authorities due acknowledgment of the obligation appears in the list.

The geology of Niagara Falls, as related to the flora, demands at this place some brief attention. On either side of the river, at a distance not very constant, glacial clays appear. At Clifton they form a considerable elevation. With them, in places, also appear the usual gravel drift of the same period. These make up the soil of the adjacent country. But, as Hall and Lyell long ago pointed out, the evidence is complete that the river once stood at a very considerable height above its present rocky boundaries. At Chippewa, in Prospect Park and at other places, the gravel deposits of the river, made whilst it was occupying its higher level, are still to be seen. But before the river formed its present bed in the gorge below the cataract, it cut through and carried away some portion of its former deposit, leaving, as it subsided, a terrace, on either side, still easily traced. This gravel deposit, once extending continuously across the present river, covered all of what is now Goat Island. It is characterized by a great abundance of fresh-water shells, of the genera *Unio*, *Melania*, *Lymnæa*, etc., identical in species with those still inhabiting the river. The vegetation of the island is that then which might be expected to luxuriate upon a deep calcareous soil, enriched with an abundance of organic matter.

On either side of the river, following its course northerly, for a long distance, the same deposit is met with, alternating with patches from which it has been removed, and in which the underlying rock has but a scanty covering of soil. Near DeVaux College, and immediately above the Devil's Hole, near Lewiston, the rock is almost naked. Yet, even in such places, growing in the crevices of the rock or fringing the verge of the chasm, are to be found some of our most interesting plants. In some places, long since the chasm was excavated, the high rocky bank has given way, through the action of small streams of water, and perhaps by the operation of other causes, forming sheltered recesses of considerable extent, in which a rich humus has accumulated, supporting a dense growth of timber, and giving a congenial situation to some of our rarer plants. Among such places may be named the Devil's Hole, Foster's Flat, above Queenston, and the whirlpool wood, on the Canadian side.

The Niagara Escarpment, as it has been called by the geologists of the State of New York, known at Lewiston as the "Mountain" and in Canada as "Queenston Heights," presents some features entitled to notice in this place. At the point where the river makes its way through it, its height above Lake Ontario is 374 feet, and above Lake Erie, 32 feet. Though it presents a northerly exposure, yet among the plants, growing upon its talus and below, are a number which belong rather to the south and southwestward, and are much more abundant in Ohio than in Western New York. The fact may be explained by the higher annual temperature at Queenston and Lewiston and thence northward to Lake Ontario, than prevails at the Falls or immediately southward.

The flora of Goat Island presents few plants which may be called uncommon in Western New York. For the rarer plants, included in the catalogue, other localities must be visited. But it is still true that Goat Island is very rich in the number of its species. Probably no tract of land in its vicinity, so restricted in area, can be found, exhibiting so large a number. Its vernal beauty is attributable, not merely to its variety of plants, conspicuous in flower, but also to the extraordinary abundance in which they are produced. Yet it seems likely that there was a time, probably not long ago, when other species of plants, of great beauty, were common upon the island, but which are not now to be found there. It is hardly possible that several orchidaceous plants and our three native lilies did not once embellish its woods and grassy places. Within a little while the Harebell (*Campanula rotundifolia*, L.), has gone and the Grass of Parnassus (*Parnassia Caroliniana*, L.), is fast going. This is undoubtedly due to careless flower-gatherers, who have plucked and pulled without stint or reason. The same fate awaits the *Sanguinaria*, the *Diclytras* and the *Trilliums*, which do so much to beautify the island, unless the wholesale spoliation is soon arrested.

The suggestion may here be made, with great propriety, that pains be taken to re-establish upon the island the attractive plants which it has lost. The effort would entail but little expense, its success would be entirely certain and to many the pleasure of a visit to the island would be greatly enhanced. It would surely be a step, and not an unimportant one, in restoring the island to the state in which nature left it.

Frequent reference has been made in the catalogue to localities in the neighborhood of the Falls, but not within the boundaries of the Reservation. For this, however, there is a sufficient reason in the fact that if the catalogue had been confined within the limits of the Reservation, it would have been of far less value to the many who are now interested in botanical science; and to them alone can such a work be of much interest or value. As already stated, some of the rarest plants of western New York and Ontario grow in the neighborhood of Niagara river, but not within the confines of the Reservation.

Thus, it may be mentioned that in the wooded grounds adjoining DeVaux College and belonging to that institution, there may be found *Rhus aromatica*, Ait., *Liatris cylindracea*, Michx., *Aster ptarmicoides*, T. & G., *Asclepias quadrifolia*, Jacq., and *Morus rubra*, L. Among its rocks, perhaps there may still be found a specimen of *Pellæa atropurpurea*, Link., *Camptosorus rhyzophyllus*, Link., and *Asphenium Trichomanes*, L. The Devil's Hole, now almost inaccessible, was once a paradise of ferns—*Phegopteris Dryopteris*, Fée, being its chief rarity. The plateau of rock, which overlooks the ravine, produces *Arabis hirsuta*, Scop., and *Selaginella rupestris*, Spreng., elsewhere in this region quite uncommon. Between the "Mountain" and Lewiston, the explorer will find *Rannunculus multifidus*, Pursh., *Xanthoxylum Americanum*, Mill., *Houstonia cærulea*, L., and *Asplenium ebeneum*, Ait., rare plants in western New York. Queenston Heights have yielded us *Anemonella thalictroides*, Spach., *Asimina triloba*, Dunal, *Lupinus perennis*, L., *Frasera Carolinensis*, Walt. and *Celtis occidentalis*, L.,—species scarcely found elsewhere in our vicinity. Of Foster's Flat, above Queenston, it may be said that the spring seems to visit its rocky fastnesses some weeks earlier than the table-land above. Among its uncommon plants, it may be mentioned that Professor Macoun and Dr. Burgess have detected *Aspidium Lonchites*, Swartz, and *Aspidium Bootii*, Tuckm.—ferns whose accustomed range is a hundred miles northward. The woods near the whirlpool, on the Canadian side, produce in abundance *Cerastium arvense*, L., *Arctostaphylos Uva-ursi*, Spreng., *Castilleia coccinea* Spreng., and the only sassafras trees known in the neighborhood of the Falls. The low land, near Clifton, on the Canadian side, only a few inches higher than the river, affords such uncommon plants as *Gentiana serrata*, Gunner, and *Parnassia Caroliniana*, Michx., *Calamintha Nuttallii*, Benth., still grows on the damp rocks, near the border of the river, and *Gerardia purpurea*, L., and *Utricularia cornuta*, Michx., appear sparingly in the wet and oozy soil near by.

In the preparation of this list, the practice has been followed, now almost universal, of indicating introduced plants (which it will be seen are a large number), by giving their names in small capitals. The later changes in the nomenclature of the species have also been adopted, adding, however, (in parentheses) the names under which they were described in the last edition of Gray's Manual.

Of the 909 species of plants named in the Catalogue, 758 are native and 151 foreign.

The following table exhibits, synoptically, the number of species and genera belonging to each natural family of plants mentioned in the Catalogue:

Names of families.	No. of genera.	No. of species.
Ranunculaceæ	10	32
Magnoliaceæ	2	2
Anonaceæ	1	1
Menispermaceæ	1	1
Berberidaceæ	4	4
Nymphaceæ	2	2
Papaveraceæ	2	2
Fumariaceæ	2	3
Cruciferæ	12	25
Capparidaceæ	1	1
Violaceæ	2	8
Polygalaceæ	1	3
Caryophyllaceæ	7	12
Portulacaceæ	2	3
Hypericaceæ	1	7
Malvaceæ	4	6

Names of families.	No. of genera.	No. of species.
Tiliaceæ	1	2
Linaceæ	1	2
Geraniaceæ	4	6
Rutaceæ	1	1
Simarubiaceæ	1	1
Ilicaceæ	2	2
Celastraceæ	2	4
Rhamnaceæ	1	1
Vitaceæ	2	4
Sapindaceæ	3	8
Anacardiaceæ	1	6
Leguminosæ	13	33
Rosaceæ	15	42
Saxifragaceæ	6	10
Crassulaceæ	2	4
Hamamlidaceæ	1	1
Haloragæ	1	3
Lythraceæ	1	1
Onagraceæ	5	11
Ficoidæ	1	1
Umbelliferæ	15	19
Araliaceæ	1	3
Cornaceæ	2	7
Caprifoliaceæ	6	14
Rubiaceæ	4	13
Valerianaceæ	1	2
Dipsaceæ	1	1
Compositæ	38	97
Lobeliaceæ	1	3

Ericaceæ	8	14
Primulaceæ	3	6
Oleaceæ	2	4
Apocynaceæ	2	3
Asclepiadaceæ	2	7
Gentianaceæ	2	3
Polemoniaceæ	2	2
Hydrophyllaceæ	1	2
Borraginaceæ	6	11
Convolvulaceæ	3	7
Solanaceæ	5	7
Scrophulariaceæ	13	26
Orobanchaceæ	3	3
Lentibulaceæ	1	2
Acanthaceæ	1	1
Verbenaceæ	2	3
Labiatæ	18	27
Plantiginaceæ	1	4
Amarantaceæ	1	2
Chenopodiaceæ	2	7
Phytolaccaceæ	1	1
Polygonaceæ	2	17
Aristolochiaceæ	1	1
Piperaceæ	1	1
Lauraceæ	2	2
Thymelaceæ	2	2
Eleagnaceæ	1	1
Santalaceæ	1	1

Names of families.	No. of genera.	No. of species.
Euphorbiaceæ	2	6
Ceratophyllaceæ	1	1
Urticaceæ	8	13
Platanaceæ	1	1
Juglandaceæ	2	6
Betulaceæ	2	4
Cupuliferæ	6	15
Salicaceæ	2	16
Coniferæ	5	6
Hydrocharidaceæ	2	2
Orchidaceæ	4	7
Iridaceæ	2	3
Smilacaceæ	1	2
Lilaceæ	16	20
Pontederiaceæ	2	2
Juncaceæ	2	9
Typhaceæ	2	4
Araceæ	4	4
Lemnaceæ	3	3
Alismaceæ	3	4
Naiadaceæ	3	18
Cyperaceæ	5	54
Graminæ	31	68
Equisetaceæ	1	6
Ophioglossaceæ	2	3
Filices	12	27
Selaginellaceæ	2	3
Hydropterides	2	2
Total	410	909

CATALOGUE.

RANUNCULACEAE.

Clematis Virginiana, L......................**Clematis. Virgin's Bower.**

On the Canadian side, near Clifton, and elsewhere.

Anemone cylindrica, Gray....................**Anemone. Wind Flower.**

Goat Island and elsewhere.

Anemone Virginiana, L....................**Anemone. Wind Flower.**

Goat Island and near DeVaux College.

var. **alba**, Wood.

Goat Island.

Anemone dichotoma, L. (*A. Pennsylvanica*, L.)

Anemone. Wind Flower.

Goat Island and elsewhere.

Anemone nemorosa, L.......................**Anemone. Wind Flower.**

Goat Island.

var. **quinquefolia**, Gray.

With the last.

Anemone Hepatica, L. (*Hepatica triloba*, Chaix.)............**Liverwort.**

Goat Island and near Lewiston. Rather rare.

Anemone acutiloba, Lawson. (*Hepatica acutiloba*, DC.).......**Liverwort.**

Goat Island and elsewhere. Less rare.

Anemonella thalictroides, Spach. (*Thalictrum anemonoides*, Michx.)

Near Brock's Monument, Ontario.

Thalictrum dioicum, L..................................**Meadow Rue.**

Goat Island, Lewiston and elsewhere.

Thalistrum Cornuti, L.

Meadow Rue.

Shores of the river above the Falls.

Thalictrum purpurascens, L.............................Meadow Rue.

Near Clifton, Canada.

var. **ceriferum, Austin.**

"Near Drummondsville, Niagara Falls." Macoun.

Ranunculus aquatilis, L..............................Water Crowfoot.

var. **trichophyllus, Chaix.**

In pools above the falls on the Canadian side.

Ranunculus multifidus, Pursh..............................Buttercup.

In a pool not far from the river's bank, above Lewiston.

Ranunculus Flammula, L..................................Spearwort.

var. **reptans, Meyer.**

At the water's edge on Strawberry and Grand Islands, and, probably, in similar situations nearer the Falls.

Ranunculus abortivus, L.

Goat Island and elsewhere.

Ranunculus sceleratus, L...........................Cursed Crowfoot.

Lewiston, and near Suspension Bridge.

Ranunculus recurvatus, Poir.......................Hooked Crowfoot.

Goat Island.

Ranunculus Pennsylvanicus, L.....................Bristly Crowfoot.

Clifton, Canada.

Ranunculus fascicularis, Muhl.......................Early Buttercup.

Near DeVaux College.

Ranunculus septentrionalis, Poir ....Buttercup.

Clifton, Canada.

RANUNCULUS BULBOSUS, L...................................BUTTERCUP.

Goat Island. Introduced.

RANUNCULUS ACRIS, L.......................................BUTTERCUP.

Goat Island and elsewhere.

Caltha palustris, L......................Cowslips. Marsh Marygold.

Lewiston, Clifton, Ontario.

Aquilegia Canadensis, L.............................Wild Columbine.

Goat Island and along the rocky banks of the river in many places.

Hydrastis Canadensis, L...................................**Hydrastis.**

Cayuga Island, and probably elsewhere nearer the falls.

Actaea spicata, L.....................**Herb Christopher. Red Cohosh.**

var. rubra, Ait.

Goat Island, near DeVaux College.

Actaea alba, Bigelow..................................**White Cohosh.**

Goat Island.

Cimicifuga racemosa, Nutt.........................**Black Snake-root.**

Probably occurs near Queenston, Ont.

MAGNOLIACEAE.

Liriodendron Tulipifera, L....................**Whitewood. Tulip-tree.**

Goat Island. Not common. One fine specimen is growing near the carriage way on the north side of the island.

Magnolia acuminata, L...............................**Cucumber tree.**

"Near the falls of Niagara." *Provancher.* Not seen by us.

ANONACEAE.

Asimina triloba, Dunal...**Papaw.**

Queenston Heights, Ontario, *Macoun.*

MENISPERMACEAE.

Menispermum Canadense, L..............................**Moon seed.**

Common both in New York and Ontario, but not noticed by us near the Falls. Doubtless overlooked.

BERBERIDACEAE.

BERBERIS VULGARIS, L...................................BARBERRY.

Goat Island, near Lewiston.

Caulophyllum thalictroides, Michx......................**Blue Cohosh.**

Goat Island. Abundant.

Jeffersonia diphylla, Pers.............**Twin Leaf. Rheumatism Root.**

Niagara Falls. *Clinton.* Very rare.

Podophyllum peltatum, L....................**May Apple. Mandrake.**

Goat Island. Abundant.

NYMPHACEAE.

Nymphaea tuberosa, Paine. . . . . White Water Lily.

Abundant in shallow places in the river, some distance above the Falls.

Nuphar advena, Ait. . . . . Yellow Water Lily.

With the last.

PAPAVERACEAE.

CHELIDONIUM MAJUS, L. . . . . CELANDINE.

Clifton, Ontario.

Sanguinaria Canadensis, L. . . . . Blood Root.

Goat Island, where it has been found producing pink flowers.

FUMARIACEAE.

Diclytra Cucullaria, DC. (*Dicentra Cucullaria*, DC.) **Dutchman's Breeches.**

Goat Island. Abundant.

Diclytra Canadensis, DC. (*Dicentra Canadensis*, DC.) . . . . . . **Squirrel Corn.**

Goat Island. Abundant. Between the two species numerous hybrids have been noticed on Goat Island.

Corydalis glauca, Pursh. . . . . Corydalis.

Occurs at Tonawanda, and should be found near the Falls.

CRUCIFERAE.

ALYSSUM CALYCINUM, L. . . . . ALYSSUM.

Near Brock's monument, Ontario.

DRABA VERNA, L. . . . . WHITLOW GRASS.

Introduced on Goat Island, but perhaps not established.

Dentaria diphylla, Michx. . . . . Pepper Root. Crinkle Root.

Goat Island.

Dentaria laciniata, Muhl.

Goat Island.

Cardamine rhomboidea, DC. . . . . Spring Cress.

var. **purpurea**, Torr.

Goat Island and elsewhere. The typical form, probably, may be found in the low ground near Clifton, Ontario.

Cardamine hirsuta, L. . . . . Bitter Cress.

Goat Island.

Arabis lyrata, L. .... Rock Cress.

Goat Island, The Three Sisters, and along the gorge to Lewiston.

Arabis hirsuta, Scop.

Near DeVaux College, and at Lewiston, and on the opposite side of the river.

Arabis laevigata, Poir.

Devil's Hole. Queenston Heights, Ontario. *Macoun.*

Arabis Canadensis, L. .... Sickle Pod.

Goat Island, and elsewhere.

Arabis perfoliata, Lam. .... Tower Mustard.

Near Clifton, Ontario.

Arabis Drummondii, Gray.

Lewiston. Not common.

BARBAREA PRAECOX, R. Br. .... EARLY WINTER CRESS.

Brock's monument, Ontario. *Macoun.*

Barbarea vulgaris, R. Br. .... Winter Cress.

Road sides near the Falls.

Erysimum cherianthoides, L. .... Worm-seed Mustard.

Margin of the river above the Falls.

SYSIMBRIUM OFFICINALE, Scop. .... HEDGE MUSTARD.

Road sides and waste places near the Falls.

BRASSICA SINAPISASTRUM, Boiss. .... MUSTARD. CHARLOCK.

Abundant on both sides of the river.

BRASSICA NIGRA, Koch. .... BLACK MUSTARD.

Between the Falls and DeVaux College.

NASTURTIUM OFFICINALE, R Br. .... WATER CRESS.

Near the river's edge above the Falls. Clifton, Ontario.

Nasturtium palustre, DC. .... Marsh Cress.

In damp places above the Falls.

Nasturtium lacustre, Gray .... Lake Cress.

In the river above the Falls.

NASTURTIUM ARMORACIA, Fries .... HORSE RADISH.

"At Niagara Falls." *Macoun.*

CAPSELLA BURSA-PASTORIS, Moench..................SHEPHERD'S PURSE.

Common everywhere.

Lepidium Virginicum, L...........................Wild Peppergrass.

Road sides, near the village.

LEPIDIUM CAMPESTRE, R. Br........................FIELD PEPPERGRASS.

"Clifton, near Niagara Falls." *Macoun.*

CAPPARIDACEAE.

Polanisia graveolens, Raf.

"Abundant on the sands at Niagara." *Macoun.*

Plentiful at the foot of Lake Erie.

CISTACEAE.

Helianthemum Canadense, Michx..........................Frost Weed.

Common in dry places in western New York, no doubt occurring near DeVaux College, and at Lewiston.

Lechea major, Michx......................................Pinweed.

Probably occurs with the last.

Lechea minor, Lam...Pinweed.

Probably occurs with Helianthemum Canadense.

VIOLACEAE.

Ionidium concolor, Benth. and Hook. (*Solea concolor*, Ging).

Green Violet.

Goat Island. Foster's Flat, Ontario.

Viola blanda, Willd...Violet

Goat Island and near Clifton, Ontario.

Viola palmata, L...Violet.

var. cucullata, Gray. (*V. cucullata*, Ait.)

Goat Island and elsewhere.

Viola canina, L..Dog Violet.

var. **Muhlenbergii**, Gray. (*Viola canina* L., var. *sylvestris*, Regel.)

Goat Island and elsewhere.

Viola rostrata, Muhl...........................Long-spurred Violet.

Goat Island.

Viola Canadensis, L...............................Canada Violet.

Goat Island.

Viola pubescens, Ait. .... **Downy Yellow Violet.**

var. **eriocarpa**, Nutt.

Goat Island.

var. **scabriuscula**, Torr. and Gray.

Goat Island.

POLYGALACEAE.

Polygala verticillata, L.

Near DeVaux College. Queenston Heights, Ontario, *Macoun*.

Polygala Senega, L. .... **Seneca Snake Root.**

Both the narrow and the broad leaved varieties are to be found near the whirlpool on both sides of the river.

Polygala incarnata, L.

Said by Douglass (1823) to have been found in rocky places on the Niagara river near the Falls. *Macoun.*

CARYOPHYLLACEAE.

DIANTHUS ARMERIA, L. .... DEPTFORD PINK.

Lewiston, scarce. In a field near Clifton, Ontario, plentiful.

SAPONARIA OFFICINALIS, L. .... SOAPWORT.

Goat Island and the mainland.

Silene stellata, Ait. .... **Starry Campion.**

Found by Douglass, in 1823, in dry, stony places on the Niagara river. *Macoun.*

Silene antirrhina, L. .... **Sleepy Catch-fly.**

Near DeVaux College and elsewhere.

SILENE NOCTIFLORA, L. .... NIGHT-FLOWERING CATCH-FLY.

Above the Falls on the American side of the river.

LYCHNIS GITHAGO, Lam. .... CORN COCKLE.

Fields on the main land.

CERASTIUM VISCOSUM, L. .... MOUSE-EAR CHICKWEED.

Goat Island and elsewhere.

CERASTIUM VULGATUM, L. .... MOUSE-EAR CHICKWEED.

Goat Island and elsewhere.

STELLARIA MEDIA, Smith. .... CHICKWEED.

Goat Island and elsewhere.

Stellaria longifolia, Muhl .... **Long-leaved Stitchwort.**

In damp, grassy places above the Falls.

ARENARIA SERPYLLIFOLIA, L. .... SAND WORT.

Road sides at Clifton and elsewhere.

Arenaria lateriflora, L.

Goat Island.

PORTULACACEAE.

PORTULACA OLERACEA, L. .... PURSLANE.

Waste places on the main land.

Claytonia Caroliniana, Michx .... **Spring Beauty.**

Goat Island.

Claytonia Virginica, L. .... **Spring Beauty.**

Goat Island.

HYPERICACEAE.

Hypericum Kalmianum, L. .... **Kalm's St. John's Wort.**

Goat Island. "Rochers au bas de la chute de Niagara." *Provancher.* (*Fl. Canad. p.* 104.)

HYPERICUM PERFORATUM, L. .... COMMON ST. JOHN'S WORT.

Goat Island and elsewhere.

Hypericum maculatum, Walt. (*H. corymbosum*, Muhl.)

Margin of the river above the Falls.

Hypericum mutilum, L.

Wet places along the river above the Falls.

Hypericum Canadense, L.

In similar places as the last.

Hypericum Ascyron, L. (*H. pyramidatum*, Ait.)... **Great St. John's Wort.**

Grand Island and probably nearer the Falls.

Hypericum Virginicum, L. (*Elodes Virginica*, Nutt.)

In swampy places along the river above Clifton, Ontario.

MALVACEAE.

ALTHAEA ROSEA, L. .... HOLLYHOCK.

Clifton, Ontario. Escaped from cultivation.

MALVA ROTUNDIFOLIA, L. .... LOW MALLOW.

Goat Island and elsewhere.

AT THE SPRING GOAT ISLAND.

MALVA SYLVESTRIS, L. . . . . HIGH MALLOW.

Near the Devil's Hole.

MALVA MOSCHATA, L. . . . . MUSK MALLOW.

Road sides on the main land.

ABUTILON AVICENNAE, Gaert. . . . . VELVET LEAF.

Waste places on the main land.

Hibiscus Moscheutos, L. . . . . **Swamp Rose Mallow.**

Probably may be found along the river above the Falls, as it grows in such situations near the foot of Lake Erie.

TILIACEAE.

Tilia Americana, L. . . . . **Linden. Basswood.**

Goat Island. An abundant and conspicuous element of its forest.

Tilia——? . . . . **Basswood.**

Goat Island. The tree, here indicated, seems to be quite distinct from the last. It may be readily distinguished by its bark, which is as white as that of the white ash, *Fraxinus Americana, L.*

LINACEAE.

Linum Virginianum, L. . . . . **Wild Flax**

"Near Niagara Falls." *Macoun.* Above Foster's Flat.

LINUM USITATISSIMUM, L. . . . . COMMON FLAX

Occasionally seen on railroad embankments.

GERANIACEAE.

Geranium maculatum, L. . . . . **Wild Geranium.**

Goat Island.

Geranium Robertianum, L. . . . . **Herb Robert.**

Goat Island.

Floerkia proserpinacoides, Willd.

Goat Island.

Impatiens pallida, Nutt. . . . . **Impatience. Wild Balsam.**

Goat Island.

Impatiens biflora, Walt. (*I. fulva*, Nutt.). . . . **Impatience. Wild Balsam.**

Goat Island and the main land.

Oxalis stricta, L. . . . . **Sorrel.**

Great Island.

RUTACEAE.

Xanthoxylum Americanum, Mill........................**Prickly Ash.**

Near Lewiston.

SIMARUIBACAE.

Ailanthus glandulosus, Desf..............**Ailanthus. Tree of Heaven.**

Spontaneous near Clifton, Ontario.

ILICACEAE.

Ilex verticillata, Gray..................................**Winter Berry.**

Chippewa. *Macoun.* Probably nearer the Falls.

Nemopanthes Canadensis, DC.

Near Clifton, Ontario.

CELASTRACEAE.

Celastrus scandens, L....................................**Bittersweet.**

Goat Island.

Euonymus atropurpureus, Jacq................**Burning Bush. Wahoo.**

Goat Island.

Euonymus Americanus, L.................**Strawberry Bush. Wahoo.**

"Niagara." *Macoun.*

var. obovatus, Torr. and Gray.

"Hills around Niagara Falls." *Macoun,* on the authority of *Dr. Maclagan.*

RHAMNACEAE.

Ceanothus Americanus, L...........................**New Jersey Tea.**

Near DeVaux College.

VITACEAE.

Vitis aestivalis, Michx................................**Summer Grape.**

Goat Island and elsewhere.

Vitis riparia, Michx. (*V. cordifolia,* Michx.)................**Frost Grape.**

Goat Island and elsewhere.

Vitis Labrusca, L......................................**Wild Grape.**

Erroneously attributed to the vicinity of the Falls, as a native, in the "*Plants of Buffalo and Vicinity*" (p. 26), and by *Provancher* in the *Flore Canadienne* (p. 112). Occasionally spontaneous.

Ampelopsis quinquefolia, Michx.....................**Virginia Creeper.**

Goat Island and elsewhere.

SAPINDACEAE.

Staphylea trifolia, L. .... Bladder Nut.

Foster's Flat, Ontario.

Aesculus glabra, Willd. .... Buckeye.

Spontaneous near Lewiston. From the west.

Acer spicatum, Lam. .... Mountain Maple.

Goat Island, near the Horse-shoe falls.

Acer saccharinum, Wang. .... Sugar Maple.

Goat Island. One of the most abundant trees.

Acer dasycarpum, Erhart .... White Maple.

Near Clifton, Ontario.

Acer rubrum, L. .... Red Maple.

Goat Island. A shade tree in the village.

Acer Negundo, L. (*Negundo aceroides*, Moench.) .... Box Elder.

Planted in Prospect park, where it is now appearing spontaneously.

ACER PLATANOIDES, L. .... NORWAY MAPLE.

A common shade tree in the village.

ANACARDIACEAE.

Rhus typhina, L. .... Stag-horn Sumach.

Goat Island.

Rhus glabra, L. .... Smooth Sumach.

Queenston Heights. *Macoun.*

Rhus venenata, DC. .... Poison Sumach.

Swampy places above the Falls, near Clifton, Ontario.

Rhus Toxicodendron, L. .... Poison Ivy.

Goat Island. Too plentiful.

var. **radicans**, Torrey .... Poison Ivy.

Goat Island.

Rhus aromatica, Ait. .... Aromatic Sumach.

Near DeVaux College and on the opposite side of the river. Common in the places named.

LEGUMINOSAE.

Lupinus perennis, L. .... Lupine.

Queenston Heights, Ontario.

MEDICAGO LUPULINA, L. .... BLACK MEDICK.

Above the Falls on the American side.

MELILOTUS OFFICINALIS, Willd........................YELLOW MELILOT.

Above the Falls on the American side.

MELILOTUS ALBA, Lam................................SWEET CLOVER.

Goat Island and the main land.

TRIFOLIUM ARVENSE, L.........................RABBIT'S-FOOT CLOVER.

Abundant near Lewiston.

TRIFOLIUM PRATENSE, L...................................RED CLOVER.

Goat Island and elsewhere.

TRIFOLIUM REPENS, L..................................WHITE CLOVER.

Goat Island and elsewhere.

TRIFOLIUM HYBRIDUM, L...............................ALSIKE CLOVER.

American side of the river above the Falls. Lewiston.

TRIFOLIUM PROCUMBENS, L.................................HOP CLOVER.

Near Clifton, Ontario.

Robinia Pseudacacia, L..............................Common Locust.

A frequent shade-tree; often spontaneous.

Robinia viscosa, Vent...............................Clammy Locust.

Spontaneous in places near Lewiston.

Astragalus Cooperi, Gray.......................Cooper's Milk Vetch.

Goat Island near the Horse-shoe Fall.

Astragalus Canadensis, L................................Milk Vetch.

Common on the islands in the river near Tonawanda. It may be looked for nearer the Falls.

Desmodium nudiflorum, DC.............................Tick Trefoil.

Near DeVaux College.

Desmodium acuminatum, DC.

Near DeVaux College.

Desmodium pauciflorum, DC.

"Woods at Niagara Falls." *Macoun.*

Desmodium rotundifolium, DC.

Near DeVaux College.

Desmodium cuspidatum, Torr. and Gray.

Queenston Heights, Ontario. *Macoun*, on the authority of *Douglass.*

Desmodium Dillenii, Darlington.

Near DeVaux College. Queenston Heights, Ontario. *Macoun.*

Desmodium paniculatum, DC.

Near DeVaux College and on the opposite side of the river.

Desmodium Canadense, DC.

Near Clifton, Ontario.

Desmodium ciliare, DC.

Queenston Heights, Ontario. *Macoun.*

Lespedeza reticulata, Pers. (*L. violacea*, Pers. var. *sessiliflora*, Torr. and Gray.)

Near DeVaux College.

Lespedeza hirta, Ell. .... Bush Clover.

"Queenston Heights and Niagara Falls." *Macoun.*

Lespedeza capitata, Michx. .... Bush Clover.

Near DeVaux College.

Vicia Cracca, L. .... Vetch. Tare.

Near Clifton, Ontario.

Vicia Caroliniana, Walt. .... Vetch.

Goat Island.

Vicia Americana, Muhl. .... Vetch.

Goat Island.

Lathyrus ochroleucus, Hook. .... Wild Pea.

Goat Island, and near the whirlpool, Ontario.

Lathyrus paluster, L. .... Marsh Pea.

var. myrtifolius, Gray.

Goat Island and margins of the river above the Falls.

Amphicarpaea monoica, Elliott. .... Hog Peanut.

Goat Island.

Apios tuberosa, Moench. .... Ground Nut.

Doubtless near the Falls, but not yet observed by us.

Gleditschia triacanthos, L. .... Honey Locust.

Spontaneous along the bank of the river near Lewiston.

ROSACEAE.

AMYGDALUS PERSICA, L. .... PEACH.

Spontaneous on Goat Island and near the Devil's Hole.

Prunus Americana, Marshall .... Wild Plum.

Goat Island and elsewhere.

PRUNUS CERASUS, L. .... COMMON RED CHERRY.

Spontaneous on Goat Island, and abundant along roadsides below the Falls.

Prunus Virginiana, L. .... **Choke Cherry.**

Near DeVaux College.

Prunus serotina, L. .... **Black Cherry.**

Near Clifton, Ontario.

Spiraea salicifolia, L. .... **Common Meadow Sweet. Spiraea.**

Above the Falls on the Canadian side.

Neillia opulifolia, Benth. and Hook. (*Spiraea opulifolia*, L.). . **Nine-bark.**

Goat Island. The Three Sisters.

Rubus odoratus, L. .... **Purple Flowering Raspberry.**

Goat Island and elsewhere.

Rubus triflorus, Rich.

Wet places near Clifton, Ontario.

Rubus strigosus, Michx .... **Red Raspberry.**

Goat Island.

Rubus occidentalis, L. .... **Black Raspberry.**

Goat Island.

Rubus villosus, Ait .... **Blackberry.**

Goat Island.

Rubus Canadensis, L. .... **Dewberry. Low Blackberry.**

Goat Island. Lewiston.

Geum album, Gmelin .... **Avens.**

Goat Island.

Geum Virginianum, L.

Goat Island. Chippewa. *Macoun*, on the authority of *Dr. Maclagan.*

Geum strictum, Ait.

Goat Island.

Geum rivale, L. .... **Purple Avens.**

Wet places above Clifton, Ontario.

Waldsteinia fragarioides, Tratt .... **Barren Strawberry.**

Goat Island.

Fragaria Virginiana, Duchesne........................Wild Strawberry.

Goat Island and the main land.

Fragaria vesca, L..................................Wood Strawberry.

Goat Island.

Potentilla Norvegica, L.

Goat Island and the main land.

Potentilla Canadensis, L..................................Cinquefoil.

Goat Island.

var. simplex, Torr. and Gray.

Near DeVaux College.

Potentilla argentea, L..............................Silvery Cinquefoil.

Near DeVaux College. Clifton, Ontario.

Potentilla Anserina, L..................................Silver Weed.

Low grounds above the Falls on both sides of the river.

Potentilla palustris, Scop........................Marsh Five-finger.

Chippewa. Clinton.

POTENTILLA PILOSA, Willd.

Near Clifton, Ontario. This is the plant called by us, *P. recta*, L., in the "*Plants of Buffalo and Vicinity*." The present determination was by *Prof. Macoun*.

Agrimonia Eupatoria, L..................................Agrimony

Goat Island.

Rosa blanda, Ait..................................Early Wild Rose.

Goat Island.

Rosa Carolina, L..................................Swamp Rose.

Wet grounds, near Clifton, Ontario.

Rosa parviflora, Ehrt. (*R. lucida*, Ehrt.)..................Shining Rose.

Goat Island.

Rosa rubiginosa, L..................................Sweet Briar.

Goat Island. Devil's Hole. Lewiston.

ROSA MICRANTHA, Smith..................................SWEET BRIER.

Goat Island.

PYRUS MALUS, L..................................APPLE.

Spontaneous on Goat Island and near the Devil's Hole and Lewiston.

PYRUS COMMUNIS, L. .... PEAR.

Spontaneous on Goat Island.

Pyrus coronaria, L. .... WILD CRAB-APPLE.

Near DeVaux College. Lewiston. Queenston. Queenston Heights, Ontario.

PYRUS AUCUPARIA, Gaertn. .... ROWAN TREE. MOUNTAIN ASH.

Within the gorge of the river, on the Canadian side near the Cantilever bridge.

Crataegus coccinea, L. .... Thorn.

Goat Island.

Crataegus tomentosa, L. .... Black Thorn.

Goat Island.

Crataegus Crus-galli, L. .... Cockspur Thorn.

Goat Island. Not common in western New York; but here, quite abundant.

Amelanchier Canadensis, Torr. and Gray .... Shad-bush.

var. Botryapium, Gray.

Goat Island.

var. oblongifolia, Gray.

Goat Island. Clifton, Ontario.

SAXIFRAGACEAE.

Saxifraga Virginiensis, Michx .... Spring Saxifrage.

Goat Island. Near DeVaux College. Lewiston.

Saxifraga Pennsylvanica, L. .... Swamp Saxifrage.

Cayuga Island, near LaSalle. *Clinton.*

Tiarella cordifolia, L. .... Mitre Wort.

Goat Island.

Mitella diphylla, L. .... Mitre Wort.

Goat Island.

Mitella nuda, L. .... Naked Mitre Wort

Near Chippewa. *Clinton.*

Chrysosplenium Americanum, Schw .... Golden Saxifrage.

Wet grounds near Clifton, Ontario.

Parnassia Caroliniana, Michx.

Goat Island near the Horse-shoe Fall. Wet grounds near Clifton, Ontario. Near the water's edge at the whirlpool on the Canadian side.

Ribes Cynosbati, L. .... **Wild Gooseberry.**

Goat Island.

Ribes oxyacanthoides, L. (*R. hirtellum*, Michx.)..... **Swamp Gooseberry.**

Wet places near Clifton, Ontario.

Ribes floridum, L. .... **Wild Black Currant.**

Along the descent to the ferry on the Canadian side.

CRASSULACEAE.

Penthorum sedoides, L. .... **Ditch Stone Crop**

Damp places near Clifton, Ontario.

SEDUM ACRE, L. .... STONE CROP

Goat Island and the main land. Abundant.

Sedum ternatum, Michx .... STONE CROP

Attributed to rocky places on Niagara river by Douglass. Not seen by us.

SEDUM TELEPHIUM, L. .... LIVE-FOR-EVER

Near DeVaux College.

HAMAMELACEAE.

Hamamelis Virginica, L. .... **Witch Hazel**

Near DeVaux College.

HALORAGEAE.

Myriophyllum spicatum, L. .... **Water Milfoil**

Niagara river above the Falls in shallow and quiet places.

Myriophyllum verticillatum, L. .... **Water Milfoil**

With the last.

Myriophyllum heterophyllum, Michx .... **Water Milfoil**

Pools near Clifton, Ontario.

LYTHRACEAE.

Nesaea verticillata, H. B. K. .... **Swamp Loosestrife**

Found in several places along the shores of Niagara river near Lake Erie. Therefore very likely to occur near the Falls, but not yet observed by us.

ONAGRACEAE.

Epilobium spicatum, L. (*E. augustifolium*, L.)..............**Willow Herb**

Near DeVaux College.

Epilobium hirsutum, L.

Introduced near Clifton, Ontario. Perhaps not established.

Epilobium palustre, L.

var. **lineare**, Gray.

Near Clifton, Ontario.

Epilobium molle, Torr.

In wet places near Clifton, Ontario.

Epilobium coloratum, Muhl.

Above the Falls on the American side.

Ludwigia palustris, Ell.

Not yet seen by us near the Falls, but may be confidently looked for.

Oenothera biennis, L..................................**Evening Primrose**

Goat Island and elsewhere.

Oenothera pumila, (*Oe. chrysantha*, Michx.)....**Dwarf Evening Primrose**

Near the cantilever bridge on the Canadian side of the river. Queenston Heights, Ontario. *Macoun*.

Gaura biennis, L..**Gaura**

Near the Devil's Hole.

Circaea Lutetiana, L..........................**Enchanter's Nightshade**

Goat Island.

Circaera alpina, L............................**Enchanter's Nightshade**

Damp and shady woods near Clifton, Ontario.

CUCURBITACEAE.

Echinocystis lobata, Torr and Gray......................**Wild Cucumber**

Sicyos angulatus, L......................................**Star Cucumber**

These two members of the Gourd family, occurring not rarely in Ontario and western New York, have not been observed by us near the Falls. They may be expected.

FICOIDEAE.

Mollugo verticillata, L....................................**Carpet Weed**

"On the railway track between Niagara Falls and Queenston." *Macoun*.

UMBELLIFERAE.

Hydrocotyle Americana, L. .... Penny Wort

Damp and shady places near Clifton, Ontario.

Sanicula Canadensis, L. .... Sanicle

Goat Island.

Sanicula Marilandica, L. .... Sanicle

Goat Island.

Conium maculatum, L. .... Poison Hemlock

Road sides on the main land.

Cicuta maculata, L. .... Water Hemlock

Goat Island.

Cicuta bulbifera, L. .... Water Hemlock

Wet places near Clifton, Ontario.

Carum Carui, Koch .... Caraway

Goat Island.

Sium cicutaefolium, Gmel., (*S. lineare*, Michx.) .... Water Parsnip

Wet places near Clifton, Ontario.

Pimpinella integerrima, Beuth. and Hooker. (*Zizia integerrima*, DC.)

Near DeVaux College.

Cryptotaenia Canadensis, DC. .... Hone Wort

Goat Island.

Osmorrhiza longistylis, DC. .... Smooth Sweet Cicely

Goat Island.

Osmorrhiza brevistylis, DC. .... Hairy Sweet Cicely

Goat Island.

Thaspium barbinode, Nutt.

Near DeVaux College. Foster Flat, Ontario. *Macoun.*

Thaspium aureum, Nutt. .... Golden Alexanders

Goat Island.

Archangelica atropurpurea, Hoffm. .... Great Angelica

Wet grounds near Clifton, Ontario.

Peucedaneum sativum, Beuth. and Hook. (*Pastinaca sativa*, L.) .. Parsnip

Along the descent to the ferry on the Canadian side.

Heracleum lanatum, Michx. .... Cow Parsnip

Goat Island. More abundant *on the* main land.

DAUCUS CAROTA, L. .... CARROT

Near the cantilever bridge on the Canadian side.

Erigenia bulbosa, Nutt. .... **Harbinger of Spring**

Goat Island. Introduced. Established.

ARALIACEAE.

Aralia nudicaulis, L. .... **Wild Sarsaparilla**

Goat Island. The Three Sisters.

Aralia quinquefolia, Gray .... **Ginseng**

Goat Island, but rare.

Aralia trifolia, Gray. .... **Dwarf Ginseng**

Goat Island.

CORNACEAE.

Cornus florida, L. .... **Flowering Dogwood**

Goat Island. Near DeVaux College.

Cornus circinata, L'Her. .... **Round-leaved Dogwood**

Goat Island near the Horse-shoe Fall.

Cornus sericea, L. .... **Silky Dogwood**

Goat Island.

Cornus stolonifera, Michx .... **Red Osier**

Goat Island.

Cornus paniculata, L'Her. .... **Dogwood**

Goat Island. Near DeVaux College.

Cornus alternifolia, L.

Not uncommon in western New York. Probably overlooked.

Nyssa multiflora, Wang .... **Pepperidge. Tupelo**

Not a rare tree in Erie and Niagara counties. Ought to be found near the Falls.

CAPRIFOLIACAE.

Sambucus racemosa, L. .... **Red-berried Elder**

var. pubens, Watson.

Goat Island.

Sambucus Canadensis, L. .... **Elder**

Goat Island.

Viburnum Opulus, L........Snowball. Guelder Rose. High Cranberry

Grand Island. Likely to be found on either side of the river near the Falls.

Viburnum acerifolium, L..................................Arrow wood

Near DeVaux College.

Viburnum pubescens, Pursh................................Arrow wood

Goat Island. Near DeVaux College. Along the descent to the ferry on the Canadian side.

Viburnum dentatum, L.....................................Arrow wood

Chippewa, Ontario. *Macoun*, on the authority of *Dr. Maclagan*. Common on Grand Island.

Viburnum nudum, L.......................................Withe rod

Probably in the wet grounds near Clifton, Ontario.

Viburnum Lentago, L....................Sheep-berry. Sweet Viburnum

Wet grounds near Clifton, Ontario.

Triosteum pedfoliatum, L...............................Horse Gentian

Goat Island. Near DeVaux College.

Symphoricarpus racemosus, Michx.........................Snowberry

var. pauciflorus, Robbins.

Goat Island. Near DeVaux College and elsewhere.

Lonicera ciliata, Muhl..............................Fly Honeysuckle

Goat Island. Near Clifton, Ontario.

LONICERA TARTARICA, L.....................TARTARIAN HONEYSUCKLE

Goat Island. Near DeVaux College. Well established.

Lonicera glauca, Hill. (*L. parviflora*, Lam., var. *Douglasii*, Gray.)

Goat Island.

Diervilla trifida, Moench.........................Bush Honeysuckle

Near Clifton, Ontario.

RUBIACEAE.

Houstonia caerulea, L.......................Bluets. Innocence

Near Lewiston.

Houstonia purpurea, L.

var. ciliolata, Gray.

Goat Island. Near the whirlpool on both sides of the river.

Cephalanthus occidentalis, L. .... **Button Bush**

Wet places near Clifton, Ontario.

Mitchella repens, L. .... **Partridge Berry**

Goat Island.

Galium Aparine, L. .... **Cleavers**

Goat Island and elsewhere.

GALIUM MOLLUGO, L. .... BEDSTRAW

Goat Island. Introduced.

Galium pilosum, Ait. .... **Bedstraw**

Near DeVaux College. Queenston, Ontario. *Macoun.*

Galium circaezans, Michx.

Lewiston. Queenston Heights, Ontario. *Macoun.*

Galium lanceolatum, Michx.

Near DeVaux College. Near the whirlpool. Ontario. *Macoun*, on the authority of *Dr. Maclagan.*

Galium boreale, L. .... **Northern Bedstraw**

Goat Island. Near DeVaux College.

Galium trifidum, L.

var. tinctorum, Gray.

Goat Island.

Galium asprellum, Michx .... **Rough Bedstraw**

Goat Island.

Galium triflorum, Michx .... **Sweet Bedstraw**

Goat Island. Woods near DeVaux College.

VALERIANACEAE.

VALERIANA OFFICINALIS, L. .... MILLEFLEUR. VALERIAN

Near the cantilever bridge on the Canadian side.

Valeriana dioica, L., var. sylvatica, Watson. (V. *sylvatica.* Rich.)

"Meadows, Niagara Falls." [Ontario.] *Macoun*, on authority of *Dr. Maclagan.* Not seen by us.

DIPSACEAE.

DIPSACUS SYLVESTRIS, Huds. .... TEASEL.

Above the Falls on the American side of the river. Near the Devil's Hole.

Eupatorium purpureum, L. .... Purple Thoroughwort

Damp grounds near Clifton, Ontario.

Eupatorium perfoliatum, L. .... Boneset. Thoroughwort

Above the Falls on the American side.

Eupatorium ageratoides, L. .... White Snakeroot

Goat Island and elsewhere.

Liatris cylindracea, Michx. .... Button Snakeroot

Near DeVaux College.

Solidago caesia, L. .... Golden Rod

Near DeVaux College.

Solidago latifolia, L. .... Golden Rod

Goat Island.

Solidago bicolor, L. .... Golden Rod

Goat Island. Near DeVaux College.

var. concolor, Torr. and Gray.

Near DeVaux College.

Solidago ulmifolia, Muhl. .... Golden Rod

Goat Island.

Solidago neglecta, Torr. and Gray .... Golden Rod

"Niagara Falls." *Macoun*, on the authority of *Dr. Burgess.*

Solidago arguta, Ait. .... Golden Rod

Chippewa. *Macoun*, on the authority of *Dr. Maclagan.*

Solidago juncea, Hook. (*S. arguta*, Ait.) .... Golden Rod

Goat Island.

Solidago serotina, Ait. .... Golden Rod

Goat Island. "Niagara District." *Macoun.*

var. gigantea, Gray.

Goat Island.

Solidago Canadensis, L. .... Golden Rod

Goat Island. Near Clifton, Ontario.

Solidago nemoralis, Ait. .... Golden Rod

Goat Island. Near Clifton, Ontario.

Solidago rigida, L. .... **Golden Rod**

Near DeVaux College.

Solidago lanceolata, Ait. .... **Golden Rod**

Margin of Niagara river above the Falls on the American side.

BELLIS PERENNIS, L. .... ENGLISH OR TRUE DAISY.

In a lawn at Clifton, Ontario, where it has maintained itself a number of years.

Aster corymbosus, Ait. .... **Aster**

Goat Island. Lewiston.

Aster macrophyllus, L. .... **Aster**

Goat Island. Near DeVaux College.

Aster Novae-Angliae, L. .... **New England Aster**

Goat Island and above the Falls on either side.

Aster patens, Ait. .... **Aster**

Near DeVaux College.

var. **phlogifolius, Nees.**

With the typical variety.

Aster azureus, Lindl. .... **Aster**

Near DeVaux College. La Salle. *Clinton.*

Aster cordifolius, L. .... **Aster**

Near DeVaux College.

Aster sagittifolius, Willd. .... **Aster**

Goat Island. "Niagara." *Macoun*, on the authority of *Dr. Maclagan.*

Aster laevis, L. .... **Aster**

Near DeVaux College.

Aster ericoides, L. .... **Aster**

Near DeVaux College.

Aster multiflorus, L. .... **Aster**

Goat Island.

Aster vimineus, Lam. (*A. puniceus*, L., var. *vimineus.* Gray .... **Aster**

Near Clifton, Ontario.

Aster diffusus, Ait. (*A. miser*, Nutt.) .... **Aster**

Goat Island. Near Clifton, Ontario.

Aster Tradescanti, L. .... **Aster**

Goat Island.

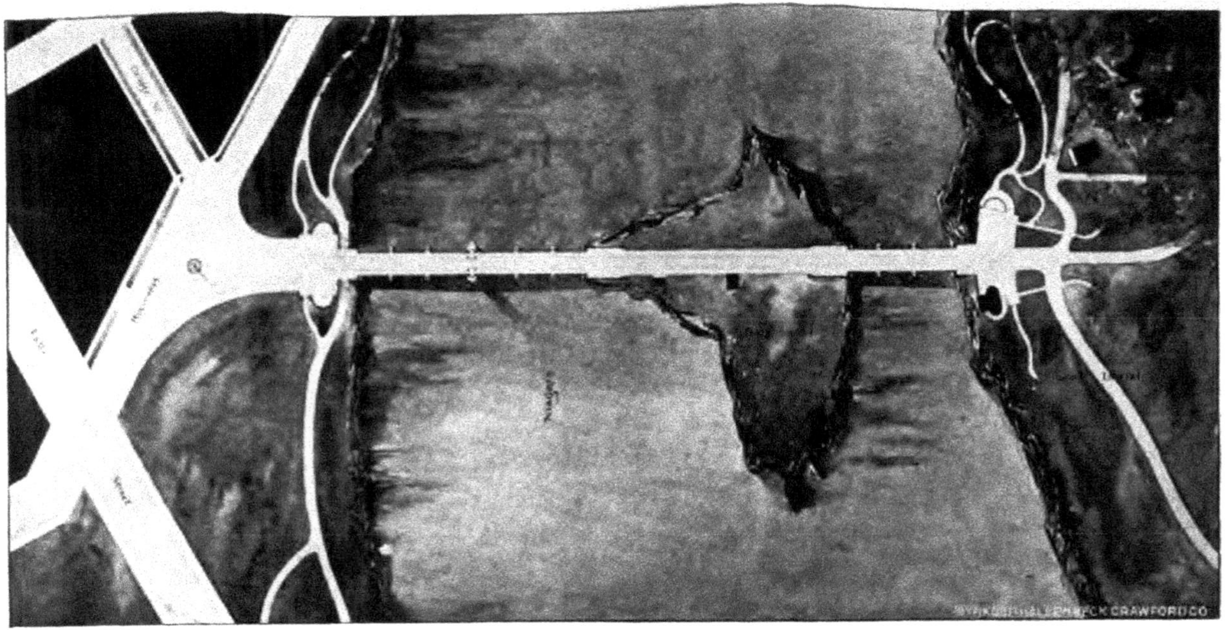

Niagara Falls. Walker & Morris, Architects, New York.

Plan for a Bridge from the Mainland to Goat Island.

Menu

Aster paniculatus, Lam. . . . . **Aster**

Goat Island. Near DeVaux College.

Aster prenanthoides, Muhl. . . . . **Aster**

Near Clifton, Ontario.

Aster puniceus, L. . . . . **Aster**

Goat Island.

Aster umbellatus, Mill. (*Diplopappus umbellatus*, Torr. and Gray.) **Aster.**

Goat Island. Near Clifton, Ontario.

Aster ptarmicoides, Torr. and Gray. . . . . **Aster**

Near DeVaux College. A species not common in our region, but here rather abundant.

Erigeron bellidifolius, Muhl. . . . . **Poor Robin's Plantain**

Goat Island.

Erigeron Philadelphicus, L. . . . . **Pink Fleabane**

Near Clifton, Ontario.

Erigeron annuus, Pers. . . . . **Fleabane**

Waste places on the main land.

Erigeron strigosus, Muhl. . . . . **Fleabane**

Goat Island.

Erigeron Canadensis, L. . . . . **Horseweed**

Main land above the Falls on the American side.

Antennaria plantiginifolia, Hook. . . . . **Everlasting**

Lewiston.

Anaphalis margaritacea, Benth. and Hook. (*Antennaria margaritacea*, R. Br.). . . . . **Pearly Everlasting.**

Near DeVaux College.

Gnaphalium polycephalum, Michx. . . . . **Everlasting**

Near DeVaux College.

Gnaphalium uliginosum, L.

Damp places along road sides on the main land.

INULA HELENIUM, L. . . . . ELECAMPANE.

Goat Island.

Polymnia Canadensis, L. .... **Leaf-cup.**

Along the descent to the ferry on the Canadian side of the river, and at the whirlpool and Foster's Flat, Ontario.

Silphium trifoliatum, L.

Attributed to the Falls, by *Torrey*, on the authority of *Dr. Eddy*. Not observed by us.

Ambrosia trifida, L. .... **Rag Weed.**

Margin of the river above the Falls.

Ambrosia artemisiaefolia, L. .... **Rag Weed**

Goat Island and with the foregoing species.

Xanthium Canadense, Mill. .... **Cockle-bur.**

Above the Falls on the American side.

var. **echinatum**, Gray.

Chippewa, Ontario. *Macoun*, on the authority of *Dr. Maclagan*.

Heliopsis laevis, Pers. .... **False Sun-flower.**

Near Clifton, Ontario.

Rudbeckia hirta, L. .... **Yellow Daisy.**

Near Lewiston.

Rudbeckia laciniata, L. .... **Cone Flower.**

Wet places near Clifton, Ontario.

HELIANTHUS ANNUUS, L. .... COMMON SUN-FLOWER.

Escaped near Lewiston.

Helianthus divaricatus, L. .... **Sun-flower.**

Goat Island.

Helianthus strumosus, L. .... **Sun-flower.**

Goat Island. Near DeVaux College.

Helianthus decapetalus, L. .... **Sun-flower.**

Damp places near Clifton, Ontario.

Bidens frondosa, L. .... **Cockle. Beggar's ticks.**

Waste places on the main land.

Bidens connata, Muhl. .... **Beggar's ticks.**

Goat Island.

Bidens cernua, L.

Wet places near Clifton, Ontario.

Bidens chrysanthemoides, Michx.

With the last and on the American side of the river above the Falls.

Bidens Beckii, Torr.

Chippewa, Ontario. *Macoun*, on authority of *Dr. Maclagan*. Niagara river near Grand Island.

Helenium autumnale, L. .... **Sneeze-weed.**

Goat Island and elsewhere in wet ground.

ANTHEMIS COTULA, L. (*Maruta Cotula, D. C.*) .... MAYWEED.

Road sides on the main land.

Achillaea Millefolium, L. .... **Yarrow.**

Goat Island.

CHRYSANTHEMUM LEUCANTHEMUM, L. .... OX-EYE DAISY.

Goat Island and elsewhere.

CHRYSANTHEMUM PARTHENIUM, Pers. .... FEVERFEW.

Lewiston. A garden scape.

TANACETUM VULGARE, L. .... TANSY.

Near the cantilever bridge on the Canadian side.

ARTEMISIA VULGARIS, L. .... MUGWORT.

Along the descent to the old landing on the American side of the river of the steamer "Maid of the Mist."

TUSSILAGO FARFARA, L. .... COLTSFOOT.

By the railroad track between the Devil's Hole and Lewiston.

Senecio aureus, L. .... **Golden Ragwort.**

Goat Island.

var. **Balsamitae,** Torr. and Gray.

Near DeVaux College. Rays sometimes wanting.

SENECIO VULGARIS, L. .... GROUNDSEL.

American side of the river above the Falls.

Erechtites hieracifolia, Raf. .... **Fireweed.**

Goat Island. Near Clifton, Ontario.

ARCTIUM LAPPA, L. .... BURDOCK.

Goat Island.

CNICUS ARVENSIS, Hoffm. .... CANADA THISTLE.

Goat Island and elsewhere.

CNICUS LANCEOLATUS, Hoffm. .... COMMON THISTLE.

Goat Island and elsewhere.

Cnicus pumilus, Torr. .... Pasture Thistle.

Near DeVaux College.

Cnicus altissimus, Willd. .... Tall Thistle.

var. **discolor**, Gray.

Goat Island.

PICRIS ECHIOIDES, L.

"Along the road side between Clifton and Niagara Falls." *Macoun.*

LAMPSANA COMMUNIS, L.

Queenston Heights, Ontario. *Macoun*, on the authority of *Millman.*

HIERACIUM AURANTIACUM, L. .... Hawkweed.

Goat Island. Introduced.

Hieracium Canadense, Michx. .... Canadian Hawkweed.

Near DeVaux College.

Hieracium paniculatum, L. .... Hawkweed.

With the last.

Hieracium venosum, L. .... Hawkweed.

Goat Island.

Hieracium scabrum, Michx. .... Hawkweed.

Near DeVaux College.

Hieracium Gronovii, L. .... Hawkweed.

Goat Island. Near Clifton, Ontario.

Prenanthes alba, L. (*Nabalus albus*, Hook.) .... White Lettuce.

Goat Island.

Prenanthes altissima, L. (*Nabalus altissimus*, Hook.)

Lewiston. Foster's Flat, Ontario.

TARAXACUM OFFICINALE, Web. .... DANDELION.

Goat Island and elsewhere.

Lactuca Canadensis, L. .... Wild Lettuce.

Goat Island.

Lactuca integrifolia, Bigelow. (*L. Canadensis* L., var. *integrifolia*, Torr. and Gray.)

Goat Island. Near Clifton, Ontario.

Lactuca leucophaea, Gray. (*Mulgedium leucophoeum*, DC.)...**Blue Lettuce.**

Goat Island.

SONCHUS OLERACEUS, L....SOW THISTLE.

Goat Island.

LOBELIACEAE.

Lobelia syphilitica, L....**Blue Lobelia.**

On the American side of the river above the Falls.

Lobelia Kalmii, L....**Kalm's Lobelia**

Goat Island. Low grounds near Clifton, Ontario.

Lobelia inflata, L....**Indian Tobacco.**

Lewiston and elsewhere.

CAMPANULACEAE.

Specularia perfoliata, A. DC....**Venus's Looking Glass.**

Niagara Falls. *Macoun*, on the authority of *Dr. Burgess.*

CAMPANULA RAPUNCULOIDES, L....BELLWORT.

Road sides on the main land. Escaped from cultivation.

Campanula rotundifolia, L....**Harebell.**

Goat Island, and thence to Lewiston.

Campanula aparinoides, Pursh...**Marsh Bellflower.**

Goat Island, and in grassy places on the edge of the river.

Campanula Americana, L....**Tall Bellflower.**

Near Clifton, Ontario.

ERICACEAE.

Gaylussacia resinosa, Torr. and Gray...**Huckleberry.**

Goat Island, and thence to Lewiston.

Vaccinium stamineum, L....**Deerberry.**

Near DeVaux College and at Lewiston.

Vaccinium vacillans, Solander...**Low Blueberry.**

Goat Island.

Vaccinium corymbosum, L....**Swamp Blueberry.**

Wet ground near Clifton, Ontario.

Arctostaphylos Uva-ursi, Spreng...**Bearberry. Killikinnick.**

Goat Island. Vicinity of the whirlpool on both sides of the river.

Gaultheria procumbens, L....**Winter-green.**

Near the whirlpool on the Canadian side.

Chimaphila umbellata, Nutt............................**Prince's Pine.**

Near DeVaux College.

Pyrola secunda, L.

Near DeVaux College.

Pyrola chlorantha, Swartz.

Niagara Falls. *Clinton.* Near the whirlpool on the Canadian side.

Pyrola elliptica, Nutt.

Goat Island.

Pyrola rotundifolia, L............................**False Winter-green.**

Near DeVaux College.

Pterospora andromedea, Nutt.

Near the whirlpool on the American side.

Monotropa uniflora, L..................................**Indian Pipe.**

Near the whirlpool on the Canadian side.

Monotropa Hypopitys, L..................................**Pine-sap.**

Goat Island. *Clinton.*

A sphagnous swamp, near Black Creek, Ontario, a few miles south of Chippewa, has produced the following named plants of this family:

Chiogenes hispidula, Torr. and Gray.............**Creeping Snowberry.**

Andromeda polifolia, L..................................**Andromeda.**

Cassandra calyculata, Don..............................**Leather Leaf.**

Ledum latifolium, Ait..................................**Labrador Tea.**

PRIMULACEAE.

Dodecatheon Meadia, L. American Cowslip.............**Shooting Star.**

Goat Island. Introduced.

Steironema ciliatum, Raf.

Goat Island.

Steironema longifolium, Gray.

"Crevices of rocks at Niagara Falls." *Macoun.*

Lysimachia stricta, Ait.

Wet grounds near Clifton, Ontario.

LYSIMACHIA NUMMULARIA, L..........................MONEYWORT.

Escaping from gardens on the main land.

Lysimachia thyrsiflora.

Wet grounds near Clifton, Ontario.

OLEACEAE.

LIGUSTRUM VULGARE, L. .... PRIVET.

Well established near Clifton, Ontario.

SYRINGA VULGARIS, L. .... LILAC.

A well-grown lilac-tree was observed in the flower in the gorge of the river, on the Canadian side, near the cantilever bridge, where it could not have been planted by man.

Fraxinus Americana, L. .... **White Ash.**

Goat Island and elsewhere.

Fraxinus sambucifolia, Lam .... **Black Ash.**

Goat Island. Near Clifton, Ontario.

APOCYNACEAE.

Apocynum androsaemifolium, L. .... **Dog Bane.**

Near DeVaux College.

Apocynum cannabinum, L. .... **Indian Hemp.**

Goat Island. Prospect Park. Lewiston.

VINCA MINOR, L. .... PERIWINKLE. CREEPING MYRTLE.

Goat Island. Introduced and spreading.

ASCLEPIADACEAE.

Asclepias tuberosa, L. .... **Butterfly Weed.**

Below the Falls on both sides of the river.

Asclepias incarnata, L. .... **Swamp Milkweed.**

Goat Island. Wet grounds near Clifton, Ontario.

Asclepias Cornuti, Decaisne .... **Common Milkweed.**

Goat Island and elsewhere.

Asclepias phytolaccoides, Pursh .... **Wood Milkweed.**

Goat Island.

Asclepias quadrifolia, L. .... **Four-leaved Milkweed.**

Near DeVaux College.

Asclepias verticillata, L. .... **Whorled Milkweed.**

Near DeVaux College.

Acerates viridiflora, Ell .... **Green-flowered Milkweed.**

"Niagara Falls." *Macoun.*

GENTIANACEAE.

Gentiana serrata, Gunner .... **Shorn Gentian.**

Goat Island. Wet grounds near Clifton, Ontario.

Gentiana Andrewsii, Grieseb .... **Closed Gentian.**

Chippewa, Ontario. *Macoun*, on the authority of *Dr. Maclagan.*

Gentiana crinita, Froel, occurs on islands in Niagara river near Lake Erie. *Gentiana Saponaria*, L., may be looked for on either side of the river above the Falls. *Gentiana quinqueflora*, Lam., probably will be found on the wooded hillsides near the whirlpool on the Canadian side.

Frasera Carolinensis, Walt .... **American Columbo.**

Queenston Heights, Ontario. *Jos. Sturdy. Macoun.*

POLEMONIACEAE.

Phlox divaricata, L .... **Blue Phlox.**

Goat Island and elsewhere.

Polemonium reptans, L .... **Polemonium.**

Goat Island. Uncommon.

HYDROPHYLLACEAE.

Hydrophyllum Virginicum, L .... **Water Leaf.**

Goat Island. Clifton, Ontario.

Hydrophyllum Canadense, L .... **Water Leaf.**

Near Clifton, Ontario.

BORRAGINACEAE.

CYNOGLOSSUM OFFICINALE, L .... HOUNDS-TONGUE.

Goat Island and elsewhere.

Cynoglossum Virginicum, L .... Wild Comfrey.

Near DeVaux College.

Echinospermum Virginicum, Lehm. (*Cynoglossum Morisoni*, DC.)

Goat Island.

ECHINOSPERMUM LAPPULA, Lehm .... STICKWEED.

Road sides on the main land.

Myosotis laxa, Lehm .... **Forget-me-not.**

Wet grounds near Clifton, Ontario.

Myosotis verna, Nutt.

Lewiston.

LITHOSPERMUM ARVENSE, L. .... CORN GROMWELL.

Goat Island and elsewhere.

LITHOSPERMUM OFFICINALE, L. .... COMMON GROMWELL.

Goat Island and elsewhere.

Lithospermum latifolium, Michx.

Near DeVaux College.

SYMPHYTUM OFFICINALE, L. .... COMFREY.

Niagara Falls. *Macoun*, on the authority of *Dr. Maclagan*.

ECHIUM VULGARE, L. .... VIPER'S BUGLOSS.

Below the Falls on the American side.

CONVOLVULACEAE.

IPOMOEA PURPUREA, Lam. .... MORNING GLORY.

Occasionally seen as a garden scape.

CONVOLVULUS ARVENSIS, L. .... BINDWEED.

Above the Falls on the American side. Lewiston.

Convolvulus spithameus, L. (*Calystegia spithamoea*, Pursh.) ...... **Hedge-Bindweed.**

Near the whirlpool, Ontario. *Clinton*. Abundant near Lewiston.

Convolvulus sepium, L. (*Calystegia sepium*, B. R.) .. **Wild Morning Glory.**

Below Prospect park. Near DeVaux College.

Cuscuta inflexa, Englemann .... **Dodder.**

Below Lewiston. Identified by *Dr. Englemann*.

Cuscuta Gronovii, Willd. .... **Dodder.**

Above the Falls on the American side.

CUSCUTA EPILINUM, Weihe. .... FLAX DODDER.

Flax fields near Lewiston. *Clinton*.

SOLANACEAE.

SOLANUM NIGRUM, L. .... COMMON NIGHTSHADE.

Waste places on the main land.

SOLANUM DULCAMARA, L. .... BITTERSWEET.

Goat Island.

Physalis Virginiana, Mill (*P. viscosa*, L.)..............**Ground Cherry.**

Below the Falls on the Canadian side.

LYCIUM VULGARE, Dunal..............................MATRIMONY VINE.

Near the lower suspension bridge on the American side. *Clinton.*

DATURA STRAMONIUM, L....................STRAMONIUM. THORN APPLE.

Waste places on the main land.

DATURA TATULA, L..............................PURPLE THORN APPLE.

Near Lewiston.

HYOSCYAMUS NIGER, L..HENBANE.

Niagara. *Judge Logie.*

SCROPHULARIACEAE.

VERBASCUM THAPSUS, L..MULLEIN.

Goat Island and elsewhere.

VERBASCUM BLATTARIA, L..............................MOTH MULLEIN.

Niagara Falls, Ontario, and between Queenston and Niagara. *Macoun.*

LINARIA VULGARIS, Mill..................BUTTER-AND-EGGS. TOAD FLAX.

Above the Falls on the American side. Lewiston.

Collinsia verna, Nutt.

Goat Island. Introduced.

Scrophularsa nodosa, M..**Figwort.**

var. **Marilandica**, Gray.

Goat Island.

Chelone glabra, L....................................**Turtle-Head.**

Wet ground near Clifton, Ontario.

Pentstemon pubescens, Solander..........................**Pentstemon.**

Goat Island and elsewhere.

Pentstemon laevigatus, Solander. (*P. Digitalis*, Nutt.)......**Pentstemon.**

Goat Island. Introduced.

Mimulus ringens, L..............................**Monkey-Flower.**

Low grounds on the American side above the Falls.

Gratiola Virginiana, L.

With the last. Gill Creek. *Clinton.*

Veronica Virginica, L..............................**Culver's Physic.**

Goat Island. Introduced.

Veronica Anagallis, L. .... Water Speedwell.

Wet grounds near Clifton, Ontario.

Veronica Americana, Schw. .... Brooklime.

Margin of river on the American side above the Falls.

Veronica scutellata, L. .... Marsh Speedwell.

Wet grounds near Clifton, Ontario.

Veronica officinalis, L. .... Speedwell.

Near Lewiston.

Veronica serpyllifolia, L. .... Thyme-leaved Speedwell.

Goat Island.

Veronica peregrina, L.

Above the Falls on the American side.

VERONICA ARVENSIS, L. .... CORN SPEEDWELL.

Goat Island.

Gerardia flava, L. .... Downy False Foxglove.

Near DeVaux College.

Gerardia quercifolia, Pursh .... Smooth False Foxglove.

Goat Island. Near DeVaux College.

Gerardia purpurea, L. .... Purple Gerardia.

Goat Island. Wet grounds near Clifton, Ontario.

Gerardia tenuifola, Vahl.

Near DeVaux College.

Castilleia coccinea, Spreng. .... Painted Cup.

Near the whirlpool on the Canadian side.

Pedicularis Canadensis, L. .... Lousewort.

Goat Island.

Pedicularis lanceolata, Michx.

Wet grounds near Clifton, Ontario.

Melampyrum Americanum, Michx. .... Cow-wheat.

Goat Island. Near DeVaux College.

OROBANCHACEAE.

Aphyllon uniflorum, Gray .... Broom-rape.

Near the whirlpool on the Canadian side.

Conopholis Americana, Wallr. . . . . **Cancer-root.**
Above the Falls on the American side. *Clinton.* "Vicinity of Niagara Falls." *Macoun*, on the authority of *Dr. Maclagan.*

Epiphegus Virginiana, Bart . . . . Beechdrops.
Goat Island. Not common.

LENTIBULARIACEAE.

Utricularia vulgaris, L. . . . . **Bladderwort.**
Shallow and quiet places in Niagara river near the American shore above the Falls.

Utricularia cornuta, Michx.
Wet grounds near Clifton, Ontario. Not common now. "Abundant on the Table Rock." [1818.] *Nuttall, Gen. Am. Pl., vol.* 1, *p.* 14.

ACANTHACEAE.

Dianthera Americana, L. . . . . **Water Willow.**
Margin of Niagara river above the Falls on the American side. Chippewa, Ontario. *Macoun*, on the authority of *Dr. Maclagan.*

VERBENACEAE.

Phryma Leptostachya, L. . . . . **Lop-seed.**
Goat Island.

Verbena urticaefolia, L. . . . . **White Vervain.**
Above the Falls on the American side.

Verbena hastata, L. . . . . **Blue Vervain.**
Goat Island and elsewhere.

LABIATAE.

Teucrium Canadense, L. . . . . **Germander.**
Goat Island. Chippewa, Ontario. *Macoun*, on the authority of *Dr. Maclagan.*

AJUGA REPTANS, L. . . . . BUGLE.
Goat Island. Introduced.

Collinsonia Canadensis, L. . . . . **Horse Balm.**
Near Clifton, Ontario.

MENTHA VIRIDIS, L. . . . . SPEARMINT.
Goat Island. *Clinton.* Niagara Falls. *Macoun*, on the authority of *Dr. Burgess.*

MENTHA PIPERITA, L. .... PEPPERMINT.

Above the Falls on the American side. Near Lewiston.

Mentha Canadensis, L. .... **Wild Mint.**

Goat Island and elsewhere.

Lycopus Virginicus, L. .... **Bugle Weed.**

Goat Island.

Lycopus sinuatus, Ell. (*L. Europaeus*, L., var. *sinuatus*, Gray.)

Shore of Niagara river on the American side above the Falls.

Pycnanthemum lanceolatum, Pursh .... **Mountain Mint.**

Near DeVaux College.

Calamintha Nuttallii, Benth. (*C. glabella*, Benth., var. *Nuttallii*, Gray.)

In wet, rocky places above Clifton, Ontario.

Calamintha Clinopodium, Benth .... **Basil.**

Goat Island.

MELISSA OFFICINALIS, L. .... LEMON BALM.

Niagara Falls. *Macoun*, on the authority of *Dr. Burgess.*

Hedeoma pulegioides, Pers .... **False Penny Royal.**

Near Lewiston.

SALVIA OFFICINALIS, L. .... SAGE.

Near DeVaux College. Well established.

Monarda fistulosa, L. .... **Horse Mint.**

Goat Island and elsewhere.

Monarda didyma, L., **Scarlet Balm**, ought to be found in the low, rich grounds along the river, near Clifton, Ontario.

Lophanthus nepetoides, Benth .... **Giant Hyssop.**

Near DeVaux College. Between Niagara Falls and Lake Ontario. *Macoun*, on the authority of *Dr. Maclagan.*

Lophanthus scrophulariaefolius, Benth.

Slopes of Queenston Heights. *Macoun.*

NEPETA CATARIA, L. .... CATNEP.

Goat Island.

NEPETA GLECHOMA, Benth .... GROUND IVY.

Goat Island.

Scutellaria lateriflora, L. .... **Skull-Cap.**

Wet grounds near Clifton, Ontario.

Scutellaria parvula, Michx. . . . .**Skull-Cap.**

Near the whirlpool on the Canadian side.

Scutellaria galericulata, Michx. . . . .**Skull-Cap.**

Goat Island and elsewhere.

Brunella vulgaris, L. . . . .**Heal-All.**

Goat Island and elsewhere.

Physostegia Virginiana, Benth.

Shores of the river above the Falls.

LEONURUS CARDIACA, L. . . . . MOTHERWORT.

Goat Island.

Stachys aspera, Michx. (*S. palustris*, L., var. *aspera*, Gray.). . . . . . .**Hedge Nettle.**

Goat Island.

PLANTAGINIACEAE.

PLANTAGO MAJOR, L. . . . . .COMMON PLAINTAIN.

Goat Island and elsewhere.

Plantago Rugellii, Decaisne. (*P. Kamtschatica*, Cham.). . . . . . .**Plaintain.**

Goat Island and elsewhere.

PLANTAGO LANCEOLATA, L. . . . . .RIB GRASS.

Near Clifton, Ontario.

PLANTAGO MEDIA, L.

"Niagara." *Provancher, Flore Canadienne, p.* 474. Not seen by us.

AMARANTACEAE.

AMARANTUS RETROFLEXUS, L. . . . . .PIGWEED.

Road sides on the main land.

AMARANTHUS ALBUS, L. . . . . .WHITE AMARANT.

Road sides on the main land.

CHENOPODIACEAE.

CHENOPODIUM ALBUM, L. . . . . .PIGWEED.

Goat Island and elsewhere.

CHENOPODIUM GLAUCUM, L.

Road sides and waste places on the main land.

CHENOPODIUM URBICUM, L. . . . . .PIGWEED.

Bath Island.

CHENOPODIUM HYBRIDUM, L.

Goat Island.

CHENOPODIUM BOTRYS, L. .... OAK-OF-JERUSALEM.

"Niagara Falls." *Macoun.*

ATRIPLEX PATULA, L. .... ORACHE.

var. HASTATA, Gray.

American side of the river above the Falls.

var. LITTORALIS, Gray.

With the last.

PHYTOLACCACEAE.

Phytolacca decandra, L. .... Poke weed.

Goat Island.

POLYGONACEAE.

POLYGONUM AVICULARE, L. .... KNOT GRASS.

Waste places on the main land.

POLYGONUM ERECTUM, L.

Above the Falls on the American side.

Polygonum incarnatum, Ell.

Shores of the river above the Falls.

Polygonum Pennsylvanicum, L.

"In the Niagara District." *Macoun*, on the authority of *Dr. Maclagan.*

Polygonum amphibium, L.

Islands of Niagara river. *Clinton.*

Polygonum Muhlenbergii, Watson. (*P. amphibium*, var. *terrestre*, Willd.)

Margin of the river above the Falls.

POLYGONUM PERSICARIA, L. .... LADY'S THUMB.

Waste places on the main land.

Polygonum Hydropiper, L. .... Smart weed.

Margin of the river above the Falls.

Polygonum acre, H. B. K. .... Smart weed.

On the American side of the river above the Falls.

Polygonum hydropiperoides, Michx.

Wet places near Clifton, Ontario.

Polygonum arifolium, L. .... Tear-thumb.

Chippewa, Ontario. *Macoun*, on the authority of *Dr. Maclagan.*

Polygonum sagittatum, L. .... **Tear-thumb.**

Doubtless to be found in the low, wet ground near Clifton, Ontario.

POLYGONUM CONVOLVULUS, L. .... BLACK BINDWEED.

Waste places and road sides of the main land.

Polygonum dumetorum, L. .... **False Buckwheat.**

var. **scandens**, Gray.

Chippewa, Ontario. *Macoun*, on the authority of *Dr. Maclagan.*

RUMEX CRISPUS, L. .... YELLOW DOCK.

Road sides on the main land.

RUMEX OBTUSIFOLIUS, L. .... BITTER DOCK.

In similar places as the last.

RUMEX ACETOSELLA, L. .... SHEEP SORREL.

Near Clifton, Ontario.

ARISTOLOCHIACEAE.

Asarum Canadense, L. .... **Wild Ginger.**

Near Clifton, Ontario. Lewiston. Only the larger form (or species?) noticed.

PIPERACEAE.

Saururus cernuus, L. .... **Lizard's Tail.**

"Charles's Island above the Falls." [Ontario.] *Macoun*, on the authority of *Dr. Burgess.*

LAURACAE.

Sassafras officinale, Nees .... **Sassafras.**

Lewiston. Near the whirlpool on the Canadian side.

Lindera Benzoin, Meisner .... **Spice bush.**

Goat Island and the Three Sisters.

THYMELACEAE.

Dirca palustris, L. .... **Moosewood.**

Niagara, Ontario. *Macoun*, on the authority of *Dr. Maclagan.*

DAPHNE MEZEREUM, L. .... MEZEREON.

Goat Island. Introduced and spreading.

ELEAGNACEAE.

Shepherdia Canadensis, Nutt .... **Shepherdia.**

Goat Island and on each side of the river to Lewiston and Queenston.

Comandra umbellata, Nutt. .... **Bastard Toad-Flax.**

Goat Island. Lewiston and elsewhere.

EUPHORBIACEAE.

Euphorbia maculata, L. .... **Spotted Spurge.**

Main land on both sides of the river.

Euphorbia platyphylla, L. .... **Spurge.**

Road sides on the main land.

EUPHORBIA HELIOSCOPIA, L. .... SPURGE.

With the last.

EUPHORBIA CYPARISSIAS, L. .... CYPRESS SPURGE.

Escaped.

Euphorbia hypericifolia, L. .... **Spurge.**

Above the Falls on the American side.

Acalypha Virginica, L. .... **Three-seeded Mercury.**

Near DeVaux College.

CERATOPHYLLACEAE.

Ceratophyllum demersum, L. .... **Hornwort.**

In shallow places in the river above the Falls on the American side. In pools near Clifton, Ontario.

URTICACEAE.

Ulmus fulva, Michx .... **Slippery Elm.**

Goat Island. Lewiston.

Ulmus Americana, L. .... **American Elm.**

Goat Island.

Ulmus racemosa, Thomas .... **Thomas's Elm.**

Bath Island. Planted.

ULMUS CAMPESTRIS, L. .... ENGLISH ELM.

Luna Island. Planted.

Celtis occidentalis, L. .... **Nettle Tree. Sugar Berry.**

"Rather common between Queenston and Niagara." *Macoun.*

Morus rubra, L. .... **Red Mulberry.**

Near DeVaux College. One small specimen observed near the ferry landing on the Canadian side. "Not uncommon from Niagara town along the river to the whirlpool." *Macoun.*

MORUS ALBA, L. .... WHITE MULBERRY.

Spontaneous near Lewiston. Niagara Falls. *Macoun.*

MORUS ———

An undetermined species has been planted on Luna Island.

Urtica gracilis, Ait. .... **Tall Nettle.**

Goat Island.

Laportea Canadensis, Gaud. .... **Wood Nettle.**

Damp grounds above Clifton, Ontario.

Pilea pumila, Gray .... **Richweed.**

Goat Island.

Boehmeria cylindrica, Willd. has probably been overlooked.

CANNABIS SATIVA, L. .... HEMP.

Waste places on the main land.

PLATANACEAE.

Platanus occidentalis, L. .... **Button Wood. Sycamore.**

Goat Island.

JUGLANDACEAE.

Juglans cinerea, L. .... **Butternut.**

American side of the river above the Falls.

Juglans nigra, L. .... **Black Walnut.**

Near DeVaux College. "Niagara Falls." *Macoun.*

Carya alba, Nutt. .... **White Hickory. Shell-bark Hickory.**

Goat Island. Near DeVaux College. "At Queenston Heights and the Falls it constitutes the greater part of the forest." *Macoun.*

Carya tomentosa, Nutt. .... **Hairy Hickory.**

"Amongst other hickories in the Niagara peninsula." *Macoun.*

Carya porcina, Nutt. .... **Pignut Hickory.**

"Queenston Heights and Niagara Falls." *Macoun.*

Carya amara, Nutt. .... **Bitter Hickory.**

Goat Island. In the village near the river. Below Lewiston.

BETULACEAE.

Betula lenta, L. .... **Black Birch.**

Goat Island.

Betula lutea, Michx. .... **Yellow Birch.**

Goat Island near the Horse-shoe Fall.

Betula papyracea, Ait. .... **Paper Birch.**

Goat Island. Below Lewiston.

Alnus incana, Willd .... **Alder.**

Goat Island. Wet grounds near Clifton, Ontario.

CUPULIFERAE.

Carpinus Caroliniana, Walt. (*Carpinus Americana*, Michx.)... **Blue Beech.**

Goat Island.

Ostrya Virginica, Willd .... **Iron Wood. Hop Hornbean.**

Goat Island. Some of the trees very large.

Corylus rostrata, Ait .... **Hazelnut.**

Near DeVaux College.

CORYLUS AVELLANA, L. .... EUROPEAN FILBERT.

Planted on Luna Island.

Fagus ferruginea, Ait .... **Beech.**

Goat Island. Abundant.

Castanea vulgaris, Lam .... **Chestnut**

var. **Americana**, A. DC.

Near DeVaux College. Lewiston. Queenston.

Quercus alba, L. .... **White Oak.**

Goat Island; but more abundant near DeVaux College.

Quercus obtusiloba, Michx .... **Post Oak.**

"Niagara Falls." *Provancher, Flore Canadienne, p.* 543.

Quercus Prinus, L. .... **Chestnut Oak.**

"Niagara." *Provancher, Flore Canadienne, p.* 543.

Quercus macrocarpa, Michx .... **Bur Oak.**

"Niagara." *Provancher, Flore Canadienne, p.* 543.

Quercus prinoides, Willd. (*Q. Prinus*, L., var. *humilis*, Gray.)

Dwarf Chestnut Oak.

Goat Island. "Common on Queenston Heights and in numerous places around Niagara." *Macoun.*

Quercus rubra, L. .... **Red Oak.**

Near Clifton, Ontario.

Quercus coccinea, Wang .... **Scarlet Oak.**

Goat Island. "In the forest along the Niagara river it is an abundant tree." *Macoun.*

Quercus tinctoria, Bartram.................................**Quercitron.**

Near DeVaux College. "Not uncommon at Niagara." *Macoun.*

Quercus palustris, Du Roi.................................**Pin Oak.**

"Wet woods below Queenston Heights." *Macoun.*

SALICACEAE.

Salix nigra, Marsh.................................**Black Willow.**

Goat Island. Moist places near Queenston, Ontario. *Macoun.*

Salix lucida, Muhl.................................**Shining Willow.**

Goat Island. Near Clifton, Ontario.

Salix discolor, Muhl.................................**Glaucous Willow.**

Goat Island.

Salix rostrata, Richardson. (*S. livida*, Wahl., var. *occidentalis*, Gray.)
Livid Willow.

Above the Falls on both sides of the river.

Salix petiolaris, Smith.

American side of the river above the Falls.

Salix cordata, Muhl.................................**Heart leaved Willow.**

Goat Island.

SALIX PURPUREA, L.................................BASKET WILLOW.

On the American side of the river above the Falls. "Between Niagara Town and Queenston." *Macoun.*

SALIX ALBA, L., and

SALIX BABYLONICA, L. are not uncommon in cultivation.

POPULUS ALBA, L.................................ABELE. WHITE POPLAR.

Near DeVaux College.

POPULUS CANESCENS, Smith.................................WHITE POPLAR.

Planted as a shade tree in places. This and the last species have been often confounded by American botanists. *P. canescens* is much the commoner.

Populus tremuloides, Michx.................................**American Aspen.**

Goat Island.

Populus grandidentata, Michx.................................**Large-toothed Aspen.**

Goat Island. The Three Sisters.

Populus monilifera, Ait. . . . . **Cotton wood.**

Goat Island and occasionally on both sides of the river, above the Falls to Lake Erie.

Populus balsamifera, L.

var. **candicans**, Gray. . . . . **Balm of Gilead.**

Lewiston. A single tree near the ferry landing on the Canadian side; probably not planted. The typical form may occur in the vicinity of the Falls as it is not uncommon on the islands in the river near Lake Erie. Not yet observed near the Falls.

POPULUS DILATATA, L. . . . . LOMBARDY POPLAR.

Planted as a shade tree on the main land and spreading by the root.

CONIFERAE.

Thuja occidentalis, L. . . . . **Arbor-vitae. White Cedar.**

Goat Island. Near DeVaux College. The most abundant of the evergreens growing near the Falls.

Juniperus communis, L. . . . . **Juniper.**

Goat Island.

Juniperus Virginiana, L. . . . . **Red Cedar.**

Goat Island. Apparently disappearing.

Taxus baccata, L., var. **Canadensis**, Gray. . . . . **Ground Hemlock American Yew.**

Goat Island.

Pinus Strobus, L. . . . . **White Pine.**

Goat Island. A few specimens. More plentiful and of larger growth near DeVaux College and below.

Tsuga Canadensis, Carriere. (*Abies Canadensis*, Michx.). . . . . . . **Hemlock.**

Goat Island. Not a prevailing tree.

HYDROCHARIDACEAE.

Elodea Canadensis, Michx. (*Anacharis Canadensis*, Planch.). . . . . . **Water Snake-Weed.**

In the old mill-race above the Falls on the American side and elsewhere.

Vallisneria spiralis, L. . . . . **Tape-grass. Ell-grass.**

With the last.

ORCHIDACEAE.

Corallorhiza multiflora, Nutt. .... **Coral root.**

Near the whirlpool, Ontario.

Spiranthes latifolia, Torr. .... **Ladies' Tresses.**

Wet places near Clifton, Ontario.

Spiranthes cernua, Rich. .... **Ladies' Tresses.**

In the same places as the last.

Habenaria Hookeri ana, Torr. (*Habenaria Hookeri*, Torr.)...**Tway Blade.**

Near the whirlpool, Ontario.

Habenaria hyperborea, R. Br.

Goat Island near the Horse-shoe Fall.

Cypripedium pubescens, Willd......**Ladies' Slipper. Moccasin Flower.**

Near DeVaux College.

Cypripedium parviflorum, Salisb....**Ladies' Slipper. Moccasin Flower.**

Near the whirlpool, Ontario.

IRIDACEAE.

Iris versicolor, L. .... **Blue Flag.**

Goat Island. In wet places above the Falls on both sides of the river.

Sisyrinchium anceps, Cav. (*Sisyrinchium Bermudiana*, L. var. *anceps*, Gray.) .... **Blue-eyed Grass.**

Near DeVaux College.

Sisyrinchium mucronatum, Mich. (*Sisyrinchium Bermudiana*, L. var. *mucronatum*, Gray.)

Not seen by us. Probably overlooked.

SMILACEAE.

Smilax herbacea, L. .... **Carrion Flower.**

Near DeVaux College.

Smilax hispida, Muhl. **(Cat-Brier),** and **Smilax rotundifolia,** L., **(Green Brier),** have not been observed, but probably may be found.

LILIACEAE.

Allium tricoccum, Ait. .... **Wild Leek.**

Goat Island. Abundant.

Allium Canadense, Kalm.

Goat Island. Not common.

Polygonatum biflorum, Ell. . . . .**Solomon's Seal.**

Goat Island.

Smilacina racemosa, Desf. . . . . **False Solomon's Seal.**

Goat Island and elsewhere.

Smilacina stellata, Desf.

Goat Island. Only the small variety.

Maianthemum Canadense, Desf. (*Smilacina bifolia*, Ker. var. *Canadensis*, Gray.) . . . . **Two-leaved Solomon's Seal.**

ASPARAGUS OFFICINALIS, L. . . . . ASPARAGUS.

Goat Island and elsewhere. Not common.

Lilium Philadelphicum, L. . . . . **Fire Lily.**

Near DeVaux College.

HEMEROCALLIS FULVA, L. . . . . DAY LILY.

Near the Cantilever bridge on the Canadian side. Escaped from cultivation.

Erythronium Americanum, Ker. . . . . **Yellow Adder's-Tongue.**

Goat Island and elsewhere.

Uvularia grandiflora, Smith. . . . . **Bellwort.**

Goat Island.

Oakesia sessilifolia, Watson. (*Uvulari asessilifolia*, L.). . **Small Bellwort.**

Goat Island.

Streptopus roseus, Michx. . . . . **Streptopus.**

Goat Island.

Prosartes lanuginosa, Don. . . . . **Prosartes.**

Goat Island.

Veratrum viride, Ait. . . . . **False Hellebore.**

Low grounds near Clifton, Ontario.

Chamaelirium Carolinianum, Wild. (*Chamaelirium luteum*, Gray.) **Devil's Bit.**

Between Stamford and the whirlpool, Ontario. Abundant.

Medeola Virginica, L. . . . . **Cucumber Root.**

Goat Island. Not common.

Trillium erectum, L. . . . . **Purple Trillium.**

Goat Island.

var. **album**, Pursh.

With the typical form.

Trillium grandiflorum, Salisb. .... **Large White Trillium.**

Goat Island. Flower with green stripes through the petals frequently produced.

PONTEDERIACEAE.

Pontederia cordata, L. .... **Pickerel Weed.**

Niagara river above the Falls in quiet places.

Schollera graminifolia, Willd. .... **Water Star Grass.**

Not uncommon along the margin of Niagara river near Lake Erie and likely to be found in places nearer the Falls.

JUNCACEAE.

Luzula pilosa, Willd. .... **Hairy Wood-Rush.**

Near DeVaux College.

Luzula campestris, DC.

With the last.

Juncus effusus, L. .... **Soft Rush.**

Above the Falls near the river on the American side.

Juncus bufonius, L. .... **Rush.**

In similar situations as the last.

Juncus tenuis, Willd. .... **Rush.**

Low places on the main land.

Juncus acuminatus, Mich. .... **Rush.**

var. debilis, Engleman.

In low grounds.

Juncus nodosus, L. .... **Rush.**

Goat Island, in wet places near the margin of the river.

var. megacephalus, Torr.

With the last.

Juncus Canadensis, J. Gay .... **Rush.**

On both sides of the river above the Falls.

TYPHACEAE.

Typha latifolia, L. .... **Cat-tail Flag.**

At the water's edge near the foot of the American staircase.

Typha angustifolia, L. .... **Narrow-leaved Cat-tail Flag.**

Niagara Falls. *Clinton.*

Sparganium eurycarpum, Engleman........................**Bur Reed.**

American side of the river above the Falls.

Sparganium simplex, Hudson..............................**Bur Reed.**

With the last.

AROIDEAE.

Arisaema triphyllum, Torr........**Indian Turnip. Jack-in-the-Pulpit.**

Goat Island. A large and small variety common.

Peltandra Virginica, Raf..................................**Arrow Arum.**

Niagara Falls. *Clinton.* Not seen by us.

Symplocarpus foetidus, Salisb..........................**Skunk's Cabbage.**

Wet places above Clifton, Ontario.

Acorus Calamus, L..........................**Calamus. Sweet Flag.**

With the last.

LEMNACEAE.

Lemna trisulca, L......................................**Duck's meat.**

Niagara river above the Falls in quiet places.

Spirodela polyrrhiza, Schleid..........................**Duck's meat.**

In similar places as the last.

Wolffia Columbiana, Karst.

Niagara river above the Falls. *Prof. Kellicott.*

ALISMACEAE.

Alisma Plantago, L..................................**Water Plantain.**

var. Americana, Gray.

Margin of Niagara river above the Falls.

Sagittaria variabilis, Engelmann.............................Arrow-Head.

Above the Falls in wet places on the Canadian side.

Sagittaria heterophylla, Pursh.

Niagara river, near La Salle. *Clinton.*

Triglochin palustre, L.

Wet ground above Clifton, Ontario. Of unusual size.

NAIADACEAE.

Naias flexilis, Rostk..**Naiad.**

Niagara river above the Falls.

Zannichellia palustris, L. .... **Horned Pondweed.**

Abundant in the upper portion of Niagara river and therefore to be expected nearer the Falls.

Potamogeton natans, L. .... **Pondweed.**

Niagara river near Strawberry Island. *Clinton.*

Potamogeton hybridus, Michx. .... **Pondweed.**

Black Creek, Ontario, opposite Grand Island.

Potamogeton rufescens, Schrader .... **Pondweed.**

Niagara river. *Rev. Thomas Morong.*

Potamogeton fluitans, Roth. (*P. lonchites*, Tuckerman.) .... **Pondweed.**

Niagara river. *Rev. Thomas Morong.*

Potamogeton lucens, L. .... **Pondweed.**

Niagara river.

Potamogeton amplifolius, Tuckerman .... **Pondweed.**

Niagara river. *Rev. Thomas Morong.*

Potamogeton gramineus, L. .... **Pondweed.**

var. **heterophyllus**, Fries.

Niagara river.

var. **elongatus**, Morong.

Niagara river. *Rev. Thomas Morong.*

Potamogeton praelongus, Wulfen .... **Pondweed.**

Niagara. *Provancher, Flore Canadienne*, p. 627.

Potamogeton perfoliatus, L. .... **Pondweed.**

var. **lanceolatus**, Robbins.

Niagara river. *Rev. Thomas Morong.*

Potamogeton zosteriaefolius, Schum. (*P. compressus*, L.) ..... **Pondweed.**

Niagara river.

Potamogeton Niagarensis, Tuckerman .... **Pondweed.**

"Rapids above the Falls." *Gray's Manual (5th Ed.)*, p. 489. Rediscovered, after many years, in 1886, in the old mill-race above the Falls, on the American side, by the *Rev. Thomas Morong.*

Potamogeton pauciflorus, Pursh .... **Pondweed.**

Niagara river.

Potamogeton pusillus, L. .... **Pondweed.**

Niagara river.

Potamogeton pectinatus, L. .... Pondweed.

Niagara river.

Potamogeton Robbinsii, Oakes .... Pondweed.

Niagara river.

[NOTE.—It is not asserted that all the species of Potamogeton, above named, have yet been detected in the immediate vicinity of the Falls, but as they all occur more or less abundantly in the upper portion of the river (except as noted), they may well be expected nearer the Falls.]

CYPERACEAE.

Cyperus diandrus, Torr. .... Galingale.

var. castaneus, Torr.

Goat Island, on the east side, near the river.

Cyperus esculentus, L. (*Cyperus phymatodes*, Muhl.) .... Galingale.

Goat Island near the river.

Cyperus strigosus, L. .... Galingale.

Wet places near Clifton, Ontario.

Cyperus filiculmis, Vahl., has probably been overlooked.

Dulichium spathaceum, Richard .... Dulichium.

Wet places near Clifton, Ontario.

Eleocharis ovata, R. Br. (*Eleocharis obtusa*, Schult.) .... Spike Rush.

Near the shores of the river above the Falls.

Eleocharis palustris, R. Br. .... Spike Rush.

With the last and growing in the water.

Eleocharis tenuis, Schult. .... Spike Rush.

Damp places above the Falls.

Eleocharis acicularis, R. Br. .... Spike Rush.

Shores of the river above the Falls on either side.

Scirpus planifolius, Muhl. .... Spike Rush.

Near DeVaux College.

Scirpus pungens, Vahl. .... Spike Rush.

Wet places above Clifton, Ontario.

Scirpus lacustris, L. (*S. validus*, Vahl.) .... Great Bull Rush.

Margin of the river, on the American side, above the Falls. Wet places above Clifton, Ontario.

Scirpus fluviatilis, Gray....................................**Club Rush.**

Wet places above Clifton, Ontario.

Scirpus atrovirens, Muhl.

Wet places along the river, above the Falls, on either side.

Scirpus lineatus, Michx.

East side of Goat Island in wet places near the river.

Eriophorum cyperinum, L. (*Scirpus Eriophorum*, Mich.).....**Wool Grass.**

Wet places above Clifton, Ontario.

Carex polytrichoides, Muhl....................................**Sedge.**

Wet places above Clifton, Ontario.

Carex Steudellii, Kunth....................................**Sedge.**

Near Clifton, Ontario.

Carex bromoides, Schkr....................................**Sedge.**

Wet places near Clifton, Ontario.

Carex vulpinoides, Michx....................................**Sedge.**

Above the Falls, on the American side, in low places.

Carex stipata, Muhl....................................**Sedge.**

With the last.

Carex rosea, Schkr....................................**Sedge.**

Goat Island in damp places.

var. **retroflexa**, Torr. (*Carex retroflexa*, Muhl.)................**Sedge.**

Goat Island in damp places.

Carex sterilis, Willd....................................**Sedge.**

Near Clifton, Ontario.

Carex scoparia, Schkr....................................**Sedge.**

With the last.

Carex tribuloides, Wahl. (*Carex lagopodioides*, Schkr.)............**Sedge.**

With the last.

var. **cristata**, Bailey. (*Carex cristata*, Schw.)..................**Sedge.**

Goat Island.

Carex straminea, Schkr....................................**Sedge.**

Near DeVaux College.

Carex aquatilis, Willd....................................**Sedge.**

Margin of the river above the Falls.

Carex torta, Boott..Sedge.

Near Clifton, Ontario.

Carex stricta, Lam...Sedge.

In wet, grassy places on the American side above the Falls.

Carex Crinita, Lam...Sedge.

Near Clifton, Ontario.

Carex aurea, Nutt...Sedge.

Near Clifton, Ontario.

Carex granularis, Muhl..Sedge.

Near Clifton, Ontario.

Carex conoidea, Schkr...Sedge.

Moist places on the American side above the Falls.

Carex grisea, Wahl...Sedge.

With the last.

Carex virescens, Muhl...Sedge.

Near DeVaux College.

Carex plantaginea, Lam..Sedge.

Near Lewiston.

Carex retrocurva, Dew..Sedge.

Goat Island.

Carex platyphylla, Carey...Sedge.

Near DeVaux College.

Carex digitalis, Willd...Sedge.

Goat Island.

Carex laxiflora, Lam...Sedge.

var. plantiginea, Boott.

Near DeVaux College.

Carex eburnea, Boott..Sedge.

Goat Island, and near DeVaux College.

Carex Pennsylvanica, Lam..Sedge.

Lewiston.

Carex varia, Muhl...Sedge.

Near Clifton, Ontario.

Carex prasina, Wahl. (*Carex miliacea*, Muhl.).........................Sedge.

Wet places *above Clifton, Ontario.*

Carex debilis, Michx. . . . . **Sedge.**

With the last.

Carex Oederi, Retz. . . . . **Sedge.**

Goat Island near the Horse-shoe Fall. Niagara. *Provancher, Flore Canadienne*, p. 658.

Carex riparia, Curtis. . . . . **Sedge.**

Eastern side of Goat Island.

Carex trichocarpa, Muhl. . . . . **Sedge.**

Near Clifton, Ontario.

Carex comosa, Boott. . . . . **Sedge.**

Near Clifton, Ontario.

Carex intumescens, Rudge. . . . . **Sedge.**

Near Clifton, Ontario.

Carex lupulina, Muhl. . . . . **Sedge.**

Wet grounds above the Falls on either side.

Carex rostrata, With. var. **utriculata**, Bailey. (*Carex utriculata*, Boott.) . . . . **Sedge.**

With the last.

[NOTE.—It is not unlikely that a number of other species of this large and difficult genus may still be found in the vicinity of the Falls; and, as our specimens have not been submitted to any one who has made the *Carices* a special study, it is quite probable that some of our determinations may prove erroneous.]

GRAMINEAE.

PANICUM GLABRUM, Gaudin. . . . . PANIC GRASS.

Road sides on the main land.

PANICUM SANGUINALE, L. . . . . CRAB GRASS. PANIC GRASS.

Above the Falls on the American side of the river.

Panicum capillare, L. . . . . **Witch Grass.**

A garden weed on the main land.

Panicum virgatum, L. . . . . **Panic Grass.**

Wet grounds near Clifton, Ontario. Dry places near DeVaux College. The latter an unusual situation.

Panicum latifolium, L. . . . . **Panic Grass.**

Near DeVaux College.

Panicum clandestinum, L. .... **Panic Grass.**

Goat Island.

Panicum dichotomum, L. .... **Panic Grass**

Near DeVaux College.

Panicum depauperatum, Muhl. .... **Panic Grass.**

Goat Island and the Three Sisters in rocky places.

Panicum Crus-galli, L. .... **Barn-Yard Grass.**

Waste places on the main land.

Setaria viridis, Beauv. .... **Green Fox-tail Grass.**

Above the Falls on the American side.

Spartina cynosurioides, Willd. .... **Cord Grass.**

Margin of the river above the Falls on either side.

Zizania aquatica, L. .... **Wild Rice.**

Niagara river above the Falls.

Leersia Virginica, Willd. .... **White Grass.**

Goat Island.

Leersia oryzoides, Swartz. .... **Rice Cut Grass.**

Wet places above Clifton, Ontario.

Andropogon provincialis, Lam. (*A. furcatus*, Muhl.) .... **Beard Grass.**

Near DeVaux College.

Andropogon scoparius, Michx. .... **Beard Grass.**

Near DeVaux College.

Chrysopogon nutans, Benth. (*Sorghum nutans*, Gray.) ..... **Indian Grass.**

Near DeVaux College.

Phalaris arundinacea, L. .... **Reed Canary Grass.**

Near Clifton, Ontario. Islands in the river above the Falls.

Oryzopsis melanocarpa, Muhl. .... **Mountain Rice.**

Near the whirlpool, Ontario.

Oryzopsis asperifolia, Michx. .... **Mountain Rice.**

Goat Island. Near DeVaux College.

Oryzopsis juncea, Michx. (*Oryzopsis Canadensis*, Torr.) ... **Mountain Rice.**

Near Lewiston.

Muhlenbergia glomerata, Trin. .... **Drop-seed Grass.**

Niagara Falls. *Clinton.*

Muhlenbergia Mexicana, Trin. .... **Drop-seed Grass.**

Niagara Falls. *Clinton.*

Muhlenbergia sylvatica, Torr. and Gray. .... **Drop-seed Grass.**

Foster's Flat, Ontario. *Clinton.*

Muhlenbergia Willdenovii, Trin. .... **Drop-seed Grass.**

Lewiston.

Muhlenbergia diffusa, Schreb. .... **Nimble Will.**

Near the whirlpool, Ontario. *Clinton.*

Muhlenbergia capillaris, Kunth. .... **Hair Grass.**

"On the talus below the Falls on the American side." *Clinton.*

Brachyelytrum aristatum, Beaud.

Goat Island. Near DeVaux College.

PHLEUM PRATENSE, L. .... TIMOTHY.

Goat Island and the main land.

Alopecurus geniculatus, L. .... **Fox-tail Grass.**

Near the lower Suspension bridge, Ontario.

Sporobolus vaginaeflorus, Vasey. (*Vilfa vaginoeflora*, Torr.). . **Rush Grass.**

Lewiston. Near the whirlpool, Ontario.

Agrostis perennans, Tuckerman. .... **Thin Grass.**

Lewiston.

Agrostis scabra, Willd. .... **Hair Grass.**

Near DeVaux College.

Agrostis alba, L. (*Agrostis vulgaris*, With.). .... **Red Top Grass.**

Grassy places. Goat Island and elsewhere.

var. **vulgaris**, Thurb. .... **Fiorin.**

Margin of the river above the Falls on the American side.

Cinna arundinacea, L. .... **Reed Grass.**

Near Clifton, Ontario.

Deyeuxia Canadensis, Beauv. (*Calamagrostis Canadensis*, Beauv.). . . **Blue-joint Grass.**

Goat Island, on the water's edge.

Deschampsia flexuosa, Griseb. (*Aira flexuosa*, L.). . . **Common Hair Grass.**

Near DeVaux College.

..................................Wild Oat.

..................................OAT.

..................................Reed.

Eatonia Pennsylvanica, Gray..................................Eatonia.

..........................**Low Spear Grass.**

and the main land.

L......................................**Wire Grass.**

and the main land.

Poa serotina, Ehrh......................................**False Red Top.**

Wet grounds above Clifton, Ontario.

Poa pratensis, L.......**Common Meadow Grass. Kentucky Blue Grass.**

Goat Island.

Poa debilis, Torr.

Near DeVaux College.

Poa alsodes, Gray.

Near DeVaux College.

Glyceria nervata, Trin............................**Fowl-Meadow Grass.**

Above the Falls on both sides of the river.

Glyceria pallida, Trin.

Niagara river above the Falls on the American side.

Glyceria arundinacea, Kunth. (*G. aquatica,* Smith.).......**Reed Meadow Grass.**

Wet places above Clifton, Ontario.

Glyceria fluitans, R.Br.

Old mill-race above the Falls on the American side.

Festuca duriuscula, L. (*F. ovina*, L., var. ***duriuscula***, Gray.)......**Sheep's Fescue.**

Goat Island.

FESTUCA ELATIOR, L.................................MEADOW FESCUE.

Goat Island.

Festuca nutans, Spreng..**Fescue.**

Near DeVaux College. Lewiston.

BROMUS SECALINUS, L...CHESS.

Fields near Lewiston.

BROMUS RACEMOSUS, L...CHESS.

Fields near Lewiston.

Bromus Kalmii, Gray....................................**Wild Chess.**

Near the whirlpool, Ontario.

Bromus ciliatus, L.......................................**Wild Chess.**

Near DeVaux College.

LOLIUM TEMULENTUM, L.............................BEARDED DARNEL.

Near the landing of the old steamer "Maid of the Mist," on the American side. *Clinton.*

Agropyrum repens, Beauv. (*Triticum repens*, L.)......**Couch, Quitch, or Quick Grass.**

Road sides near Clifton, Ontario.

Agropyrum caninum, R. and S. (***Triticum caninum***, L.)...**Awned Wheat Grass.**

Goat Island.

Elymus Virginicus, L...................................**Lyme Grass.**

On either side of the river above the Falls.

Elymus Canadensis, L...................................**Lyme Grass.**

Near the whirlpool, Ontario.

Elymus striatus, Willd..................................**Lyme Grass.**

Near De Veaux College.

Asprella Hystrix, Willd. (*Gynostichum Hystrix*, Schreb.)...**Bottle-Brush Grass.**

EQUISETACEAE.

Equisetum arvense, L. .... **Horse-Tail.**

Goat Island and elsewhere.

Equisetum limosum, L. .... **Horse-Tail.**

Islands of Niagara river. *Clinton.*

Equisetum palustre, L. .... **Horse-Tail.**

Wet places above Clifton, Ontario.

Equisetum hyemale, L. .... **Scouring Rush.**

Near the whirlpool, Ontario, and elsewhere.

Equisetum variegatum, Schleicher.

Goat Island. Wet places above the Falls, on the Canadian side.

Equisetum scirpoides, Michx.

Near the whirlpool, Ontario.

OPHIOGLOSSACEAE.

Botrychium Virginianum, Swartz. (*B. Virginicum*, Swartz.).. **Moonwort.**

Goat Island and elsewhere.

Botrychium ternatum, Swartz .... **Moonwort.**

var. intermedium, D. C. Eaton. (*B. lunarioides*, D. C. Eaton.) Near Clifton, Ontario.

Ophioglossum vulgatum, L. .... **Adder's Tongue.**

Occurs on Grand Island, and may, therefore, be confidently looked for in favorable situations near the Falls.

FILICES.

Onoclea Struthiopteris, Hoffm. (*Struthiopteris Germanica*, Willd.)

Ostrich Fern.

Goat Island and elsewhere.

Onoclea sensibilis, L. .... **Sensitive Fern.**

Goat Island and elsewhere.

Osmunda regalis, L. .... **Royal Fern.**

Goat Island. Not common. Near Clifton, Ontario.

Osmunda Claytoniana, L. .... **Interrupted Fern.**

Near Clifton, Ontario.

Osmunda cinnamomea, L. .... **Cinnamon Fern.**

Near Clifton, Ontario.

Cystopteris fragilis, Swartz.

Between Lewiston and Youngstown. Near DeVaux College.

Cystopteris bulbifera, Bern.............................**Bladder Fern.**

Goat Island and along the American staircase. Near the whirlpool, Ontario.

Aspidium Noveboracense, Swartz......................**New York Fern.**

Goat Island.

Aspidium Thelypteris, L..................................**Shield Fern.**

Goat Island.

Aspidium spinulosum, Swartz............................**Shield Fern.**

var. intermedium, D. C. Eaton.

Near Clifton, Ontario.

Aspidium cristatum, Swartz...............................**Shield Fern.**

Foster's Flat, Ontario.

Aspidium Goldianum, Hook................................**Shield Fern.**

Near the whirlpool, Ontario.

Aspidium marginale, Swartz..............................**Shield Fern.**

Goat Island. Devil's Hole. Foster's Flat, Ontario.

Aspidium Lonchitis, Swartz ....**Shield Fern.**

"Sparingly at Foster's Flat," Ontario. *Burgess.*

Aspidium achrostichoides, Swartz.....................**Christmas Fern.**

Goat Island. Near DeVaux College. Lewiston.

Phegopteris Dryopteris, Fée..............................**Beech Fern.**

Devil's Hole.

Camptosorus rhizophyllus, Link........................**Walking Fern.**

Not uncommon near the whirlpool, Ontario. Foster's Flat, Ontario. Near DeVaux College, but rare.

Scolopendrium vulgare, Smith....................**Hart's Tongue Fern.**

Introduced in one place in the gorge of the river by Judge Clinton several years ago.

Asplenium Trichomanes, L...............................**Spleen wort.**

Near De Vaux College. Foster's Flat, Ontario.

Asplenium ebeneum, Ait...................................**Spleen wort.**

Near DeVaux College, but rare. Near Lewiston, abundant.

Asplenium achrostichoides, Swartz. (*Asplenium thelypteroides*, Michx.) **Spleen wort.**

Near the whirlpool, Ontario.

Asplenium Filix-foemina, Bern............................**Lady-Fern.**

Foster's Flat, Ontario.

Pellaea gracilis, Hook....................................**Cliff-Brake.**

"Crevices of rocks at Foster's Flat," Ontario. *Burgess.* Not seen by us.

Pellaea atropurpurea, Link.................................**Cliff-Brake.**

Formerly on Goat Island and the Three Sisters. Not lately seen by us. Probably extirpated. Near DeVaux College. Foster's Flat, Ontario. Rare.

Pteris aquilina, L....................................**Common Brake.**

Near DeVaux College at Lewiston.

Adiantum pedatum, L.............................**Maiden-Hair Fern.**

Goat Island. Not abundant. Near Clifton, Ontario.

Polypodium vulgare, L....................................**Polypody.**

Goat Island. The Three Sisters. Near DeVaux College. Lewiston.

LYCOPODIACEAE.

Lycopodium lucidulum, Michx.

Lycopodium dendroideum, Michx.

Lycopodium complanatum, L.

No species of Lycopodium (or Ground Pine) has been seen by us in the immediate vicinity of the Falls. Yet it is likely that the three species here mentioned may be found, as they are not rare elsewhere in western New York.

SELAGINELLACEAE.

Selaginella rupestris, Spring........................**Rock Selaginella.**

Cliffs of Niagara river near the Devil's Hole.

Selaginella Apus, Spring.................................**Selaginella.**

Goat Island near the Horse-shoe Falls. Damp grounds on both sides of the river above the Falls.

Isoetes echinospora, Duiren................................**Quill wort.**

var. Braunii, Englemann.

Not seen by us; but occurring in the river above the Falls. *Clinton.*

HYDROPTERIDES.

Marsilia quadrifola, L. .... **Marsilia.**

Introduced into the Niagara river above the Falls and in a pool near Lewiston.

Azolla Caroliniana, Willd. .... **Azolla.**

"At a saw-mill half a mile above the village." Niagara. *Dr. Robbins.*

NOTE.—The unnamed species of *Tilia*, herein referred to, seems to be only a variety of *Tilia Americana*, L. July 26, 1888.

FIFTEENTH ANNUAL REPORT

OF THE

COMMISSIONERS

OF THE

ate Reservation at Niagara.

RANSMITTED TO THE LEGISLATURE FEBRUARY 1, 1899.

WYNKOOP HALLENBECK CRAWFORD CO.,
STATE PRINTERS,
NEW YORK AND ALBANY.
1899.

OLD MILL RACE, WILLOW ISLAND.

FIFTEENTH ANNUAL REPORT

OF THE

COMMISSIONERS

OF THE

State Reservation at Niagara.

TRANSMITTED TO THE LEGISLATURE FEBRUARY 1, 1899.

WYNKOOP HALLENBECK CRAWFORD CO.,
STATE PRINTERS,
NEW YORK AND ALBANY.
1899.

STATE OF NEW YORK.

No. 37.

ASS M ,

FEBRUARY 1, 1899.

FIFTEENTH ANNUAL REPORT

OF THE

of the State Reservation

NEW YORK, *January* 27, 1899.

To the Honorable the Speaker of the Assembly:

Sir.—I transmit herewith for presentation to the Legislature the Fifteenth Annual Report of the Commissioners of the State Reservation at Niagara for the fiscal year ending September 30, 1898.

Respectfully,

ANDREW H. GREEN,

President.

FIFTEENTH ANNUAL REPORT

OF THE

COMMISSIONERS

OF THE

STATE RESERVATION AT NIAGARA

FOR THE FISCAL YEAR FROM

OCTOBER 1, 1897, to SEPTEMBER 30, 1898.

Commissioners:

ANDREW H. GREEN, President.

GEORGE RAINES, CHARLES M. DOW,

THOMAS P. KINGSFORD, ALEXANDER J. PORTER.

Treasurer and Secretary:

HENRY E. GREGORY.

Superintendent:

THOMAS V. WELCH.

REPORT.

To the Honorable the Legislature of the State of New York:

The Commissioners of the State Reservation at Niagara, as required by law, submit their report for the fiscal year begun October 1, 1897, and ended September 30, 1898, being their fifteenth annual report.

The Commissioners have work of a two-fold nature to carry on at Niagara — first, that of maintenance; second, that of restoration and improvement. Of the former, it is only necessary to say that it has proceeded with its usual regularity, a detailed account of it being contained in the report of the superintendent which is appended to this report.

The work of restoration and improvement also has been continued during the year, in accordance with the plan and purpose of the Commissioners, and with gratifying results. It has consisted largely of grading, filling, riprapping, sodding and planting. The approach to Goat Island, for example, has been graded and riprapped, the shore of Bath Island (now Green Island) has been graded, and filled with soil where that treatment was needed; the inner margin of the loop-driveway at Port Bay has been graded and sodded, while its outer shore or bank has been riprapped.

There has been considerable filling and grading along the riverway, especially below Fourth street. It may be stated that the construction of the loop-driveway at Port Bay, and the filling in that has been done at different points within the Reservation has somewhat increased the acreage of the State's property at Niagara.

There is more land requiring care and attention than there was in 1885, when the Reservation was opened to the public, and consequently more money is needed for maintenance.

The Hall building, or pavilion, so long a conspicuous object in Prospect Park, was destroyed by fire in January, probably by an incendiary. There can be no doubt that the destruction of this building has improved the appearance of the grove in which it was situated. The debris has been removed, and the cellar space filled, graded and sodded. Prospect Park is a grove of very limited area, and the presence of cheap and unornamental structures within its boundaries can only be regarded as unsuitable and obtrusive. Convenience and necessity alone justify the retention of those structures that still remain. The Inclined Railway building is not tolerable from an aesthetic point of view, but it has been found convenient for the office of the Superintendent and necessary for the railway. It can never be anything but an ill-looking structure, whether seen from the grove, from the International bridge, or from the Queen Victoria Niagara Falls Park. Located as it is, in close proximity to the American Fall, it obstructs the finest view that could otherwise be had of both the American and Horseshoe Falls from the grounds It is the one unsightly building remaining.

A small house that was occupied by the Inclined Railway operator, situated near the northern boundary of the grove, and overlooking the gorge, has also been removed, and the ground raised so as to afford a better prospect of the lower river.

An elevation of earth and rock has been substituted at Hennepin's View for the observation platform that had been there for some years.

The most considerable and noteworthy improvement in Prospect Park has been effected by removing the old parapet wall at Prospect Point. It is at this point that visitors are wont to congregate

Gorge View, Prospect Park.

in order to gaze at the American Fall, since the nearest and best view of it is here obtainable. The wall afforded ample security to those who wished to approach as nearly as possible to the edge of the cataract. But it was not in accord with the surroundings. Photographs of this portion of the Reservation taken about 1845 show it in its natural state. To restore Prospect Point to something like this natural state, was the purpose of the Commissioners in directing the removal of the parapet wall. Another consideration was the condition of the bluff upon which the wall rested. The constant soaking which it receives from the water and spray of the falls made necessary frequent observations of the soil. Portions of the bluff to the northward had occasionally, through disintegration, become loosened and fallen, and it was not unnatural to apprehend some such occurrence at the Point itself. In other words, the removal of the parapet wall could not much longer have been postponed without endangering the safety of visitors.

The soil now slopes gradually down to the margin of the river, and the former artificially is succeeded by a pleasing natural appearance. An iron railing provides a safeguard for visitors.

The Commissioners were able, during the summer, to begin work upon a stone arched bridge from Goat Island to the First Sister Island, an improvement contemplated for some years. While the wooden structure that had been there so long, was safe, it was not suitable, and could not be regarded as permanent.

In 1893 the Commissioners had ordered the preparation of plans of a stone bridge to be substituted for the wooden one when the requisite amount should be forthcoming.

At the last session of the Legislature a sufficient sum of money was appropriated for the purpose. Advertisements for bids were inserted in the local and neighboring newspapers, and in due time the contract for the construction of the bridge was awarded, the

Commissioners agreeing to furnish the stone to the contractor. Although the bridge was not completed prior to the close of the fiscal year, it may not be inappropriate to give a brief description of it.

It is built of native sandstone of rustic finish, with a clear span sixty feet in width, the pathway being twelve feet wide. The total length of the bridge, including abutments, is one hundred and twenty feet; the rise of the arch is nine feet, and the extreme heighth of the pathway above the river bed is seventeen feet.

On the Goat Island side a semi-circular bay gives variety to the lines of the abutment and additional space for visitors to linger and enjoy the view of the rapids without obstructing the pathway. There is a similar bay on both sides of the Sister Island approach.

The parapet walls are of large stones, rock faced. The pathway is asphalted. The abutments are built on the solid rock, and the stone work already has a dark gray coloring which harmonizes with the surroundings. The bridge altogether seems to be exceptionally well built. Although at present rather massive in appearance, when partially overgrown with vines it will be an exceedingly picturesque and graceful object.

The State, in establishing the Reservation, did something more than put an end to the mis-treatment of the Niagara scenery; it did something more than provide a pleasuring ground for the people; it made a noteworthy contribution to the cause of popular education. By terminating private ownership of the islands and shore of the Niagara, and proceeding to restore them to a more natural condition, the State gave to individuals the opportunity of contemplating and the privilege of being influenced by a great natural phenomenon in the midst of agreeable surroundings. A stimulus was thus given to the appreciative study of Nature in one of her most wonderful manifestations.

ther the people in sufficient numbers have taken advantage of this opportunity and made the most of this privilege is a ques- may be differently answered by different persons. It must be admitted, however, that the stay of most visitors at the falls is to be followed by the best results. Those who obtain the benefit from a visit to Niagara are they who tarry long to become intimately acquainted, so to speak, with all the of the scenery. It is impossible to know nature well without long-continued study. A recent writer has pointed out that the first requisite in the study of nature is contemplation, which implies deliberation, leisure, receptivity. The appreciation and enjoyment of natural scenery are not, save in exceptional instances, parts of one's natural endowment. Cultivation is absolutely essential.

The importance of natural scenery in the life of the people is not likely to be over-estimated. In this age, when the tendency of so large a proportion of the population is towards cities and the artificialities of urban existence, when the exactions of business life are so incessant, when the dominance of the practical and materialistic forces is so recognizable, the importance of natural scenery as an educating and restoring agency in the popular life is greater than ever before.

"If, as seems likely," says a contemporary writer, "we can bring into definite shape, by educative means, the emotions which lead to pleasure in the landscape, we shall thereby add another important art to those which serve to dignify our lives."

The Reservation has been the means of attracting the attention of individuals to the desirability and advantage of the acquisition by the State, or people, of places remarkable for striking natural scenery, or interesting historical associations.

There are in this State not a few such places and objects that might with general approval be taken by the State and made public possessions for all time. Money expended in fostering a love for natural scenery and stimulating a popular interest in national or local history must be regarded as money well invested.

Niagara as a great producer of power, in the development of manufacturing and electrical industries, is not likely to be neglected; it is as an educating force that the river and falls of Niagara are in danger of being disregarded. The current of the upper river has a commercial value with difficulty computed, but recognizably great which the State through its law-makers has seen fit to dispose of to corporations without exacting compensation.

As the city of Niagara Falls is to be, if it is not already, one of the most conspicuous and important manufacturing centres in the country, with the consequent development of the commercial and materialistic element, it would seem more than ever desirable that the Reservation should receive legislative, as well as popular attention, and that its natural attractions, freed from artificialities, should stand out in more striking and vivid contrast to the factories and other useful but unornamental structures that are so distinctly and obtrusively visible both above and below the Falls.

It should not be forgotten that the Reservation really belongs to the State, to the whole State and not to any portion or section of it. The organized movement for the protection of the scenery of the Falls had its origin in the great city at the mouth of the Hudson. The same city is assessed for more than one half of the State taxes. Local interest in the Reservation is entirely subordinate to the interest of the State.

The volume of the river and cataract at Niagara is of course dependent upon the water supply of the Great Lakes. The Niagara river is but the overflow of Lake Erie, into which flows the waters of the other lakes. The lowering of the level of these lakes would diminish the flow into Lake Erie, and reduce the volume of the Niagara river. Any very large withdrawal or diversion of water from one or more of the Great Lakes would scarcely fail to be noticeable in a reduced flow at the cataract.

The Commissioners deem it advisable that the National government be requested to appoint a commission to confer with a Canadian commission as to the means to be devised to prevent any excessive diversion of the waters of the Great Lakes, and to consider the whole subject of the uses and control of these waters, and to report its conclusions to Congress, with such recommendations as it may desire to submit.

The Commissioners have been careful to avoid anything approaching extravagance, and by adopting prudent and business-like methods they have endeavored to secure an adequate return for moneys paid out. No expenditure of theirs, so far as is known, has ever been unfavorably criticised in the public prints, nor has any voucher been rejected by the State Comptroller, upon whom is imposed by law the duty of auditing the bills and approving the accounts of the Commissioners. They have, by giving in their annual reports full information of their receipts and expenditures, invited attention to their disbursement of public money and to their management of the Reservation.

They have endeavored to keep in view the interests and benefits of the people. The people paid for the Reservation, and they should have the fullest enjoyment of it.

MONTHLY RECEIPTS FROM THE RESERVATION.

	Inclined railway.	Rentals.	Interest on balances in bank.	Dividends.	Conscience.
1897.					
October	$284 80	....	....	....	$6 00
November	49 50	....	....	....	....
December	28 60	....	$11 00	....	....
1898.					
January	37 40	....	....	....	....
February	50 70	....	....	$14 17	....
March	64 00	....	24 97	....	....
April	56 80	....	....	....	....
May	218 50	....	....	....	....
June	288 35	....	25 81	14 17	....
July	1,096 85	$302 10	....	....	....
August	1,367 50	800 00	....	....	....
September	1,022 35	700 00	15 34	....	....
	$4,565 35	$1,802 10	$77 12	$28 34	$6 00

Inclined railway	$4,565 35
Rentals	1,802 10
Interest on balances in bank	77 12
Dividends	28 34
Conscience	6 00
	$6,478 91

Monthly pay-rolls were as follows:

1897.	
October	$1,796 03
November	1,448 70
December	1,320 66
1898.	
January	1,396 02
February	1,333 66
March	1,495 67
April	1,780 16
May	1,795 12
June	1,798 48
July	1,799 80
July, supplemental	496 37
August	1,497 56
September	1,376 23
Total	$19,334 46

Maintenance expenditures, as per abstract, were as follows:

Abstract CV	$2,386 77
Abstract CVI	3,414 64
Abstract CVII	5,933 11
Abstract CVIII	2,272 99
Abstract CIX	2,460 34
Abstract CX	9,155 97
	$25,623 82

Improvement abstracts were as follows:

Abstract II, series H	$2,406 39
Abstract III, series H	1,296 19
Abstract IV, series H	1,944 07
Abstract V, series H	1,333 85
	$6,980 50
Abstract I, series I	5,063 03

The receipts and earnings of the Reservation have been sent to the State Treasurer monthly, and interest on balances in the Manufacturers & Traders' Bank, Buffalo, has been remitted to the same official quarterly.

The Comptroller has advanced to the Commissioners quarterly a fourth part of the $25,000 appropriated by chapter 306, Laws of 1897.

Of the treasurer's report herewith submitted, exhibiting in detail all receipts and disbursements for the fiscal year ending September 30, 1898, the following is a summary:

Balance on hand October 1, 1897		$758 53
Receipts.		
Inclined railway	$4,565 35	
Rentals	1,802 10	
Interest	77 12	
Dividends	28 34	
Conscience	6 00	
		6,478 91
From the State treasury (chapter 306, Laws of 1897).		25,000 00
From the State treasury (chapter 790, Laws of 1897).		6,980 50
From the State treasury (chapter 606, Laws of 1898).		5,063 03
		$44,280 97

Looking Northward from Gorge View.

Payments.

Pay rolls at Niagara (maintenance)	$19,334 46	
Repairs, materials, superintendent's expenses, etc	4,878 93	
Treasurer's salary and expenses	1,163 85	
Commissioners, traveling expenses, etc	246 58	
		$25,623 82
Remitted to State Treasurer		6,478 91
Improvements (chapter 790, Laws of 1897)		6,980 50
Improvements (chapter 696, Laws of 1898)		5,063 03
Balance September 30, 1898		134 71
		$44,280 97

Total receipts since organization of the Commission, 1883	$534,493 02
Total disbursements	534,358 31
Balance	$134 71

The Legislature has made appropriations for maintenance, current expenses and salaries, as follows:

By chapter 336, Laws of 1883	$10,000 00
By chapter 656, Laws of 1887	20,000 00
By chapter 270, Laws of 1888	20,000 00
By chapter 569, Laws of 1889	25,000 00
By chapter 84, Laws of 1890	20,000 00
By chapter 144, Laws of 1891	20,000 00
By chapter 324, Laws of 1892	20,000 00
By chapter 414, Laws of 1893	25,000 00
By chapter 654, Laws of 1894	25,000 00

By chapter 807, Laws of 1895	.....	$25,000 00
By chapter 948, Laws of 1896	.....	25,000 00
By chapter 306, Laws of 1897	.....	25,000 00
By chapter 593, Laws of 1898	.....	25,000 00
Total	.....	$285,000 00

For special improvements appropriations have been made as follows:

By chapter 570, Laws of 1889	.....	$15,000 00
By chapter 302, Laws of 1891	.....	15,000 00
By chapter 356, Laws of 1892	.....	15,000 00
By chapter 726, Laws of 1893	.....	25,000 00
By chapter 358, Laws of 1894	.....	20,000 00
By chapter 932, Laws of 1895	.....	20,000 00
By chapter 950, Laws of 1896	.....	10,000 00
By chapter 790, Laws of 1897	.....	15,000 00
By chapter 606, Laws of 1898	.....	15,000 00
Total	.....	$150,000 00

In compliance with statutory directions, the Commissioners have remitted to the State treasury the receipts from the Reservation, as follows:

From October 1, 1887, to September 30, 1888	.....	$9,331 55
From October 1, 1888, to September 30, 1889	.....	7,393 77
From October 1, 1889, to September 30, 1890	.....	7,670 29
From October 1, 1890, to September 30, 1891	.....	9,327 6[illegible]
From October 1, 1891, to September 30, 1892	.....	9,823 0[illegible]
From October 1, 1892, to September 30, 1893	.....	10,923 8[illegible]
From October 1, 1893, to September 30, 1894	.....	9,251 [illegible]

From October 1, 1894, to September 30, 1895	$6,179 09
From October 1, 1895, to September 30, 1896	7,448 01
From October 1, 1896, to September 30, 1897	7,551 83
From October 1, 1897, to September 30, 1898	6,478 91
Total	$91,379 43

The following is "an estimate of the work necessary to be done and the expenses of maintaining said Reservation for the ensuing fiscal year," ending September 30, 1899:

Construction.

Administration building and alteration of inclined railway building	$20,000 00
Grading, stairways, approaches, guard railings, and necessary improvements at Goat Island and Luna Island	5,000 00
Water pipes	2,000 00
Electric lighting	2,000 00
Planting	1,000 00
	$30,000 00

Maintenance.

Salaries, office and traveling expenses	$4,750 00
Reservation police and caretakers	5,400 00
Labor	7,500 00
Materials, tools, etc	6,000 00
Miscellaneous	1,350 00
Total	$25,000 00

Estimated receipts from October 1, 1898, to September 30, 1899:

Inclined railway	$5,500 00
Cave of the Winds	1,200 00
Ferry and steamboat landing	500 00
Carriage service	100 00
Interest	100 00
	$7,400 00

The Niagara river has a particular interest for scientific persons. Affording, as it does, an index of the lapse of geologic time, it is important that accurate measurements by trained scientists should from time to time be made, in order that the recession of the falls may be noted and other changes marked.

ANDREW H. GREEN.
GEORGE RAINES.
ALEXANDER J. PORTER.
THOMAS P. KINGSFORD.
CHAS. M. DOW.

Commissioners of the State Reservation at Niagara.

Report of the Superintendent

OF THE

STATE RESERVATION AT NIAGARA

FOR THE

Fiscal Year Ending September 30, 1898.

To the Commissioners of the State Reservation at Niagara:

Gentlemen.— The most important work of the year was the construction of the stone arched bridge from Goat Island to the First Sister Island, according to plans prepared by Vaux & Emery, architects, New York city.

STONE ARCHED BRIDGE TO FIRST SISTER ISLAND.

The span of the bridge is 60 feet. Inside width 12 feet, outside width 16 feet. The bridge is constructed of native stone, of the locality.

Early in the month of July the old wooden bridge was removed and a temporary bridge constructed further up the stream. The contract for the construction of the stone arched bridge was let to W. A. Shepard, of Niagara Falls. The contract price was $5,044. The stone for the construction of the bridge was furnished by the Commissioners. The work upon the bridge was commenced July 18, 1898, and the last stone was laid in place November 4, 1898. The total cost of the bridge was $7,831.04, being less than one-half of the estimated cost.

Since the completion of the bridge the approaches have been filled and graded and the side walls banked up in readiness for planting. In the spring the grading will be completed and vines and shrubs planted about the bridge.

Photographs of the stone arched bridge are herewith submitted.

DEEPENING THE CHANNEL.

While the bridge was being built a cofferdam was placed by the contractor in the channel, above the bridge; advantage was taken of this to deepen the channel by blasting out the rock in the stream from one to two feet, so as to insure a larger volume of water upon the completion of the bridge, which spans the stream just below the Hermit's Cascade. For several years the water between Goat Island and the First Sister Island has been very low. The deepening of the channel will tend to restore the former volume of water.

PROSPECT POINT.

In the last report of the Superintendent attention was called to the artificial constructions at Prospect Point, on the brink of the American Falls, the principal point of view within the Reservation, where a stone wall enclosure, with cut-stone coping, stone stairs, a plank walk and a plank platform marred the natural beauty of the scene. During the past year all of these artificial constructions have been removed, and a great change made in the appearance of the locality.

The parapet wall has been taken down to the surface of the ground, and the wall covered with ledges of rock, laid in cement. On the arch of the Inclined Railway, where the wall bulges out, an excavation was made, and a bed of rock, laid in cement, to prevent further movement by the action of the frost. The stone steps at Prospect Point were removed. Inside the parapet wall, at Prospect Point, an excavation was made down to the solid rock and large rustic stones laid in Portland cement. The wall was then removed on the Rapids side. The outlet from the Inclined Railway was covered with flag stones. Prospect Point was widened up

REMOVING STONE WALL, PROSPECT POINT, SEPTEMBER, 1898.

stream, a shore line of large rustic stones laid, and the enlarged space at Prospect Point filled with gravel and a floor of ledges of rock laid in, so as to restore, as nearly as possible, the natural appearance of the place.

Inside the stone wall, stone posts, 12x14 inches, 4 feet long, were placed (care being taken to avoid the straight lines of the stone wall), and a substantial iron railing, of the pattern designed by Mr. Calvert Vaux, erected.

The plank walk along the stone wall was removed, and a spacious walk of ledges of rock, laid in sand, substituted. A few rustic stones were placed, the gravel walks at Prospect Point rearranged, and the low ground adjacent to the stone walk filled and covered with grass sod.

Photographs showing the changes made at Prospect Point are herewith submitted.

HENNEPIN'S VIEW.

Hennepin's View is a projecting point upon the high bank, about midway between the American Falls and the northern boundary of the Reservation, commanding the best general view of the Falls from the American side. It was occupied by a plank platform, reached by ascending a flight of wooden stairs. During the past year the wooden platform and steps have been removed and an elevation of earth and rock made upon the locality, resembling as closely as possible, a natural formation. A space has been arranged for seats for visitors upon the point of view and the walk carried over the elevation, around which a substantial guard railing has been erected.

A photograph of Hennepin's View is herewith submitted.

GORGE VIEW.

The northern boundary of the Reservation was marked by a wire fence. In the northwest corner was a frame cottage occupied by one of the employes. During the past year the cottage, fences and sheds at this point have been removed. The site of the cottage commands a fine view of the Niagara Gorge, and the international bridges to the northward. This point has been raised, the high bank riprapped with large rustic stones, and a new and desirable view opened to visitors. The walks at Gorge View have been rearranged, a space for seats for visitors provided, and an iron guard railing erected around the point of view. Gorge View has already become a favorite place of observation for visitors. The northern boundary of the Reservation is now marked by rustic stones placed at irregular intervals.

PROSPECT PARK.

The northwestern portion of Prospect Park was low and wet, and not in keeping with the other portion of the grounds. During the past year an opportunity occurred of obtaining a large quantity of earth for filling, without expense of hauling. The low ground in Prospect Park has been drained, filled, graded and covered with grass sod, making a remarkable change in the appearance of the locality.

ROADS.

A portion of the old road in Prospect Park has been discontinued, and a loop driveway constructed, making a continuous and easy drive through the grounds. The loop drive in Prospect Park commands the prospect at Gorge View and gives access to the pavilion in the picnic grounds.

Repairs have been made upon the roads in Prospect Park and on Goat Island. Extensive repairs of the macadam roads within the Reservation will be needed during the coming year.

BUILDINGS.

January 28, 1898, the Hall building in Prospect Park was destroyed by fire of incendiary origin, as no fire had been kept in the building for two months previous to that time. A number of seats and stage fixtures were destroyed with the building. The greatest loss caused by the fire was the destruction of several fine trees adjacent to the building, showing the mistake of locating destructible frame buildings in such places.

The Hall building was long, low and unsightly; it obstructed the view, and its removal added much to the appearance of the grounds. It was, however, very convenient for public gatherings, and accessible to visitors in case of sudden rain-storms, and the need of a building for such purposes has been very great during the past summer.

During the year the library or art gallery building, one gate house, two of the refreshment booths, three closets, as well as the dwelling house and adjoining structures, all frame buildings, have been removed from Prospect Park, relieving it of a cluttered appearance, opening the view of the scenery and restoring to the place a grove-like character.

Two frame structures have been removed from the riverway, one from Green Island and one from Goat Island.

A storehouse, made necessary by the removal of the buildings in Prospect Park, has been built in the lumber yard on Goat Island. Extensive repairs were made upon the copper roof of the lower terminal station of the Inclined Railway. The open pavilion in Prospect Park and the office of the Commissioners on Green Island were renovated and an additional building provided at the cottage on Goat Island.

PLANTING.

The inner shore of the pond at Port Day, which was prepared for planting last year, has been planted with shrubs and vines, which have been carefully pruned and cultivated during the year. In addition to the stock in the nursery, 250 dogwoods, 250 honeysuckles, 100 yellow willows, 100 creeping roses, 100 Virginia creepers and 100 Japanese ivy have been procured for planting.

As soon as the weather will permit in the spring some planting should be done along the Riverway, on Green Island, adjacent to the stone arched bridge from Goat Island to the First Sister Island, and at the approach to Goat Island. Some tree planting should also be done in Prospect Park to replace the natural growths, some of which are showing signs of decay.

THE RIVERWAY.

During the past winter the outer shore of the Loop Driveway at Port Day, was damaged by erosion, caused by high water and floating ice. During the year the entire outer shore of the driveway has been riprapped with large stones, obtained near by, at only the cost of hauling.

Since riprapping the shore has withstood the action of the ice and water, and it now presents a more permanent and natural appearance.

The neighborhood of the Loop Driveway has become one of the most attractive portions of the Reservation.

REMOVAL OF RETAINING WALL.

On the Riverway, below Fourth street, the slope adjacent to the roadway was sustained by a high retaining wall, artificial and unsightly in appearance. During the year the wall has been removed and the stones carted to Goat Island for use in the construction of the new stone arched bridge.

The slope on the Riverway has been riprapped with large rustic stones to sustain the bank, upon which is a number of fine trees. The bank has been graded and covered with grass sod and now presents a sloping and natural appearance.

FILLING AND GRADING ON THE RIVERWAY.

The shore on the Riverway has been filled and graded from the Boulder bridge up to Fourth street, adding considerably to the land area in that locality and restoring, approximately, the original shore line. The filling and grading are not yet completed. In the spring new walks may be constructed and some planting done in that locality.

The bluff or terrace on the Riverway commands unequalled views of the rapids, the islands, and the upper Niagara river. The walks upon the bluff should be extended and improved so as to make the locality more accessible to visitors.

THE WING.

The locality immediately above Willow Island is known as "The Wing." A massive stone wall has been constructed at that point to deflect waters from the river into "The Race," a channel which conveyed it to mills which have been removed. The water inside the wall is a foot or two higher than the water in the river. The outlet from the Wing is in a dilapidated condition and should be reconstructed in a suitable manner. Parts of the Wing wall have fallen away, and if the wall is to be retained some repairs should be made. The wall at the upper end might be drawn in by an easy curve and connected with the mainland below Fourth street by a small stone arched bridge, and a loop walk or promenade thus obtained somewhat similar to the Loop Driveway at Port Day. It would command a fine view of the rapids at that point and the upper Niagara river, and would undoubtedly be a favorite resort for visitors.

Suitable stone for this work can now be obtained at Port Day, probably without expense except for hauling, an opportunity which may never occur again.

WALKS.

The Riverway, from the northern boundary of Prospect Park to the south line of Niagara street, is included in the Reservation. The plank walk on the west side of the Riverway in that locality should be removed, and a spacious gravel walk with broad margins of green sward substituted, similar to those to the southward on the Riverway. This work is made more necessary by recent changes made in that locality.

APPROACH TO GOAT ISLAND.

The retaining walls, of quarry stone on either hand, at the approach to Goat Island have been removed, and rustic stone work substituted. Rustic stones have also been placed in the banks, ahead and to the right and left in the slopes, formed when the road on Goat Island was constructed.

The planting of the slopes may be done as soon as the weather will permit in the Spring.

THE SPRING ON GOAT ISLAND.

The frame platform at the Spring on Goat Island has been removed, and the space floored with ledges of natural rock, laid in sand. The Board of Health of the city of Niagara Falls has called attention to the probable pollution of the water by persons dipping pails and other vessels into the spring to obtain water. To guard against possible contamination, it may be well to close the opening in the stone canopy of the Spring, so that the water may be obtained only through a tube.

SYSTEM OF WATER PIPES.

A two-inch water main has been laid from Falls street to the Inclined Railway building, and connections made so that the toilet rooms in the building may be kept open during the winter season, which has not been possible heretofore, because the water pipes in the grounds were surface pipes.

A general plan for an adequate system of water pipes for the grounds is needed.

During the year, an additional Drinking Fountain has been placed near the bridge to the Three Sister Islands. A water pipe should also be laid to the Cave of the Winds building.

ELECTRIC LIGHTING.

The present arrangement for lighting the grounds is not adequate. If the grounds are to be lighted at all, they should be well lighted.

A comprehensive plan for adequate lighting of the grounds is needed.

LANDSLIDES.

Two landslides occurred during the year on the high bank in Prospect Park, extending back as far as the stone post supporting the iron guard railings.

The cavities have been filled and planted with willows to retain the bank.

NUMBER OF VISITORS.

The number of visitors during the year is estimated at about 600,000. The "excursionists" to the Reservation aggregated 5,563 cars, bringing an estimated number of 356,765 passengers.

No disorders occurred, and no serious breach of the ordinances of the Reservation was committed by visitors.

LICENSED CARRIAGE DRIVERS.

Seventeen complaints have been made against licensed carriage drivers. Four drivers have been excluded from the Reservation for violation of the ordinances of the Commissioners.

Two public carriage stands, one on Bridge street and one on Falls street below the Riverway, have been abolished.

THE RESERVATION CARRIAGE SERVICE.

The Reservation carriage service has been in operation during the year. Since the completion of the Loop Driveway the route of the service has been extended to Port Day and return. One or more of the carriages will remain in the service during the winter season.

THE STEAMBOAT LANDING.

The landing has been used by the Maid of the Mist Steamboat Company. Only one boat has been in operation.

The level of the water in the river has been somewhat higher than in recent years.

ELEVATOR AT THE CAVE OF THE WINDS.

As this locality is between the American and the Horseshoe Falls, the view is unparalleled, but the winding stairway leading to it is so unsuitable and fatiguing that very few persons are enabled to enjoy the beauty of the scenery below the high bank, where a walk can be easily made along the edge of the water from the American to the Horseshoe Falls.

The waiting rooms and winding stairway of the Cave of the Winds have received the usual repairs during the year. The dressing rooms are inadequate and unsuitable. Better accommodations should be provided for the traveling public.

The Biddle staircase has been examined and found safe, but it is old and dilapidated in appearance, and affords but a fatiguing

PROSPECT POINT, NOVEMBER 1, 1898.

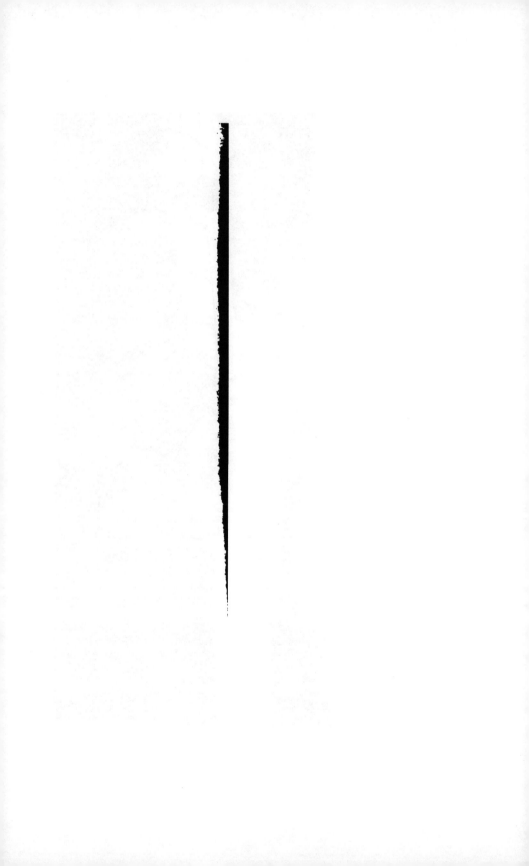

method of obtaining the view of the Falls from below. It should be replaced by a commodious elevator, with a free stairway attached. Such an elevator, operated for a nominal fee of five cents up or down, would furnish an estimated revenue of $5,000 a year, which, with the present revenue, mainly from the Inclined Railway, would almost render the Reservation self-sustaining.

ADMINISTRATION BUILDING.

The removal of the frame buildings from Prospect Park has given rise to the urgent necessity of a substantial administration building in that locality. It should be located convenient to the main entrance to the grounds, and in addition to the administration offices should contain waiting rooms, lavatories, toilet rooms, store rooms, a parcel room, and other conveniences for the accommodation of a large number of people. Such a building being provided, the present Inclined Railway building might be removed and an underground structure substituted, which would leave the view of the falls unobstructed.

The office building, which now shows so prominently on Green Island, then might also be removed.

SHELTER BUILDINGS.

Shelter buildings are needed at the American Fall and at the Horseshoe Fall on Goat Island, where visitors congregate in large numbers. At times of sudden rain-storms the present shelters are too far apart for public convenience.

EMPLOYES.

The regular force employed, exclusive of laborers, consists of 10 men, to wit, 1 superintendent, 1 clerk, 6 police gatemen and caretakers, and 2 Inclined Railway men, one of whom is employed during the summer season only.

Statement of number of employes on maintenance and improvement pay-rolls for the year ending September 30, 1898.

	MAINTENANCE.				CHAPTER 790, LAWS 1897.				CHAPTER 606, LAWS 1898.			
	Foreman.	Assistant foreman.	Laborers.	Teamsters.	Foreman.	Assistant foreman.	Laborers.	Teamsters.	Foreman.	Assistant foreman.	Laborers.	Teamsters.
1897.												
October	......	1	24	1	1	2	12	5	......	......	......	......
November	......	2	19	1	1	......	4	4	......	......	......	......
December	1	......	15	1	......	......	3	3	......	......	......	......
1898.												
January	1	......	14	1	......	......	5	1	......	......	......	......
February	1	......	15	1	......	......	......	......	......	......	......	......
March	1	......	21	......	......	1	11	5	......	......	......	......
April	1	......	26	......	......	2	10	2	......	......	......	......
May	1	......	24	......	......	3	17	3	......	......	......	......
June	1	......	24	......	......	3	23	3	......	......	......	......
July	1	2	33	1	......	......	......	......	......	2	24	3
August	......	1	16	......	......	......	......	......	1	3	48	7
September	......	1	16	......	......	......	......	......	1	3	42	8

TABULAR STATEMENTS.

The receipts from the Inclined Railway during the fiscal year re $4,565.05; from rentals and other sources, $1,836.44. Total eipts, $6,401.49.

Vouchers paid for material and labor, $11,795.36. Pay-rolls for aintenance and improvements, $28,593.91. Total expenditures Superintendent, $40,389.27.

Detailed statements of the receipts and expenditures of the uperintendent, the amount of the pay-rolls for each month and e classification of the pay-rolls and accounts are hereto appended.

Respectfully submitted.

THOMAS V. WELCH,
Superintendent.

EXCURSIONS 1897–98.

1897.		Cars.	Persons.
Oct.	3. Hampden Hose Company, Reading, Pa	1	60
	Cleveland via C. and B	27	1,620
	4. Students Holy Angels Academy, Buffalo, N. Y	2	120
	5. Seneca and Cattaraugus Indians Temperance Convention	1	60
	6. Ancient and Honorable Artillery Company, Boston	6	360
	7. National Association Photo-Engravers	1	60
	9. Oswego Normal School	3	180
	Representatives N. Y. Life Insurance Company, Chicago, Ill	1	60
	10. Buffalo, N. Y	10	600
	Almond and Johnsburg, Pa	9	540
	Carbondale and Port Jervis	9	540
	13. Philadelphia, Pa	1	60
	Brotherhood of St. Andrew	5	300
	Washington and Baltimore	4	240
	15. Brotherhood of St. Andrew	3	180
	Hendrickson party, Brooklyn, N. Y.	1	60
	Delegates Brotherhood Locomotive Engineers	3	180
	18. Delegates Brotherhood Locomotive Engineers	4	240

97.		Cars.	Persons.
	18. Brotherhood of St. Andrew	4	240
	New York, via Wabash	2	120
	19. Delegates Railway Convention (Electric)........................	10	600
	20. Delegates Railway Convention (Electric)........................	10	600
	31. Cook County Democracy, Chicago, Ill	6	360
898.			
›.	7. President S. B. Dole, of Hawaii, and party	1	60
.	10. Hotel Men's Benefit Association, Erie, Pa	7	420
	7. Postponed Arbor Day excursion, Lockport, N. Y	11	660
	19. Exempt Firemen, Wilmington, Del.	1	60
	21. Central High School of Buffalo	4	240
	22. Main line Erie R. R..............	6	360
	Buffalo, via N. Y. C. and trolley ...	6	360
	24. Queen's birthday excursion........	25	1,800
	29. New York city via West Shore	10	600
	Erie Bicycle Club, Erie, Pa........	5	300
	Main line Erie R. R..............	10	600
	Buffalo, via N. Y. C. and trolley ...	20	1,200
	30. Memorial Day excursion, all sources.	40	2,400
	31. Delegates to Convention African M. E. Church	1	60
une	2. American Society Mechanical Engineers	4	240
	Buffalo street railway employes	1	60

1898.		Cars.	Pas
June	5. From all sources	80	1,
	8. County officials and friends, York and Ontario counties, Canada	2	
	11. West Shore half-holiday excursion..	6	
	Buffalo Kindergarten training class.	1	
	American Axle Manufacturers' Association	2	
	Erie R. R.	4	
	12. Canadian militia from Niagara on the Lake	4	
	Sunday excursions	35	2,
	14. County superintendents of the poor.	2	
	East York Farmers' Institute and Bonar Camp Sons of England	4	
	15. Delegates to American Water Works Association	5	
	16. Wholesale and Retail Grocers, Dayton, Ohio	4	
	Iroquois Indian children	......	
	St. Thomas, Ont	5	
	18. Syracuse and Rochester	6	
	Employees Northey Co., Toronto...	6	
	I. O. O. F., Monroe, Ont	4	
	Reunion of Long family...........	1	
	Saturday half-holiday excursions....	4	
	Toronto, Ont., by boat.............	......	1
	19. German Social Society, Rochester, N. Y.	1	
	Lehigh Valley line	6	
	Sunday excursion, all sources......	20	1,

1898.		Cars.	Persons.
ne 20.	Employees Kilgore Paperware Mfg. Co., Toronto, Ont..............	4	240
21.	Shelbourne Street Methodist Church, Toronto.......................	3	180
	Special party Minneapolis Daily Journal, Minneapolis............	1	60
24.	Akron and Avon, via Erie R. R....	8	480
	Warren, Pa., via Dunkirk.	7	420
25.	Brotherhood Railway Trainmen, Toronto.......................	4	240
	Graduating class, School No. 13, Buffalo	1	60
	Thursday Club, Lockport, N. Y....	1	60
	Foresters' Union Picnic, Lindsay, Ont	6	360
	Saturday half-holiday excursions, all sources	10	600
26.	Sunday excursion, all sources.......	40	2,400
27.	Graduating class, School No. 38, Buffalo	1	60
	Employees McLaughlin Carriage Works, Oshawa, Ont............	3	180
29.	City employees, Toronto, Ont......	3	180
	Junior Department Y. M. C. A., Cleveland, Ohio	13	780
30.	Presbyterian Church, Erie, Pa......	4	240
y 1.	Dominion Day excursions, all sources	50	3,000
2.	Saturday half-holiday excursion, all sources	20	1,200
	Royal Canadian Yacht Club.......	1	60

1898.		Cars.	Persons.
July	3. Sunday excursions, all sources.....	85	5,100
	4. from all sources..................	250	15,000
	5. New excursion, via W. N. Y. & P.	6	360
	6. Century Club, New Orleans, La....	2	120
	St. Louis Church Sunday School, Buffalo	2	120
	St. Francis Xavier Church, Black Rock.........................	2	120
	7. Cumberland Valley State Normal School, Shippensburg, Pa.......	1	60
	Holy Trinity Church, Buffalo, N. Y.	5	300
	8. Kentucky Press Association.......	3	180
	Union churches, Caledonia, N. Y...	4	240
	Raymond and Whitcomb party....	4	240
	Railway surgeons, Toronto, Ont....	2	120
	Governor Wolcott and staff, Massachusetts......................	2	120
	N. Y. State Teachers, Rochester, N. Y.........................	4	240
	9. American Library Association Jamestown, N. Y..............	6	360
	Employees Coulter Clothing Mfg. Co., Hamilton, Ont............	8	480
	Saturday half-holiday excursions, all sources	10	600
	10. Sunday excursions, all sources.....	30	1,800
	11. Employees biscuit-makers, Toronto, Ont	4	240
	12. First Presbyterian Church, Lockport, N. Y....................	5	300

PROSPECT POINT, SEPTEMBER 1, 1898

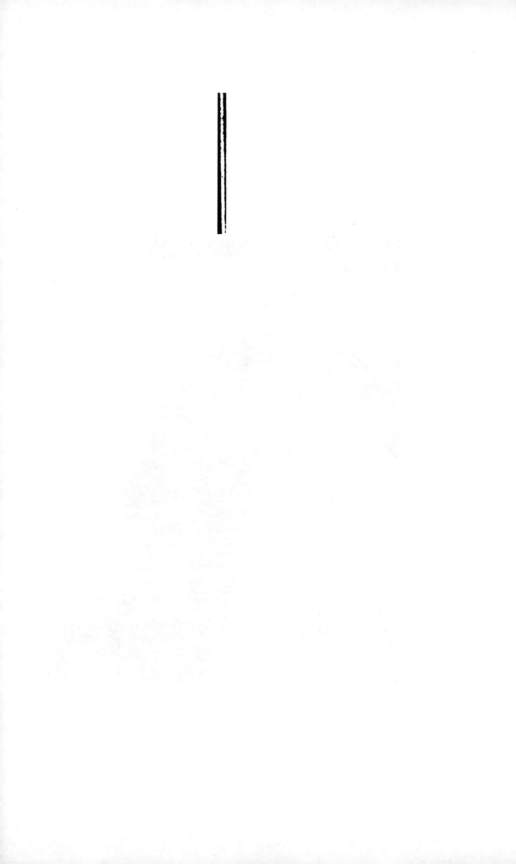

896.		Cars.	Persons.
y 12.	Orange lodges, Toronto, Ont......	5	300
13.	Annual excursion, Baltimore and Washington..................	20	1,200
	Baptist Young People...........	6	360
14.	International Bill Posters' Convention	3	180
	Baptist Young People's Union.....	6	360
	Episcopal Church, Lockport, N. Y.	6	360
	Methodist Church, Erie, Pa.	7	420
15.	Baptist Young People's Union.....	25	1,500
16.	Baptist Young People's Union.....	60	3,600
	Newton Chautauqua party........	7	420
	Employees Cornell and Co. Blank Book and Stationery, Toronto, Ont.	3	180
	Employees G. T. R., Toronto, Ont.	10	600
	Half-holiday excursions, all sources.	20	1,200
17.	Sunday excursions, all sources......	50	3,000
18.	Baptist Young People's Union.....	6	360
	St. Vincent de Paul Society, Toronto, Ont	5	300
20.	Employees wholesale and retail grocers, Hamilton, Ont....... ..	10	600
	Baptist Young People's Union, Florida delegates.	2	120
	Agents American Railway Association	6	360
	St. Margaret's Church, Toronto, by boat	......	400
	Buffalo, by boat..................	......	600
21.	Washington and Baltimore, via Lehigh Valley.	12	720

1898.		Cars.
July 21.	Railway Y. M. C. A., via R., W. and O., Oswego N. Y..........	12
	Western New York and Pennsylvania........................	7
	Columbus, Ohio, via C. and B......	10
	N. Y. State Bankers' Association....	10
22.	Employees Empire Clothing Co., Toronto, Ont..................	5
	Cleveland, via C. and B. line......	8
23.	Saturday excursions, all sources....	385
	Buffalo, by boat..................	......
24.	Sunday excursions, all sources......	84
25.	Olean, via W. N. Y. and P.... ...	7
26.	Dundas and Niagara on the Lake, Ont.........................	6
	Buffalo, by boat..................	......
27.	Hamilton and St. Catherines, Ont, via M. C......................	8
	London, Ont....................	4
	Delaware, L. and W.............	8
	Cleveland, O., via C. and B.	5
	Buffalo, by boat	......
28.	Presbyterian Sunday School, Kenmore, N. Y.	2
	Order of Foresters, Toronto, Ont...	5
	Canastota, via N. Y. C............	7
30.	Chautauqua party, via W. N. Y. and P	7
	I. O. O. F., Erie, Pa., Nickel Plate,	8

189[illegible]			Cars.	Persons.
ly	30.	Port Jervis and Binghamton, via Erie R. R.	18	1,080
		Employes C. Brown Biscuit Mfg. Co., Toronto, Ont	18	1,080
		C. P. R	12	720
		General half-holiday excursions, all sources	20	1,200
		Buffalo, by boat	......	350
	31.	Cleveland, O., via C. and B.	7	420
		Buffalo	60	3,600
		Buffalo, by boat	......	300
Aug.	1.	Foresters, Guelph, Ont	8	480
		Civic holiday, Guelph, Ont., by boat	......	700
		Delegates to Labor Convention	7	420
		Detroit, Lansing and Saginaw, via G. T. R	10	600
	2.	Special trolley party, Buffalo	3	180
		Toledo, via C. and B. line	7	420
		Rochester, via Erie R. R	6	360
	3.	Public school, Buffalo	5	300
		Rochester, via Lehigh Valley R. R.	8	480
		Mechanics and iron workers, Toronto, Ont	8	480
		Lindsay, Ont., via Grand Trunk R. R.	5	300
		St. Catherines, Ont., Sunday school	4	240
		Buffalo, by boat	......	80
	4.	Akron, Ohio	17	1,020
		Hop Growers' Association, Auburn, N. Y	25	1,500
		Toledo, Ohio	10	600

1898.		Cars.	Persons.
Aug.	4. Michigan city and western points...	30	1,800
	Apple Shippers' Association.	2	120
	Erie R. R., main line.............	20	1,200
	W. N. Y. and P.................	10	600
	Buffalo, Rochester and Pittsburg...	10	600
	Buffalo, by boat.....	......	100
	5. Michigan, Indiana and Ohio, via Wabash R. R..................	34	2,000
	Erie and western roads............	40	2,400
	Nickel Plate line.................	82	4,920
	Michigan Central R. R..	65	3,900
	Buffalo, by trolley...............	25	1,500
	Buffalo, by boat..................	......	150
	6. Farmers' picnic, via R. W. and O ..	22	1,320
	Chautauqua party, via W. N. Y. and P.........	9	540
	Half-holiday excursion...........	20	1,200
	7. Cleveland, Ohio, via C. and B. R. R.	10	600
	N. Y. C. and H. R. R. R..........	20	1,200
	N. Y., L. E. and W	12	720
	Jamestown, N. Y	10	600
	Buffalo, Rochester and Pittsburg...	10	600
	Buffalo, via steam and trolley......	50	3,000
	Port Jervis	12	720
	Carbondale, Pa	11	660
	Buffalo, by boat	......	250
	8. Civic holiday, Hamilton, Ont......	16	960
	Civic holiday, Simcoe, Ont	10	600
	Civic holiday, Toronto, Ont	12	720
	League Cycle Club, Philadelphia, Pa.	2	120

1898.		Cars.	Persons.
Aug. 8.	Lehigh Valley R. R.	12	720
	Woodstock, Ont., via G. T. R.	12	720
	Buffalo, by boat	......	50
9.	Youngstown, via L. S. and M. S.	10	600
	Erie, main line	10	600
	Chautauqua party, via W. N. Y. and P.	8	480
	Buffalo, Rochester and Syracuse, via N. Y. C.	12	720
	Milton, via G. T. R.	9	540
	Buffalo, by boat	......	200
Aug. 10.	Officers and general agents New York Life Ins. Co.	2	120
	Scandinavian Picnic, Cleveland, Ohio	8	480
	Special trolley party, Buffalo, N. Y.	5	300
	Newmarket, Ont., by boat	......	1,500
	Buffalo, N. Y., by boat	......	50
11.	Railway Y. M. C. A., Rochester, N. Y.	14	840
	National Hay Association	5	300
	Dunkirk and Erie, via L. S. and M. S.	8	480
	I. O. O. F., Hamilion, Ont.	6	360
	Meadville, Pa.	3	180
	Buffalo, N. Y.	3	180
12.	Indiana, Michigan and Illinois	32	1,920
	Buffalo, Rochester and Pittsburg	8	480
	Baltimore, Washington and Philadelphia	12	720
	Lake Shore and Michigan Southern	10	600
	Buffalo by trolley	3	180

1898.		Cars.	Persons.
Aug. 13.	R., W. and O. picnic	10	600
	Masonic picnic, Norwich, Ont.	10	600
	Dundas Tool Works, Dundas, Ont.	10	600
	Chautauqua party, W. N. Y. and P.	8	480
	Saturday half-holiday	20	1,200
	Buffalo, by boat	......	100
14.	N. Y. C., West Shore and M. C.	30	1,800
	L. A. W., Rochester, N. Y.	10	600
	C. and B. line and trolley, Buffalo	50	3,000
15.	Agricultural Implement Mfg. Co., Brantford, Ont	4	240
	Union excursion, colored people, Toronto	6	360
	Raymond and Whitcomb excursion	2	120
	Order of Foresters, Brantford, Ont.	8	480
	Buffalo by boat	......	500
16.	Ohio, Indiana, Michigan and Illinois points	70	4,200
	Jamestown, N. Y	9	540
	Hornellsville, N. Y	9	540
	St. Thomas, Ont	7	420
	Burlington, Ont	7	420
17.	Big Four excursion, Illinois, Michigan, Indiana and Ohio	80	4,800
	Buffalo, by boat	......	200
18.	Hamilton, Dayton and Cincinnati	70	4,200
	Orangetown, Ont	7	420
	Buffalo, by boat	......	100
19.	Employees Massey-Harris Co., Toronto and Brantford, Ont	133	7,980

	Other sources....................	50	3,000
	Buffalo, by trolley...............	100	6,000
	Buffalo, by boat..................	......	300
22.	Reunion Eighth Heavy Artillery...	8	480
	Buffalo, by boat..................	......	250
23.	Yunger Maennerchor Club, Philadelphia..........................	2	120
	English Lutheran Church, Lockport, N. Y........................	8	480
	Lake Shore and M. S.............	12	720
24.	Western points, via Nickel Plate...	20	1,200
	Union excursion, lodges from Canada	20	1,200
	Corning and Elmira, via Erie R. R..	10	600
	Masonic excursion, Dansville, N. Y..	10	600
	D., L. and W...	12	720

1898.		Cars.	Persons.
Aug. 25.	Alliance, Youngstown and Cincinnati, Ohio	22	1,320
	Union churches and schools, Auburn, Syracuse, etc., N. Y.	14	840
	Order of Foresters, Cleveland, Ohio.	10	600
	County officials York county, Ont.	8	480
	Buffalo, by trolley	10	600
	Buffalo, by boat	......	75
	Toronto, by boat	......	400
26.	Special Erie excursion, via C. and B. Line	18	1,080
	Lehigh Valley	12	720
27.	Knights of Pythias returning from Indianapolis	10	600
	General half-holiday excursion	25	1,500
28.	Baltimore, Washington, Wheeling, etc	22	1,320
	Cleveland, Ohio, via C. and B.	20	1,200
	Buffalo, by trolley	15	900
	Buffalo, by boat	......	120
29.	All sources	15	900
	Buffalo, by boat	......	100
30.	Lehigh Valley	13	780
	Lafayette Ave. Sunday School, Buffalo, N. Y.	3	180
	Buffalo, Rochester and Pittsburg	10	600
	Buffalo, by boat	.....	150
	Toronto, by boat	......	420
31.	Rochester, N. Y., via West Shore R. R.	12	720

STONE ARCHED BRIDGE TO FIRST SISTER ISLAND, ERECTED 1898.

[illegible]		Cars.	Persons.
Aug. 31.	Cleveland and Akron, Ohio	14	840
Sept. 1.	Annual reunion Society Army of the Potomac	12	720
	Annual reunion Niagara County Veteran Ass'n....	6	360
	Newton Chautauqua party	8	480
	Hornellsville, N. Y., via Erie R. R.	12	720
	Dunkirk, N. Y., via Erie R. R.....	10	600
	Carbondale, Pa., via Erie R. R.....	12	720
	Buffalo, by boat..................	......	450
	Toronto, by boat..........	......	150
2.	Lehigh Valley	12	720
	Washington, Baltimore and Philadelphia....	14	840
	Corning, Elmira, etc., via Erie R. R.	12	720
	Western New York and Pennsylvania R. R..................	12	720
3.	Cleveland, Ohio...........	20	1,200
	Toronto, Ont....................	30	1,800
	General half holiday excursions....	25	1,500
	Buffalo, N. Y., by boat...........	......	100
	Toronto, Ont....................	......	500
4.	New York city, via West Shore R. R..........	14	840
	Main line Erie R. R...........	12	720
	Main line Lehigh Valley..	12	720
	Main line D., L. & W...........	11	660
	Cleveland, Ohio	20	1,200
	Rochester and Avon, via Erie R. R.	11	660

1898.			Cars.
Sept.	4.	Buffalo	60
		Buffalo, by boat	......
	5.	Labor Day excursions, all sources	167
	6.	Jamestown, N. Y	12
		Bradford, Pa	13
		Knights of Pythias convention, Buffalo, N. Y	5
		Hendrickson party, Brooklyn, N. Y.	3
		Raymond and Whitcomb party	3
		Toronto fair	30
		Toronto, by boat	......
	7.	Dunkirk and Allegany	12
		Toronto, via G. T. R	50
		Toronto, via boat	......
	8.	Cleveland, Akron, Cincinnati	15
		Salem, Ohio	12
		Toronto fair, via G. T. R.	50
		Toronto fair, via boats	......
	9.	Public Press Ass'n, Brooklyn, N. Y.	2
		G. A. R. delegates from Cincinnati	2
		Toronto, Ont., via G. T. R	67
		Toronto, Ont., via boats	....
		Buffalo, N. Y	42
		Lewiston, N. Y	45
	10.	New York Central R. R	30
		Buffalo, N. Y	30
		Buffalo, via boat	....
		Toronto, via boat	....
		Toronto, via Gorge Road	40

1898.	Cars.	Persons.
Sept. 11. N. Y. C., L. V. and W S., R, W. & O. and Erie R. Rs	203	12,180
Cleveland, Ohio	17	1,020
12. Foresters' convention	10	600
Return of Co. E, 3d N. Y. V. I., excursion	20	1,200
15. Traveling Locomotive Engineers	2	120
Locomotive Firemen's Ass'n	4	240
16. Wabash line	12	720
Locomotives Firemen's Ass'n, Toronto	3	180
Lakewood, N. J	3	180
17. Olean, Salamanca, and Bradford	9	540
18. Main line N. Y. C., W. S. and R., W. & O	60	3,600
Main line Erie and D., L. & W	22	1,320
Cleveland, Ohio, via C. & B	10	600
20. City officials Milwaukee, Wis	1	60
24. St. Luke's church and Sunday school, Brockport, N. Y	2	120
25. Sunday excursions, via N. Y. C., Lehigh Valley and Erie	150	9,000
27. Newark, N. J., Grocery Men's Association	5	300
28. Newark, N. J., Grocery Men's Association	6	360
30. Baltimore and Washington	14	840
Total	5,563	356,765

RECAPITULATION.

	Cars.	Persons.
1897.		
October excursions	123	7,380
1898.		
February excursions	1	60
April excursions	7	420
May excursions	139	8,340
June excursions	266	17,280
July excursions	1,448	92,240
August excursions	2,071	130,265
September excursions	1,508	100,780
	5,563	356,765

Report of the Treasurer

FOR THE

Year beginning Oct. 1, 1897, and ending Sept. 30, 1898.

Report of the Treasurer.

1897.			
Oct. 1.	Balance on hand this date		$758 53
	Receipts (Maintenance).		
Nov. 4.	Quarterly advance from the State Comptroller	$6,250 00	
1898.			
Jan. 26.	Quarterly advance from the State Comptroller	6,250 00	
Apr. 20.	Quarterly advance from the State Comptroller	6,250 00	
July 18.	Quarterly advance from the State Comptroller	6,250 00	
			25,000 00

Special appropriation as per Chap. 790, Laws of 1897.

1897.		
Oct. 8.	Payment by State Comptroller on account	$921 46
12.	Payment by State Comptroller on account	226 29
Nov. 8.	Payment by State Comptroller on account	904 30
Dec. 10.	Payment by State Comptroller on account	354 34
1898.		
Jan. 12.	Payment by State Comptroller on account	130 75

1898.			
Feb. 14.	Payment by State Comptroller on account	$89 50	
Apr. 12.	Payment by State Comptroller on account	410 18	
May 11.	Payment by State Comptroller on account	582 51	
12.	Payment by State Comptroller on account	83 25	
17.	Payment by State Comptroller on account	500 00	
June 7.	Payment by State Comptroller on account	1,444 07	
July 8.	Payment by State Comptroller on account	1,333 85	
			$6,!

Special appropriation as per Chap. 606, *Laws of* 1898.

July 8.	Payment by State Comptroller on account	$22 00	
1897.			
Aug. 8.	Payment by State Comptroller on account	984 56	
23.	Payment by State Comptroller on account	400 35	
Sept. 7.	Payment by State Comptroller on account	1,450 00	
15.	Payment by State Comptroller on account	2,206 12	
			5,0
Nov. 1.	Draft on Bank of Niagara for October receipts	$290 80	

The Wing, Riverway.

Dec. 1.	Draft on Bank of Niagara for November receipts	$49 50	
31.	Draft on Bank of Niagara for December receipts	28 60	
1898.			
Feb. 1.	Draft on Bank of Niagara for January receipts	37 40	
Mar. 1.	Draft on Bank of Niagara for February receipts..........	50 70	
31.	Draft on Bank of Niagara for March receipts	64 00	
May 2.	Draft on Bank of Niagara for April receipts..........	56 80	
June 1.	Draft on Bank of Niagara for May receipts	218 50	
July 2.	Draft on Bank of Niagara for June receipts..............	288 35	
Aug. 1.	Draft on Bank of Niagara for July receipts.......... ...	1,398 95	
Sept. 1.	Draft on Bank of Niagara for August receipts............	2,167 50	
30.	Draft on Bank of Niagara for September receipts.........	1,722 35	
			$6,373 45
Feb. 18.	Dividend on deposits in Cataract Bank.....................	$14 17	
June 27.	Dividend on deposits in Cataract Bank.....................	14 17	
			28 34
1897			
Dec. 31.	Interest on balances in Manufacturers and Traders' Bank....	$11 00	

1898.			
Mar. 31.	Interest on balances in Manufacturers and Traders' Bank	$24 97	
June 30.	Interest on balances in Manufacturers and Traders' Bank	25 81	
Sept. 30.	Interest on balances in Manufacturers and Traders' Bank	15 34	
			$77 12
	Total		$44,280 97

EXPENDITURES.

Date.	No. of abstract.	No. of voucher.	Items.	Amount.
1897.				
Nov. 1..	CV	1481	Pay-roll for October	$1,796 03
		1482	Thomas V. Welch, superintendent, office expenses	35 59
4..		1483	J. E. Perkins, repairs	3 90
		1484	Globe Ticket Co., Inclined Railway tickets	22 50
		1485	Power City Lumber Co., lumber	24 02
		1486	Niagara Sand Co., gravel	94 57
		1487	John Sullivan, stone	112 00
		1488	Timothy Horan	13 00
		1489	Coleman Nee	144 00
		1490	John Gordon, tools	9 80
		1491	Cataract Ice Co., ice	22 80
		1492	N. F. Hydraulic Power Manufacturing Co., electric lighting	50 00
11..		1493	Henry E. Gregory, treasurer and secretary, office expenses	28 19
12..		1494	Alex. Henschel, clerk to president.....	13 62
		1495	National Press Intelligence Co	16 75
			Total	$2,386 77

Date.	No of abstract.	No. of voucher.	Items.	Amonnt.
1897.				
Dec. 1	CVI	1496	Pay-roll for November	$1,448 70
		1497	Thomas V. Welch, superintendent, office expenses	47 85
6		1498	W. A. Shepard, mason work	30 25
		1499	N. F. Hydraulic Power and Manufacturing Co., electric lighting	50 00
		1500	Hardwicke & Co., tools, etc	8 04
		1501	McDonald & Welch, coal	30 50
		1502	Walker & Paterson, tools, etc	18 12
		1503	Walker & Paterson, tools, etc	15 73
		1504	Cornelius Burns, broken stone	9 00
		1505	H. Hugerman, plumbing	3 80
		1506	Jas. O'Brien, broken stone	89 00
		1507	C. Nee, broken stone	14 00
		1508	John Clifford, tools	6 94
Dec. 31		1509	Henry E. Gregory, treasurer and secretary, salary October, November and December	275 00
		1510	Pay-roll for December	1,320 66
		1511	Thomas V. Welch, superintendent, office expenses	47 05
			Total	$3,414 64
1898.				
Jan. 8	CVII	1512	P. C. Flynn & Son, painting	$94 48
		1513	J. W. Mead, brooms	12 75
		1514	O. V. Sage, agent and warden, letter heads, etc	6 25
		1515	J. E. Perkins, repairs	8 56
		1516	P. J. Davy, plumbing	110 26
		1517	Power City Lumber Co., repairs	12 08
		1518	Power City Lumber Co., repairs	21 14
		1519	Walker & Paterson, tools, etc	14 84
		1520	N. F. Hydraulic Power and Manufacturing Co., electric lighting	50 00

Date. 1898.	No. of abstract.	No. of voucher.	Items.
Feb. 1..		1521	....Pay-roll for January..................
		1522	....Thomas V. Welch, superintendent, office expenses, etc....................
9..		1523	....Walker & Paterson, tools, etc........
		1524	....Jas. Davy, stationery...................
		1525	....G. Chormann, tools....................
		1526	....N. F. Hydraulic Power and Manufacturing Co., electric lighting ....
		1527	....Braas Bros., raising ceiling ....
		1528	....Braas Bros., shelves, etc.............
Feb. 15..		1529	....Wm. Hamilton, Commmissioner, traveling expenses..........................
Mch. 1..		1530	....Pay-roll for February ....
		1831	....Thomas V. Welch, superintendent, office expenses ....
8..		1532	....Thos. E. McGarigle, repairing water wheel ....
		1533	....Power City Lumber Co., repairing Incline Railway building ....
		1534	....P. C. Flynn & Son, painting..........
		1535	....Walker & Paterson, tools, etc.........
		1536	....McDonald & Welch, coal.............
		1537	....Maloney & McCoy, ice................
		1538	....N. F. Hydraulic Power and Manufacturing Co., electric lighting.........
31..		1539	....Henry E. Gregory, treasurer and secretary, salary January, February and March ....
		1540	....Pay-roll for March....................
		1541	....Thomas V. Welch, superintendent, office expenses. ....
		1542	....Alex. Henschel, clerk to president, services.............................
		1543	....Hardwicke & Co., repairs.............
		1544	....Braas Bros., carpentering.............

Date. 1898.	No. of abstract.	No. of voucher.	Items.	Amount.
		1545	Braas Bros., carpentering	$148 34
		1546	Jas. Davy, stationery	3 25
		1547	Peter Lammerts, tools	40 20
		1548	Walker & Paterson, tools, etc.	33 60
		1549	N. F. Hydraulic Power and Manufacturing Co., electric lighting	50 00
		1550	Fred Batchelor, seed	18 00
				$5,933 11
April 9	CVIII	1551	Omar V. Sage, agent and warden, stationery	$13 76
May 2		1552	Pay-roll for April	1,780 16
		1553	Thomas V. Welch, superintendent, office expenses	48 49
9		1554	W. A. Shepard, repairs to conduit	44 31
		1555	Thos. E. McGarigle, repairs and tools	12 63
		1556	Power City Lumber Co., lumber	12 13
		1557	Hardwicke & Co., repairs	43 93
		1558	Peter Henderson & Co., scythes	12 00
		1559	P. J. Davy, repairs	64 67
		1560	Walker & Paterson, tools, etc	22 44
		1561	P. C. Flynn & Son, painting	77 36
		1562	Jas. Davy, stationery, etc	10 20
		1563	J. McDonald, coal	15 00
		1564	N. F. Hydraulic Power and Manufacturing Co., electric lighting	50 00
		1565	Wm. Hamilton, Commissioner, traveling expenses	30 25
		1566	Henry E. Gregory, treasurer and secretary, office expenses	35 66
				$2,272 99
June 1	CIX	1567	Pay-roll for May	$1,795 12
		1568	Thomas V. Welch, superintendent, office expenses, etc	49 03

Date.	No. of abstract.	No. of voucher.	Items.	Amount.
1898.				
June 2..		1569	J. Mackenna & Son, shades, etc	$5 50
		1570	Hardwicke & Co., hose, etc	18 23
		1571	J. H. Ellenbaum, plumbing	29 21
		1572	Braas Bros., carpentering	25 34
		1573	P. C. Flynn & Son, painting	62 47
		1574	F. W. Oliver & Co., tools	25 02
		1575	P. J. Davy, plumbing	48 87
		1576	Walker & Paterson, hardware, etc	58 91
		1577	Niagara Falls Gas Light Co., gas	17 64
		1578	N. F. Hydraulic Power & Mfg. Co., electric lighting	50 00
30..		1579	Henry E. Gregory, treasurer and secretary, salary April, May and June	275 00
			Total	$2,460 34
July 1..	CX	1580	Pay-roll for June	$1,798 48
		1581	Thomas V. Welch, superintendent, office expenses	49 27
5..		1582	Walker & Paterson, tools, etc	35 45
		1583	P. J. Davy, plumbing	126 77
		1584	Braas Bros., repairs	26 61
		1585	Braas Bros., repairs	208 94
		1586	P. C. Flynn & Son, painting	159 45
		1587	Hardwicke & Co., repairs	139 18
		1588	Charlotte Haeberle, tools	5 28
		1589	Power City Lumber Co., lumber	34 64
		1590	Omar V. Sage, agent and warden, stationery	6 25
		1591	N. F. Hydraulic Power and Mfg. Co., electric lighting	50 00
		1592	N. F. Gas Light Co., gas	13 32
Aug. 1..		1893	Pay-roll for July	1,799 80
		1594	Thomas V. Welch, superintendent, office expenses	37 52
3..		1895	Supplemental pay-roll	496 37

No. of abstract.	No. of voucher.	Items.	Amount.
3..	1596	Jas. McCarthy, tools	$25 00
	1597	Charlotte Haeberle, repairs	11 18
	1598	J. H. Cook & Co., repairs	10 28
	1599	N. F. Gas Light Co., gas	10 08
	1600	N. F. Hydraulic Power and Manufacturing Co., electric lighting	50 00
	1601	Niagara Wagon Works, tools	6 45
	1602	Carl Steinbrenner, repairs	2 16
	1603	Belmont Iron Works, park seats	181 50
	1604	Nat. Tent and Awning Co., canopy	50 00
	1605	Walker & Paterson, tools, etc	44 06
	1606	Braas Bros., bridge repairs	19 45
	1607	Braas Bros., bridge repairs	21 58
1..	1608	Pay-roll for August	1,497 56
	1609	Thomas V. Welch, superintendent, office expenses	19 11
	1610	W. F. Wall, cable for Inclined Railway	212 64
	1611	Globe ticket Co., tickets	45 00
	1612	P. C. Flynn & Son, painting	76 20
	1813	P. J. Davy, plumbing	19 71
	1614	Walker & Paterson, tools, etc	19 27
	1615	N. F. Gas Light Co, gas	4 14
	1616	N. F. Hydraulic Power and Manufacturing Co., electric lighting	50 00
	1617	W. A. Shepard, raceway repairs	37 93
	1618	Charlotte Haeberle, tools, etc	6 09
	1619	Pay-roll for September	1,376 23
	1620	Thomas V. Welch, superintendent, office expenses	14 29
	1621	Thomas P. Kingsford, Commissioner, traveling expenses	83 73
	1622	Henry E. Gregory, treasurer and secretary, salary July, August and September	275 00
		Total	$9,155 97

Payments out of $15,000 *as per chapter* 790, *Laws of* 1897.

Date.	No. of abstract.	No. of voucher.	Items.	Amount.
1897.	Series II.			
Oct. 14..	II........	13....	Pay-roll, walks and grading...........	$851 71
		14....	Samuel Parsons, Jr., traveling expenses.	25 50
		15....	Fred Batchelor, seed................	9 00
		16....	Fred Batchelor, seed................	9 00
		17....	W. A. Shepard, gutters	26 25
		18....	Niagara Sand Co., gravel.............	226 29
Nov. 10..		19....	Pay-roll, roads and grading...........	904 30
Dec. 10..		20....	Pay-roll, roads	354 34
			Total	$2,406 39
1898.				
Jan. 12..	III.......	21....	Pay-roll, walks.	$130 75
Feb. 15..		22....	Pay-roll, walks and grading	89 50
April 12..		23....	Pay-roll, grading....................	410 18
		24....	Pay-roll, grading....................	582 51
		25....	W. McCulloch, survey, etc............	58 50
		26....	Samuel Parsons, Jr.................	24 75
			Total	$1,296 19
May 17..	IV.......	27....	Walker & Morris, plans on bridge....	$500 00
June 7..		28....	Sam'l Parsons, Jr., plans for planting, etc.........................	400 00
		29....	Pay-roll, rods and planting...........	787 57
		30....	Elizabeth Nursery Co., shrubs.........	19 00
		31....	Ellwanzer & Barry, vines.	29 50
		32....	Andorra Nurseries, shrubs............	208 00
			Total	$1,944 07
July 11..	V	33....	Pay-roll, roads, grading, etc..........	$1,187 60
		34....	Parsons & Sons Co., shrubs...........	35 75
		35....	Sam'l C. Moon, shrubs................	40 00
		36....	George Cook, planting................	43 50
		37....	Sam'l Parsons, Jr., landscape architect	27 00
			Total	$1,333 85

Payments out of $15,000, *as per chapter* 606, *Laws of* 1898.

Date.	No. of abstract.	No. of voucher.	Items.	Amount.
1898.	Series I.			
July 11..	I........	1....	Cataract Publishing Co., advertising..	$2 50
		2....	Wm. Pool, advertising................	1 75
		3....	Buffalo News, advertising............	5 60
		4....	Gazette Publishing Co., advertising...	2 50
		5....	Buffalo Commercial, advertising	4 20
		6....	E. T. Williams, advertising...........	1 25
		7....	The Times, advertising..............	4 20
Aug. 9..		8....	Buffalo Courier, advertising	3 50
		9....	Buffalo Express, advertising	3 50
		10....	Pay-roll, bridge and grading..........	977 56
24..		11....	Pay-roll, grading....................	400 35
Sept. 8..		12....	W. A. Shepard, stone arched bridge...	1,450 00
16..		13....	Pay-roll............................	1,464 04
		14....	B. Messing, stone arched bridge	140 00
		15....	David Phillips, stone arched bridge ...	101 67
		16....	John D. Dietrich, tools	42 40
		17....	Conway & Munson, stone arched bridge	458 01
			Total	$5,063 03

Remittances to the State Treasurer.

1897.		
Nov.	1..Draft for October receipts..................................	$290 80
Dec.	1..Draft for November receipts	49 50
	31..Draft for December receipts	28 60
1898.		
Feb.	1..Draft for January receipts.................................	37 40
Mar.	1. Draft for February receipts	50 70
	31..Draft for March receipts	64 00
May	2..Draft for April receipts	56 80
June	1..Draft for May receipts	218 50
July	2..Draft for June receipts	288 35
Aug.	1..Draft for July receipts	1,398 95

Sept. 1..	Draft for August receipts	$2,167 50
30..	Draft for September receipts	1,722 35
	Total	$6,373 45
Mar. 1..	Dividend on deposits in Cataract Bank	$14 17
June 30..	Dividend on deposits in Cataract Bank	14 17
	Total	$28 34
1897.		
Dec. 31..	Interest on balances in Manufacturers and Traders' Bank.	$11 00
1898.		
Mar. 31..	Interest on balances in Manufacturers and Traders' Bank.	24 97
June 30..	Interest on balances in Manufacturers and Traders' Bank.	25 81
Sept. 30..	Interest on balances in Manufacturers and Traders' Bank.	15 34
	Total	$77 12
	Cash balance in bank	$134 71
	Grand total	$44,280 97

HENRY E. GREGORY,

Treasurer.

We, the undersigned, hereby certify that we have examined the foregoing report of the treasurer for the fiscal year ended September 30, 1898, the vouchers and other papers; and we find the report and accompanying documents correct, and that the treasurer has properly accounted for all moneys received and disbursed by him during the fiscal year ended September 30, 1898.

THOMAS P. KINGSFORD,

CHAS. M. DOW,

Commissioners of the State Reservation at Niagara.

....	$166 21
expenses............	63 85
t)..................	575 83
and clerk)................	98
....	00
....	1,571 27
....	3,196 53
....	2,315 34
....	2,423 62
....	1,178
....	105 50
....	550 00
....	83 80
....	18 00
lings	2,573 05
....	362 91
ges	142 78
surer and secretary..........................	1,100 00
r pipes..............................	282 92
nse......................................	80 87
bing	3 80
iture	81 88
ires..	5 50
ers	7 00
....	181 50
king fountains..........................	54 55
rs..	26 38
Day pond............................	397 50
	$25,623 82

Improvements under chapter 790, *Laws of* 1897.

Planting	$459 10
Roads	1,810 26
Grading	2,731 46
Walks	835 93
Bridges	558 50
Filling	208 75
Seed	18 00
Gutters	26 65
Shrubs and vines	332 25
	$6,980 50

Improvements under chapter 606, *Laws of* 1898.

First Sister Island bridge	$2,787 24
Grading	1,341 60
Deepening channel	365 05
Iron railing	77 2[illegible]
Hennepin's View	449 [illegible]
Tools	42 [illegible]
	$5,063 [illegible]

DEPARTMENT OF THE INTERIOR—U. S. GEOLOGICAL SURVEY
CHARLES D. WALCOTT, DIRECTOR

RECENT EARTH MOVEMENT

IN THE

GREAT LAKES REGION

BY

GROVE KARL GILBERT

EXTRACT FROM THE EIGHTEENTH ANNUAL REPORT OF THE SURVEY, 1896-97
PART II—PAPERS CHIEFLY OF A THEORETIC NATURE

CONTENTS.

ILLUSTRATIONS.

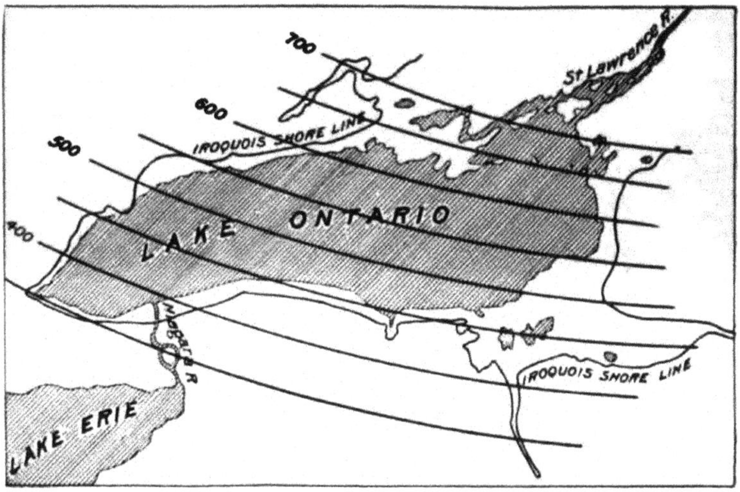

FIG. 94.—Map of the Iroquois shore line. Modern water bodies are shaded. A line shows the boundary of the ancient lake. The parallel curves are isobases.

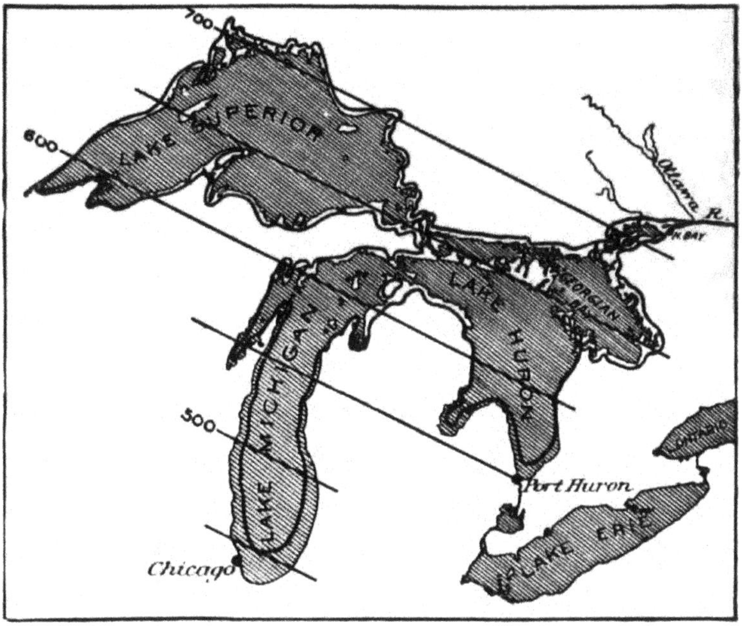

FIG. 95.—Map of the shore line of Great Lake Nipissing. Modern water bodies are shaded. A line shows the boundary of the ancient lake. The parallel lines are isobases.

Mrol

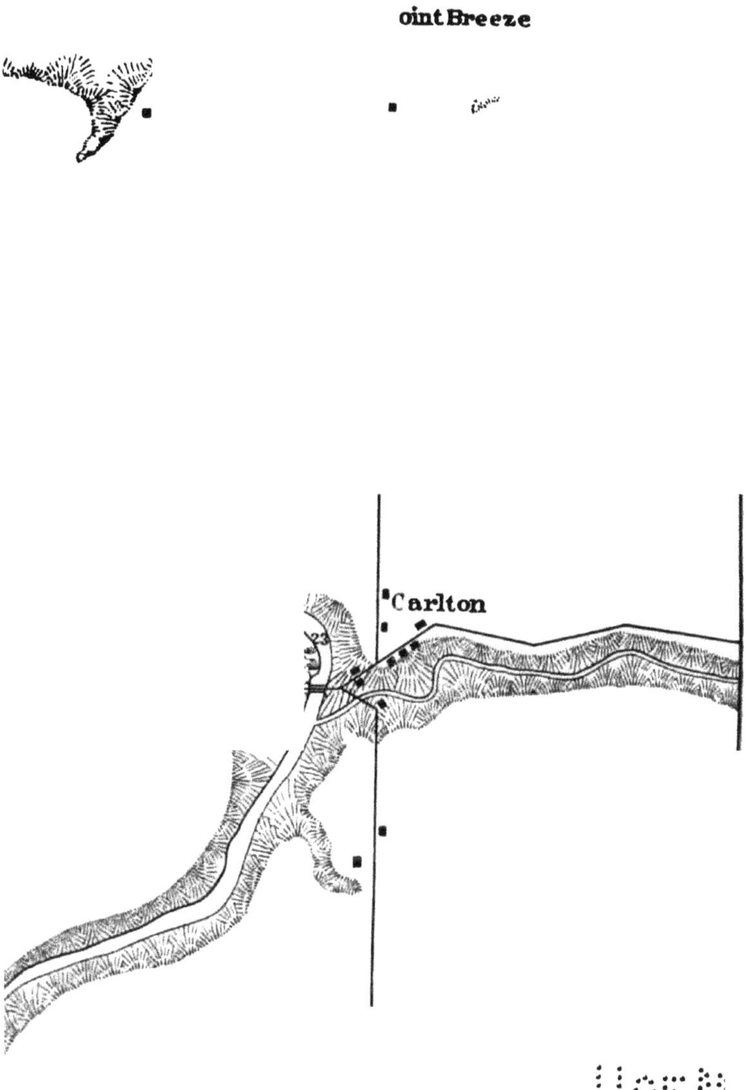

UARY AT THE MOUTH OF OAK ORCHARD CREEK, ORLEANS COUNTY, NEW YORK.
Scale, 1 inch=1,500 feet. Figures give soundings in feet.
—ays are sharply incised in the plain. Partial refilling is shown by r
'r at lake level reaches to Carlton, above which the creeks are sha

M7ol

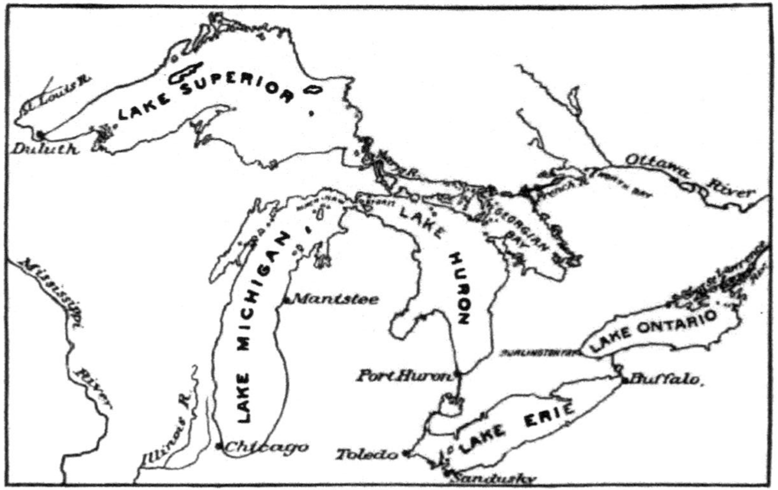

FIG. 93.—Map of Laurentian lakes.

MroU

RECENT EARTH MOVEMENT IN THE GREAT LAKES REGION.

By G. K. Gilbert.

INTRODUCTION.

The geologic history of the earth shows that in all parts of its surface there have been great oscillations of level. Modern history also records upward and downward movement of the land at various points. The modern movements are of small amount, but it is believed that they are of the same kind as the ancient, and that the great changes of the geologic past were effected slowly. Nearly all discoveries of modern change have been made at the seashore, but there is no reason to suppose that the land is now more stable in the interior of the continents than along their coastal borders. Observations are restricted to the coast because the sea level affords the best available datum plane for comparison. The present paper discusses the stability of the region of the Laurentian lakes, and uses the surfaces of the lakes as datum levels or planes of reference.

OBSERVATIONS BY MR. STUNTZ.

In 1869 there was presented to the American Association for the Advancement of Science a paper by G. R. Stuntz, a land surveyor of Wisconsin, describing certain observations on Lake Superior made by him in 1852 and 1853. He states that in those years a certain mill race at the falls of St. Marys River was entirely dry. As St. Marys River is the outlet of Lake Superior, its volume and the supply of water for the mill race depend on the height of water in the lake, and he therefore inferred that at that end of the lake the water was low. He also states that a small stream at Pindle's mill, entering Lake Superior not far from the outlet, runs with swift current to the lake, and has no widening, marshes, or other indication that its valley overflows by the lake setting back into it. He then describes the strongly contrasted condition of streams entering the lake near its western end.*

As you go westward, the Ontonagon River exhibits a slight filling up. The valley near the mouth shows that at the time it was excavated the surface of the lake was lower than at present. The same is also apparent at the mouth of Bad River, still farther west. At the mouth of Boise Brulé the same thing is exhibited, only to a greater extent. From this to the west end of the lake not only does the lake set back into the valleys of the streams, but the waters are making rapid encroachments on the banks. So rapidly is the filling back, that the deposits of the streams do not keep pace with the filling up. The consequence is, that there is a large marsh and pond in the mouth of the valley of Boise Brulé and Aminecan River. But nowhere is this filling up more apparent than in the bay above the mouth of the St. Louis River. In several parts submerged stumps, several feet below the present water level, are found. The numerous inlets surrounding the main bay, when we consider the nature of the soil and the formation (a tough, red clay), in all of which the water is deep, could not have been excavated in the natural course of events with the water at its present level. The testimony of the Indians also goes to strengthen the same conclusions. At the time of running the State line above mentioned, the Indians, ever jealous of their rights, called me to a council to inquire why I ran the line through Indian land. In the explanation, I gave, using the language of the law, as a starting point, the lowest rapid in the St. Louis River. The chief immediately replied that formerly there was a rapid nearly opposite the Indian village. Start, said he, from that place, and you will be near the treaty line. After he had been further questioned, I

* On some recent geological changes in northeastern Wisconsin: Proc. Am. Ass. Adv. Sci. Vol. XVIII, 1870, pp. 206-207.

learned that it was only a few years since the river was quite rapid at the Indian village. At the time the said line was run the first rapid was about one mile by the stream above the village. From these facts I conclude that a change is taking place gradually in the level of this great valley.

From these data Stuntz infers "the gradual rise of water at this [west] end of the lake and the falling of the same at the east," and it is evident from the context that he refers this change of the water to a westward canting of the basin, the western part becoming lower as compared with the eastern.

So far as I am aware, this paper broaches for the first time the idea of differential elevation in the Great Lakes region, and it contains the only observations that have ever been cited as showing recent changes of that character. In later years the subject has been approached from the geologic side, and Dr. J. W. Spencer has expressed his opinion that a warping or tilting of the whole region is now in progress.

EARTH MOVEMENTS DURING THE CLOSING EPOCHS OF THE PLEISTOCENE PERIOD.

The Great Lakes came into existence in the latest of the geologic periods, the Pleistocene. Their number and position underwent numerous and important changes during the latter part of the period, and their area and drainage systems have been greatly modified even within the time to which human history belongs. In late Pleistocene time, the great Laurentide ice field, which just before had covered the entire lake basin, was slowly growing less through the melting away of its edges, there were a series of lakes along its southern margin. These were held in at the north by ice and on other sides by uplands, and they found outlet southward over the lowest passes of the divide between the Laurentian basin and the basins of the Mississippi, Susquehanna, and Hudson. With changes in the position of the ice barrier, individual lakes were from time to time divided or drained and separate lakes united, so that the lacustrine geography had a complex history. After the ice had wholly disappeared from the region, the drainage did not at once assume its present system, for Lake Huron, instead of overflowing to Lake Erie, discharged its surplus water over the pass at North Bay, Canada, and thence down the Mattawa and Ottawa Rivers to the St. Lawrence.

In the decipherment of this history much use is made of the shore lines of the vanished lakes. These consist of sand and gravel terraces that were once deltas, of cliffs and strands carved from hillsides by the waves, and of spits and beach ridges thrown up by the same agency. A number of these lines have been traced for great distances, and wherever thus traced it is found that they are no longer level, but are gently inclined. When formed they were of course horizontal, for they were made by waves generated on a water surface, and the fact that they are not now level shows that the land on which they are marked has undergone changes of relative height. The general direction of inclination of the shore lines is toward the south-southwest, showing that the basin of the lakes has been canted in that direction. The amount of change has not been everywhere the same, and it is probable that the direction of the canting varies somewhat from place to place. Where several shore lines are traced on the same slope the first made are usually more steeply inclined than the last made, and hence it is inferred that the general change of relative altitude was in progress through the whole epoch of the glacial lakes. The plane of the Iroquois shore line in the basin of Lake Ontario descends toward the south-southwest at an average rate of 3½ feet a mile, the slope being steeper at the north than at the south.* The Oswego shore

*J. W. Spencer, Trans. Roy. Soc. Canada, Section IV, 1889, pp. 121-131; G. K. Gilbert, Sixth Ann Rept. Commissioners of the State Reservation at Niagara, Albany, 1890.

line, in the same basin, slopes in the same direction at the rate of more than 3 feet a mile. The Warren shore line, traced from Lima, New York, about the sides of the Ontario, Erie and Huron basins, to Pompeii, Michigan, is nearly level in the Maumee basin, but rises northeastward with a rate gradually increasing to 2 feet a mile. Its northward rise in Michigan is 1½ feet a mile.* The present southward inclination of the water plane of Lake Algonquin, which occupies the Superior, Michigan, and Huron basins, ranges from a few inches to 3 feet a mile.† Great Lake Nipissing, which occupied the same basins after the disappearance of the ice and had its outlet at North Bay, conformed more nearly to the present slopes, the general inclination of its water plane being about 7 inches to the mile.‡

On the accompanying maps of Lake Iroquois and Great Lake Nipissing (figs. 94 and 95) the character of the tilting is shown by means of isobases, or lines drawn at right angles to the direction of tilting. All points on one of these lines have been uplifted the same amount since the time of the corresponding lake. If we think of the plane of the water surface of one of the old lakes as having been deformed by uplift or warping, then the isobases are contours, or lines of equal present height, on the deformed plane.

Other evidence of the tilting of the land is found in the character of stream channels as they approach lake shores. The streams reaching Lake Superior from the southwest have already been described in the quotation from Stuntz, and similar characters are found in the basins of Lake Erie and Lake Ontario. Considerable tracts of land along the southern shores and about the western ends of these lakes are smooth plains, there surfaces having been leveled by deposits of fine sediment from the Pleistocene lakes just men-

* F. B. Taylor, Bull Geol. Soc. America, Vol. VIII 1897, p. 55.
† F. B. Taylor, A Short History of the Great Lakes, Terre Haute, 1897.
‡ Ibid.

tioned. The creeks and rivers traversing the plains have readily cut the soft deposits, carving out narrow valleys. In the upper parts of these valleys the streams are shallow and descend with lively current, but on approaching the lake they become deep and sluggish, the change usually occurring several miles from the lake shore. Stated in another way, each stream, instead of debouching into the lake, enters the head of a long, narrow bay or estuary. The origin of such estuaries is well understood. They are found on all sinking coasts, and their meaning in this region is that the land has gone down or the lake level has risen, so that the waters of the lake occupy portions of the channels carved by the streams in the lowland plain. This description applies to the greater number of streams entering Lake Ontario between the Genesee and Don Rivers and to those entering Lake Erie between Cuyahoga River and Maumee Bay. Individual mention may be made of Oak Orchard, Eighteenmile and Twelvemile Creeks in New York, of Twelvemile and Twentymile Creeks and the Credit and Humber rivers in Ontario, and of Rocky River, Black River, Vermillion River, Old Womans Creek, Pike Creek, Turtle Creek and Ottawa River, in Ohio. Even the largest rivers of the district, including the Genesee, Niagara, Cuyahoga and Maumee, have features indicative of the same history.*

By reference to the map (fig. 93) it will be seen that these outlets of these lakes are at their northernmost points, and this fact is related to the conditions of the stream channels. The water level of a lake is maintained by the balance between inflow and outflow. It is just high enough to enable the outflowing stream to carry off the excess from inflow, and the height of water on all shores is thus determined by the height of the outlet. So if these basins are canted northward the outlets are thus lowered with reference to

* F. B. Taylor mentions a few other localities on the same lakes: Am. Geologist. Vol. XV, 1895, pp. 174-176.

other parts, and the waters recede on the southern shores. If they are canted southward, the outlets are raised and the waters are made to advance on the southern shores. Reasoning from effect to cause, the fact that the lake water invades the new-made stream channels on the southern shores is evidence of the southward canting.

It should not be assumed that the "drowning" of stream channels is restricted to the tracts mentioned above. Those tracts are specified because they fall within the range of the writer's personal observation and are known to exhibit the phenomena in a striking way. It is believed that similar features may be found wherever the local conditions are favorable throughout the whole coast lines of Lake Ontario and Lake Erie, about the head of Lake Michigan from Manistee, Michigan, to Kewaunee, Wisconsin, and about the whole of the American shore of Lake Superior.

REASONS FOR REGARDING A PROGRESSIVE MODERN CHANGE AS PROBABLE.

Independent of the phenomena described by Stuntz, there are various considerations tending to direct attention to the question of the stability or instability of the Laurentian area at the present time. The first to be mentioned is purely geologic. The epoch during which the overflow from the upper lakes followed the valleys of the Mattawa and Ottawa is definitely associated with a certain stage of the Niagara River. The cataract of Niagara is at the present time increasing the length of the Niagara gorge at a somewhat rapid rate. The formation from which the water leaps is a firm limestone 60 feet thick, and beneath this are shales which are comparatively soft and weak. The cataract, by eroding the shale, undermines the limestone, which falls away in blocks, and these blocks are in turn utilized by the water as an instrument with which to grind the shale. Whirled about by the water, the blocks not only wear away the face of the shale cliff, but drill down deeply, so that beneath the cataract there is a pool nearly or quite 200 feet deep. Working in this way, the cataract has extended the gorge several hundred feet since the first accurate measurements were made, the average annual rate being between four and five feet.

With the present arrangement of the drainage system the Niagara carries the surplus water from the basins of lakes Huron, Erie, Michigan and Superior; but when the upper lakes sent their overflow to the St. Lawrence by way of the Ottawa, the Niagara carried only the discharge from the Erie basin. Its volume was only one-eighth of the present volume and its power was correspondingly less. It could not move the great blocks of limestone which fell from the cliff, and, instead of scooping out a deep pool, as now, it excavated a comparatively shallow channel, whose bottom was cumbered with limestone débris. Owing to this difference in method of erosion it is possible to discriminate the parts of the gorge excavated when the river was small and when it was large, and thus to determine the place of the cataract when the outlet of Lake Huron was shifted from North Bay and the Ottawa River to Port Huron and the St. Clair and Detroit Rivers. That place is at the head of the Whirlpool Rapids, 11,600 feet from the present cataract. Assuming that the cataract worked at its present rate through this distance, we may compute the time consumed. At four and one-half feet a year, it would be about two thousand six hundred years. F. B. Taylor, making allowance for various qualifying factors, estimates the time to have been not less than five thousand years.*

When Lake Huron changed its outlet, the plane of its water surface extended from the pass at North Bay to the pass at Port

---

* A short history of the Great Lakes.

Huron, but the North Bay pass now stands 140 feet higher than the Port Huron. This difference of altitude, amounting to six inches a mile, has, therefore, been wrought within the period of about five thousand years. In view of the gradual nature of such movements, this is not a long period to assign to the measured change, and it is natural to inquire whether the movement is not still in progress.

Dr. J. W. Spencer, who has devoted much time to the study of the Niagara gorge and the glacial lakes, is confident that change of level has not yet ceased, and that it will eventually turn the water of the upper lakes southward to the Illinois and Mississippi rivers, leaving the Niagara channel dry. Addressing the American Association for the Advancement of Science in 1894, he said: *

The end of the falls seems destined, if we read the future by the past, to be effected, not by the erosion expending itself on the rocks, but by terrestrial deformation turning the drainage of all the upper lakes into the Mississippi, by way of Chicago, just as the Huron waters were lately turned from the Ottawa into the Niagara drainage; and at the recent rate it would seem that about five thousand or six thousand years at most will be needed. The change of drainage should arrive before the cataract shall have receded to Buffalo.

Another consideration of the same tendency is found in the condition of the estuaries described in the last section. Most of the streams flowing into these rise in districts of unconsolidated drift and carry forward in flood time a considerable load of detritus. This is deposited in the estuaries, the coarser part making deltas at their heads, and the finer settling as mud in the deeper water. The process tends to convert the estuaries, first to marshes and then to dry land, but in most instances little

*Proc. Am. Ass. Adv. Sci., Vol. XLIII, 1894, p. 246.

progress in that direction has been made. There are a few creeks rising in sandy districts which have succeeded in filling their estuaries, changing them to marshes; but as a rule the delta at the head of the estuary invades it but a short distance, and the marshes which border it here and there at points sheltered from the flood currents are impassable except by boats, and have the appearance of submerged flood plains. These characters, from their close resemblance to the features observable along the subsiding parts of the Atlantic coast, give the impression that a slow flooding of the stream valleys is still in progress.

A third consideration is connected with the record of recent changes on the coasts of the continent. It has long been known that the Atlantic coast south of Connecticut is subsiding, and Prof. G. H. Cook was able to determine the rate in New Jersey as about two feet a century.* Dr. Robert Bell has recently collated a variety of facts tending to show that the land has risen in the region about Hudson and James bays,† and he estimates the rate at from five to seven feet a century. If these two movements are parts of a general movement affecting the northeastern part of the continent, then the Great Lakes region, approximately intermediate in position between the rising and sinking areas, should be found to exhibit a southward tilting.

These various facts, all tending in one direction, are sufficient warrant for the working hypothesis that the tilting of the lake region demonstrated by the slopes of the old shore lines is still in progress; and the writer, who has for many years been interested in the problems of the Great Lakes, has made repeated efforts to secure an investigation by which the hypothesis might be tested.

---

* Am. Jour. Sci., 2d series, Vol. XXIV, 1857.
†Am. Jour. Sci., 4th series, Vol. I, 1896.

The mode of investigation first suggested was the establishment of elaborate observation stations at three points — Port Huron, Chicago and Mackinac. By a suitable series of observations at these points, the relative heights of benches might be established with high precision, the water surface being used as a leveling instrument. Then, after an interval of one or two decades, the observations might be repeated and any changes in the heights of benches due to differential uplift detected. The matter was submitted in 1890 to the Superintendent of the United States Lake Survey and to the Superintendent of the United States Coast and Geodetic Survey, but, though it was received favorably by the latter officer, the work was not undertaken.

Other plans were then considered, and it was finally decided to make a study of existing records of lake level, and, if necessary, supplement them by additional observations. The results of this investigation are set forth in the following pages.

GENERAL PLAN OF INVESTIGATION.

Variations in the height of the ocean level at any place depend chiefly on tides, winds, and atmospheric pressure. By means of long series of observations the effect of these disturbing factors can be eliminated and a mean level obtained which is practically uniform from year to year and decade to decade. The height of the water surface must depend also on the quantity of water in the ocean, but the actual variations of volume are so small as compared to the extent of the ocean surface that the resulting variations of level may be neglected and the mean level used as a standard for the discussion of differential movements of the earth's crust. With the Great Lakes the case is materially different. There are variations due to wind, atmospheric pressure, and tides, but when these have been eliminated by long series of observations the resulting mean level is far from constant, varying from season to season and year to year with the volume of water. In each lake there is an annual change of more than a foot, depending on the seasonal inequality between gain by precipitation and loss by evaporation (fig. 96), and there is a still greater change resulting from the cumulative effect of series of dry and series of moist years. The records show that the water surface in each lake has been several feet higher in some years than in others. (See fig. 97.)

For this reason the water surface of a lake does not afford a datum plane by reference to which the elevation or subsidence of coasts can be directly determined. Fortunately, however, there is an indirect method by which practically the same result may be attained. If the mean level of a lake surface be determined for two parts of the coast at the same time, these two planes may be regarded as parts of the same level surface, and, through reference to this common datum, fixed objects on the land at the two localities can be compared with each other so as to determine their relative altitudes. If, then, after an interval of time, the measurements are repeated, a change in the relative height of the fixed objects may be discovered and measured. The investigation described in the following pages made use of this method.

The fundamental principle of the method is illustrated by the diagram, fig. 98, in which A C B is the cross profile of a lake basin. At a certain time the mean plane of the water surface occupies the position XX′. By means of the engineer's level it is ascertained that a bench mark A has a certain height above the water plane at X, and that a bench mark B has a certain height above X′. The difference between these two measurements is the difference in altitude between A and B. After an interval of years the measurements are repeated. The water plane then stands at some different level, say YY′. The height of A above Y is measured, and the height of B above Y′; the difference between the two measurements gives the relative height of A and B. If earth movements have occurred during the interval between the two sets of measurements the second determination of the comparative height of A and B will differ from the first determination, and the amount of difference will measure the differential earth movement.

AVAILABLE DATA.

Gage readings.—In order to eliminate the temporary effects of disturbing factors, it is necessary to have a series of observations of the height of the water surface at each of the localities compared. The gages by means of which such observations are made are of various kinds. One of the simplest is a graduated plank, fixed vertically, by attaching it to a dock or other structure, so that one end is above water and the other below. Sometimes the plank is omitted and the graduation marked upon the side of a dock or pier. The height of the water surface is ascertained by direct comparison with the lines of the graduation. Another form of gage which has been extensively used in the lakes consists of a graduated rod, not fixed, but held in the hand; with this the distance from the water surface to a fixed point is measured. Usually the fixed point chosen is above the water surface, but at one station, Port Colborne, it is the submerged sill of a canal lock. Another form of gage includes a float to which a graduated vertical rod is fixed, and the graduations of the rod are compared with a fixed point on the land; or a chain attached to the float may pass over a pulley and carry a counterpoise, in which case an index, fastened to some part of the chain or counterpoise, moves up and down past a stationary graduated scale. There are also automatic gages making periodic or continuous records.

Previous to the year 1859 records of lake level are meager, and not of such nature as to be suited to the purposes of this investigation. A general account of them is given by Col. Charles Whittlesey in Volume XII of the Smithsonian Contributions to Knowledge, and a fuller account in the Report of the United States Deep Waterways Commission for 1896. In 1859 the investigation of lake levels was undertaken by the United States Lake Survey. Several stations were established on each lake, and at these regular observations were made, usually three times a day. From time to time stations were discontinued and others were established, and after the close of the field work of the Lake Survey the observations were, in many cases, continued by officers of the Engineer Corps in charge of harbor improvements. With reference to the present investigation, I have examined United States Lake Survey and other United States engineer records for the following stations:

On Lake Superior: Superior City, Duluth, Ontonagon, Marquette and Sault Ste. Marie.

On Lake Michigan-Huron: Chicago, Milwaukee, Grand Haven, Lockport, Sand Beach, Port Austin, Ponte aux Barques, Tawas, Escanaba, Thunder Bay.

On Lake Erie: Monroe, Rockwood, Cleveland, Ashtabula, Erie, Buffalo.

On Lake Ontario: Port Dalhousie, Niagara, Charlotte, Oswego, Sacketts Harbor.

These records are for the most part published in the form of monthly means, but the individual observations are preserved in the Engineer Office at Washington, and these have been made accessible to me through the courtesy of Gen. William P. Craighill, Chief of Engineers. By the Canadian Department of Railways and Canals I have been enabled to make use of a long series of observations at Port Colborne, on Lake Erie, the head of the Welland canal; observations at Toronto, on Lake Ontario, have been furnished me by the city engineer, and observations at Collingwood, on Lake Huron, by Mr. Frank Moberly.

Benches.—As gages at the water side are subject to various accidents, it is rarely possible to maintain their zeros for long periods at a constant level, and unless they are connected by leveling with bench marks of a permanent character their records have little value for purposes of comparison. Previous to 1871 such connection with benches was not made by the United States Lake Survey, or if made, the records are lost. There were, however, certain stations, notably Chicago, Milwaukee, Cleveland, Port Colborne, Buffalo, Charlotte and Oswego, at which this matter had received attention. The structures at Chicago on which the early bench marks were made are thought to have afterwards settled.* At Milwaukee the early bench marks no longer exist, and although there is reason to believe that other benches were substituted with care, my researches have not discovered a satisfactory record. The same remark applies to Buffalo; and the record of the original bench at Charlotte has been lost. At Port Colborne and Oswego the zero of gages are permanent structures, which have probably suffered no change; and at Cleveland, although the oldest benches no longer exist, it is believed that the record of transfer is complete and satisfactory.

In 1870 Gen. C. B. Comstock was placed in charge of the United States Lake Survey, and the scientific methods introduced by him included the establishment of a complete system of benches in connection with the gages. From 1872 until the completion of the field work of the Lake Survey there was an annual inspection of the gages, and the relations of their zeros to the bench marks were

---

* Report on Chicago City Datum and City Bench Marks, by W. H. Hedges, Chicago, 1895.

redetermined as often as seemed necessary. From 1871 to 1878 the supervision of gages and the reduction of records were in charge of Mr. O. B. Wheeler, and from 1879 to 1882 of Mr. A. R. Flint. The results of the present investigation are largely indebted to the care and thoroughness with which these engineers performed their work.

SELECTIONS OF STATIONS AND YEARS.

Under the general method outlined above, the first step was the selection of suitable pairs of stations on the shores of the various lakes. As the geologic data indicated a tilting of the land toward the south-southwest, or, more precisely, in the direction S. 27° W., it was desirable to have each pair of stations separated by a long distance in that direction. As the hypothetic change was exceedingly slow, it was desirable to compare observations separated by the longest practicable time intervals. It was essential that the gage readings before and after the time intervals be accurately connected with the same benches. Consideration was also given to the fact that the results might be vitiated if use were made of observations taken during the prevalence of storms, when the water is sometimes driven by the wind so as to stand abnormally high on certain shores; and in order that the use of such observations might be avoided it was important to select years during which the force of the wind was daily recorded. With these considerations in view the available data were examined, and the following selection was made of stations (see fig. 99) and years:

For Lake Ontario, Charlotte and Sacketts Harbor, 1874 and 1896.

For Lake Erie, Cleveland and Port Colborne, 1858 and 1895.

For Lake Michigan-Huron, Milwaukee and Port Austin, 1876 and 1896, and Milwaukee and Escanaba, 1876 and 1896.

No comparison was undertaken for stations on Lake Superior.

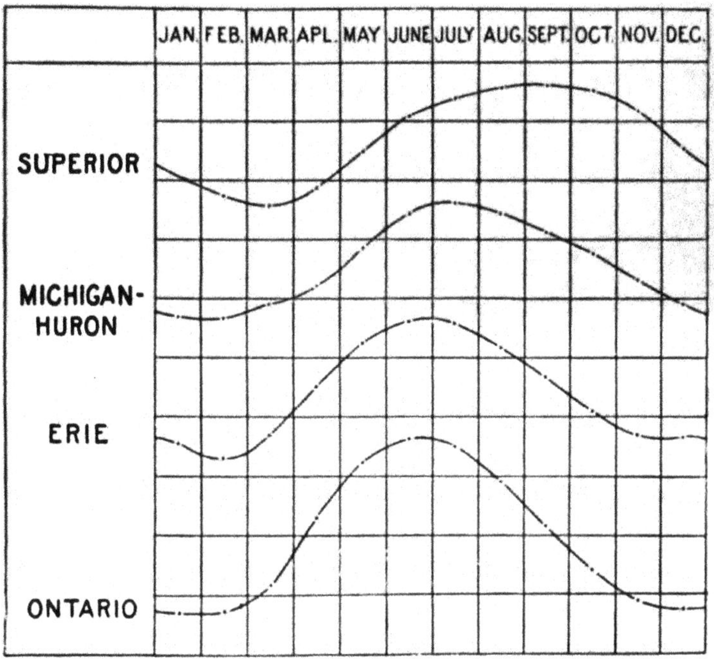

FIG. 96.—Annual oscillations of the surfaces of the Laurentian lakes. Compiled from monthly means published by the Chief of Engineers, U. S. A. Each vertical space represents six inches. The observations for Lake Superior cover the period 1862-1895, for Michigan-Huron, 1860-1895, for Erie, 1855-1895, for Ontario, 1860-1895.

١

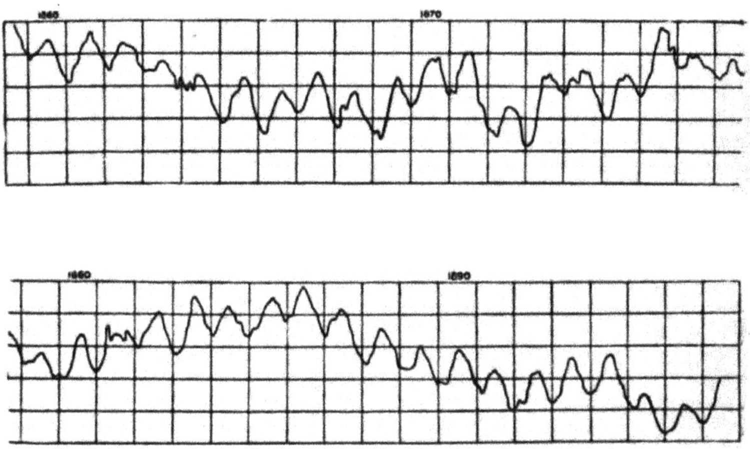

Fig. 97.—Oscillations of the surface of Lake Michigan due to changes in the volume of the lake. Compiled under the direction of the Chief of Engineers, U. S. A., from gage readings at Milwaukee, Wisconsin, from August, 1859, to June, 1897. Each horizontal space represents a calendar year; each vertical space, one foot.

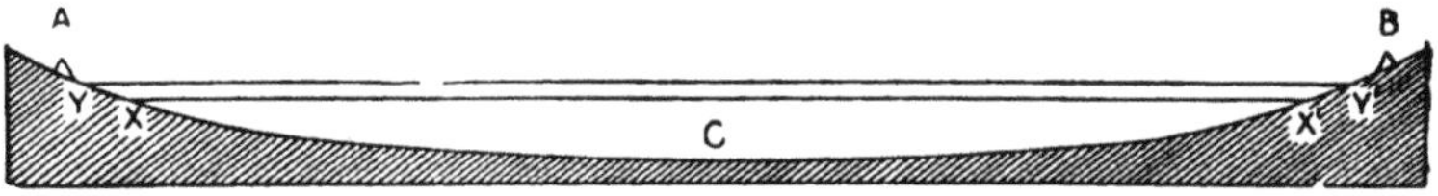

Fig. 98.—Diagram illustrating method of measuring earth movements.

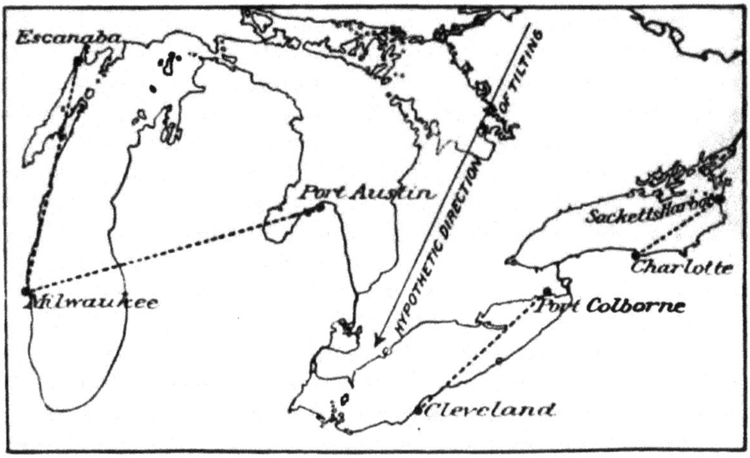

FIG. 99.—Arrangement of selected stations.

SPECIAL OBSERVATIONS IN 1896.

Certain of the selected stations are not now maintained by the United States engineers, and in order to complete the data it was necessary to make special observations. This was done in the summer of 1896, during the months of July, August, September and October. The necessary attention was also given to bench marks, and provision was made for observations of a special character at the regular engineer stations.

At Sacketts Harbor use was made of a gage which had been established for temporary purposes by Maj. W. S. Stanton, U. S. E. It was connected, by leveling, with an old bench mark, and an observer was employed. At Charlotte the relation of the gage zero to a bench mark was determined by leveling, and special series of observations were made by the observer employed by the United States engineers. At Port Austin a new gage was established and a special observer employed. At Milwaukee and Escanaba the relative heights of gages and bench marks were determined under direction of Capt. George A. Zinn, U. S. E., and special observations were made by the observers regularly employed by the United States engineers.

The "special" observations at these stations consisted of series of readings intended to eliminate the effect of the oscillations called "seiches." The equilibrium of a lake surface is disturbed not only by winds, which blow the water toward the lee shore, but by inequalities of atmospheric pressure occurring during thunderstorms and during the passage of cyclones; and the impulses thus received are not quickly dissipated, but cause a long-continued swaying of the water. In large lakes these oscillations are so enduring as to cover the interval from one disturbing impulse to another, and keep the water perpetually in motion. Near the ends of the lakes and in bays with gradually converging sides the range of oscillation may be as great as 1 foot, and it [illegible] from 1 inch to ½ inch. For this reason a single observation may not approximate closely to the mean level of the water, and the actual mean level was determined [illegible] observations at short intervals. In arranging the work of [illegible] the observers were instructed to record the water level every five minutes for an hour each morning and evening of all days when the wind was light; and at Sackets Harbor, where the seiche has an exceptionally long period, the length of the series was afterwards increased.

DISCUSSION OF DATA FROM PAIRS OF STATIONS.

SACKETTS HARBOR AND CHARLOTTE.

In 1874 the zeros of gages at these stations were points marked on docks, and readings were made by means of graduated vertical rods attached to floats. They give the distance of the water surface below the gage zeros. At the time of each observation record was also made of the direction and force of the wind. The work was under the direction of the United States Lake Survey. Mr. A. Wilder was the observer at Charlotte, and Mr. Henry Metcalf at Sacketts Harbor.

The gage at Charlotte was put in place in November, 1871, and the measurements showed its zero to be 32.7 feet below a bench mark. In January, 1873, its zero was found to be 32.959 feet below the same bench mark. On May 11, 1874, it was again compared with the bench mark, and the difference was found to have increased to 33.003 feet. It is probable that this change of .044 foot was occasioned by the settling of the dock to which the gage was attached. A manuscript report dated February 3, 1875, says: "The bank is here partly of timbers and partly of earth. The earth has been washed out and has fallen away from the timber in some places." The gage at Sacketts Harbor was also found unstable. The report of an inspection in May, 1874, states that the zero of gage "has been lowered 0.555 foot;" and a report dated February 3, 1875, says: "This gage is fastened to the timbers of an old and unused dock. The instability of gages determined the selection of time for the comparison of stations. Both gages having been compared with benches in May, 1874, that at Charlotte on the 11th and that at Sacketts Harbor probably on the 14th, the computations were based on a period including these dates. Within this period selection was made of those times of observation when the wind force at both stations was less than 3 on a scale of 10. Thus treated, the observations of 54 days gave 51 comparisons.

Computation of the height of the gage zero at Sacketts Harbor, New York, above the gage zero at Charlotte, New York, in the spring of 1874.

DAY.	Hour.	Gage Reading.		Difference.
		Sacketts Harbor.	Charlotte.	
1874.		Feet.	Feet.	Feet.
April 17	9 p. m.	5.43	3.15	2.28
19	9 p. m.	5.33	3.01	2.32
22	9 p. m.	5.23	2.98	2.25
23	7 a. m.	5.16	2.80	2.36
	2 p. m.	5.18	2.78	2.40
24	7 a. m.	5.20	2.80	2.40
25	7 a. m.	5.18	2.80	2.38
	2 p. m.	5.11	2.81	2.30
27	7 a. m.	5.08	2.78	2.30
	9 p. m.	4.88	2.78	2.10
28	2 p. m.	5.13	2.87	2.26
	9 p. m.	5.06	2.88	2.18
May 3	7 a. m.	5.06	2.95	2.11
4	7 a. m.	5.43	2.95	2.48
	2 p. m.	5.09	2.94	2.15
5	9 p. m.	5.16	2.94	2.22
6	9 p. m.	5.18	2.91	2.27
7	9 p. m.	5.12	2.85	2.27
8	7 a. m.	5.06	2.87	2.19
11	9 p. m.	5.17	2.85	2.32
14	9 p. m.	5.17	2.82	2.35
15	9 p. m.	5.12	2.83	2.29
	7 a. m.	5.17	2.84	2.33
	2 p. m.	5.25	2.82	2.43
18	2 p. m.	5.13	2.82	2.31
	9 p. m.	5.18	2.83	2.35
20	7 a. m.	5.05	2.83	2.22
	2 p. m.	5.10	2.82	2.20
21	7 a. m.	5.02	2.81	2.21
22	9 p. m.	5.02	2.82	2.20
24	7 a. m.	5.12	2.95	2.17
	2 p. m.	5.08	2.92	2.16
	9 p. m.	5.08	2.91	2.17
26	9 p. m.	5.08	2.86	2.2[illegible]
27	7 a. m.	5.00	2.86	2.1[illegible]
	9 p. m.	5.10	2.84	2.2[illegible]
28	7 a. m.	4.98	2.83	2.1[illegible]
	9 p. m.	5.02	2.83	2.1[illegible]

Computation of the height of the gage zero at Sacketts Harbor, New York, above the gage zero at Charlotte, New York, in the spring of 1874—(Continued).

DAY.	Hour.	Gage Reading.		Difference.
		Sacketts Harbor.	Charlotte.	
1874.		Feet.	Feet.	Feet.
May 30	7 a. m.	5.00	2.81	2.19
	9 p. m.	5.03	2.82	2.21
June 1	9 p. m.	4.97	2.82	2.15
2	7 a. m.	5.00	2.83	2.17
	9 p. m.	5.02	2.85	2.17
4	9 p. m.	5.10	2.82	2.28
5	9 p. m.	5.00	2.78	2.22
6	7 a. m.	5.00	2.78	2.22
	9 p. m.	5.09	2.79	2.30
7	7 a. m.	5.10	2.79	2.31
	2 p. m.	5.00	2.81	2.19
8	9 p. m.	4.97	2.81	2.16
9	7 a. m.	4.97	2.82	2.15
Mean				2.247 ±.008

In 1896 the gage at Charlotte was a graduated plank spiked to a pile just north of the western abutment of the Rome, Watertown and Ogdensburg railroad bridge. The readings give the distance of the water surface above the zero of the gage. At Sacketts Harbor the arrangement was similar, the gage being spiked to an unused dock. The observer at Charlotte was Mr. J. W. Preston, harbor master; at Sacketts Harbor, Mr. Wilbur S. McKee. Observations were made morning, noon and night, the morning and evening observations being extended into series whenever the water was so little agitated by waves that the position of its surface could be determined with precision. As the times selected for these periods of observation were also comparatively free from atmospheric disturbances, and therefore favorable to a general equilibrium of lake surface, the computations were restricted to such times. In the four months of observations there were but five occasions when series were made at both stations.

Computation of the height of the gage zero at Sacketts Harbor, New York, above the gage zero at Charlotte, New York, in the summer of 1896.

DATE.	HOUR OF COMMENCING OBSERVATION.		NUMBER OF FIVE MINUTE READINGS.		MEAN OF READINGS.		DIFFERENCE.
	Sacketts Harbor.	Charlotte.	Sacketts Harbor	Charlotte.	Sacketts Harbor.	Charlotte.	Charlotte minus Sacketts Harbor.
1896.					Feet.	Feet.	
Aug. 8....	7.15 a. m.	7 a. m.	13	13	0.984	0.962	—0.022
8....	6.30 p. m.	6 p. m.	13	12	0.912	0.934	0.022
Sept. 9....	5.30 a. m.	7 a. m.	13	13	0.351	0.428	0.077
14....	5.00 p. m.	6 p. m.	13	13	0.270	0.386	0.098
Oct. 27....	8.15 a. m.	7 a. m.	45	11	—0.148	—0.048	0.100
Mean							+0.055 ±0.014

The bench at Charlotte is a mark on the upper surface of the water table of the old light-house. The walls of the building show no cracks, and there is every reason to believe the bench stable. On May 11, 1874, the zero of gage was found by Mr. E. S. Wheeler, assistant engineer United States Lake Survey, to be 33.003 feet below this bench mark. On June 30, 1896, I leveled from the zero of the present gage to the bench mark, obtaining 38.950 as the mean of two measurements. On July 11, 1897, Mr. Warner W. Gilbert obtained 38.954 feet as a mean of two measurements.

The only bench mark existing at Sacketts Harbor in 1874 and 1896 is a point on the upper outer edge of the water table at the northeast corner of the stone building known as the Masonic Temple. In May, 1874, this was determined by Mr. Wheeler to be 12.225 feet above the zero of gage. On June 28, 1896, by duplicate measurements, I found it to be 20.425 feet above the zero of the present gage. The building bearing this mark rests on a foundation of bed rock, but nevertheless has yielded to such extent that its walls are cracked. I was informed that the cracking and repairing of the walls took place some years previous to 1874, and regard it as probable that there has been no change since that date in the height of the bench mark.

These several data are combined in the following table:

Computation of the height of the bench mark at Charlotte, New York, above the bench mark at Sacketts Harbor, New York, in 1874 *and* 1896.

	1874. Feet.	1896. Feet.
Charlotte bench mark above Charlotte gage zero	+33.003	+38.950
Charlotte gage zero above Sacketts Harbor gage zero...........................	— 2.247	— 0.055
Sacketts Harbor gage zero above Sacketts Harbor bench mark......	—12.225	—20,425
Sum of above — Charlotte bench mark above Sacketts Harbor bench mark..	+18.531	+18.470
Difference	—0.061	

The results of the computations indicate that the height of the Charlotte bench mark above the Sacketts Harbor bench mark has diminished in twenty-two years to the extent of 0.061 foot. This quantity is the algebraic sum of six other quantities, two measurements through water leveling and four measurements by the engineer's level. The probable errors of the water levelings are ±0.008 and ±0.014 foot; the probable errors of my own instrumental levelings were each ±0.01 foot. Assigning the same precision to the earlier levelings, we obtain for the resulting quantity (0.061 foot) a probable error of about ±.03 foot.

This probable error attempts to express only such deviations from accuracy as are exhibited by the discordance of observations it does not include errors of the class called constant. The result may be vitiated by the instability of either bench or by river freshets in 1874, and there are qualifications related to tides and cyclonic gradient.

The data at Sacketts Harbor are not subject to errors from stream floods. The gages at Charlotte were on the bank of the Genesee River near its mouth. The channel is deep, and at ordinary river stages the current is so gentle that river level and lake level are the same, but in time of flood the river level is somewhat higher. In 1896 no flood periods were included, but the records for 1874 are not full enough to insure freedom from flood influences. If the Charlotte data include errors due to that cause, their correction would increase the computed change of relative height.

The tides of the Great Lakes are so small as to be masked by the seiches, but they are nevertheless of sufficient magnitude to affect an investigation of this sort. Lieut. Col. J. D. Graham determined a lunar tide of Lake Michigan at Chicago amounting to $1\frac{3}{4}$ inches and a spring tide amounting to $3\frac{1}{2}$ inches.* Gen. C. B. Comstock determined a lunar tide of Lake Michigan at Milwaukee of 1 inch and a solar tide of one-half inch; and a tide of $1\frac{1}{2}$ inches was found at the west end of Lake Superior.† The tides of Lake Ontario have not been investigated, and therefore a correction for them can not be applied. It would be quite possible to eliminate their effect by making the periods of observation include complete tidal cycles; but the local conditions gave greater importance to other criteria for the selection of times. An inspection of dates with reference to tidal cycles shows that the observations are so distributed that the influence of tide can not be great.

A complete comparative discussion of lake levels should also take account of differences of atmospheric pressure. It is evident that

*Ann. Rept. Chief of Engineers, U. S. A., for 1860, p. 296.

†Ann. Rept. Chief of Engineers, U. S. A., for 1872, pp. 1033, 1035, 1040; 1876, pp. 1173, 1192, 1194.

in a condition of equilibrium the level water surface must be deformed by local inequalities of atmospheric pressure, and the effect of pressure differences of course coexists with inequalities due to other causes. In planning these computations the intention was to apply corrections for barometric gradient, but this intention was afterwards relinquished because of the difficulty of properly discussing the available barometric data. Such examination as was given to the subject led to the opinion that during the stormless periods selected for the comparison of gage readings the error arising from the neglect of the pressure correction is small.

PORT COLBORNE AND CLEVELAND.

The character of the gage used at Cleveland in 1858 is not described in the records I have seen. Neither is the name of the observer given, but various circumstances indicate that the readings were made either by Col. Charles Whittlesey or under his immediate direction. The readings give the distance of the water surface below the high-water level of 1838, and that level was adopted by the United States Lake Survey as the plane of reference for all observations on Lake Erie. At Port Colborne the upper sill of Lock No. 27 of the Welland Canal was the zero of measurement, and the measurement was made by the lock master, Mr. John McGillivray, by thrusting a graduated pole into the water until the end rested on the lock sill. As the reference point at Cleveland was above the water surface and that at Port Colborne below, their difference in height is obtained by adding the two readings. Most of the observations at Cleveland were made at 8 a. m. and the observations at Port Colborne at noon. At Port Colborne the direction of the wind was recorded; at Cleveland, the direction and force. I do not know the scale of force employed, but the record numbers range from 0 to 5. All observations at both stations were rejected when the wind force at Cleveland was recorded as greater than 1.

The gage zero used at Cleveland in 1895 was the upper edge of a cleat nailed to a plank forming one wall of a well in a wharf. From this the. observer measured to the water surface in the well with a graduated rod. The gage zero was set at the level of high water in 1838, which is mentioned in the records as "the plane of reference." Three observations were made daily, at 7 a. m., 1 p. m. and 7 p. m., the work being under the direction of the United States Engineers. At Port Colborne observation was made by means of a float connected through a chain with a counterpoise, and was therefore indirect; but the readings were checked by occasional observations with a pole, after the method of 1858. An index on the counterpoise was so adjusted as to indicate on a scale the depth of water on the lock sill. I inspected the gage in 1896, finding it in close adjustment, except that an error in either direction of a fraction of an inch might arise from friction. The observer in 1895 was Mr. John Henshaw. In the following table the readings at Port Colborne, which were recorded in feet and inches, have been converted to feet and hundredths. The record of wind at the two stations was the same as in 1858, and there were also available the wind and pressure observations of the United States Weather Bureau. From an inspection of these data three periods were selected for comparison: June 28 to July 3, July 18 to 28, and August 3 to 18. These periods are so related to the tidal cycle as nearly to eliminate tidal error.

Computation of the height of the "plane reference" at Cleveland, Ohio, above the sill of Lock Welland Canal at Port Colborne, Ontario, in 1858, and 1895.

DATE.	Reading at Cleveland.		Reading at Port Colborne.		Sum.		DATE.	Readings at Cleveland. 7 a.m.	1 p.m.	7 p.m.	Mean.	Reading at Port Colborne.	Sum.
1858.	Ft.	in.	Ft.	in.	Ft.	in.	1895.	Feet.	Feet.	Feet.	Feet.	Feet.	Feet.
Aug. 20	0	9.0	14	7	15	4.0	June 28	......	3.30	3.50	3.40	10.92	14.32
24	0	8.0	14	6	15	2.0	29	3.55	3.50	3.53	3.53	10.75	14.28
25	0	8.5	14	6	15	2.5	30	3.55	3.50	3.53	3.53	11.08	14.61
28	0	5.0	14	0	14	5.0	July 1	3.30	3.40	3.50	3.40	10.75	14.15
31	0	8.0	14	3	14	11.0	2	3.47	3.45	3.55	3.49	10.83	14.32
Sept. 1	0	7.2	14	4	14	11.2	3	3.57	3.60	......	3.58	10.92	14.50
3	0	8.0	14	0	14	8.0	18	3.57	3.80	3.83	3.73	11.00	14.73
6	0	9.5	13	11	14	8.2	19	3.85	3.50	3.70	3.68	10 75	14.43
9	0	9.4	14	1	14	10.4	20	3.80	3.68	3.53	3.67	11.25	14.92
12	0	8.7	14	0	14	8.7	21	3.33	3.87	3.63	3.61	11.00	14.61
14	0	10.7	13	8	14	6.7	22	3.90	3.50	3.60	3.67	10.75	14.42
15	0	6.0	13	11	14	5.0	23	3.60	3.70	3.70	3.67	10.83	14.50
16	0	2.3	13	9	13	11.3	24	3.68	3.60	3.70	3.66	10.83	14.49
18	0	6.7	14	3	14	9.7	25	3.72	3.73	3.82	3.76	10.67	14.43
20	0	9.3	14	0	14	9.3	26	3.70	3.73	3.63	3.69	11.00	14.69
23	1	0.8	13	11	14	11.8	27	3.75	3.70	3.32	3.59	10:92	14.51
24	1	1.5	13	5	14	6.5	28	3.50	3.43	3.80	3.58	10.75	14.33
25	1	0.9	14	0	15	0.9	Aug. 3	......	3.79	3.80	3.80	11.00	14.80
27	1	1.8	14	2	15	3.8	4	3.80	3.73	3.73	3.75	10.75	14.50
30	1	2.5	14	5	15	7.5	5	3.73	3.68	3.68	3.70	10.58	14.28
Oct. 4	1	1.6	14	0	15	1.6	6	3.90	3.73	3.70	3.78	11.33	15.11
10	1	3.0	13	10	15	1.0	7	3.93	3 95	3.76	3.83	11.75	15.58
11	1	4.0	13	6	14	10.0	8	3.75	3.60	3.80	3.72	10.67	14.39
14	1	2.2	14	3	15	5.2	9	3.72	3.70	3.59	3.67	10.83	14.50
15	0	11.0	14	3	15	2.0	10	3.90	3.75	3.70	3.78	10.75	14.53
16	0	10.5	14	5	15	3.5	11	3.80	3.50	3.64	3.65	10.67	14.32
18	1	5.0	14	2	15	7.0	12	3.54	3.78	3.68	3.67	11.00	14.67

[illegible]	[illegible]	[illegible]	[illegible]	[illegible]	[illegible]	[illegible]
[illegible]	1	4 9	12	7	13	11 9
[illegible]	1	4 3	12	11	14	3.3
[illegible]	1	6 1	13	3	14	9 1

Mean [illegible] .... 11.800 / .057

				Mean	[illegible]	[illegible]
	Feet	Feet	Feet	Feet	Feet	Feet
Aug. [illegible]	3 6[illegible]	3 6[illegible]	3 60	3 [illegible]	1[illegible] 8[illegible]	11 1[illegible]
14	3 68	3 6[illegible]	3 7[illegible]	3 68	1[illegible]	11 68
15	3 6[illegible]	3 7[illegible]	3 6[illegible]	3 66	11 [illegible]	11 [illegible]
16	3 [illegible]	3 50	3 67	3 [illegible]	10 83	11 4[illegible]
17	3 70	3 68	3 70	3 69	10 7[illegible]	11 41
18	3 80	3 83	3 70	3 78	11 17	11 93

[illegible] 11.901 / [illegible]

Height of Cleveland plane of reference above Port Colborne lock sill

1858	11.800	[illegible]
1895	11.901	[illegible]
Difference	0.[illegible]	[illegible]

The zero of gage at Port Colborne, being submerged masonry, is of unquestioned stability. The canal was constructed in 1833, and if any settling followed construction it was doubtless complete before 1858; but the appearance of the masonry above the water gives no suggestion of yielding.

The earlier work at Cleveland was connected with several bench marks, all of which have been destroyed, but before the disappearance of the last one the datum was transferred by leveling to other points. The chain on which the record depends is as follows:

1. "Top of coping of the northeast wall of the Ohio Canal lock where it joins the river." The high water of 1838 was directly compared with this bench, and Whittlesey states that it is 6.30 feet above that high water plane.* As the observations in 1858 were made near the lock, and as Whittlesey, who reports them, was a civil engineer whose writings show that he appreciated the importance of precise bench marks, it is probable that the observations were properly connected with the bench. Explicit statement, however, is lacking; the record merely refers the lake level to the high water of 1838. The bench was destroyed in 1877 or 1878.

2. "Cross on water table, northeast corner of Johnson House block, southwest corner of Front and East River streets." On June 15, 1875 (as shown by manuscript records in the office of the Chief of Engineers, U. S. A.), Assistant Engineer T. W. Wright, United States Lake Survey, leveled from this bench mark to the canal lock coping (1), finding the difference (1 above 2) to be 3.67 feet. This bench mark is still in existence. The walls of the building are cracked in such manner as to indicate some settling of the northeast corner, and the broad flagstone on which the bench is marked stands (in 1897) 0.04 foot lower than the next stone of the water table toward the west. As the lower stone supports part of

* Canadian Naturalist, Vol. VII, 1875, p. 412.

the building and the higher stone carries no load, the latter may be assumed to show the original level of the former. It is impossible to say whether this settling affects the record of water levels. The building was erected in 1842, and is, therefore, 55 years old; it was 83 years old in 1875 when the datum of levels was transferred to it. The datum remained with it eighteen years, until 1893. If settling has progressed at a uniform rate, the datum was affected 0.013 foot, but it is equally possible that the settling belonged to the early history of the building, and that a condition of practical stability was reached prior to 1875.

3. "Bottom of west angle iron, on bottom of north longitudinal plate girder, middle of first full-depth bent, close to stone pier, new L. S. & M. S. R. R. drawbridge, now [1893] being finished." As the bridge is symmetric and reversible, this description applies to two different points, but measurement shows that they have the same height. It was copied from manuscript records in the United States Engineer's office at Cleveland, courteously placed at my service by Col. Jared A. Smith. The records show that in June, 1893, the bridge bench (3) was connected by leveling with the Johnson House bench (2) and also with the gage zero, and that the gage zero was checked by the bridge in 1896 and found correct. The gage readings in 1895 (used in our computations) are thus referred to the bridge bench. The height of the bridge bench is given as 4.84 feet above the "plane of reference," and by implication as 1.71 feet above the Johnson House bench (2). The drawbridge rests on a stone pier many years older than the present bridge, and there can be little question of its stability.

In these records of bench marks and levelings in Cleveland there is certainly much to be desired, but the presumption is, nevertheless, in favor of good work.

It appears from the computation that the ground at Port Colborne has risen, as compared to the ground at Cleveland, 0.239 foot, or about 2⅞ inches in thirty-seven years. The probable error of this measurement, as indicated by the discordance of gage data, is three-fourths of an inch.

As a check upon this result, a third computation was made from gage readings in the summer of 1872, a year in which the gage zero at Cleveland was connected with the canal-lock bench mark by instrumental leveling. That computation gives for the height of the plane of reference at Cleveland above the lock sill at Port Colborne 14.714 feet. If we assume a gradual change from 1858 to 1895, and interpolate between 14.800 feet, the determination for 1858, and 14.561, the determination for 1895, we obtain for the summer of 1872 the value 14.710 feet, which differs from the result of that year's observations by only 0.004 foot. The observations on Lake Erie thus accord well with the theory of a progressive southward tilting of the land.

The Port Colborne gage is not so related to streams as to subject its readings to error from floods. The Cleveland gage, like the Charlotte, is on a river estuary, and the readings are subject to influence by floods. The records include no systematic account of the condition of the river, and it is, therefore, possible that some of the observations were made when the river level was above the lake level.

PORT AUSTIN AND MILWAUKEE.

At each of these stations automatic gages were maintained for several years, and their tracings give the height of water level with an amount of detail permitting the complete elimination of seiches and tides; but there was, unfortunately, some uncertainty as to the position of the zeros, and the danger of thus introducing constant errors led me to avoid the automatic records and choose times when other gages were employed. The earlier period selected for the comparison was the summer of 1876, and the gages then used were floats, carrying graduated vertical rods. The force and direction of the wind were recorded at Port Austin by the gage observer, and at Milwaukee by the United States Weather Bureau. From an inspection of these records, together with the Weather Bureau records of barometric gradient, selection was made of the periods July 11 to 19 and August 16 to 24, excepting only certain hours when the force of the local wind was recorded as greater than 3 in a scale of 10. This gave 46 separate comparisons, from which the difference in height of the gage zeros was computed. The chosen periods are well disposed with reference to tides. The readings at Milwaukee were made at 7 a. m., 1 p. m. and 6 p. m. by Mr. John McCabe; at Port Austin the hours were 7 a. m, 2 p. m. and 9 p. m., and the observer was Mr. J. W. Kimball. In the computations the midday observations, though one hour apart, and the evening observations, though three hours apart, were treated as simultaneous.

DATE.	Readings at Milwaukee.			Readings at Port Austin.			Differences.		
	7 a. m.	1 p. m.	6 p. m.	7 a. m.	2 p. m.	9 p. m.			
1876.	Feet.	Feet.	Feet.	Feet.	Feet.	Feet.	Feet.	Feet.	Feet.
July 11	2.17	2.23	2.12	7.22	7.31	7.47	5.05	5.08	5.35
12	2.12	2.03	2.26	7.54	7.25	7.22	5.42	5.22	5.06
13	2.20	2.05	2.35	7.35	7.37	7.35	5.15	5.32	5.00
14	1.95	2.12	2.10	7.50	7.41	7.30	5.55	5.29	5.20
15	2.16	2.15	2.06	7.33	7.35	7.30	5.17	5.20	5.24
16	2.13	2.12	2.11	7.34	7.25	7.40	5.21	5.13	5.29
17	2.15	2.20	2.21	7.37	7.37	7.30	5.22	5.17	5.09
18	2.20	2.07	2.20	7.21	7.25	7.28	5.01	5.18	5.08
19	2.18	......	......	7.34	......	......	5.16	......	......
Aug. 16	2.29	......	2.20	7.50	......	7.35	5.21	......	5.15
17	2.19	......	2.21	7.33	......	7.50	5.14	......	5.29
18	2.23	2.02	2.27	7.50	7.35	7.40	5.27	5.33	5.13
19	......	2.22	......	......	7.45	......	......	5.23	......
20	......	2.18	2.23	......	7.62	7.37	......	5.44	5.14
21	2.25	2.20	2.24	7.41	7.34	7.37	5.16	5.14	5.13
22	2.32	2.18	2.33	7.38	7.51	7.40	5.06	5.33	5.07
23	2.26	1.91	2.10	7.38	7.60	7.59	5.12	5.69	5.49
24	2.25	......	2.42	7.49	......	7.49	5.24	......	5.07
Mean									5.210 ±.013

In 1896 the gage at Milwaukee consisted of a graduated rod held in the observer's hand in measuring down to the water from a fixed point or zero. At Port Austin a board, carrying a graduated scale, was spiked to the side of a timber crib, and the position of the water surface on the scale was noted by the observer. At each of these stations a series of 12 observations, at five minute intervals, was made every morning and evening when the surface of the water was nearly smooth. The mean of a series was afterwards treated as one observation, and the computation was based on the simultaneous pairs of observations—53 in number. The selection of times was thus determined by conditions favorable for the elimination of seiches, but it appears by inspection that tidal influences also are very nearly eliminated. The observers were: At Milwaukee, Mr. John McCabe; at Port Austin, Mr. John P. Smith.

As the zero at Milwaukee was above the water, and the zero at Port Austin below, the sum of the readings gives the height of one zero above the other.

Computation of height of gage zero at Milwaukee, Wisconsin, above gage zero at Port Austin, Michigan, in the summer of **1896**.

DATE.		Time.	Readings (Means of Series). Milwaukee.	Port Austin.	Sum.
1896.			Feet.	Feet.	Feet.
July	20	A. M.	5.528	1.271	6.799
	24	A. M.	5.483	1.333	6.816
	28	A. M.	5.703	1.385	7.088
	29	A. M.	5.470	1.354	6.824
		P. M.	5.560	1.375	6.935
Aug.	1	A. M.	5.436	1.297	6.733
	2	P. M.	5.620	1.425	7.045
	7	P. M.	5.447	1.420	6.867
	9	A. M.	5.519	1.448	6.967
	11	A. M.	5.575	1.433	7.008
	14	A. M.	5.338	1.455	6.793
	20	A. M.	5.587	1.340	6.927
	20	P. M.	5.571	1.406	6.977
	21	A. M.	5.558	1.330	6.888
		P. M.	5.588	1.391	6.979
	22	A. M.	5.505	1.259	6.764
		P. M.	5.540	1.229	6.769
	24	P. M.	5.595	1.375	6.970
	25	A. M.	5.721	1.221	6.942
	25	P. M.	5.792	1.268	7.060
	28	P. M.	5.721	1.279	7.000
	30	P. M.	5.797	1.239	7.036
Sept.	1	P. M.	5.725	1.259	6.984
	2	P. M.	5.748	1.203	6.951
	4	P. M.	5.720	1.248	6.968
	5	P. M.	5.515	1.134	6.649
	7	P. M.	5.739	1.275	7.014
	8	A. M.	5.649	1.203	6.852
		P. M.	5.595	1.139	6.734
	9	P. M.	5.585	1.077	6.662
	14	A. M.	5.584	1.208	6.792
	15	A. M.	5.560	1.181	6.741
	20	P. M.	5.892	1.281	7.173
	23	A. M.	5.791	1.307	7.098
	25	A. M.	5.932	0.803	6.735
	28	P. M.	5.755	1.167	6.922
	29	A. M.	5.615	1.013	6.628

Computation of height of gage zero at Milwaukee, Wisconsin, above gage zero at Port Austin, Michigan, in the summer of 1896—(Continued.)

DATE.		Time.	READINGS (MEANS OF SERIES).		Sum.
			Milwaukee.	Port Austin.	
1896.			Feet.	Feet.	Feet.
Oct.	2	A. M.	5.566	1.250	6.816
	3	A. M.	5.594	1.465	7.059
		P. M.	5.574	1.186	6.760
	4	P. M.	5.632	1.101	6.733
	5	P. M.	5.705	1.085	6.790
	10	P. M.	5.506	0.889	6.395
	15	A. M.	5.784	0.769	6.553
	17	A. M.	5.642	1.444	7.086
	18	A. M.	5.720	1.398	7.118
	19	A. M.	5.846	1.215	7.061
	22	A. M.	6.182	1.212	7.394
	25	A. M.	6.139	0.800	6.939
	26	P. M.	5.960	0.858	6.818
	27	A. M.	5.918	0.750	6.668
		P. M.	5.802	0.722	6.531
	29	A. M.	5.864	0.724	6.588
Mean					6.875 ± .019

Milwaukee is well provided with engineer bench marks, and it is probable that thorough research would establish the connection of the gage zeros at each epoch with several of the bench marks; but after inspection of the data readily accessible, I thought it best to make use of only one bench, that called the "check point." This consists of the top of a copper bolt leaded into the north side of the center pier of the swing bridge over the river between Chestnut and Division streets. The gage observer is required at stated intervals to check the stability of the zero of his gage by means of this check point. Using two rods, with the aid of an assistant he makes a series of simultaneous measurements from the check point and from the gage zero down to the water level, and from these measurements the relation of the gage zero to the check point is determined. Their relation has also been determined by means of the engineer's level at various times, and was so determined on August 8, 1876, by Assistant Engineer L. L. Wheeler, who found the check point 0.843 foot above the gage zero. In 1896 the check observations by the observer were very thorough, series of twenty simultaneous readings being made every fortnight, and from five of these series the relation of the two points is computed as follows:

Computation of height of Milwaukee check point above Milwaukee zero of gage in the summer of 1896.

	Feet.
July 12 (mean of twenty comparisons by simultaneous readings)	1.203
July 26	1.212
August 14	1.200
August 28	1.203
September 16	1.206
Mean	1.205
	± .002

In response to a letter of inquiry as to the stability of the Milwaukee check point, Capt. George A. Zinn, United States engineer in charge of harbor improvements, writes as follows:

The Chestnut Street Bridge, on the center pier of which the check point is established, was built in 1872.

Mr. G. H. Benzenberg, city engineer, states that the pier rests on a pile foundation; that to his knowledge the drawbridge has never been releveled since put in place, and that if any appreciable settlement had taken place in the center pier it would have interfered with the operating of the swing bridge. He stated positively that no settlement had occurred.

The principal bench mark used in 1876 at Port Austin, called the Wisner bench mark, was a copper bolt leaded into bed rock; but in 1896 I was unable to find it, and, as at Milwaukee, I had recourse to a bench mark originally established and used as a check point. It is the top of an iron bolt driven into a vertical face of bed-rock on the west side of a promontory opposite the residence of Mr. J. W. Kimball. In July, 1875, and October, 1876, Assistant Engineer T. W. Wright found the check point 7.424 feet below the Wisner bench mark; in June, 1896, I found the gage zero 5.125 feet below the check point, this quantity being the mean of two measurements.

Manuscript records in the archives of the Lake survey state that the Port Austin gage zero was originally placed on a level with the Wisner bench mark, but that in July, 1875, it was 0.003 foot too low, and that on October 18, 1876, it was 0.040 foot too low, having settled during the interval. As the observations used fall within this interval, it was necessary to make some assumption in regard to this settling, and the assumption made was that it had been at uniform rate through the whole period. The correction interpolated for the time of observation was 0.034 foot. Combining this correction with data from leveling in 1875 and 1876, I obtained as the height of the gage zero above the check point in July and August, 1876, 7.460 feet. The various data thus described are combined in the following table:

Computation of height of Milwaukee check point above Port Austin check point in the summers of 1876 and 1896.

	1876. Feet.	1896. Feet.
Milwaukee check point above Milwaukee gage zero	0.843	1.205
Milwaukee gage zero above Port Austin gage zero	—5.210	6.875
Port Austin gage zero above Port Austin check point	7.460	—5.125
Sum of above=Milwaukee check point above Port Austin check point	3.093	2.955
Difference	—0.138	

This result indicates that the ground at Milwaukee, as compared to the ground at Port Austin, has subsided 0.138 foot in the twenty years from 1876 to 1896. It is the algebraic sum of six measurements, of which three are levelings by water surface and three by the engineer's level. The probable errors of the water-level measurements are ±0.019, ±0.013, and ±0.002. The probable errors of the Port Austin levelings in 1896, as indicated by the discordance of two independent results, is ±0.008. If the probable error of each of the other measurements was ±0.010, the probable error of the result is less than ±.03 foot. There is also an uncertainty arising from the possibility that the stone pier to which the Milwaukee check mark is attached has settled, another uncertainty due to the possibility of river floods, and a third involved in the assumption that the settling of the Port Austin gage zero in 1876 was at a uniform rate. If all the settling of the Port Austin zero took place before the period of observation, the assumption makes the result too large by 0.006 foot; if all the settling took place after the observations, the assumption makes the result too small by 0.031 foot. The Port Austin record is free from stream-flood influences, but the Milwaukee gage station is on a narrow estuary, like the stations at Charlotte and Cleveland.

ESCANABA AND MILWAUKEE.

In comparing Escanaba with Milwaukee the same general periods of observation were employed as in comparing Port Austin with Milwaukee, but the individual days, though selected in the same manner, were in part different. Fifty-one separate comparisons were made in 1876, and 52 in 1896. The selection of times was controlled by conditions favorable for the elimination of seiches, but the combination of days chosen was found to approximately eliminate tidal effects also.

The observations at Escanaba in 1876 were conducted in the same manner as at Milwaukee and Port Austin, the hours being 7 a. m., 2 p. m., and 9 p. m., and the observer Mr. George Preston. In 1896 the system was the same as at Milwaukee, the observer being Mr. Clinton B. Oliver. The following tables give the computations for the two years:

DATE.	7 a. m.	2 p. m.	9 p. m.	7 a. m.	1 p. m.	6 p. m.	Differences.		
1876.	Feet.	Feet.	Feet.	Feet.	Feet.	Feet.	Feet.	Feet.	Feet
8 July 11	1.86	1.90	1.78	2.17	2.23	2.12	0.31	0.33	0.34
12	1.90	1.75	1.78	2.12	2.03	2.16	.22	.28	.38
13	2.00	2.07	1.95	2.20	2.05	2.35	.20	—.02	.40
14	2.10	2.15	2.05	1.95	2.12	2.10	—.15	—.03	.05
15	1.96	1.95	2.00	2.16	2.15	2.06	.20	.20	.06
16	1.90	1.95	1.85	2.13	2.12	2.11	.23	.17	.26
17	1.89	1.80	1.85	2.15	2.20	2.21	.26	.40	.36
18	2.05	1.85	1.95	2.20	2.07	2.20	.15	.22	.25
19	1.75	....	....	2.18	....	....	.43	....	....
Aug. 16	2.07	1.94	1.90	2.29	2.30	2.20	.22	.36	.30
17	2.00	1.95	1.95	2.19	2.13	2.21	.19	.18	.26
18	1.70	1.93	2.00	2.23	2.02	2.27	.53	.09	.27
19	1.78	2.00	....	2.13	2.22	....	.35	.22	....
20	1.95	1.90	1.85	2.23	2.18	2.23	.28	.28	.38
21	1.85	2.05	2.10	2.25	2.20	2.24	.40	.15	.14
22	1.83	2.05	1.90	2.32	2.18	2.33	.49	.13	.43
23	1.91	1.83	1.75	2.26	1.91	2.10	.35	.08	.35
24	1.95	1.85	1.90	2.25	2.10	2.42	.30	.25	.52
Mean									0.255 ±.012

Computation of height of gage zero at Escanaba, Michigan, above gage zero at Milwaukee, Wisconsin, in the summer of 1896.

DATE.		Time.	Readings (Means of Series).		Difference.
			Milwaukee.	Escanaba.	
1896.			Feet.	Feet.	Feet.
July	2	P. M.	5,465	5,917	0.452
	7	A. M.	5,434	5,907	.473
	8	A. M.	5,505	5,920	.415
	9	A. M.	5,348	5,837	.489
		P. M.	5,356	5,765	.409
	10	A. M.	5,442	5,771	.329
		P. M.	5,567	5,694	.127
	11	P. M.	5,576	5,771	.195
	13	A. M.	5,411	5,869	.458
		P. M.	5,493	5,776	.283
	14	A. M.	5,574	5,750	.176
	17	A. M.	5,431	5,865	.434
	19	A. M.	5,524	6,007	.483
		P. M.	5,496	5,887	.391
	20	A. M.	5,528	5,803	.275
	21	A. M.	5,573	5,973	.400
	23	A. M.	5,645	5,908	.263
	25	A. M.	5,601	5,856	.255
	28	A. M.	5,703	5,857	.154
	31	A. M.	5,446	5,938	.492
Aug.	1	P. M.	5,360	5,859	.499
	4	A. M.	5,654	5,912	.258
	8	P. M.	5,347	5,954	.607
	9	A. M.	5,519	5,658	.139
	10	P. M.	5,328	5,546	.218
	13	A. M.	5,273	5,616	.343
		P. M.	5,378	5,752	.374
	14	A. M.	5,338	5,670	.332
	15	A. M.	5,360	5,730	.370
		P. M.	5,402	5,710	.308
	19	A. M.	5,414	5,878	.464
	21	A. M.	5,558	5,935	.377
		P. M.	5,588	5,872	.284
	22	A. M.	5,505	5,698	.183

nputation of height of gage zero at Escanaba, Michigan, above gage zero at Milwaukee, Wisconsin, etc.—(Continued).

DATE.	Time	Readings (Means of Series).		Difference.
		Milwaukee.	Escanaba.	
1896.		Feet.	Feet.	Feet.
t. 4	A. M.	5,734	6,028	.294
13	P. M.	5,452	5,848	.396
14	A. M.	5,584	5,762	.178
16	A. M.	5,500	5,937	.437
18	A. M.	5,701	6,047	.346
26	A. M.	5,914	6,187	.273
28	P. M.	5.755	6,224	.469
29	P. M.	5,510	6,133	.623
7	A. M.	5,514	6,164	.650
14	A. M.	5,731	6,270	.539
	P. M.	5,813	6,157	.344
17	P. M.	5,622	6,160	.538
19	A. M.	5,846	6,287	.441
20	A. M.	5,857	6,346	.489
22	A. M.	6,182	6,539	.357
	P. M.	6,148	6,540	.392
26	A. M.	6,080	6,471	.391
	P. M.	5,980	6,544	.564
Mean				0.374 ±.012

he bench employed at Milwaukee has already been described. Escanaba there were three bench marks in good standing, as ows: No. 1, the top of the water sill on the southeast corner of Adler building, northwest corner of Ludington street and ıseman avenue; No. 2, the top of the water sill of the Escanaba ıt-house at the north side of front door, against the brick wall; . 3 is described in 1876 as the "center of a copper bolt set hori- ıtally in the foundation of the light-house, west side, north cor- , 3 feet north from steps." In a description by Capt. George Zinn, dated June 30, 1896, the top of the bolt is specified. I informed by Mr. Clinton B. Oliver, the gage observer, that the

diameter of the bolt is three-eighths inch. The relative he
two or more of these bench marks have been determined in
six different years, the measurements being made indep
with the engineer's level. It is advantageous to compa
measurements, not only to learn what confidence is to be
in the individual benches, but for the sake of whatever lig
be cast on the general precision of such data.

Comparison of Escanaba bench marks with one another.

Year.	Above Zero of Gage.			Difference Between Bench Marks.			Deviation from Mean.		
	No. 1.	No. 2.	No. 3.	1—2.	1—3.	2—3.	1—2.	1—3.	2—3.
	Feet.	Feet.	Feet	Feet.	Feet.	Feet.	Feet.	Feet.	Feet.
1874	4.135	—2.100	....	a6.235	....	....	....	....	....
1875	7.859	2.382	....	5.477	....	....	+0.003	....	....
1876	7.874	2.409	1.392	5.465	6.482	1.017	— .009	+0.002	+0.005
1880	7.878	2.402	1.035	5.476	a6.843	a1.367	+ .002	....	....
1887	7.753	2.285	1.280	5.468	6.473	1.005	— .006	— .011	— .007
1896	7.780	2.297	1.283	5.483	6.497	1.014	+ .009	+ .013	+ .002
Mean	....	....	....	5.474	6.484	1.012	....	....	....

a Not used in computing means.

In this table the reading of the height of bench mark No. 3 in 1876 is corrected for the distance between center and top of bolt. In the first division of the table the benches are referred to zero of gage, but as the gage was not constant in position these numbers differ widely from year to year. In the second division the relations of the gages one to another are given, being deduced by subtraction from the numbers of the first division, and these figures are more accordant. It appears, however, that the difference between benches 1 and 2 in 1874 departs widely from differences found in other years, and it is therefore probable that a blunder of measurement or record was made in that year. It appears further, by inspection, that the difference between benches 1 and 3 and the difference between benches 2 and 3 in 1880 are not in accord with the differences found in other years, and it is evident that some blunder was made in the measurement or record of the height of bench 3 for that year. These figures were accordingly thrown out and not used in the computation of the means. The numbers of the third division were obtained by subtracting the means from the several numbers of the second division, and they show the deviations from mean after rejecting the records showing gross errors. Inspection of the table of deviations shows that their signs are irregularly distributed, and discovers no evidence of progressive change from year to year. It is therefore probable that all three of the benches are stable, and that the deviations of the measurements from uniformity represent ordinary errors of observation. They may accordingly be used as a rough measure of the precision, barring blunders, of the instrumental leveling on which the results of this investigation largely depend. Their mean is 0.006 foot, and the computed probable error of a single measurement is ±0.008 foot. In combining various data for the comparison of Escanaba with Milwaukee, bench mark No. 1 of Escanaba was first used.

Computation of the height of Escanaba bench mark No. 1 *above Milwaukee check point in the summers of* 1876 *and* 1896.

	1876. Feet.	1896. Feet.
Escanaba bench No. 1 above Escanaba gage zero	7.874	7.780
Escanaba gage zero above Milwaukee gage zero	—0.255	.375
Milwaukee gage zero above Milwaukee check point	—0.843	—1.205
Sum of above = Escanaba bench No. 1 above Milwaukee check point	6.776	6.949
Difference	+ .173	

The result indicates that the ground at Escanaba, as compared with the ground at Milwaukee, has risen 0.173 foot in twenty years. This quantity is the algebraic sum of six measurements, of which three were made through water leveling and three by instrumental leveling. The probable errors of the water levelings are ±0.012, ±0.012 and ±0.002 foot; the estimated probable error of the instrumental levelings at Milwaukee is ±0.010 foot, and of the two levelings at Escanaba each ±0.008 foot. This gives as the probable error of the result ±0.022 foot.

A similar computation, using bench mark No. 2 instead of bench mark No. 1, gives 0.155 foot instead of 0.173, and a computation based on bench mark No. 3 gives 0.156. The mean of the three results is 0.161 foot, with a probable error of ±0.022 foot. The only important uncertainties to which this result is subject, besides those indicated by the discordance of measurements, arise from the possibility of the settling of the bridge pier to which the Milwaukee check point is attached and the possibility of river floods.

DISCREPANCY NOTED BY CAPTAIN MARSHALL.

In the later work of the United States Lake Survey all determinations of lake level were referred to the high-water level of 1838, which is called the "plane of reference." That plane was directly observed by Dr. I. A. Lapham, the geologist, and with the aid of a bench mark on his house at Milwaukee was permanently recorded. For other stations on Lake Michigan-Huron its position was determined by assuming that the level of 1838 had everywhere the same height above the mean lake level as determined by long series of ob ervations. For the determination of this plane at Escanaba use was made of observations for the period from January 1, 1860, to December 31, 1875. In 1887 Capt. W. L. Marshall, U. S. E., under whose direction the gage readings at Milwaukee and Escanaba were then made, detected a discrepancy, which he reported to the Chief of Engineers in a letter dated October 1.*

In former reports the zero of Escanaba gage has been assumed as 0.76 foot above the plane of reference, but a comparison of corrected readings at Milwaukee and Escanaba shows that the determinations of the plane of reference at Milwaukee and Escanaba vary 0.187 foot, the Escanaba plane being too high or the Milwaukee determination too low.

In the light of present knowledge it seems probable that the discrepency thus noted by Captain Marshall as an error was occasioned either wholly or in chief part by the progressive tilting of the land. This conclusion is difficult of verification, because little record survives of such checks as may have been made upon the heights of gage zeros during the period 1860-1875; but the indicated change agrees in direction, and approximately in rate, with the change deduced from the present investigation. From the middle of the period 1860-1875 to the summer of 1887 was an interval of twenty

* Ann. Rept. Chief of Engineers, U. S. A., for 1887, part 3, p. 2417.

years, equal to the interval 1876–1896 here used, and the discrepancy of 0.187 foot discovered by Captain Marshall differs from the change of 0.161 foot here deduced by a quantity little greater than the probable error ascribed to the latter determination.

SUMMARY OF RESULTS.

In the following table are assembled the numerical results as to changes in relative height of the four pairs of stations. Besides the measured changes, the table includes the periods intervening between dates of measurement and distances between the stations of each pair. The lines connecting pairs of stations have a southwesterly direction (fig. 99, op. p. 88), and it is the northeastern station of each pair that appears to have risen as compared to the other.

The results thus show a general agreement with the working hypothesis, that the latest change recorded by geologic data is still in progress. To make the comparison quantitative there should be substituted for the direct distances between stations the corresponding distances in the assumed direction of tilting, S. 27° W., and the measured results for various distances and various time intervals should be reduced to a common basis. In the third column of the table are the reduced distances, and in the sixth the reduced rates of change. Assuming the change to have a uniform rate and to be the same for all parts of the region, the measurements at the different pairs of stations give for a distance of 100 miles and a period of a century the quantities of the sixth column. The seventh column contains the probable errors of quantities in the sixth, and is based on the probable errors of the measured changes in pairs of stations:

Summary of distances, time intervals, and measurements of differential earth movements.

PAIRS OF STATIONS.	Direct distance.	Distance in direction S. 27° W.	Interval between dates of measurement.	Change in relative height.	Change per 100 miles per century.	Probable error of quantities in last column.
	Miles	Miles.	Years.	Feet.	Feet.	Feet.
Sacketts Harbor and Charlotte	88	76	22	0.061	0.37	0.18
Port Colborne and Cleveland	158	141	37	0.239	0.46	0.11
Port Austin and Milwaukee	259	176	20	0.137	0.39	0.09
Escanaba and Milwaukee	192	186	20	0.161	0.43	0.06
Mean					0.41	±0.044
Weighted mean					0.42	

IS THE LAND TILTING?

With the numerical results of the investigation before us we may now recur to the main subject and ask whether the evidence warrants the conclusion that a general, gradual tilting of the basin is in progress. In the discussion of the data used in comparing the several pairs of stations it has been found that, taken at their face value, they indicate a tilting in the hypothetic direction, but it has also been found impossible to resolve all doubts as to the stability of the gages and benches and the accuracy of the measurements. By reason of these doubts the result from no single pair of stations is conclusive, but when assembled they exhibit a harmony which argues strongly for their validity. As tabulated, there are four results, but these are not all independent, since observations and measurements at Milwaukee are used twice. There are, however, three results wholly independent and a fourth partly independent. To these may be added a fifth partly independent, namely, the determination of change between Port Colborne and Cleveland for the shorter period, 1872-1895. Not only do all these results indicate a change of the same sort, but they agree fairly well as to quantity. The computed change for 100 miles in a century ranges only from 0.37 to 0.46 foot, and the greatest deviation of an individual result from the mean of four is 12 per cent. This measure of harmony appeals strongly to the judgment, and is also susceptible of approximate numerical expression. If the four determinations tabulated in the sixth column are, in fact, measures of the same quantity—that is, if the tilting has been uniform throughout, as we have assumed—then the probable error of the determined value of that quantity (0.42 foot) is less than ±0.05 foot.

The most important factors tending to throw doubt on the conclusion are the possibilities of accidental change in the various benches to which the measurements are referred. The bench at

Port Austin, being a mark on bedrock, is trustworthy, and the agreement between the three benches used at Escanaba is good evidence of their stability; but the bench at Milwaukee, with which both are compared, is a pier of a bridge in daily use and may, perhaps, be slowly settling. If it is settling, the comparisons with benches at Escanaba and Port Austin may merely reveal that fact and not measure the subsidence of the land. The fact that the swing bridge on the pier has not required re-leveling is certainly favorable to the stability of the pier, especially when it is considered that a change of fully 1½ inches is to be accounted for; and there is further confirmation in the discovery of a discrepancy between Milwaukee and Escanaba by Captain Marshall, whose data are probably independent of the check mark. Of the benches on Lake Erie, the one at Port Colborne is satisfactory, but those at Cleveland may have settled at critical times, and if so their change would influence the result in the direction found. Of the benches on Lake Ontario, the one at Charlotte is eminently stable; the only practical question affects the bench at Sacketts Harbor, which is on a building that has not been wholly stable since its construction, although presumably so since the making of the bench. If the building at Sacketts Harber settled between 1874 and 1896, the effect of the lowered bench was to produce, not such a change as appears from the measurements, but one with the opposite sign.

It seems to me that the harmony of the measurements and their agreement with prediction from geologic data make so strong a case for the hypothesis of tilting that it should be accepted as a fact, despite the doubts concerning the stability of the gages.

RATE OF MOVEMENT.

The deduced mean rate of change—0.42 foot to the 100 miles in a century—depends on assumptions which are convenient rather than probable. These are: (1) that the whole region moves together as a unit, being tilted without internal warping, and (2) that the direction of its present tilting is identical with the direction of the total change since the epoch of the Nipissing outlet of the upper lakes. What we know of the general character of earth movements gives no warrent for such assumptions of uniformity, but no better assumptions as to this region are now available. Under the law of probabilities, the close agreement of four measurements, three of which are wholly independent, gives a good status to their mean, but there are other considerations tending to weaken this status. The probable errors of the individual measurements are rather high, ranging from 14 to 50 per cent., and this suggests the possibility that the closeness of their correspondence may be accidental. It should be remembered also that at two or three stations there was reason to believe that the gage zeros were settling during the period in which the observations were made, and the results involve the doubtful assumption that the rate of settling was uniform. There is room for doubt as to the precision of the instrumental leveling; in only a few instances is the fact of duplicate measurements recorded, and single measurements are notoriously insecure. Error was doubtless admitted by ignoring the effects of barometic gradient. River floods may have introduced errors. In the absence of flood records the records of rainfall at Rochester (near Charlotte), Cleveland, and Milwaukee were compared with the gage readings, the results showing only that if flood errors are involved they must be small. There may also be personal equations of observers, especially as the gages at pairs of stations were not in every case of the same type. For all these reasons I am disposed to ascribe only a low order of precision to the deduced rate of change, and regard it as indicating the order of magnitude rather than the actual magnitude of the differential movement.

The rate of change indicated by Stuntz's observations is more rapid. As already quoted, he states that at a time when Lake Superior was exceptionally low at its outlet, it was nevertheless so high at its western extremity as to obliterate from the St. Louis River a rapid which had been visible only a few years before. This statement involves no definite measures, but it implies that the change within the memory of individuals involves feet rather than the inches deduced from the studies in the other lakes. Similar inferences may be drawn from his statement as to submerged stumps. The recorded range of water level in Lake Superior is about 5 feet, and trees would grow little if any below high-water mark. If, then, with low stage at the east end, stumps are submerged at the west, a change of 5 feet or more would seem to have occurred during the period covered by the growth of a tree and the survival of its stump. The differences between the inferences drawn from this evidence and the result based on gage readings on the other lakes is so wide as to suggest the possibility of error in the Lake Superior observations. It is certainly important that they be verified. Unfortunately I have not been able to visit the region, and the gage records accessible to me are not so connected with bench marks as to give a satisfactory basis for computation. The United States Lake Survey made observations of lake level at Superior City from 1859 to 1871, and then transferred the station to Duluth, where it was continued for two or three years. No bench mark at Duluth is described, and the only recorded bench mark at Superior City is upon a wooden structure, Johnson & Alexander's sawmill. If this bench survives, a good test could be made by renewing the gage station at Superior City. At the other end of the lake, at Sault Ste. Marie, there are authentic benches

If we assume that the rate of 0.42 foot per 100 miles per century is uniform and secular, and project it backward to the time when the drainage of Lake Huron was shifted from North Bay to Port Huron, we obtain for the period since that change about 10,000 years. From studies at Niagara, Taylor has estimated the same period as between 5,000 and 10,000 years;* and the comparison indicates that the rate of modern change is of such magnitude as to accord well with the idea that it merely continues the geographic change.

It is to be hoped that eventually a better measure of the rate of tilting and a surer indication of its direction may be obtained, but even with present knowledge there is interest and profit in considering the economic and geographic consequences of the tilting.

GEOGRAPHIC CHANGES RESULTING FROM THE MOVEMENT.

Assuming that the general result of this investigation is substantially correct — that the whole lake region is being lifted on one side or depressed on the other, so that its plane is bodily canted toward the south-southwest, and that the rate of change is such that the two ends of a line 100 miles long and lying in a south-southwest direction are relatively displaced four-tenths of a foot in 100 years — certain general consequences may be stated. The waters of each lake are gradually rising on the southern and western shores or falling on the northern and eastern shores, or both. This change is not directly obvious, because masked by temporary changes due to inequalities of rainfall and evaporation and various other causes, but it affects the mean height of the lake surface. In Lake Ontario the water is advancing on all shores, the rate at any place being proportional to its distance from the isobase through the outlet

* Bull. Geol. Soc. America, Vol. IX, 1898, p. 83.

(AA, fig. 100). At Hamilton and Port Dalhousie it amounts to 6 inches in a century. The water also advances on all shores of Lake Erie, most rapidly at Toledo and Sandusky, where the change is 8 or 9 inches in a century. All about Lake Huron the water is falling, most rapidly at the north and northeast, where the distance from the Port Huron isobase (CC, fig. 100) is greatest; at Mackinac the rate is 6 inches, and at the mouth of French River 10 inches, a century. On Lake Superior the isobase of the outlet (DD, fig. 100) cuts the shore at the international boundary; the water is advancing on the American shore and sinking on the Canadian. At Duluth the advance is 6 inches, and at Heron Bay the recession is 5 inches, a century. The shores of Lake Michigan are divided by the Port Huron isobase. North of Oconto and Manistee the water is falling; south of those places it is rising, the rate at Milwaukee being 5 or 6 inches a century, and at Chicago 9 or 10 inches. Eventually, unless a dam is erected to prevent, Lake Michigan will again overflow to the Illinois River, its discharge occupying the channel carved by the outlet of a Pleistocene glacial lake. The summit in that channel is now 8 feet above the mean level of the lake, and the time before it will be overtopped (under the stated assumption as to rate of tilting) may be computed. Evidently the first water to overflow will be that of some high stage of the lake, and the discharge may at first be intermittent. Such high-water discharge will occur in 500 or 600 years. For the mean lake stage such discharge will begin in about 1,000 years, and after 1,500 years there will be no interruption. In about 2,000 years the Illinois River and the Niagara will carry equal portions of the surplus water of the Great Lakes. In 2,500 years the discharge of the Niagara will be intermittent, falling at low stages of the lake, and in 3,500 years there will be no Niagara. The basin of Lake Erie will then be tributary to Lake Huron, the current being reversed in the Detroit and St. Clair channels.

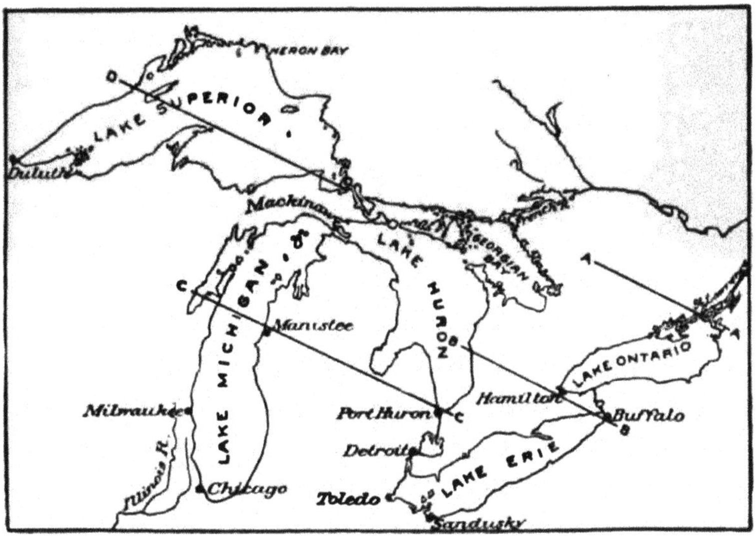

FIG. 100.—Relations of the shores of the Great Lakes to the isobases drawn through their outlets.

The most numerous economic bearings of this geographic change pertain to engineering works, especially for the preservation of harbors and regulation of water levels. But the modifications thus produced are so slow as compared to the growing demands of commerce for depth of water that they may have small importance. It is a matter of greater moment that cities and towns built on lowlands about Lakes Ontario, Erie, Michigan and Superior will sooner or later feel the encroachment of the advancing water, and it is peculiarly unfortunate that Chicago, the largest city on the lakes, stands on a sinking plain that is now but little above the high-water level of Lake Michigan.

PLANS FOR PRECISE MEASUREMENT.

While it is believed that the general fact of earth movement has been established by the present investigation, the measurement of its rate and the determination of its direction fall far short of the precision which is desirable. For the purposes of science the order of magnitude of the change is more important than its precise measurement, but there are involved great economic interests, and these demand more definite information. The account of the present investigation is therefore supplemented by an outline plan of the more elaborate investigation which appears necessary to give measurements of the precision that is desirable.

Existing data are neither full enough nor exact enough to give satisfactory measures of the small quantities sought. Doubtless a more elaborate discussion would yield better results than I have obtained, but the improvement could not be great. Observations by the Lake Survey were conducted for purposes not demanding a high order of precision, and high refinement was not attempted. The supplementary work done in 1896 attempted only to be good enough for use in combination with the work of 1874 and 1876, and can not serve as the first term of a new comparison. The problem requires a new set of high-grade observations at each station of a carefully planned system, to be followed, after an interval of at least a decade, by a second set of observations at the same stations.

Foreseeing no opportunity to undertake such a work myself, I have formulated in the following paragraphs a plan embodying the results of my experience—a plan intended to afford useful suggestions to some investigator by whom the work may be actually undertaken.

Selection of stations.—To measure the rate of change in any given direction, observations at two stations suffice; but to determine also the direction of change, it is necessary to use three stations grouped in the form of a triangle. The longer the sides of the triangle the better the measurement of rate, and the larger its smallest angle the better the determination of direction. A brief inspection shows that the shores of Lake Michigan and Lake Huron give the best opportunity for the planning of a well-conditioned triangle. Though the narrowness of their connecting strait has led to the giving of separate names, they are really a single lake, and the stretch of their water surface is in every direction greater than that of Lake Superior.

For the purpose in view the point of first importance is the outlet of the lake at Port Huron. This is peculiar in that the plane of the mean water level has here a constant relation to the adjacent land, a relation altogether independent of the progressive deformation of the basin. This station should not be on the St. Clair River, but on the shore of the lake near by.

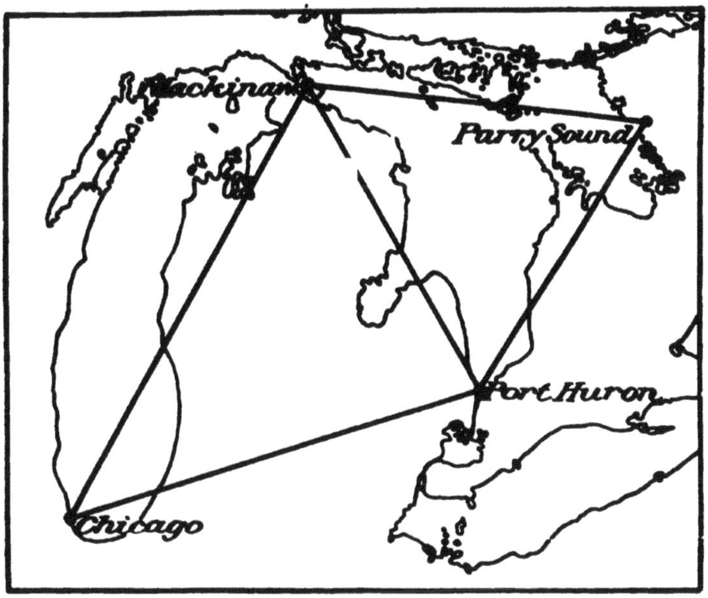

FIG. 10[illegible].—Proposed systems of stations for the precise measurement of earth movements.

M701:

The second point of vantage is Chicago. As economic interests are more seriously affected by the geographic change at that point than elsewhere, it is desirable to determine directly, by comparison with Port Huron, the rate at which the lake is encroaching on the land.

A third point of prime importance is the Strait of Mackinac. Although the equilibrium levels of the surfaces of the two lakes are the same, there are considerable periods when their equilibrium is disturbed, and during such periods a current flows in one direction or the other through the strait. Only when this current is nil is the whole water body in perfect equilibrium, and it is essential to precise leveling through the water surface either that times of equilibrium be chosen or that due allowance be made for the gradients associated with flow. Observations must therefore be made on the current in the strait, and it is best to connect them with the work of a complete station.

As appears by the annexed diagram (fig. 101), the triangle formed by these three stations is well conditioned as to size and form; the lengths of its sides are approximately 225, 275 and 310 miles, and its smallest angle is about 45 degrees.

While the proper use of these three stations will give answer to the questions of greatest economic and scientific importance there will be material scientific advantage in adding a fourth station to the system. It should be placed somewhere on the north shore of Georgian Bay, and, giving consideration to accessibility as well as geographic position, it is probable that Parry Sound should be selected. By adding this station another well-conditioned triangle would be completed, and there would result an additional determination of the rate and direction of tilting. If rate and direction vary from place to place the fact will probably be brought out. There would be additional advantage in the fact that Parry Sound and Chicago are separated by the greatest practicable distance in the direction of maximum change, so that a comparatively short period of time might afford a valuable measurement. The approximate results of the present investigation indicate that the change in the relative height of Parry Sound and Chicago in ten years would be about 2 inches.

Conditions controlling equipment.—In order to plan intelligently the system of observations, full consideration should be given to the conditions affecting the problem, and provision should be made for all possible sources of error. Prominent among these are the various factors which modify the water level at points on the lake shore. Such factors have been considered in the preceding discussion of gage data, but they are assembled here in a more systematic way.

The lake continually receives water from streams and from rain, and continually parts with water by discharge at its outlet and by evaporation. In the long run gain and loss are equal, but for short periods they are usually unequal; so that from day to day from season to season, and from year to year the volume of the lake and the consequent mean level of its surface are continually changing.

In bays and estuaries there are local temporary variations occasioned by the floods of tributary streams.

There are solar and lunar tides, small as compared to those of the ocean, but not so small that they may be neglected.

The wind pushes the lake water before it, piling it up on lee shores and lowering the level on weather shores. During great storms these changes have a magnitude of several feet, and the effect of light wind is distinctly appreciable. Even the land and sea breezes, set up near the shore by contrasts of surface temperature, have been found to produce measurable effects on the water level.

There is also an influence from atmospheric pressure. When the air is in equilibrium, if that ever occurs, the pressure is the same on all parts of the lake surface, and the equilibrium of the lake is not disturbed; but when the air pressure varies from point to point this variation of pressure is a factor in the equilibrium of the water surface, the surface being comparatively depressed where the air pressure is greater and elevated where it is less.

When a storm wind ceases, the water not merely flows back to its normal position but is carried by momentum beyond, and an oscillation is thus set up which continues for an indefinite period. A similar oscillation is started whenever the equilibrium is disturbed by differences of atmospheric pressure; and these swaying motions, called seiches, analogous to the swaying of water in a tub or hand basin, persist for long periods. In fact, they bridge over the intervals from impulse to impulse, so that the water of the Great Lakes never comes to rest.

Every disturbance which causes the water to rise on one shore of the lake and fall on the other interferes with the equilibrium between the two lakes at the Mackinac Strait. If a strong wind blows the water eastward, raising the level on the east shores of the lakes and lowering it on the west shores, there is high water at the west end of the strait and low water at the east, producing a current toward the east; and when the wind ceases the water that has poured from Lake Michigan to Lake Huron must return, producing a current in the opposite direction. Theoretically, analogous effects should be produced by tides and barometric gradients, and there can be little question of their detection if the phenomena at the strait are studied.

These various influences work independently but simultaneously, and their effects are blended in the actual oscillations of the water surface at any point. In using the water surface for the purpose of precise leveling, it is necessary to take account of all such factors and make provision for the avoidance or correction of the errors they tend to produce.

Equipment.—In view of the complexity of the phenomena to be analyzed, it is desirable that most of the instruments employed be of the automatic kind, giving continuous record. While such instruments accomplish much more than could be done by an observer alone, they do not dispense with his services. They are complex as compared to the apparatus for personal observation, and can be successfully employed only by a man of scientific training. The first essential, therefore, at each of the stations is an expert observer.

The gage employed for the determination of water height should be of some automatic type, giving a continuous record. This is necessary in order that the study of the record may furnish data for the complete elimination of errors from tides, seiches, and land and sea breezes. The gage should be protected, not only from the direct shock of waves, but from all secondary agitation of the water due to wave shock. It should be so installed as to be secure from settling. The height of its zero should be readily verifiable.

Near each station there should be at least three benches, constructed with special reference to permanence and stability. They should be independent of one another and independent of other structures.

Pressure of the air should be continuously recorded by a barograph, carefully standardized. A wind vane and anemometer should give automatic records.

At Mackinac there should also be means for securing a record of the direction and velocity of water currents.

Treatment of observations.—Stress having already been laid on the importance of putting the work in expert hands, it would be unwise to attempt the formulation of a code of instructions for either the making or the reduction of observations; but there may be advantage in a few suggestions based on experience acquired in the present investigation.

While it is doubtless possible to deduce from a study of currents in Mackinac Strait a theory of the relation of those currents to the equilibrium conditions of the two lakes, it will probably be found best to use the current observations chiefly for the discrimination of favorable and unfavorable times, and to compare the lake-level observations only for times when the current at Mackinac is gentle. It will also be better to avoid the use of observations during the prevalence of strong winds or high barometric gradients than to attempt the application of corrections for those factors.

The times not barred by high winds, high gradients and currents will ordinarily not be found to have such duration and distribution that tidal effects can be eliminated by including complete tidal cycles. It will therefore be necessary to discuss the solar and lunar tides for each station and prepare tables of correction to be applied to all observations employed. The same treatment will be necessary for the effects of land and sea breezes. Barometric gradient of amount too great to be ignored nearly always exists, and this, as determined by observations at the stations themselves, should be the subject of computation and correction.

Seiches should be fully discussed for each station, and the observations finally used should be grouped in periods of sufficient length to eliminate the seiche effect.

SUPPLEMENT.—INVESTIGATION BY MR. MOSELEY.

The main body of manuscript for this paper was prepared in June and July, 1897. An abstract was communicated in August to the American Association for the Advancement of Science, meeting in Detroit, and a fuller abstract was printed in the September number of the National Geographic Magazine.* As a result of this publication I became acquainted with a cognate investigation by E. L. Moseley, of Sandusky, Ohio. His data and results were communicated to the Ohio Academy of Sciences in December, 1897, and printed soon afterwards in a Sandusky newspaper. They received more permanent as well as fuller presentation in an article contributed to the Lakeside Magazine,† and this article reaches me while the proof sheets of the present paper are in hand. As will be readily understood from the following abstract, the data he has gathered constitute an important contribution to the subject.

North of Sandusky Bay, near the west end of Lake Erie, is a cluster of islands, of which the five largest are each several miles in extent.‡ About them the water is shallow, and if the lake were lowered 30 to 35 feet they would all be connected with the mainland. On these islands grow many species of wild plants, and the origin of this flora is related to the geologic history of the islands. There was a time during the ice retreat when the whole basin was covered by a glacial lake. If the water was gradually lowered from the plane of the glacial lake to the present plane of Lake Erie, the islands were at first barren and were eventually occupied only by such plants as were in some way conveyed across the intervening straits, from 2½ to 3 miles wide. As Moseley points out, there are many modes of such adventitious introduction, but they could not be expected to give to the islands a flora so varied as that of the adjacent mainland.

If, on the other hand, as inferred from the slopes of the old shore lines and other data, the attitude of the land was different when the

* Modification of the Great Lakes by earth movement: Nat Geog. Mag., September, 1897, Vol. VIII, pp. 233–247.

† Lake Erie enlarging; the islands separated from the mainland in recent times; by E. L. Moseley: The Lakeside Magazine, Lakeside, Ohio, April, 1898, Vol I, pp. 14–17.

‡ For the relation of these islands to the lake and the isobase of its outlet, see fig. 100, p. 640.

glacial lake was drained away, the original Lake Erie occupied only the eastern part of the Erie basin, and the western part, including the district of the islands, was dry land. Subsequently, from the tilting of the land, the lake waters advanced westward so as to flood the straits and convert the lowland hills into the present islands. In connection with such a geologic history the islands would have acquired their flora at the same time with the mainland, and should now present the same variety of species, so far as local conditions permit. Moseley has carefully compared the insular flora with that of the mainland, and finds that the only mainland species which do not occur on the islands are such as do not find there a congenial soil. The botanic evidence thus supports the geologic, and verifies the conclusion that the land has been tilted toward the southwest since the birth of Lake Erie.

The islands are composed largely of limestone and are surrounded by limestone cliffs. In South Bass or Put-in-Bay Island there are caves opening at the water's edge and partly occupied by lake water. Exploring these, Moseley finds stalactites extending from the roof down into the water, and stalagmites lying 3 or 4 feet below the present surface of the lake. Comparing the present water level with the lowest levels known in recent times, it appears that these stalagmites have not been above water during the present century, and as stalagmites are formed only in the air, it is clear that the lake has encroached on the land since they were made.

These data show only that a change has occurred, and ascribe no date, but other phenomena observed in the neighborhood of Sandusky indicate clearly that change is now in progress. A tract of land on which hay was made in 1828 is now permanently under water. A tract of land one-half mile square, surveyed in 1809, has since become marsh, with water and mud 12 to 18 inches deep. Various parts of Sandusky Bay where rushes grew within the memory of men still living are now covered with open water. "By the high water that prevailed in 1858 to 1860 large trees were killed in many places where the waves could not reach them." "Hundreds of walnut stumps are still standing on the border of the marshes east of Sandusky where even now, although the water is lower than usual, it is too wet for walnut trees to grow. One that stood recently on ground only 6 inches above the present lake level measured 5 feet 4 inches in diameter. We may infer from this that during the life of this tree, probably over 300 years, the water was not so high as in the present century." Many stumps and prostrate trunks with roots and branches attached are found from 1 to 4 feet below the present lake level, and in one locality it is inferred that the lake during the life of the trees must have been as much as 8 feet lower than "during much of the time for the last forty years."

These various facts, and others of the same tenor enumerated by Moseley, are in complete accord with the qualitative results derived from the discussion of gage readings, but, like the data gathered by Stuntz, they suggest a more rapid rate of change than do those results.

AN ACCOUNT OF THE RESEARCHES RELATING TO THE GREAT LAKES.

By Dr. J. W. Spencer, Toronto.

PREFACE.

The following "Account of the Researches Concerning the Great Lakes" was read before the Detroit meeting of the American Association for the Advancement of Science in August, 1897.* Although two summers have been spent in field work in the mountains of New York, New England and Canada, only one or two notes † have appeared, so that the only important contribution by the author on this questien since the writing of the Detroit paper is that on "Another Episode in the History of Niagara," ‡ which is a sequel to the former paper on "The Duration of Niagara Falls." An abstract of this recent paper will be appended.

Twenty years ago very little was known of the history of the Great Lakes and of Niagara Falls. Since then the origin of the lake basins has been explained by the discovery of the buried and drowned ancient Laurentian River and its tributaries, the valleys of which have been obstructed by drift, their altitude above the sea now greatly reduced, and their respective barriers in part raised up by the recent unequal tilting of the earth's crust. These physical changes have also diverted the drainage of western Pennsylvania, so that now the upper Ohio and Allegheny Rivers discharge into the Mississippi in place of into the St. Lawrence, through the Erie Basin, as they formerly did.

The after-history of the lakes has been partly studied, for their old and now deserted shore lines have been approximately mapped over a large area, and the amount of subsequent tilting of these old water lines has been measured. The most interesting feature in their subsequent history is the change of outlets, both past and in prospect. Thus the three uppermost lakes discharged through Lake

* American Geologist, Vol. XXI, pp. 110-123, 1898.

† Ib., Vol. XXII, p. 262, 1898.

‡ Am. Jour. Sc., series IV, Vol VI, pp. 639-450, 1899.

Huron towards the east, so that the Niagara did not receive any more than the waters of the Erie Basin until in a recent period. This discovery was first made by the writer in 1888,* although more recently Mr. F. B. Taylor, by a remarkable method of reasoning, has gone out of his way to ascribe it to another,† although this other has not so claimed it, so far as is known to the writer. So also the rate of rise of the earth's crust throughout the lake region, and the consequent hypothesis announced at the same time that the Falls of Niagara would probably cease to exist in the near geological future, owing to the diversion of the waters of the four upper Great Lakes to the Mississippi by way of Chicago, was first announced by the writer in 1894.‡ This has been confirmed by the subsequent researches of Prof. G. K. Gilbert,§ with almost identical results when reduced to the same standard. In his paper (p. 602) Prof. Gilbert says: "So far as I am aware this paper broaches for the first time the idea of the differential elevation of the lake region, and it contains the only observations that have been cited as showing the recent changes of that character. In late years the subject has been approached from the geologic side, and Dr. J. W. Spencer has expressed his opinion that the warping or tilting of the whole region is now in progress." From the standpoint of measuring the changes of level of the lake waters, I believe that Prof. Gilbert is correct in his claim, but they only confirm the correctness of the previous geological determinations. The tilting of the beaches has been measured by several of us, but the direction of the tilting was determinable only after my surveys on the Canadian side of the lake enabled us to triangulate the direction of the rise, and in my papers between 1888 and 1891 this deter-

* Proc. Am. As. Ad. Sc., Vol. XXVII, pp. 198-200, 1888.

† Bull. Geol. Soc. Am., Vol. IX, p. 180, 1898. See also Am. Jour. Sc., series IV, Vol. VI, p. 439 (foot-note).

‡ Am. Jour. Sc., Vol. XLVIII, pp. 455-472, 1894.

§ Eighteenth Annual Report U. S. Geological Survey, p. 639, 1898.

mination of the direction of the rise was calculated and shown only to be confirmed by all subsequent measurements, including the recent paper of Prof. Gilbert. But the difficulty awaiting us was that we did not know the rate of warping, for, if found for one district, it could be calculated for the whole lake region. Of course it was not an absolute rate of rise, but the differential rise of the lake region. However, I found that during the last 1500 years the mean rise in the Niagara district was from 1¼ to 1½ feet per century, as compared with Chicago. To make it a measurement of the absolute rise would be to add or subtract the changes of the earth's crust at Chicago, which are not known. But the rate of rise north of the Adirondack district was three times as great, and so on for other regions. With this mean rate of rise discovered as in progress for the last 1500 years, it becomes no bold prediction to apply it to the diversion of the Niagara drainage to the Mississippi and the consequent extinction of the Falls. This hypothesis was first announced in March, 1894,* or three and a half years before it was confirmed by Mr. Gilbert's estimates of the fluctuations of the lakes. The results of the latter methods almost exactly confirm the earlier geological discoveries of a rise of the earth's crust, and the consequent confirmation of the hypothesis of the extinction of Niagara, at about the same date, when the calculations are reduced to the same basis.

The history of the Falls is so intimately connected with that of the lakes, and in their archives we find that there have been several changes in the height of the Falls as well as great variations in the amount of water that passed over them during the different episodes. All of these studies are gradually leading us to more and more nearly approach the correct determination of the age of the Falls.

* Am. Jour. Sc., Vol. XLVIII, pp 455-472, 1894.

[*From The American Geologists, Vol XXI, February, 1898.*]

AN ACCOUNT OF THE RESEARCHES RELATING TO THE GREAT LAKES*.

By DR. J. W. SPENCER.

An old text-book upon geology briefly says that the lake basins are due to movements of the earth's crust. What the movements were and how they affected the history of the great lakes was left a subject of discovery for recent years. In the meanwhile, theories arose as to their origin, the disposal or modification of which was fraught with difficulties as great as those of discovering the history itself. Ramsay had attributed the origin of the American lakes to glacial excavation;† Hunt, Newberry, Carll and many others had collected the evidence of buried channels occurring in the lake region. Gen. G. K. Warren‡ had followed up the observations of Prof. H. Y. Hind § in the history of the Winnepeg basin, and proposed the northeast warping as closing the Ontario basin, to such a degree that he may be considered the father of lacustrine geology. But the great impetus towards the investigation of the great lakes is due to Prof. J. S. Newberry, || whose contribution was followed by one from Prof. E. W. Claypole.¶ To give a full account of the researches concerning the great lakes, and to tell how each author had contributed to the subject would make a very long chapter. As the present writer has been so closely connected with the pioneering study of the subject, and has announced progress

---

* Read at the Detroit meeting of the A. A. A. S., 1897.

† Quart. Jour. Geol. Soc., Lond. vol. XVIII, pp. 185-204, 1862.

‡ Appendix J., Rep. of the Chief of Engineers, U. S. A., 1875; Am. Jour. Sci. (3), vol. XVI, 1878, pp. 416-431.

§ Report on the Assiniboine and Saskatchewan Exploring Expedition. By Henry Youle Hind. Toronto, 1859, pp. 1-20.

|| Geology of Ohio, vol. II, 1897, pp. 72-80.

¶ On the pre-Glacial Geography of the region of the Great Lakes, E. W. Claypole. Can. Nat., vol. VIII, 1877, pp. 187-206.

from time to time before the American Association, it seems a fitting opportunity to tell how his investigations have been influenced by his co-workers, leaving to others the narration of the most recent studies.

Newberry followed up on the lines of Ramsay in attributing the basins of the lakes to glacial excavations, yet there was a counter current in his writings which finally advocated that the glacial excavation had taken place only after their courses had been predetermined by river action. Adopting the teachings of Agassiz and Newberry, and going much farther, an influential school was developed which attributed the superficial features of the northern regions almost entirely to the action of continental ice—in spite of the teachings of Lesley, Dawson, Whitney and others. The extreme views, as represented by Dr. G. J. Hinde,* made the ice plough dig out the St. David's, Dundas and other valleys, irrespective of their direction, as compared with that of the ice flow. Such speculations were most common at the close of the eighth decade of the century, when the writer commenced his studies upon lacustrine history—concerning which his first paper was on the "Discovery of the Outlet of the Basin of Lake Erie," etc.† (1881). The appearance of this "avant courier" was due to the enthusiastic reception given by Prof. J. P. Lesley to the writer's discovery of the reduction of rocky barriers beneath the superficial drift between Lake Erie and the Dundas Valley, at the head of Lake Ontario, indicating an outlet for the Erie basin by a channel, the lower end of which is deeply buried by drift deposits. Prof. Lesley pointed out that this discovery satisfied the necessity

* Glacial and interglacial strata of Scarboro Heights, etc. Canadian Journal, April, 1877, p. 24.

† Discovery of the Preglacial Outlet of the Basin of Lake Erie into that of Lake Ontario; with notes on the Origin of our Lower Great Lakes. By J. W. Spencer; Proc. Amer. Phil. Soc., XIX, 198, 2n., March 30, 1881, pp. 300-337.

for some such outlet to the Erie Basin, as Hunt and Newberry had found buried channels beneath the lake, and Mr. J. F. Carll had discovered that the drainage of the Upper Allegheny and other streams had been reversed, having flowed northward into the Erie Basin in preglacial days.

The writer's paper referred to not only described the outlet of the Erie Basin, but also showed that the Niagara River was not needed in ancient times. Shortly afterwards this idea was confirmed by Dr. Julius Pohlman* who found that the Niagara channel was not sufficiently deep for the drainage of the buried valleys in the vicinity of Buffalo.

In the same paper the valley-like features beneath the lake waters were analyzed and established. But at that time the course of the ancient drainage could not be traced beyond the meridian of Oswego. The writer also objected to the theory of the glacial excavation of the basins on account of the stream-like sculpturing of the land and the sub-lacustrine escarpments; and on account of the glaciation of the region being everywhere at sharp angles to the escarpments, whether above or below the surface of the lakes. These views and the discovery of the outlet for the ancient Erie Basin confirmed the teachings of Prof. J. P. Lesley, who, from being a progenitor of the science of topography, became the father of geomorphy, of which the lake history is one of the phases. In speaking of the origin of the lake valleys, Prof. Lesley† says: "For a number of years I have been urging upon geologists, especially those addicted to the glacial hypothesis of erosion, the strict analogy existing between the submerged valleys of Lakes Michigan, Huron and Erie and the whole series of dry Appalachian 'valleys of VIII,' stretching from the Hudson River to Alabama; also of Green Bay, Lake Ontario and

---

* The Life history of Niagara. By Julius Pohlman. Trans. Am. Inst. Min. Eng.

† Report Q4 of the Geological Survey of Pennsylvania, 1881, pp. 399–406.

Lake Champlain, with all the 'valleys of II and III.' One single law of topography governs the erosion of them all, without exception, whether at present traversed by small streams or great rivers, or occupied by sheets of water, the only agency or method of erosion common to them all being that of rainwater, not in the form of a great river, because many of them neither are nor ever have been great waterways."

Notwithstanding the shortcomings and what are now known to be errors of detail, the paper on the preglacial outlet of Erie attracted considerable attention as a new departure, and at the time Prof. James Geikie, who is well known to be one of the leading glacialists, expressed himself as follows, under date June 21, 1881: "I have always had misgivings as to glacial erosion of the Great Lakes, * * * and now your most interesting paper comes to throw additional doubt upon the theory in question. Possibly those who have upheld that view will now give in. Your facts seem, to me at least, very convincing. I never could understand how those great lakes of yours could have been ground out by ice. The physical conditions of the ground seem to me very unfavorable." Prof. G. K. Gilbert, on June 15, 1881, wrote: "My first geological field work was in the drift of the Erie basin, and the problem of the origin of the basins of the great lakes has always had great attraction for me. Had I been able to understand its solution, my working hypothesis would have been that which you have demonstrated so thoroughly. * * * The matter has certainly never received a demonstration until your paper appeared. * * *"

At this time the writer was struggling to find the outlet of the basins, and looked in every possible direction for buried channels without avail. While the St. Lawrence valley, beyond the outlet of lake Ontario, was evidently only a continuation of the drowned valley occupied by the lake, and while the lower St. Lawrence indicated an elevation of the continental region to more than 1,200 feet (when the cañon of the Saguenay was being excavated), the evidence of the local oscillation of the earth's crust was not yet forthcoming. The deep cañon of the Dundas valley, and the observations of Prof. Gilbert that the Irondequoit bay was drowned to a depth of 70 feet, was taken as evidence of terrestrial oscillation, but later the writer found that the St. Lawrence, after leaving Ontario, was in part flowing over a valley buried or drowned to a depth of 240 feet; accordingly the Dundas and Irondequoit valleys were no evidence of local oscillation, which had to be found elsewhere.

In concluding a notice of this early work,* the modern aspect of the Niagara River was emphasized, and the valley of St. Davids was regarded as of inter-glacial origin—in deference to the prevailing theories of the time—in place of being, as is now known, the channel of an insignificant stream of greater antiquity. The Finger lakes of New York were explained as closed up valleys which had formerly drained the rivers of the highlands of New York, as for example Seneca lake, which has since been found to be the ancient course of Chemung and its tributaries. About this time the writer, from the data collected by the Geological Survey of Pennsylvania, pointed out the probability that the Monongahela and upper Ohio had formerly been reversed and drained into the Erie valley.* This hypothesis was afterward amplified by Dr. P. Max Foshay,† disputed by Prof. I. C. White; modified and confirmed by Mr. F.

*A short study of the Features of the Great Lakes, etc. J. W. Spencer. Proc. A. A. A. S., vol. XXX, 1881, pp. 131-146; and Surface Geology of the Region about the western end of Lake Ontario. J. W. Spencer, Can. Nat., vol. X, 1882, pp. 213-236, and 265-312.

*On the ancient upper course of the Ohio river emptying into lake Erie. Proc. Am. Phil. Soc, Phil. vol. XIX, 1881.

†Preglacial Drainage and recent Geological History of western Pennsylvania. Am. Jour. Sci, vol XL, 1890, pp 397-403.

Leverett,* and finally, with some modifications, reconfirmed by Prof. I. C. White.† In order to test the validity of his objections to the hypothesis of glacial excavation, the writer visited Switzerland and Norway for the purpose of personally observing the mechanical effects of modern glaciers, with the result that he saw in them only the agents of abrasion—the ice moulding itself round obstructions, or smoothing off irregularities, and not ploughing out channels.‡ Indeed, in a more recent visit to Norway, it became apparent that the great glacial valleys still preserve many base levels of erosion—the doctrine of which has not been applied to them, and consequently their history is as yet unwritten. The extreme views concerning glacial erosion, held a decade ago, are now greatly modified and do not belong to the present day.

In 1882, fragments of great beaches, and others which were delta deposits, were described as occuring about the western end of Lake Ontario at various elevations from 500 feet above the lake down to its present level.§ Other fragments of beaches had been known for many decades, the most notable of which were the ridge roads of New York state, that Prof. James Hall, as early as 1842, found to be rising gently upon proceeding eastward;|| and the same was found to be true at the eastern end of Lake Ontario. About this time Prof. Gilbert was studying the beaches of the western lakes, and Mr. Warren Upham those of the Winnipeg basin. The beaches in both places were found to record the evidences of gentle terrestrial movements. Following up his investigations, Prof. Gilbert connected the various fragments of a great beach upon the

*Pleistocene fluvial plains of western Pennsylvania. Am. Jour. Sci., vol. XLII, 1891, pp. 200-212; and Further studies of the Upper Ohio basin. Am. Jour. Sci., vol. XLVII, 1894, pp. 247-283.

†American Geologist, vol. XVIII, 1896, pp. 368-379.

‡The erosive power of glaciers as seen in Norway. Geol. Mag. Lond., Dec. iii, vol. IV, 1887, pp. 167-173.

§Surface Geology about the region of the western end of Lake Ontario, cited before.

||Geology of New York. Vol. IV. 1843, p. 351.

southern and eastern sides of Lake Ontario, as far as Adams Centre, near Watertown, N. Y.,* and found that the old waterline was deformed to the extent of several hundred feet in proceeding north-eastward. This was an admirable piece of work, which was invaluable to the writer, who extended the observations further† and made use of them in measuring the amount of the long sought for terrestrial deformation as the outlet of Lake Ontario, and found that these post-glacial movements were sufficient to account for the rocky barrier across the Laurentian valley, producing the basin which retains the waters of Lake Ontario. The channels across this rocky barrier, however, were closed with drift deposits reaching to a depth of 240 feet. In thus establishing the ancient drainage of the Ontario basin, after years of observation, often representing but little progress, the phenomena of the basin were discovered without the glacial theory of erosion. Then the writer found that the drowned channels across Lake Huron, and passing through Georgian bay, continued beneath hundreds of feet of drift, eastward of the Niagara escarpment, and joined the Ontario valley a few miles east of Toronto. A similar channel (the Huronian) crossed the State of Michigan, passed through Saginaw bay, and over the sub-lacustrine escarpment, to the deeper channel of the Huron basin.‡ The Erie (Erigan river) drainage had been found to pass into the head of the Ontario basin. Thus was discovered the course of the ancient Laurentian river and its tributaries of antiquity. These upper basins were also affected by the terrestrial tilting recorded in the beaches, as well as by the drift obstructing them.

Prof. Gilbert, who had, many years before, mapped beaches the head of lake Erie§ afterwards measured the deformation recor

* Report of the meeting of the Am. Assc. Adv. Sci , Science, Sept., 1885, p. 222.

† The Iroquois Beach: a Chapter in the Geological History of Lake Ontario, by J. W. Sp Trans. Roy. Soc Can., 1889, pp. 121-134. (First read before Phil. Soc., Wash., March, 188

‡ Origin of the Basins of the Great Lakes. Q. J. G. S. (Lon.), vol. XLVI, 1890. pp. 523-5

§ See Geology of Ohio, vol. II, 1874.

in the deserted shore at the eastern end of the lake;* while the writer surveyed the old water margins across Michigan, and on the Canadian sides of Lakes Ontario, Erie and Huron, and in portions of New York.† After this, very little work was done upon the deserted shores for several years, when Mr. F. B. Taylor commenced his researches about the northeast portion of Georgian bay, Lake Michigan, etc.,‡ and Dr. A. C. Lawson carried on similar observations north of Lake Superior,§ and Prof. H. L. Fairchild in New York. The deserted beaches show but little terrestrial oscillation about the western end of lake Erie, but it increases towards the northeast and amounts from four to seven feet per mile.

With the surveys of the deserted beaches, new questions arose concerning the history of the lakes and of Niagara River, which forms an inseparable chapter. At the same time, opposing hypotheses presented themselves.

None of the beaches have been fully surveyed. They occur at various altitudes from near the greatest elevation of the land down to the levels of the lakes, and they have not always been separated from other Pleistocene deposits. While there are questions as to the higher forms, those from lower levels have undoubtedly been accumulated about extensive bodies of water—the character of which is the subject of disagreement. The writer has regarded them as accumulations at sea-level, and other observers as margins of glacial lakes, irrespective of their elevation. The theoretical aspect is not one likely to be settled speedily. Those who advocate the

* The History of the Niagara River. 6th Rept. Com. State Res. Niag., Albany. 1890, pp. 61-84.

† The Iroquois Beach, etc., cited before. Deformation of the Iroquois Beach and Birth of Lake Ontario, Am. Jour. Sci., vol. XL, 1890, pp. 443-451; Deformation of the Algonquin Beach and Birth of Lake Huron, Ib., vol. XLI, 1891, pp. 11-21; High Level Shores in the Region of the Great Lakes, and their Deformation, Ib., vol. XLI, 1891, pp. 201-211; Deformation of Lundy Beach and Birth of Lake Erie, Ib., vol. XLVIII, 1894, pp. 207-212.

‡ Numerous papers recently published in Am. Jour. Sci., American Geologist, and Bul. Geol. Soc. Am.

§ Sketch of the Coastal Topography of the North Side of Lake Superior. 20th Report of the Geol. Sur. Minnesota, for 1891, pp. 181-289.

glacial character of the lakes have sought to terminate the beaches against morainic deposits to the northeast, but their ice dams have been frequently thrown along lines beyond which the beaches have subsequently been traced. Thus Prof. Claypole* made ice dams in Ontario where open water, bounded by beaches, was afterwards found to prevail. At Adams Centre, Prof. Gilbert drew an ice dam for the Ontario basin, beyond which, however, the writer found that the old shore line extended, and this was later confirmed by Prof. Gilbert. Mr. Leverett made an ice dam at Cleveland, beyond which the writer has been informed by two observers that the beach extends, and Prof. Gilbert and Mr. Leverett described another glacial dam near Crittenden, N. Y., beyond which the beaches have been discovered by Prof. Fairchild. Another diagnosis of the glacial lake is the occurrence of gravel floors over low divides, which are regarded as the outlets of them, and upon this feature alone many such lakes have been named. But the advocates of these glacial outlets have not explained how the terraces (at hundreds of feet above the drainage) upon the southern side of them are indistinguishable in character from those upon the northern side.† If these supposed outlets be evidence *per se* of glacial dams then the most perfect which the writer has ever seen may be found within 16° of the equator, at an altitude of less than 800 feet, suggesting that the Mexican gulf had a glacial dam, discharging into the Pacific ocean across the Isthmus of Tehuantepec — a suggestion which no one would seriously consider. The writer has also presented the hydrostatic objections‡ to the impossible long continuance of some of the supposed dams, the location of which demands their drainage across ice itself, which

*Report of the meeting Am. Assoc. Adv. Sci. Science, Sept., 1895, p. 222.

†Channels over divides not evidence *per se* of glacial dams. J. W. Spencer. Bull. Geol. Soc. Am., vol. III, 1891, p. 491.

‡Post-Pliocene continental subsidence versus ice dams, by J. W. Spencer. Bull. Geol. Soc. Am., vol. II, pp. 465–476, 1890.

would soon be penetrated by the warmer waters so as to reduce their level. By straightening out the deformation recorded in the deserted shore-lines, some of the beaches are shown to have undoubtedly been formed at sea level.* While recent surveys report the discovery of additional glacial lakes, or the splitting up of those first described under new names, the survey of the high level terraces in the mountain regions has suggested to the writer counterbalancing evidence of the occurrence of glacial dams, but this is a study which has been postponed, partly on account of the prejudice against post-glacial subsidence and partly on account of the writer's absorption in other questions of physical changes. Whatever may be the ultimate fate of the theory of glacial dams, the opposing hypotheses have given zest to the investigations to the degree of advancing our knowledge of the lake history.

In the survey of the beaches, besides the terrestrial deformation recorded, there seems to be no more important discovery than when the writer found how the Huron, Michigan and Superior waters (the Algonquin gulf or lake) originally emptied to the northeastward of the Huron basin in place of discharging into Lake Erie; after which, by the northeastern tilting of the land, "the waters were backed southward and overflowed into the Erie basin, thus making the Erie outlet of the upper lakes to be of recent date."† This conclusion was established by the survey of the Algonquin beach which recorded the necessary tilting. The first survey was suspended near Balsam lake, where an overflow was found; and, accordingly, in the original announcement, the generalizations were not carried farther; although there was a lower depression in the vicinity of Lake Nipissing, which was shortly afterwards made use

*The Iroquois Beach, etc., cited before; and. Deformation of the Iroquois Beach, cited elsewhere.

†Proc. A. A. A. S., vol. XXXVII, 1888, p. 199.

of by Prof. Gilbert* and the writer. With the further elevation of the land, the lower beaches — partly measured at that time (1887–8)—represented the surface of the Algonquin water discharging by the Nipissing route alone.† This has since been worked out by Mr. Taylor.‡

Co-existing with the Algonquin gulf or lake was the Lundy gulf or lake, occupying part of the Erie basin, and extending into the Ontario, having substantially the same level. Both of these bodies of water extended much farther towards the northeast than their successors, although more contracted in the opposite directions — the effect of the more recent tilting of the land. Prior to the existence of these separate bodies of water, higher shore-lines were formed, and the great gulf or lake bounded by them was called the Warren water, which name the writer has defined as applicable to the great open water of the region, until after the formation of the Forest Beach — its most perfect episode — after which it was dismembered into the Algonquin and Lundy waters.§

During the changing stages of Warren water, its configuration was somewhat varied but not sufficiently to call the water by a multiplicity of names, according to the changing levels. The old shore lines form prominent features, requiring nomenclature for the most important. And additional naming only adds confusion. Some of the beaches have been renamed by Mr. Leverett,|| contrary to the usage of naturalists.

With the continued elevation of the land, the Algonquin water sunk to the level of the Nipissing beach (of Taylor) and the Lundy became dismembered, and formed an insignificant Lake Erie.¶ In

*The history of the Nipissing River.

†Deformation of the Algonquin Beach, cited before.

‡The Ancient Strait of Nipissing. F. B. Taylor. Bull Geol. Soc. Am., vol. V, 1893.

§ High-level shores in the region of the Great Lakes, etc., cited before.

|| On the correlation of the New York moraines with the raised beaches of lake Erie, by Frank Leverett. Am. Jour. Sci., vol. L, 1895, pp. 1-20.

¶Proc. A. A. A. S., 1888, p. 199.

the Ontaria basin, the water sunk to tho Iroquois beach and lower levels, and Niagara Falls had their birth, after the river had first been a strait. Remnants of beaches of that time were long ago observed, not only in the vicinity of Niagara, but also at the head of the lake. With the temporary pauses recorded, the waters of the upper level were speedily lowered to that of the Iroquois beach, and then the Niagara river descended only 200 feet, in place of 326 feet, as at present. The effect of this diminished descent upon the excavating power of the falls was first pointed out by the writer in 1888* and published in 1889. With the continued lowering of the waters in Ontario basin, the descent of the Niagara increased to 80 feet more than at present, as first shown by Prof. Gilbert; but later, by the tilting of the earth's crust north of the Adirondack mountains, the outlet of the Ontario basin was raised, causing the backing of the waters, so as to reduce the descent of Niagara river to its present amount.

In 1886, after the third survey of Niagara Falls (by Prof. Woodward), the rate of recession was found to be much greater than had formerly been supposed. Prof. Gilbert then made a short study of the falls, the conclusions concerning which are summed up as follows by that author: † "The problem admits of expression in an equation:

Age of gorge equals $\dfrac{\text{Length of gorge.}}{\text{Rate of recession of falls.}}$
—Effect of antecedent drainage.
—Effect of thinner limestone.
—Effect of thicker shales.
—Effect of higher fall.
—Effect of more floating ice.
±Effect of variation of detrital load.
±Effect of chemical changes.
±Effect of changes of river volume.

* The Iroquois Beach, etc. Trans. Roy. Soc. Can., 1889, p. 132.

† The Place of Niagara Falls in Geological History. G. K. Gilbert. Proc. Am. Adv. Sci., Vol. XXXV, 1886, pp. 222-223.

"The catchment basin was formerly extended by including part of the area of the ice sheet; it may have been abridged by the partial diversion of Laurentian drainage to other courses." He had divided the length of the gorge by the maximum rate of recession, finding the product to be 7,000 years. If the equation be carefully examined, together with the cited quotation, all the important changing effects in the physics of the river would lessen the estimated age of the cataract below 7,000 years, except the effect "by partial diversion of the Laurentian drainage to other courses," of which no evidence was suggested; nor was any lengthening of time shown as necessary, by the long interior hight of the falls, Henceforth, Prof. Gilbert was naturally quoted as an authority that the age of the falls was only 7,000 years. This conclusion did not satisfy the writer, who from the evidence of the beaches, especially the Iroquois,* found that the rate of recession must have been for long ages much less than now, on account of the interior height of the falls; and also on account of the greatly diminished volume of water, owing to the overflow of the upper lakes to the northeast, until in recent days. But how much of the work of the falls had been done before the upper lakes were turned into the Niagara drainage, for a long time seemed undeterminable, until the features of Foster's flats were used for measuring the amount of work performed in that early episode. This standard has since been confirmed by other phenomena not yet published; and from a different standpoint the distance of the early recession has been agreed to by Prof. Gilbert, who now considers the age of the falls far greater than that formelry suggested by his paper in 1886. From all the available data up to 1894, the writer computed the age of Niagara Falls at 32,000 years.† Of the various episodes that of the cataract passing the narrows of the whirlpool rapids sti

*See Trans. Roy. Soc. Can., 1889, p. 132; and Proc. A. A. A. S., 1888, p. 199.

† *Duration of Niagara Falls.* Am. Jour. Sci., vol. XLVIII, 1894, pp. 455-472.

seems the most difficult of explanation; but the writer has recently found that the narrows record a second reduction in the amount of fall in the river, before the present descent was established, thus retarding the recession along this section of the gorge, and increasing in part the time compensation for the reduced amount of work performed. However, further discoveries are necessary to fully explain the phenomenon of the narrows. It now seems probable that the error in determining the time required for the recession of the falls through the section of the whirlpool rapids would not affect the computation of the whole age of the river by more than a few per cent.

No less important than the determination of the age of the river was that of the date when the waters of the Algonquin basin (Huron, Michigan and Superior) were first turned into the Niagara drainage, owing to the warping of the land, with the greatest rise occuring along an axis trending N. 28° E.* The date of the diversion of the waters of the upper lakes from the Ottawa to the Niagara valley has been computed by the writer at 7,200 years. This result was obtained from the mean of three distinct methods of computation, varying from 6,500 to 7,800 years.† Mr. F. B. Taylor's more recent estimate gives the range of from 5,000 to 10,000 years.

Niagara as a time piece would be incomplete without indicating the changes in the near future. From the northeastward tilting of the lake region, it was computed that in 5,000 years, not merely Niagara Falls would cease to exist, but also that the drainage of the deepest part of the Niagara river at Buffalo (45 feet) would be reversed and turned into lake Erie, whose outlet would then be through lakes Huron and Michigan into the Mississippi river by

*This direction occurs east of Georgian bay, while at the end of Lake Ontario the direction of rise is N. 25° E. See papers by the writer cited before

†See Duration of Niagara Falls, cited before.

way of Chicago. This inference was based upon the long delayed discovery of the rate at which the earth's crust has been rising in the lake region, — which was found to be for the Niagara district 1.25 feet per century more than the rate of rise at Chicago.* With this determination it was easy to calculate the rate of terrestrial deformation for other regions, — thus northeast of lake Huron the rise has been found to be two feet per century, and north of the Adirondacks, the warping is progressing at 3.75 feet in a hundred years.

The rate of deformation of 1.25 feet per century, in the Niagara district, was the minimum calculation, with a possible maximum of about 1.5 feet per century. The approximate correctness of the determination has just been confirmed by a paper presented to the American Association, by Prof. G. K. Gilbert, immediately before this communication was read.† He had used the bench-marks at various localities where the fluctuations of the lake levels have been registered the last 20-37 years. While the recorded measurements vary from about one to two and a half inches during the periods of observation, they have been extended over the lake region, with results closely agreeing with the previous determinations of the writer. This will be better understood using Professor Gilbert's application — namely, — that in 500-600 years, the Erie waters would be on a level with those of lake Huron — in 1,000 years they would overflow the natural divide near Chicago — in 2,500 years, the waters would cascade into the Niagara gorge only during high water — and in 3,000 years, the falls would be entirely drained. These changing conditions, based upon the writer's previously discovered rate of terrestrial deformation, would take — 720 years for the Erie and Huron waters to be on the same level; 1,280

* See Duration of Niagara Falls, cited before.

† Modification of the Great Lakes by earth movements. Nat. Geog. Mag., vol. VIII, 1897, pp. 233-247.

years for the overflow into the Mississippi drainage (the artificial canal would reduce this estimate to 720 years); and 2,600 years for the general drainage of the lakes into the Mississippi. In 5,000 years the whole river as far as Buffalo would be drained towards the south.

In spite of taking the minimum rate of recession and the probable errors, the closeness of these results satisfactorily confirms many of the calculations based upon Niagara as a geological chronometer.

This paper, giving the principal results of investigations into the lake history, thus shows the writer to have been greatly affected by the studies of his co-workers. Indeed, all of the researches by the different observers have been very much dove-tailed, so that our present knowledge of the history of the Great Lakes and Niagara Falls is the result of the labors of many individuals. Besides the names of those already mentioned, we should add those of Shaler, Tarr, Wright, Russell, Upham, Kibbe, Lincoln, Brigham and Scovill with the names of Hall and Lyell, too well known to need special mention.

To complete the review, mention should be made of the writings of Mr. F. B. Taylor, in connection with his important survey of the Nipissing outlet of the Algonquin basin, and of the dissected shore lines of the upper lakes; and of the important investigation of Central New York by Professor Fairchild.

APPENDIX.

The recent paper on "Another Episode in the History of Niagara"* by the writer is important in its showing that after the waters of Lake Ontario had reached their lowest state, and the Niagara its highest fall, the waters of the Ontario basin were backed up into the then made gorge, owing to the progressive tilting of the earth's crust at the outlet of the lake, so that the waters rose in the Niagara gorge to a height of 75 feet above their present level, and thus reduced the efficiency of the Falls, while they were receding past the section of the whirlpool rapids; thus this diminished height helps to explain the shallowness of the Niagara river along this section. The waters were again lowered to their present level by the St. Lawrence cutting down its channel more deeply into the rim of the water basin. In this paper there is also a revision of the episode of the Falls, correcting and reducing greater accuracy the previous investigations.

* Read before the Am. As. Ad. Sc., Boston, Aug., 1898. Published in Am. Jour. Sc., Ser. IV, Vol. VI, pp. 439-450, 1898; also, Proc. Can. Inst., Vol. I, pp. 101-103, 1898. (This abstract was the first announcement of the collecting of this episode.)

SIXTEENTH ANNUAL REPORT

OF THE

COMMISSIONERS

OF THE

State Reservation at Niagara.

FROM OCTOBER 1, 1898, TO SEPTEMBER 30, 1899.

TRANSMITTED TO THE LEGISLATURE JANUARY 31. 1900.

ALBANY:
JAMES B. LYON, STATE PRINTER.
1900.

STATE OF NEW YORK.

No. 40.

IN ASSEMBLY,

JANUARY 31, 1900.

SIXTEENTH ANNUAL REPORT

OF THE

Commissioners of the State Reservation of Niagara.

NEW YORK, *January* 20, 1900.

To the Honorable the Speaker of the Assembly:

Sir: I herewith transmit for presentation to the Legislature the sixteenth annual report of the Commissioners of the State Reservation at Niagara, for the fiscal year ended September 30, 1899.

Respectfully yours,

ANDREW H. GREEN,

President.

SIXTEENTH ANNUAL REPORT

OF THE

COMMISSIONERS

OF THE

STATE RESERVATION AT NIAGARA.

FOR THE FISCAL YEAR FROM

OCTOBER 1, 1898, TO SEPTEMBER 30, 1899.

Commissioners.

ANDREW H. GREEN, President.

GEORGE RAINES, CHARLES M. DOW,
THOMAS P. KINGSFORD, ALEXANDER J. PORTER.

Treasurer and Secretary.

RICHARD F. RANKINE.

Superintendent.

THOMAS V. WELCH.

REPORT.

To the Honorable the Legislature of the State of New York:

The Commissioners of the State Reservation at Niagara, as required by law, submit their report for the fiscal year begun October 1, 1898, and ended September 30, 1899.

During the past year many substantial improvements have been made upon the State Reservation.

The stone arched bridge from Goat Island to the First Sister Island has been completed, the approaches filled and grading and planting done about the bridge, so as to harmonize the structure as much as possible with the surrounding natural conditions.

The bridge is massive, safe and commodious, and in a few years the shrubs and vines, planted about the approaches, will give it a natural and graceful setting, in keeping with the rustic features of the locality.

The channel between the islands above the bridge has been deepened, and a larger volume of water, usual in former years of higher water in the great lakes, obtained, thus restoring the beauty and attractiveness of that part of the Reservation.

This year a new macadam road has been constructed along the rapids from Prospect park to the bridge to the Islands. It furnishes a delightful drive along the rapids, and is one of the works of improvement most appreciated and enjoyed by visitors.

The Loop driveway, in Prospect Park, has also been completed during the year, and now a continuous drive is provided for visitors along the river bank from the bridge to the Islands, to the bridge leading to the Canadian side of the river.

Four more of the old frame buildings have been removed from Prospect park during the year just closed. In all thirteen frame buildings have been removed from a territory about twelve acres in extent. Perhaps no improvement carried out since the establishment of the Reservation has been so noticeable as that caused by the removal of the parapet wall, the platform at Hennepins view and the frame buildings and the fences at Prospect Park. That part of the Reservation once more presents a grove-like appearance, with vistas commanding views of the rapids and the islands, and views of the Falls and of the gorge of the river, heretofore obscured by numerous unsightly and incongruous buildings. This change has been mainly wrought during the past two years, and the result has been a revelation of beauties of Niagara that have been shut out from visitors by artificial structures for many years.

Another improvement has been made during the past year on the riverway between the Bowlder bridge and the Loop driveway, where the shore in some places was low and marshy, and at others marred by erosions caused by high water and floating ice. Material has been obtained without expense to the State, and a strip of land made from 50 to 100 feet in width, and more than a quarter of a mile long, every part of it commanding fine views of the rapids, the islands and the upper Niagara River. The water's edge has been riprapped with large stones to prevent future erosions.

The old shore line has been substantially restored, and, when viewed from Goat Island, is in keeping with the shore along the rapids, and presents a natural appearance which will continue and improve as time goes by.

The enlargement of the hydraulic canal, adjacent to the Reservation, affords a fortunate opportunity for obtaining the large stones used in riprapping the river shore, without expense except for hauling.

August 16th these improvements were viewed by Governor Roosevelt, who made an official visit to the State Reservation, and on August 24th the works of improvement were examined by Senator Ellsworth, President pro tem. of the Senate; Senator Higgins, chairman of the Finance Committee, and Senator Stranahan.

The details of the works of improvement and maintenance are set forth in the report of the Superintendent, which is appended to this report. It will be seen that the territory planted and laid out with roads and walks that now have to be kept in order is several times as great as formerly, while the appropriation for maintenance has remained the same. As a result, an increased appropriation is desired to keep the grounds in order and the buildings, bridges, stairways and railings in a safe and proper condition.

The plans for a shelter and administration building, for which an appropriation was made at the last session of the Legislature, are being made by the State Architect. The building will be erected in the spring as soon as the weather will permit.

When this shelter building is erected the inclined railway building, which stands near the brink of the Falls in Prospect Park, may be removed. It is inadequate and unsuitable and greatly obstructs the view. It is proposed to replace it by an underground structure for the operation of the inclined railway, which will leave the view of the Falls unobstructed and greatly improve the appearance of the Reservation from Goat Island and the Canadian side of the river.

As the inclined railway cannot be operated by water power at times during the winter season when there are ice jams in the river, it is proposed, when the new underground structure is constructed, to operate the inclined railway by electrical power, which can be obtained free of charge by the Commissioners of the Reservation, as provided by chapter 513 of the Laws of 1892. It is also provided in said law for furnishing water and electricity, for light, free of charge, for the Reservation grounds and buildings, and as an improved system of lighting the grounds and buildings is greatly needed, and an adequate water supply for sprinkling the lawns, the Commissioners are considering the propriety of also availing themselves of these provisions of the law for the benefit of the property of the State.

The estimated cost of constructing an underground station for the operation of the inclined railway, and installing electrical power, and for providing a proper system of electric lighting and an adequate water supply, are included in the estimate of work necessary to be done for the ensuing fiscal year.

The Pan-American Exposition, to be held in Buffalo in 1901, will bring greatly increased numbers of visitors to Niagara.

The State Engineer and Surveyor has been requested to make an examination and report the cost of placing and maintaining the bridges, buildings and other structures in a safe and proper condition.

The work of substituting more commodious and substantial structures for the remaining wooden bridges to the Three Sister Islands should be continued and completed during the coming year.

It will be necessary to make extensive repairs upon the other bridges and upon the inclined railway and the Biddle stairs.

For these reasons the appropriations for maintenance and improvements should be considerably increased for the coming year.

Mr. Henry E. Gregory, treasurer and secretary of the board since 1888, presented his resignation November 21, 1898, and was succeeded February 1, 1899, by Mr. Richard F. Rankine.

The following is an estimate of the work necessary to be done and of the expenses of maintaining said Reservation for the ensuing fiscal year, ending September 30, 1901:

Construction.

Underground station for the inclined railway and new structure for covering inclined railway......	$15,000 00
Electrical installation for same and lighting........	5,000 00
System of water pipes for grounds and buildings...	5,000 00
Bridge from First to Second Sister island..........	10,000 00
Bridge to Horseshoe Falls at Terrapin Point.......	1,000 00
Approaches to, and guard railings, Luna island....	5,000 00
Bridges	100,000 00

Maintenance.

Salaries office and traveling expenses..............	$4,600 00
Reservation police and caretakers.................	5,400 00
Labor..	15,000 00
Materials, tools, etc..............................	15,000 00
	$40,000 00

Estimated receipts from October 1, 1899, to September 30, 1900:

Inclined railway	$6,500 00
Cave of the Winds............................	1,200 00

Ferry and steamboat landing....................	$300 00
Carriage service	100 00
Interest...	50 00
	$8,150 00

The report of the treasurer for the fiscal year is herewith submitted, showing the receipts and expenditures for maintenance and improvements, the remittances to the State Treasurer, and the classification of the accounts in detail.

From the statement of the remittances to the State Treasurer it will be seen, that the receipts from the inclined railway and rentals, received and paid into the State Treasury, amount to $8,563.90—more than one-third of the amount appropriated for the maintenance of the State Reservation during the past fiscal year.

The island between the mainland and Goat Island has been known as Bath Island. In honor of the Hon. Andrew H. Green, who has been a zealous and efficient member of the Board of Commissioners of the State Reservation at Niagara, since its establishment in 1883, and almost continuously the president of the Board, on November 16, 1898, the Commissioners by resolution changed the name of Bath Island to Green Island. As the island is a sloping green lawn, the name of Green Island is doubly appropriate.

Chapter 710 of the Laws of 1889 provides that "the commissioners of the State Reservaton at Niagara, by and with the consent of the commissioners of the land office, may construct, without expense to the state, street railroad tracks, upon and along that part of the riverway, so called, between Falls and Niagara streets, in the city of Niagara Falls, and in their discretion may grant revocable licenses, to street surface railroad companies, to use such tracks, upon such terms as said commissioners may prescribe."

A petition having been received from the several local street surface railroad companies for the construction of such a track in that portion of the riverway, the subject was referred to a special committee consisting of Commissioners Dow, Kingsford and Raines, to carefully examine into the matter and report. September 1st, a report in favor of the construction of such a track was submitted and adopted, President Green dissenting. An opinion by the Hon. Frank W. Stevens, upon the legal questions involved, was submitted and approved.

September 23d, a revocable license, as provided by law, was granted to the Niagara Falls and Suspension Bridge Railway Company to use such track, as provided in chapter 710, Laws of 1899. The report of the special committee, the legal opinion and the revocable license and agreement in the matter are hereto appended.

In the year 1890 a topographical survey of the Falls of Niagara was made under the direction of the State Engineer and Surveyor, and a valuable and exhaustive report submitted. During the ten years that have elapsed since that time, great masses of rock have fallen from the crest of the falls, causing many changes in the contour of the Cataract. The recession of the falls is a subject of great interest to the public, especially to scientists at home and abroad, and it is very desirable that a resurvey should be made during the season of 1900, so as to show the changes of the past ten years, and have recorded the exact form, location and other topographical features of the Cataract, at the close of the nineteenth century.

The estimated cost of the resurvey is included in the estimates for work to be done during the coming fiscal year.

A bill has been introduced in the Congress of the United States providing for the construction of a dam or jetties at the head of the Niagara River, at Lake Erie, in order to raise the level of the water in the great lakes.

As such an obstruction would be liable to diminish the volume of water flowing over the falls, and thus to injure the beauty of the natural scenery of Niagara, which the State of New York has expended its means and exerted its authority to protect, the proposed legislation by Congress is a matter of concern to the Legislature and the people of the State, and to the public generally, at home and abroad. The necessity for legislation upon the subject, if any such necessity exists, is probably due to the excessive diversion of the waters of the great lakes, thus referred to in the previous report of this Commission.

"The volume of the river and cataract at Niagara is of course dependent upon the water supply of the Great Lakes. The Niagara River is but the overflow of Lake Erie, into which flows the waters of the other lakes. The lowering of the level of these lakes would diminish the flow into Lake Erie and reduce the volume of the Niagara River. Any very large withdrawal or diversion of water from one or more of the Great Lakes would scarcely fail to be noticeable in a reduced flow at the cataract.

"The commissioners deem it advisable that the National government be requested to appoint a commission to confer with a Canadian commission as to the means to be devised to prevent any excessive diversion of the waters of the Great Lakes, and to consider the whole subject of the uses and control of these waters and to report its conclusions to Congress, with such recommendations as it may desire to submit.

"Any measure of this kind, which threatens even remotely the investment by the people, for the preservation of the beauty of the natural scenery of Niagara, should be closely questioned. Through the timely and providential intervention of the State, the great cataract once more has a graceful and becoming environment.

"Nearly fifteen 'years' have passed since the establishment of the reservation, and the people of the State have had time to judge of the merits or demerits of the new departure in State policy. It is now evident that the removal of the buildings, toll gates and fences, and allowing free access to Niagara, was a people's movement of the broadest kind.

"The Niagara experiment has shown the wisdom of setting aside great natural features of forest, cataract, mountain and seashore for the common enjoyment of the people.

"The State Reservation will be an ever-increasing source of gratification and enjoyment. The Falls of Niagara are the crowning natural feature of our Commonwealth. As the years go by, the people of the State will appreciate more and more the value of their most noted possession."

Respectfully submitted,

ANDREW H. GREEN,

President.

GEO. RAINES,

ALEXANDER J. PORTER,

CHAS. M. DOW,

THOMAS P. KINGSFORD,

Commissioners of the State Reservation at Niagara.

REPORT OF THE TREASURER

FOR THE

YEAR BEGINNING OCTOBER 1, 1898, AND ENDING SEPTEMBER 30, 1899.

Report of the Treasurer.

1898.				
Oct.	1.	Balance on hand this date		$134 71
		Receipts (Maintenance).		
Oct.	14.	Quarterly advance from the State Comptroller	$6,250 00	
1899.				
Jan.	23.	Quarterly advance from the State Comptroller	6,250 00	
April	17.	Quarterly advance from the State Comptroller	6,250 00	
July	19.	Quarterly advance from the State Comptroller	6,250 00	
				25,000 00
		Special appropriation as per Chapter 606, Laws of 1898:		
1898.				
Oct.	7.	Payment by State Comptroller on account	$1,970 75	
Oct.	11.	Payment by State Comptroller on account	1,700 00	
Oct.	18.	Payment by State Comptroller on account	2,017 71	
Oct.	26.	Payment by State Comptroller on account	565 00	

1898.			
Nov. 12.	Payment by State Comptroller on account	$490 61	
Dec. 8.	Payment by State Comptroller on account	3,060 12	
Dec. 16.	Payment by State Comptroller on account	132 68	
			$9,936 87
Special appropriation as per Chapter 569, Laws 1899:			
1899.			
July 19.	Payment by State Comptroller on account	$1,179 44	
Aug. 5.	Payment by State Comptroller on account	1,424 42	
Sept. 9.	Payment by State Comptroller on account	739 25	
Sept. 15.	Payment by State Comptroller on account	75 50	
			3,418 61
1898.			
Nov. 1.	Draft on Bank of Niagara for October receipts	$315 70	
Dec. 1.	Draft on Bank of Niagara for November receipts	56 35	
1899.			
Jan. 3.	Draft on Bank of Niagara for December receipts	28 85	
Feb. 1.	Draft on Bank of Niagara for January receipts	85 05	

1899.				
March	1.	Draft on Bank of Niagara for February receipts	$501 35	
April	1.	Draft on Bank of Niagara for March receipts	69 15	
May	1.	Draft on Bank of Niagara for April receipts	54 80	
June	1.	Draft on Bank of Niagara for May receipts	272 85	
July	1.	Draft on Bank of Niagara for June receipts	549 85	
Aug.	1.	Draft on Bank of Niagara for July receipts	1,714 50	
Sept.	1.	Draft on Bank of Niagara for August receipts	3,144 05	
Sept.	30.	Draft on Bank of Niagara for September receipts	1,746 40	
				$8,538 90
May	1.	Dividend on deposits in Cataract Bank		99 22
1898.				
Dec.	31.	Interest on balance in Manufacturers and Traders' Bank	$13 64	
1899.				
March	31.	Interest on balances in Manufacturers and Traders' Bank	14 36	
				28 00
1898.				
Dec.	1.	Sales from Reservation		25 00
		Total		$47,181 31

Expenditures.

Abstract CXI.

1898.	No. of voucher.			
Oct. 11.	1623..	Niagara Falls Hydraulic Power Manufacturing Co., electric lighting.....	$50 00	
	1624..	Walker & Paterson, tools and supplies............	13 04	
	1625..	Electric City Ice Co., ice...	13 10	
	1626..	Irwin Electrical Co., Inclined Railway..........	9 50	
	1627..	James Davy, stationery....	9 90	
Oct. 13.	1628..	Alex. J. Porter, Commissioner, traveling expenses.................	10 00	
Oct. 24.	1629..	Andrew H. Green, President, traveling expenses..	32 70	
Nov. 1.	1630..	Pay-roll for October.......	1,983 75	
	1631..	Thomas V. Welch, superintendent's office, expenses, etc....................	49 92	
				$2,171 91

Abstract CXII.

1898.			
Nov. 22.	1632..	Pay roll..................	$403 62
	1633..	Thos. E. McGarigle, tools..	43 12
	1634..	Grout Electric Co., Inclined Railway...............	27 50

Date	No. of voucher.		
1898.			
Nov. 22.	1635..	Power City Carting Co., buildings	$22 75
	1636..	P. J. Davy, buildings	26 13
	1637..	P. C. Flynn & Son, buildings, Inclined Railway	80 02
	1638..	A. C. Kugel, roads	1 02
	1639..	N. F. Hydraulic Power and Mfg. Co., electric lighting	50 00
	1640..	Walker & Paterson tools, buildings	14 27
Dec. 1.	1641..	H. E. Gregory, treasurer and secretary, office and traveling expenses	82 34
	1642..	Pay-roll for November	1,595 28
	1643..	Thomas V. Welch, Superintendent, office expenses, etc	43 76
10.	1644..	Pay-roll	66 00
31.	1645..	H. E. Gregory, Treasurer and Secretary, salary for October, November and December	275 00
	1646..	Pay-roll for December	1,380 08
	1647..	Alex. Henschel, clerk, April to December 31	50 00
			$4,160 89

Abstract CXIII.

1899.	No. of voucher.			
Jan. 6.	1648..	Thomas V. Welch, superintendent, office expenses, etc.	$47 36	
25.	1649..	P. J. Davy, plumbing	229 75	
	1650..	N. F. Hydraulic Power and Mfg. Co., electric lighting	50 00	
	1651..	Wm. Young, carting	14 10	
	1652..	Walker & Paterson, tools and supplies	14 35	
				$355 56

Abstract CXIV.

1899.			
Feb. 3.	1653..	Pay-roll for January	$1,195 67
	1654..	Pay-roll, supplemental, January	447 50
	1655..	H. E. Gregory, treasurer and secretary, salary January	91 66
	1656..	H. E. Gregory, treasurer and secretary, office expenses	24 64
	1657..	H. E. Gregory, treasurer and secretary, commissioner, traveling expenses	61 76
	1658..	James Davy, stationery	19 20

1899.		No. of voucher.		
Feb.	3.	1659.	.Gazette Publishing Co., office	$17 00
		1660.	.N. F. Hydraulic Power and Mfg. Co., electric lighting	50 00
		1661.	.N. F. Hydraulic Power and Mfg. Co., electric lighting	50 00
		1662.	.Walker & Paterson, inclined railway, tools, buildings	25 26
		1663.	.P. J. Davy, buildings	16 53
	7.	1664.	.Thomas V. Welch, superintendent, office expenses	39 82
	13.	1665.	.Thomas V. Welch, superintendent, office expenses, February	49 92
		1666.	.A. J. Porter, Commissioner, traveling expenses	24 55
	18.	1667.	.R. F. Rankine, treasurer and secretary, office and traveling expenses	35 85
March	1.	1668.	.R. F. Rankine, treasurer and secretary, salary February	91 67
		1669.	.Pay-roll for February	1,182 28
		1670.	.Thomas V. Welch, superintendent, Commissioner's and superintendent's office expenses	49 31

1899.		No. of voucher.			
March	8.	1671..	Pay-roll for February supplemental	$402 87	
		1672..	Grout Electrical Co., inclined railway	8 80	
		1673..	Walker & Paterson, tools, inclined railway, buildings	8 92	
		1674..	P. J. Davy, buildings	11 16	
		1675..	Ellwanger & Berry, vines	12 25	
		1676..	John D. Dietrick, tools	14 20	
		1677..	T. E. McGarigle, inclined railway	7 80	
		1678..	N. F. Hydraulic Power and Mfg. Co., electric lighting	50 00	
		1679..	Blackstone Co., treasurer, office expenses	7 60	
	25.	1680..	Nat. Press Intelligence Co., treasurer, office expenses	2 45	
	30.	1680..	T. V. Welch, superintendent, office and Commissioner's expenses	40 97	
	31.	1681..	R. F. Rankine, treasurer and secretary, salary for March	91 67	
		1682..	Pay-roll for March	1,483 54	
					$5,614 85

Abstract CXV.

1899.	No. of voucher.		
April 4.	1683.	Charlotte Haeberle, Inclined Ry	$3 64
	1684.	Maloney and McCoy, ice	50 75
	1685.	Wm. S. Humbert, deepening channel	3 29
	1686.	F. W. Oliver Co., Inclined Ry	4 10
	1687.	A. J. Porter, commissioner, traveling expenses	20 65
17.	1688.	A. J. Porter, commissioner, office furniture	90 00
4.	1689.	Payroll, deepening channel	106 62
17.	1690.	A. J. Porter, commissioner, traveling expenses	23 00
29.	1691.	Pay-roll for April	1,796 21
May 2.	1692.	Pay-roll, supplemental, April	696 19
	1693.	N. F. Hydraulic Power and Mfg. Co., electric lighting	50 00
5.	1694.	Wm. S. Humbert, deepening channel	3 20
	1695.	J. H. Cook & Co., ice	2 25
	1696.	T. E. McGarigle, Inclined Ry	82 08
	1697.	Walter Latta, trees	15 75
	1698.	Walker & Paterson, tools, buildings, Inclined Ry	21 70
	1699.	Braas Bros., buildings	182 54

		No. of voucher.			
1899.					
May	5.	1700..	A. J. Porter, commissioner, traveling expenses	$68 16	
		1701..	P. C. Flynn and Son, buildings	106 60	
		1702..	N. F. Hydraulic Power and Mfg. Co., electric lighting	50 00	
		1703..	T. V. Welch, superintendent, office expenses	36 41	
	15.	1704..	Pay-roll, Prospect Park and Goat Island	419 14	
April	29.	1705..	R. F. Rankine, treasurer and secretary, salary April	83 33	
May	11.	1706..	A. J. Porter, commissioner, traveling expenses	20 65	
	31.	1707..	R. F. Rankine, treasurer and secretary, salary, May	83 33	
		1708..	Pay-roll for May..........	1,943 85	
June	30.	1709..	Pay-roll for June.........	786 66	
					$6,750 10
July	1.	1710..	R. F. Rankine, treasurer and secretary, salary, June..	$83 33	
		1711..	T. V. Welch, superintendent, office expenses......	46 83	
May	13.	1712..	A. J. Porter, commissioner, traveling expenses	3 21	
July	8.	1713..	P. J. Davy, water pipes, buildings	294 32	
	10.	1714..	Miller and Brundage Co., damages	15 00	

1899.		No. of voucher.		
May	20.	1715.	.Wm. S. Humbert, roads....	$4 49
	31.	1716.	.Charlotte Haeberle, Inclined Ry	2 60
April	28.	1717.	.Charlotte Haeberle, roads..	4 00
June	30.	1718.	.N. F. Power and Mfg. Co., electric lighting	50 00
April	17.	1719.	.Paterson and Grout. Inclined Ry	7 80
May	9.	1720.	.F. W. Oliver Co., tools.....	13 08
	18.	1721.	.Hardwicks & Co., water pipes	137 68
	15.	1722.	.J. McDonald, coal.........	119 47
July	13.	1723.	.Gazette Printing Co., stationery	6 00
May	27.	1724.	.Gazette Printing Co., superintendent's office expenses	2 00
	23.	1725.	.J. H. Cook & Co., walks...	39 61
	31.	1726.	.N. F. Hydraulic Power and Mfg. Co., electric lighting	50 00
	13.	1727.	.P. C. Flynn & Son, buildings	139 00
July	18.	1728.	.Walker & Paterson, water pipes, tools, buildings, etc.	62 67
	8.	1729.	.James Day, superintendent's office expenses	16 45
	20.	1730.	.National Press Intelligence Co., treasurer's office expenses	2 90
	31.	1731.	.T. V. Welch, superintendent, tools, office, commissioner's expenses	32 57

1899.	No. of voucher.			
July 31.	1732..	Pay-roll for July.........	$1,520 14	
	1733..	R. F. Rankine, treasurer and secretary, July salary and office expenses	86 58	
Aug. 9.	1734..	Pay-roll July, supplemental	19 12	
				$2,758 85

Abstract CXVII.

Sept. 1.	1735..	R. F. Rankine, treasurer and secretary, salary August.	$83 33
	1736..	Milton C. Johnson & Co., treasurer's office expenses	9 00
	1737..	Pay-roll, August..........	1,601 56
	1738..	Thomas V. Welch, superintendent, office expenses..	48 90
	1739..	Wm. S. Humbert, bridges..	6 42
	1740..	Wm. S. Humbert, roads...	3 09
	1741..	Arthur C. Kugel, roads....	16 34
2.	1742..	N. F. Hydraulic Power and Mfg. Co., electric lighting	50 00
	1743..	N. F. Power and Mfg. Co., electric lighting	50 00
	1744..	J. H. Cook & Co., bridges..	24 94
	1745..	P. J. Davy, bridges, iron railings and pipes.......	38 48
	1746..	Globe Ticket Co., Inclined railway	45 00
	1747..	Walker & Paterson, tools, buildings and Inclined railway	48 89

1899.

	No. of voucher.			
Sept. 2.	1748	Alex. J. Porter, Commissioner, traveling expenses.	$11 00	
30.	1749	Pay-roll, September	1,126 67	
	1750	Thomas V. Welch, superintendent, office expenses	48 09	
	1751	R. F. Rankine, treasurer and secretary, salary, September	83 33	
				$3,295 04

Payments out of $15,000, as per chapter 606, Laws of 1898.

Series 1. Abstract II.

	No. of voucher.		
Oct. 7.	18	Pay-roll, grading, etc	$1,970 75
11.	19	W. A. Shepard, First Sister Island bridge	1,700 00
18.	20	Wm. S. Humbert, deepening channel	29 55
	21	P. J. Davy, iron railings	171 16
	22	Thomas Dunlavey, broken stone	32 00
	23	Timothy Horan, broken stone	20 00
	24	Coleman Nee, broken stone	35 00
	25	Braas Bros., repairs	49 68
	26	Braas Bros., carpenter work	163 92
	27	Braas Bros., signs	45 32
	28	Dobbie Foundry and Machine Co., iron railings	172 50

1899.		No. of voucher.			
Oct.	18.	29..	A. C. Kugel, cement	$35 98	
		30..	D. Phillips, inspector First Sister Island bridge	86 67	
		31..	C. Haeberle, stakes	3 75	
		32..	Conway & Munson, stone	365 25	
		33..	Schumacher & Mayer, carpenter work	248 99	
		34..	M. F. Saxton & Co., signs	31 00	
		35..	L. B. Ackley, grass sod	446 58	
		36..	R. D. Young, mason's work	80 36	
	26.	37..	Pay-roll, grading and road building	565 00	
Nov.	12.	38..	Pay-roll, grading and road	490 61	
					$6,744 07

Series I. Abstract III.

Dec.	8.	39..	Vaux & Emory, Hennepin's view and buildings	$147 50
		40..	D. Phillips, inspector First Sister Island bridge	18 75
		41..	L. B. Ackley, grass sod	18 00
		42..	P. J. Davy, iron railing	99 37
		43..	Conway & Munson, stone	495 22
		44..	W. S. Humbert, cement	23 80
		45..	W. A. Shepard, First Sister Island bridge	288 48
		46..	W. A. Shepard, First Sister Island bridge	1,894 00

No. of voucher.				
1899.				
Dec. 8.	47..	Vaux & Emory, First Sister Island bridge	$75 00	
16.	48..	Pay-roll, grading	113 98	
	49..	W. A. Shepard, deepening channel	18 70	
				$3,192 80

Payments out of $30,000, as per chapter 569, Laws of 1899.

Series K. Abstract I.

No. of voucher.				
1899.				
July 19.	1..	Pay-roll, filling, grading and riprapping	$1,179 44	
Aug. 5.	2..	Ellwanger & Berry, shrubs	94 06	
	3..	Pay-roll, roads, walks and grading	1,330 36	
Sept. 10.	4..	Pay-roll, roads and grading	739 25	
	5..	Thos. Dunlavey, stone	8 00	
	6..	Coleman Nee, stone	4 00	
	7..	Dean & Hoffman, roads	63 50	
				$3,418 01

Remittances to the State Treasurer:

1898.		
Nov. 1.	Check on Bank of Niagara for October receipts	$315 70
Dec. 1.	Check on Bank of Niagara for November receipts	81 35
1899.		
Jan. 3.	Check on Bank of Niagara for December receipts	28 85

	No. of voucher.			
1899.				
Feb.	3.	Check on Bank of Niagara for January receipts	$85 05	
March	2.	Check on Bank of Niagara for February receipts	501 35	
April	4.	Check on Bank of Niagara for March receipts	69 15	
May	6.	Check on Bank of Niagara for April receipts	54 80	
June	1.	Check on Bank of Niagara for May receipts	272 85	
July	1.	Check on Bank of Niagara for June receipts	549 85	
Aug.	2.	Check on Bank of Niagara for July receipts	1,714 50	
Sept.	2.	Check on Bank of Niagara for August receipts	3,144 05	
Sept.	30.	Check on Bank of Niagara for September receipts	1,746 40	
				$8,563 90
May	6.	Dividend on deposits in Cataract Bank		99 22
1898.				
Dec.	31.	Interest on balances in Manufacturers and Traders' Bank	$13 64	
March	31.	Interest on balances in Manufacturers and Traders' Bank	14 36	
				28 00
		Cash balance on hand in bank		27 51
				$47,181 31

Classification of Accounts.

Maintenance.

Commissioners' expenses	$353 23
Treasurer and secretary, office expenses	112 93
Niagara Falls office expenses (superintendent)	466 28
Salaries (superintendent and clerk)	2,899 98
Police	5,325 00
Inclined railway	1,219 60
Prospect Park	4,669 07
Goat Island	2,490 72
Roads	1,552 76
Walks	1,503 85
Coal	119 47
Electric lighting	600 00
Ice	66 10
Buildings	1,422 05
Tools	181 38
Bridges	104 86
Iron railing	33 45
Deepening channel	113 11
Treasurer and secretary's salary	1,049 98
Treasurer and secretary's traveling expenses	82 20
Water pipes	544 08
Expense, president's office	50 00
Trees	15 75
Furniture	90 00
Damages	15 00
Vines	12 25
Carting	14 10
	$25,107 20

Improvements under chapter 606, Laws 1898:

Walks	$271 07
First Sister Island bridge	5,061 37
Grading	1,538 39
Deepening channel	30 75
Iron railing	801 78
Hennepin's view	50 00
Roads	973 16
Buildings	560 09
Signs	76 32
Prospect Point	109 36
Grass sod	464 58
	$9,936 87

Improvements under chapter 569, Laws 1899:

Walks	$159 37
Filling	524 20
Shrubs	94 06
Grading	720 26
Riprapping	393 14
Roads	1,527 58
	$3,418 61

We, the undersigned, hereby certify that we have examined the foregoing report of the treasurer for the fiscal year ended September 30, 1899, the vouchers and other papers, and we find the report and accompanying documents correct, and that the treasurer has properly accounted for all moneys received and disbursed by him during the fiscal year ended September 30, 1899.

T. P. KINGSFORD,
CHAS M. DOW,
Commissioners of the State Reservation at Niagara.

GENERAL—ALL COUNTIES.

CHAPTER 710.

AN ACT to amend the railroad law regarding construction of a street surface railroad track by the commissioners of the state reservation at Niagara.

Became a law, May 25, 1899, with the approval of the Governor. Passed, three-fifths being present.

The People of the State of New York, represented in Senate and Assembly, do enact as follows:

Section 1. Section one hundred and eight of the railroad law is hereby amended to read as follows:

§ 108. Road not to be constructed upon ground occupied by public buildings or in public parks.—No street surface railroad shall be constructed or extended upon ground occupied by buildings belonging to any town, city, county or to the state, or to the United States, or in public parks, except in tunnels to be approved by the local authorities having control of such parks. Provided however that the commissioners of the state reservation at Niagara, by and with the consent of the commissioners of the land office, may construct, without expense to the state, street railroad tracks upon and along that part of the riverway, so called, between Falls and Niagara streets in the city of Niagara Falls, and in their discretion may grant revokable licenses to street surface railroad companies to use such tracks upon such terms as said commissioners may prescribe.

§ 2. This act shall take effect immediately.

To the Commissioners of the State Reservation at Niagara:

Your committee appointed to investigate and report concerning the desirability of taking action under section 108 of the Railroad Law upon the proposition of the Niagara Falls and Suspension Bridge Railway Company, respectfully report:

1. That we have carefully examined the local situation with reference to the desirability of street railroad tracks in Riverway, and have also taken advice relative to legal objections which have been suggested.

2. That we are advised in the opinion herewith submitted that no legal objections exist, in which conclusion we concur.

3. That in our opinion the construction of one street railroad track in Riverway between Falls street and Niagara street, with suitable connections to a station on the easterly side of the street, is eminently desirable to the interests of the public and in furtherance of the general purposes of the State Reservation for the reason that it will greatly accommodate visitors, facilitate economical, easy and convenient arrival and departure and will in no way impair the safety and beauty of the Reservation.

We recommend, That the proposition of the said Niagara Falls and Suspension Bridge Railway Company to furnish the money for the construction of such track be accepted upon such terms and conditions as to you shall seem proper and we suggest the following:

1. That a single track be constructed from Falls street to Niagara street in the center of Riverway with four turn-outs from the main track to the station of the said company on the east side of the street.

2. That the work of construction be carried on under the direct supervision of the commissioners and subject to such regulations as they may make from time to time.

3. That upon the completion of said track a license revocable at the pleasure of the commissioners be issued to the said Niagara Falls and Suspension Bridge Railway Company, to operate over said track, under and subject to all regulations which the commissioners may from time to time adopt.

4. That a license fee of $1,000 per annum in advance be charged for the privileges afforded by said license.

5. That no license fee be charged said company until the expiration of such time as a license fee at the foregoing rate would equal the amount advanced by the company in constructing said track and turn-outs.

6. That said track and all structures within the street be and remain at all times the property of the State.

7. That this board select a competent engineer to advise concerning the proper and desirable construction of said track.

8. That the said company enter into a proper and sufficient agreement for the construction of a station as proposed by it.

9. That the license to said company shall express that it in no way limits the right of the commissioners to license other companies to use said track upon terms solely within their discretion.

10. That all expenses incurred by the commissioners for engineering or legal advice or otherwise in connection with said track be treated as a part of the construction thereof to be paid by the company.

All of which is respectfully submitted.

(Signed) CHAS. M. DOW,
GEORGE RAINES,
T. P. KINGSFORD.

Dated *September* 1, 1899.

IN THE MATTER OF THE PROPOSED LAYING OF STREET RAILROAD TRACKS ALONG RIVERWAY IN THE STATE RESERVATION AT NIAGARA.

Section 108 of the Railroad Law, as amended by Chapter 710 of the Laws of 1899, authorizes the Commissioners of the State Reservation at Niagara to construct street railroad tracks along that part of Riverway between Falls and Niagara streets and to grant revocable licenses to street railway companies to use such tracks upon such terms as the Commissioners may prescribe.

Upon the assumption that the fee of the soil of Riverway is in the state free from all easements and rights of individuals no discussion is needed to demonstrate that the Commissioners may properly and safely proceed under this act without apprehension of trouble or interference. The questions for examination, therefore, are whether the title of the State is in any way qualified or limited by rights or easements of abutting owners or others and if so qualified or limited whether the owners of such rights or easements can in any way legally object to the proposed work.

Examination of the records discloses that Riverway was prior to the establishing of the State Reservation known as Canal street; that the part of Canal street in question was designated by the Commissioners as a portion of the lands to be taken absolutely for the reservation; that in the condemnation proceedings such street was included and it must be assumed as true beyond question that the State in these proceedings acquired absolute and unqualified title unless the same be in some way limited in the following facts.

It appears upon the condemnation proceedings that the east side of Canal street, between Falls and Niagara streets, was abutted by eleven village lots owned by individuals; that these lots fronted upon Canal street, having in the rear an alley seventeen feet in width; and that the taking of the street and closing the same to public travel would obviously destroy in great part their value.

It appears to have been assumed on all hands that the owners of these lots owned the fee of the soil to the center of the street. Such conclusion was probably correct and will be adopted by me as the one most favorable to the abutting owners.

The destruction of the value of these lots and the consequent damage which would necessarily have to be paid by the State the Commissioners sought to avoid by the adoption of a resolution the substance of which is set out in the printed record of the condemnation proceedings at page 219, Part II, as follows: "Whereas, the purposes for which the land is being taken for the State Reservation will not involve the closing of Canal street, between Niagara street and the southeast line of Mill slip, that *the* Commission did not intend in locating the State Reservation, and does not now intend, to close up said street, *but that it does intend to leave the said Canal street substantially as it now stands the easterly boundary thereof being unchanged, simply reserving such jurisdiction over all the said street as will be necessary to carry out the purposes of the State Reservation;* and they further resolve that the Commissioners of Appraisement be and they are hereby requested to base their appraisal upon the facts stated in this resolution."

This resolution was introduced in evidence by counsel for the Commissioners with the obvious purpose of reducing the award of damages to the abutting owners. What effect was given it by the appraisers is difficult to determine. Neither in their report nor the order of confirmation is it referred to. The taking, so far as disclosed by the record, was absolute and unqualified and the award of damages for each of the eleven lots was $500.00, which did not correspond with any evidence given. Only two witnesses testified on the question of damages, and while differing as to the value of the lots, they practically concurred in the view that the absolute taking of the street would diminish such value from two-thirds to three-fourths which would largely exceed the awards made. It also seems to have been practically conceded that leaving the street open for public travel, the taking of the fee for reservation purposes would not diminish the value, and one witness testified that the value under such conditions would be enhanced (Vol. III, p. 290).

The award did not follow either view, being for an amount equal to one-fifth to one-third of the value of each of the greater part of the lots.

Under these circumstances what legal or moral effect should be given to the resolution is somewhat uncertain. It is not clear that the Commissioners had power to bind either their successors or the State. It is not clear that if they possessed such power it was acted on to such an extent as to give the abutting owners an easement in the street. The view most favorable to the abutting owners is that they have such easement in Riverway as is appurtenant to lands abutting upon a public highway in which the fee of the soil is owned by the public. It cannot be claimed that they have

Assuming this to be their right I do not think it presents any objection to the construction of the street railroad. Treating Riverway as a public highway it is undoubtedly competent for the Legislature to authorize a street railroad therein subject only to the restriction contained in section 18, article 111 of the Constitution of the State which provides that "no law shall authorize the construction or operation of a street railroad except upon the condition that the consent of the owners of one-half in value of the property bounded on, and the consent also of the local authorities having the control of, that portion of a street or highway upon which it is proposed to construct or operate such railroad be first obtained, etc." The State owns the west side of Riverway, opposite the proposed railroad and its value unquestionably is equal to the value of the east frontage and I assume very much greater. If there were any doubt on this point it would be dispelled by the consent of the street railroad company which owns a part of the east frontage. The State, by the law in question, had given its consent as owner and the Commissioners are the local authorities having control of the street whose consent is also required.

But the right to construct a railroad in a street does not necessarily debar abutting owners who have an easement in the street from asserting a claim for damages. Such a claim could not be maintained by an abutting owner in this case for the reason that the fee of the soil in the street is unquestionably in the State, and it is well settled by repeated adjudications of the Court of Appeals that in such case the person or corporation constructing the railroad is not liable to the abutting owner for any consequential damages to his adjoining property arising from a reasonable use of the street for railroad purposes without substantially changing its grade and which is not exclusive in its nature but leaves the passage across and through the street free and unobstructed for public use. This is the law even for steam railroads and *a fortiori* must be for those operated by electricity. (Forbes vs. R., W. & O. R. R. Co., 121 N. Y., 505; Kellinger vs. R. R. Co., 50 N. Y., 206; People vs. Kerr, 27 N. Y., 188.) The cases cited are so directly in point and the principle so well established that no discussion is needed to make clear the proposition that the abutting owners are not entitled to damages. Since they can neither prevent the construction of the railroad nor recover damages arising from the construction I cannot see any violation of their legal rights and am wholly unable to perceive any ground for the contrary view. As to any violation of the moral limitations placed upon the rights of the State by the resolution quoted, that, of course, is exclusively for the Commissioners to determine, but it may not be inappropriate for me to observe that an inspection of the ground convinces me that the proposed railroad will necessarily enhance to a very material extent the value of the abutting property, and if this conclusion is just, the supposed moral objection is obviously dissipated.

I might well leave the discussion here, but I deem it desirable to express my judgment that the assumption made as to the effect of the resolution is too broad, and that the abutting owners have not such an unqualified easement in the street as they would possess were it a public highway.

At the time of including this street within the boundaries of the Reservation, it was a public highway, and obviously, in seeking to condemn it for reservation purposes the Commissioners sought to change its status. They did not wish it to remain a public highway; the State paid considerable damages for the change, and yet the assumption is there has been no change except in transferring the control of it as a public highway from the village of Niagara Falls to the Commissioners. Something more than this must have been intended. The Commissioners obviously desired to make it a street, and their resolution so declares, and in making it a street they necessarily opened it to the travel of all the world, including the abutting owners and those who might desire to visit the abutting property. But they go further and say that they propose to reserve such jurisdiction over it and the right to regulate travel thereon *as may be necessary to carry out the purposes of the State Reservation.*

This is meaningless unless they thereby reserve the right to establish such regulations regarding the travel as may be reasonable, having in mind the purposes of reservation. They might prohibit obnoxious vehicles, rapid driving and establish any and all regulations necessary or proper for the comfort and convenience of the people visiting the Reservation. Possessing this power, they can, obviously, place any structure in the street, unless restricted by the Legislature, which will facilitate travel, add to the convenience of the visiting public and assist in carrying out the general plan of making the Reservation easily accessible. That a street railroad might do this and yet not take away the general character of the street cannot be debated.

By their action in taking the land, and at the same time passing the resolution, the Commissioners committed themselves to the propositions that the status of the street was to be changed, that they were to be the sole authority, subject, of course, to the Legislature, in its management, and that in such management they were to be at liberty to carry out their own ideas as to the purposes of the Reservation. We may concede that they could not close the street to public travel and not in the least affect the foregoing propositions. It seems to me clear that if the Commissioners are of the opinion that the construction of the proposed railroad will subserve the purposes of the Reservation by affording greater facilities and accommodations to visitors, something beyond the resolution under discussion must be invoked to restrain them either morally or legally. That resolution, in my judgment, was carefully drafted so as to allow just such action as future commissioners should deem wise and which would not divest the street of the general character of a park road. If this be not so, then, no useful purpose can be discovered in the inclusion of Canal street within the Reservation limits. Merely to burden the State with its care, rather than the village, cannot have been the object.

I will say in conclusion that I have carefully perused the printed argument of Mr. Ansley Wilcox submitted to the Governor in opposition to the bill and think the foregoing considerations irrefutably answer the suggestions made by him at page six of such brief.

Respectfully submitted.

(Signed) FRANK W. STEVENS.

THIS AGREEMENT made the twenty-third day of September, 1899, between The Niagara Falls and Suspension Bridge Railway Company, party of the first part, and the Commissioners of the State Reservation at Niagara, party of the second part, WITNESSETH:

1. The party of the first part agrees to construct upon and along Riverway upon the State Reservation at Niagara, between Falls and Niagara streets, a street railroad track with turnouts in complete conformity to the maps, plans and specifications hereto attached and made a part hereof, and pay all expenses connected therewith incurred by the Commission.

2. Such work shall be commenced at once and prosecuted diligently to completion.

3. The work of construction shall be carried on under the direct supervision of said commissioners and subject to such regulations and directions as they may from time to time make and give, and in giving such instructions and directions, the superintendent of the Reservation shall be treated and obeyed as the agent and representative of the commissioners.

4. The party of the first part shall keep an exact and itemized statement of all the cost and expense of such construction, including expenses incurred by the commissioners in connection therewith for council, engineering or otherwise, and shall furnish to the commissioners a detailed written statement thereof, and when the work is completed the commissioners shall determine and certify the reasonable cost of such work, having power to determine what are proper expenses, payments and charges.

5. The said track and all work, material and constructions connected therewith within the limits of the State Reservation shall at all times be and remain the property of the State, free from all lien, claim or charges of the party of the first part, which shall have no claim or charge whatever against the commissioners or either of them, or the State, for the work done, or payments made hereunder.

6. Upon the completion of said track and when the same is ready for use, the said first party may use the same for street

railroad purposes pursuant to the terms of a license therefor, a copy of which is hereto annexed.

7. This agreement is made subject to the provisions of section 108 of the Railroad Law and shall not be of force until the Commissioners of the Land Office have given the consent therein required.

8. The construction of said work and the payment of the expenses connected therewith shall not constitute any claim or ground for claiming that said license shall not be revocable at the pleasure of the commissioners.

9. The party of the first part, in consideration of said license, agrees to indemnify and save harmless the State of New York, the Commissioners of the State Reservation at Niagara and each of them, and their and each of their successors in office against all costs, payments, expenses, suits and damages whatsoever, arising from or growing out of the improper condition of said tracks in said Riverway, or the use thereof, by the first party or any of its allied roads named in the said license during the construction of said tracks and the continuance of said license.

IN WITNESS WHEREOF the parties hereto have executed this agreement in duplicate the day and year first above written.

(Seal.)

THE NIAGARA FALLS AND SUSPENSION BRIDGE RY. CO.,

(Signed by) W. CARYL ELY, *President.*
ALEXANDER J. PORTER.
THOMAS P. KINGSFORD.
CHAS. M. DOW.
GEO. RAINES.

STATE OF NEW YORK, }
COUNTY OF NIAGARA. } *ss.:*

On this twenty-third day of September, in the year of our Lord one thousand eight hundred and ninety-nine, before me personally came W. Caryl Ely, of the city of Buffalo, N. Y., to me personally known, who being by me duly sworn, did depose and say that he resides in the city of Buffalo, N. Y., and is the president of the Niagara Falls and Suspension Bridge Railway Company; that he knows the corporate seal of said company; that the seal hereunto affixed is said corporate seal, and that it was hereunto affixed by order of the board of directors of the said company, and he signed his name thereto as president by like order.

(Signed) GEO. G. SHEPARD,
Notary Public, Niagara County, N. Y.

KNOW ALL MEN, That the Commissioners of the State Reservation at Niagara hereby license The Niagara Falls and Suspension Bridge Railway Company to use the street railroad tracks upon and along that part of Riverway, so called, in the State Reservation at Niagara, between Falls and Niagara streets, in the city of Niagara Falls, in the carrying on of its business.

This license is revocable and determinable absolutely at and in the pleasure of the said commissioners and their successors in office and also upon the following terms and conditions:

1. That the said railway company pay to the said commissioners for the State of New York a license fee of $1,000 per annum in advance, the first payment of said license fee to be due and payable at the expiration of such time as license fees at said rate, computed from this date would equal the amount expended or advanced by said company in the construction of said street railroad tracks.

2. If this license shall be revoked and annulled by the said commissioners after the payment of a license fee by said company, pursuant to the proceeding paragraph, and before the expiration of the year for which said license fee is paid, the said company shall be entitled to have the unearned portion of said license fee refunded by the State.

3. The motive power used shall be electricity exclusively.

4. The use of said tracks shall be at all times subject to such rules and regulations as the said commissioners may from time to time adopt or prescribe, and the company and its agents and servants shall carefully observe and obey each and every

of said rules and regulations and also observe and obey all directions and instructions which shall at any time be given either orally or in writing by the superintendent of the Reservation which are not inconsistent with said rules and regulations.

5. The said company shall at all times keep said tracks, the space between the same, and the roadway for two feet outside the rails in such condition and repair as the commissioners shall prescribe.

6. This license in no way limits the right of the said commissioners to license other railroad companies to use said tracks, and if such other license or licenses be issued, the use of said tracks by said Niagara Falls and Suspension Bridge Railway Company shall be such only as is consistent with the use by the other company or companies licensed.

7. This license extends to and permits the use of said tracks by street railroad cars owned or operated by either of the following named railroad companies whenever using the connecting tracks of said Niagara Falls and suspension Bridge Railway Company, they being allied roads and under the same general management, to wit: The International Traction Company, Buffalo and Niagara Falls Electric Railway, Buffalo and Lockport Railway, the Niagara Falls Park and River Railway Company, Buffalo Railway Company, and all the terms hereof shall apply to each of said companies, it being understood that the license fee of $1,000 per annum shall cover the use by all of said companies.

8. The terms and conditions of this license may at any time be added to or varied by the commissioners.

In Witness Whereof, The said commissioners have hereunto set their hands the 23d day of September, 1899.

(Signed) ALEXANDER J. PORTER,
THOMAS P. KINGSFORD,
CHAS. M. DOW,
GEO. RAINES.

Report of the Superintendent

OF THE

STATE RESERVATION AT NIAGARA

FOR THE

Fiscal Year Ending September 30, 1899.

Report of the Superintendent.

To the Commissioners of the State Reservation at Niagara:

Gentlemen.—The principal improvements made during the year have been the completion of the new stone arched bridge to the First Sister Island, the completion of the loop driveway in Prospect Park, the construction of a new road from Prospect Park along the rapids to the bridge to the islands and the filling and riprapping on the riverway from the bowlder bridge to the loop driveway at Port Day.

STONE ARCHED BRIDGE TO FIRST SISTER ISLAND.

The stone arched bridge was completed Nov. 4, 1898. The approaches to the bridge have been filled and graveled, the retaining walls banked up, graded and planted. The locality has been treated so as to harmonize as much as possible with the surrounding natural conditions.

THE HERMITS CASCADE.

Owing to the low level of the water in the lakes, for some years very little water has been flowing over the little fall just above the new stone arched bridge known as the Hermits Cascade. At times it has been entirely dry. The deepening of the channel above the cascade which has been completed during the past year has secured the former volume of water and restored the beauty and attractiveness of the cascade.

PROSPECT PARK.

The loop driveway in Prospect Park has been completed, thus providing a loop driveway at each terminus of the Reservation. The gate houses at Falls street and at the rapids have been removed. The removal of seven frame buildings from Prospect Park has resulted in a remarkable improvement in the appearance of the grounds which now present a natural grove-like appearance. If the Inclined Railway building were removed the falls could be viewed from any part of the grounds. The Inclined Railway building stands near the falls and greatly obstructs the view of visitors in the grounds.

THE NEW RAPIDS DRIVE.

The new road constructed during the past year along the rapids from Bridge street to Falls street has proved to be a great convenience to the public. It obviates the necessity of following an up-hill round about route from Prospect Park to Goat Island, and running as it does along by the rapids it is, for its extent one of the finest of drives. The new rapids drive and the new loop driveway in Prospect Park provide a continuous route along the river bank from the bridge to the islands to the bridge to the Canadian side of the river.

FILLING AND RIPRAPPING ALONG THE RIVER SHORE.

The low shore of the river above the bowlder bridge was wet and marshy, caused by crowding the water upon the shore when the old wing dam was constructed many years ago. At times of high water considerable erosion of the river bank took place. During the past year a large amount of filling has been obtained and the original shore line restored. The river shore has been riprapped with large stones to prevent further erosion and the made land graded and sown with lawn grass. Next year some planting of trees and shrubs should be done in that locality.

THE WING.

The locality immediately above Willow Island is known as "The Wing." A massive stone wall has been constructed at that point to deflect waters from the river into "The Race," a channel which conveyed it to mills which have been removed. The water inside the wall is a foot or two higher than the water in the river. The outlet from the Wing is in a dilapidated condition and should be reconstructed in a suitable manner. Parts of the Wing wall have fallen away, and if the wall is to be retained some repairs should be made. The wall at the upper end might be drawn in, by an easy curve and connected with the main land below Fourth street by a small stone arched bridge, and a loop walk or promenade thus obtained somewhat similar to the loop driveway at Port Day. It would command a fine view of the rapids at that point and the upper Niagara river, and would undoubtedly be a favorite resort for visitors.

Suitable stone for this work can now be obtained at Port Day probably without expense except for hauling, an opportunity which may never occur again.

PLANTING.

During the year some planting has been done in Prospect Park and along the river way, between Prospect Park and the bridge to the islands. The approaches to the new stone arched bridge have been planted with shrubs and vines and some planting done about the rustic stone arched bridge and the bowlder bridge. The stock in the nursery has been largely increased for fall planting and a large number of native plants from the woods on Goat Island have been transplanted into the nursery. The plantations already made have been carefully pruned, watered and cultivated, materials have been accumulated for fall planting at the approach to Goat Island, on Green Island, in Prospect Park and along the riverway.

BUILDINGS.

During the year two frame toll houses and two frame toilet rooms have been removed from Prospect Park. Two frame closets have been removed from Goat Island and toilet rooms built in the shelter building at the Three Sister Islands. The frame buildings at the Cave of the Winds have been enlarged and improved and a water pipe laid to the buildings. The roof of the cottage on Goat Island has been reshingled. The interior woodwork of the shelter building at the entrance to Goat Island has been revarnished, and the roof of the lower terminal station of the Inclined Railway, which had been damaged by ice, repaired.

INCLINED RAILWAY.

The water wheel at the Inclined Railway has been rebuilt, the sheaves for the cable repaired, the railroad track leveled, a new manila cable attached to the cars, and new hand wheels attached to the cable at the cars.

The building enclosing the Inclined Railway was built twenty-six years ago, and is now in need of extensive repairs. It may be well to replace it with a new structure; many of the track timbers should be renewed.

The almost constant moisture by the spray from the Falls causes the Inclined Railway structure to decay very rapidly.

BRIDGES.

One of the protecting cribs of a pier under the bridge from the main land to Green Island was damaged by floating ice during the past winter. The sheet piling of the crib has been renewed, and the crib refilled with stone. Repairs have also been made upon the retaining walls at the weir bridge over the overflow at First street, which were being undermined by the stream, and large stones have been placed at the new bridge to the first Sister Island so as to complete the grading of the structure. Repairs have been made upon the bridges to the second and third Sister Islands and the wooden structure leading to Terrapin Point, at the Horseshoe Falls. These bridges are old and inadequate, and should be replaced by more substantial and commodious structures.

ROADS AND WALKS.

A new carriage turn-out, with stone curbing, has been constructed on Goat Island at a point commanding a view of the Horseshoe Falls. The carriage turn-outs at the American Falls, the Three Sister Islands and the Inclined Railway building have been filled and graded, and extensive repairs made on the roads on Goat Island and in Prospect Park.

A gravel walk has been constructed on the east side of the riverway, adjacent to the International Theater, and the margin filled and sodded and planted with trees.

The gravel walk along the riverway from Prospect Park to the bridge to the islands has been narrowed, drainage tile laid and two catch basins constructed. Stone gutters have been laid along the walks above Bridge street at points where steep grades caused the gravel to wash out at times of heavy rainstorms.

MISCELLANEOUS.

A large amount of compost, collected during the year, was spread upon the lawns in winter. The compost is collected in the gravel pit on Goat Island, and turned and mixed from time to time, during the year. Grading has been done on the terrace at First street, a large number of dead trees and stumps have been removed, worthless material has been removed from the lumber yard on Goat Island, the water pipes repaired and extended to the Cave of the Winds building and an additional drinking fountain (the gift of Dr. Coleman Sellers, of Philadelphia) erected.

SYSTEM OF WATER PIPES.

A plan for a complete system of water pipes for the grounds is being made, also a map of the grounds, showing the made land, the roads, walks and the planting.

ELECTRIC LIGHTING.

The present arrangement for lighting the grounds is inadequate. If the grounds are lighted at all, they should be well lighted. A comprehensive plan for lighting the grounds is being prepared.

LICENSED CARRIAGE DRIVERS.

Twenty-one complaints have been made against licensed carriage drivers for violation of the ordinances. Four drivers have been excluded from the Reservation for such violations of the rules and regulations.

THE STEAMBOAT LANDING.

The steamboat landing has been leased by the Maid of the Mist Steamboat Company. On January 22nd the ice bridge moved out, destroying the steamboat dock and the steamboat waiting room on the landing. January 26th a violent wind arose, which raised the water twenty feet, seriously threatening the terminal station of the Inclined Railway in which the water stood four feet deep. No damage was done to the building aside from the bursting of the doors and the destruction of a few panes of glass.

The steamboat landing and the shelter building on the dock have been rebuilt by the Maid of the Mist Steamboat Association.

ELEVATOR AT THE CAVE OF THE WINDS.

As this locality is between the American and the Horseshoe Falls, the view is unparalleled, but the winding stairway leading to it is so unsuitable and fatiguing that very few persons are enabled to enjoy the beauty of the scenery below the high bank, where a walk can be easily made along the edge of the water from the American to theHorseshoe Falls.

During the year the Cave of the Winds building has been much enlarged and improved, but it is still unworthy of the locality. The buildings for the guides and the clothes yard have been rearranged and present a better appearance than heretofore.

The waiting rooms and winding stairway of the Cave of the Winds have received the usual repairs during the year. The dressing rooms are inadequate and unsuitable. Better accommodations should be provided for the traveling public.

The Biddle staircase has been examined and found safe, but it is old and dilapidated in appearance, and affords but a fatiguing method of obtaining the view of the Falls from below. It should be replaced by a commodious elevator, with a free stairway attached. Such an elevator, operated for a nominal fee of five cents up or down, would furnish an estimated revenue of $5,000 a year, which, with the present revenue, mainly from the Inclined Railway, would almost render the Reservation self sustaining.

The Pan-American Exposition in Buffalo will draw numbers of people to Niagara in 1901, and the Biddle stairs should be extensively repaired and strengthened during the coming year.

SHELTER BUILDINGS.

Shelter buildings are needed at the American Falls and at the Horseshoe Falls on Goat Island, where visitors congregate in large numbers. At times of sudden rain storms the present shelters are too far apart for public convenience.

STAIRWAYS AND GUARD RAILINGS.

The stairway leading to Luna Island is too narrow to accommodate the great number of visitors to that point. It might be removed and a broad walk, winding down the slope, substituted. The balcony at the foot of the stairway, commanding the finest view of the American Falls, is also inadequate and should be replaced by a permanent structure of larger size.

The iron guard railing on Luna Island is unsuitable, and should be replaced by a railing similar to that along the high bank on Goat Island.

NUMBER OF VISITORS.

The number of visitors during the year was greater than usual, and is estimated at 750,000. Of these 403,016 were excursionists, coming largely from places in the State of New York. No disorder occurred, and no damage was done to the property of the State. A statement of the excursions is herewith submitted.

EMPLOYEES.

The regular force employed, exclusive of laborers, consists of 9 men, to wit: 1 superintendent, 1 clerk, 5 police gatemen and caretakers and 2 Inclined Railway men.

STATEMENT OF NUMBER OF EMPLOYEES.

Maintenance Rolls.

1898.	Foreman.	Asst foreman.	Laborers.	Teamsters.
October	....	3	29	....
November	1	2	19	2
December	1	2	15	2
1899.				
January	....	2	8	1
February	....	1	11	....
March	....	3	17	1
April	1	3	25	1
May	1	3	21	2
June	....	....	....	....
July	....	....	21	....
August	1	2	13	2
September	....	2	6	....

Improvement Rolls.

Chapter 606, Laws of 1898.

1898.	Foreman.	Asst. foreman.	Laborers.	Teamsters.
October	1	1	33	7
November	....	....	4	....

Chapter 569, Laws of 1899.

1899.	Foreman.	Asst. foreman.	Laborers.	Teamsters.
June	1	3	20	2
July	1	3	31	5
August	....	1	40	4
September	1	....	9	1

TABULAR STATEMENTS.

The receipts from the Inclined Railway during the fiscal year were $6,938.90. From rentals and leases, $1,600. Total receipts, $8,538.90.

Vouchers paid for material and labor, $6,826.57.

Pay-rolls for maintenance, $20,156.76.

For improvements, $4,817.89.

Total expenditures by superintendent, $31,801.22.

Detailed statements of the receipts and expenditures of the superintendent, the amount of the pay-rolls for each month, and the classification of the pay-rolls and accounts are hereto appended.

Respectfully submitted,

THOMAS V. WELCH,

Superintendent.

EXCURSIONS.

1888.		No. of cars.	Persons.
Oct.	2. Akron and Cleveland, Ohio.......	30	1,800
	Buffalo, N. Y....................	35	2,100
	Rochester, N. Y., via N. Y. C. & W. S.........................	20	1,200
	Rochester, N. Y., via Lehigh Valley	15	900
	Rochester, N. Y., via Erie R. R....	10	600
	3. Buffalo, N. Y., Sisters and Scholars Holy Angels Academy.........	2	120
	5. Pennsylvania R. R. Officials......	1	60
	7. Columbia, Pa., Fire Co. No. 1.....	3	180
	8. Boston, Mass., Knight Templars..	14	840
	Buffalo, N. Y., Teachers and Scholars Central High School........	4	240
	Lockport, N. Y., Special Party....	5	200
	9. Maine, New Hampshire and Vermont Knight Templars.........	45	2,700
	12. Hornellsville and Caledonia, via Erie R. R.....................	8	480
	Brooklyn, N. Y., Special Party....	1	60
	14. Washington, D. C., and Baltimore, Md.........................	8	480
	15. Oswego, N. Y., Normal School....	2	120
	16. Camden, N. J., Republican Club...	1	60
	Buffalo, Rochester and Lockport..	15	900

		No. of cars.	Persons.
1898. Oct.	18. New Market, Ont., American Christian Association..........	6	360
	19. Twenty-sixth Annual Convention American Gas Light...........	6	360
1899. Jan.	22. Buffalo, N. Y., Ice Bridge Excursion	10	600
Feb.	2. Buffalo, N. Y., Ice Bridge Excursion	12	720
	3. Buffalo, N. Y., Ice Bridge Excursion	8	480
	4. Buffalo, N. Y., Ice Bridge Excursion	14	840
	5. Buffalo, N. Y., and Local Points Ice Bridge Excursion..........	50	2,500
	7. Buffalo, N. Y., and Local Points Ice Bridge Excursion..........	10	600
	12. Buffalo, N. Y., and Local Points Ice Bridge Excursion..........	20	1,200
	16. Buffalo, N. Y., and Local Points Ice Bridge Excursion..........	15	900
	19. Buffalo, N. Y., Ice Bridge Excursion	20	1,200
	26. Buffalo, N. Y., Ice Bridge Excursion	8	480
May	6. Lockport, via N. Y. C., Arbor Day Excursion	21	1,260
	7. Local Excursion	20	1,200
	American Machinists' Association.	2	120
	14. Local Excursion	15	900
	International Brotherhood Railway Conductors	2	120

1899.		No. of cars.	Persons.
May	17. International Brotherhood Railway Conductors	2	120
	19. International Brotherhood Railway Conductors	8	480
	20. Buffalo High School	5	300
	21. Buffalo High School	15	900
	22. Reunion 28th N. Y. Vols	1	60
	24. Toronto, Hamilton, etc., Queen's Birthday Excursion	25	1,500
	28. New York Hotel Cooks Excursion.	9	540
	Local Excursion	20	1,200
	30. All sources, Memorial Day Excursion	65	4,000
June	3. All sources, General Half-holiday Excursion	12	720
	4. All sources, Sunday Excursion	30	1,800
	6. South American Representatives to Pan-America	2	120
	7. National Credit Men's Association.	4	240
	Daughters of American Revolution	4	240
	9. Association Steam and Hot Water Fitters	8	480
	Canada, via G. T. R., Farmers' Excursion	8	480
	10. All sources, Half-holiday Excursion	12	720
	11. All sources, Sunday Excursion	50	3,000
	12. Chicago, Ill., Nobles of the Mystic Shrine	3	180

1899.		No. of cars.	Persons.
June	13. Chicago, Ill., Nobles of the Mystic Shrine	6	360
	Via Main Line Erie R. R.	6	360
	American Dancing Masters' Asso.	6	360
	14. American Wholesale Coal Dealers.	6	360
	Nobles of the Mystic Shrine	10	600
	York County, Ont., Farmers	6	360
	15. Nobles of the Mystic Shrine	15	900
	Grand Trunk & Michigan Central.	22	1,320
	16. Alabama State Press Association.	3	180
	International Accident and Insurance Association	6	360
	Nobles of the Mystic Shrine	20	1,200
	17. Employes Water Engine Works, Brantford, Ont.	10	600
	Saturday half holiday (local)	20	1,200
	18. All sources	40	2,400
	19. American Car Service Association.	4	240
	20. Improved Order of Heptasophs	4	240
	New York State Press Association.	5	300
	Toronto, Ont., by boat	......	585
	21. Improved Order Heptasophs	4	240
	Passayank Tribe, Improved Order of Red Men	4	240
	Methodist Church, Hamilton, Ont.	9	540
	Toronto by boat	......	118
	22. Minnesota Banking Association	4	240
	23. Toronto by boat	......	360
	24. Candy Makers and Brewers, Toronto, Ont.	9	540

1899.			No. of cars.	Persons.
June	25.	Rochester Special	10	600
		All sources	65	3,900
		Half holiday	25	1,500
		Foresters via C. P. R. Line	12	720
	27.	Union Methodist Churches, Erie Pa.	6	360
	28.	All sources	17	1,020
	29.	Children Immaculate Conception, Buffalo, N. Y.	3	180
		Union Churches, Rochester, N. Y.	6	360
		Mrs. Nichols' Sunday school class (city)	......	100
		City officials, Lockport, N. Y.	1	60
July	1.	Christian Endeavor Society, Brantford, Ont.	26	1,560
		Baptist Church, Dandus, Ont.	12	720
		Foresters, St. Thomas, Ont.	25	1,500
		All other sources	25	1,500
	2.	Cleveland, O., and other sources	120	7,200
	3.	Saengerbund Society, Berlin, Ont.	8	480
		Jersey City, N. J., Athetic Club	6	360
		Cooks, tourists	8	480
		All other sources	50	3,000
	4.	Independence Day, all sources	333	20,000
	6.	Toronto, by boat	......	50
	7.	Hamilton, Ont	2	120
	8.	Toronto, Ont	9	540
		Saturday half holiday	20	1,200
	9.	All sources	65	3,900
	11.	National German Teachers Ass'n.	7	420

1899.		No. of cars.	Persons.
July	11. International Longshore Men's Association	5	300
	Detroit Christian Endeavor Ass'n.	20	1,200
	12. Detroit Christian Endeavor Ass'n.	25	1,500
	13. Detroit Christian Endeavor Ass'n.	6	360
	First Presbyterian Church, Erie, Pa.	8	480
	German Lutheran Chuch, Lockport, N. Y.	9	540
	Union churches, Lancaster, N. Y.	7	420
	City officials, Lockport, N. Y.	1	60
	International Fire Underwriters' Convention	8	480
	14. Caledonia societies, Toronto and Guelph, Ont.	2	120
	All Saints Church, Toronto, Ont.	6	360
	15. Employees Michigan Central R. R.	60	3,600
	Pennsylvania and Maryland School Teachers	10	600
	Saturday half holiday (local)	20	1,200
	16. Sunday, all sources	60	3,600
	17. Special, Michigan Central party	12	720
	18. Glassford, Ont., Grocery Men	16	960
	Hamilton, Ont., Grocery Men	36	2,160
	Hornellsville, N. Y.	8	480
	19. All sources	125	7,500
	20. Union churches, Brampton, Ont.	16	960
	Syracuse, N. Y.	8	480
	Cleveland, Ohio	8	480
	Toronto, Ont., by boat	....	350

1899.		No. of cars.	Persons.
July	21. Lewiston, N. Y., bridge opening...	13	780
	Toronto, by boat................	....	275
	22. Pekin and Sanborne, N. Y., Sunday School	6	360
	New Orleans, La.................	8	480
	Seneca Falls, N. Y...............	9	540
	Saturday half holiday (local)......	25	1,500
	G. T. R. employees...............	14	840
	Toronto, by boat.................	......	2,000
	23. Syracuse and Rochester, N. Y.....	150	9,000
	24. Olean, N. Y., Labor Picnic........	16	960
	Other sources....................	10	600
	25. Hamilton, Dayton & Cincinnati, Ohio	60	3,600
	U. S. League Building and Loan Associations	2	120
	Other sources....................	10	600
	26. Local C. M. B. A. Society........	......	2,000
	Other sources....................	42	2,520
	27. Oil City and Pittsburg, Pa........	15	900
	Cleveland, Ohio..................	10	600
	Buffalo, N. Y....................	8	480
	28. Big Four excursion...............	54	3.240
	Toronto, Ont....................	10	600
	Other sources....................	5	300
	29. London and Hamilton, Ont........	40	2,400
	Batavia and Chautauqua, N. Y....	68	4,080
	Philadelphia, Pa., special..........	10	600
	Saturday half holiday (local)......	15	900

1899.			No. of cars.	Persons.
July	30.	All sources	100	6,000
	31.	Elks Street Fair	45	2,700
		National Dental Association	6	360
		Guelph, Ont	4	240
		Toronto, Ont., by boat	......	300
Aug.	1.	Port Arthur, Ont., Civil roliday	10	600
		Via Erie R. R	9	540
		From Buffalo, by trolley	10	600
		Brooklyn, N. Y	5	300
		Guelph, Ont	3	180
	2.	Toronto and Hamilton, Ont	8	480
		National Jobbing Confectioners	4	240
		Albion and Lyndonville Sunday schools	20	1,200
		Elks, from all points	25	1,500
		Welland farmers, in wagons	......	300
		Toronto, by boat	......	600
	3.	Lockport, N. Y., Second Presbyterian Church	4	240
		Buffalo Street Railway Employees	12	720
		Western points (Big Four)	100	6,000
		Odd Fellows, Toronto, by boat	......	550
	4.	Toronto, by boat	......	45
		Buffalo Union schools	6	360
		Buffalo Street Railway Employees	6	360
		Big Four	235	14,100
		Grand Trunk railway	10	600
	5.	Chautauqua (Newton party)	9	540
		Rome, Watertown & Ogdensburg R. R	20	1,200

1899.		No of cars.	Persons.
Aug.	5. Half holiday (local)	30	1,800
	Toronto, by boat	......	500
	6. York City and Pittsburg, Pa	41	2,460
	N. Y. C., W. S. and trolleys	120	7,200
	7. C. T. A. U. of America	6	360
	Employees G. T. R.	10	600
	Toronto, Ont	12	720
	Hamilton, Ont	11	660
	Toronto, by boat	....	300
	8. Brockville, Ont., Odd Fellows	6	360
	Chautauqua (Newton party)	9	540
	Buffalo Street Railway Employees	4	240
	Toronto, Ont., by boat	......	45
	Welland, Ont., St. John's school, in wagons	......	48
	9. Big Four	67	4,020
	Clover Leaf	10	600
	Toronto, Ont	3	180
	Grimsby, Ont	15	900
	Aurora and Newmarket, Ont	15	900
	Other sources	35	2,100
	10. Western N. Y. and Pennsylvania	12	720
	Allegany valley and city	11	660
	Akron & Pea Nut road	10	600
	Survivors of 151st N. Y. V. and 2d Mounted Rifles	5	300
	Buffalo Street Railway Employees	5	300
	Buffalo, by trolley	20	1,200
	11. Empire Association Deaf Mutes	4	240
	Buffalo Street Railway Employees	4	240

1899.		No. of cars.	Persons.
Aug. 11.	Cincinnati, Ohio	10	600
	Canton, Ohio	12	720
	Supreme Council Honor and Temperance	4	240
	American Foundrymen's Association	3	180
12.	Chautauqua (Newton party)	9	540
	National Fire Underwriters' Association	6	360
	Erie, Pa., employees iron works	10	600
	Toronto, Ont., Masey Harris employees	6	360
	General half holiday	35	2,100
	Brantford, Ont	16	960
13.	Cleveland, Ohio	14	840
	Philadelphia and Washington	22	1,320
	Cleveland Gymnasium Association	12	720
	Grand Rapids and western points	20	1,200
	Woodstock, Brantford and Lincoe, Ont	34	2,040
	Buffalo, N. Y	120	7,200
14.	Employees Anderson Furniture Co., Brantford, Ont	18	1,080
	Erie, Pa., Y. M. C. A	12	720
	Buffalo, N. Y	22	1,320
15.	Chautauqua (Newton party)	14	840
	Dunkirk, German Lutheran Church	10	600
	Dansville	11	660
	Buffalo	26	1,560
	New York Central R. R. Co	103	6,180

1899.			No of cars.	Persons.
Aug.	15.	Erie R. R.	45	2,700
		Buffalo, N. Y	45	2,700
		Toronto, Ont., by boat	......	600
	17.	Albany, N. Y., Gov. Roosevelt and party	1	60
		Rochester and Syracuse	20	1,200
		Tilsonburg, Ont., (Farmers' picnic)	9	540
		Buffalo, N. Y	25	1,500
	18.	Evansville and Southern Indiana	16	960
		Buffalo, N. Y	25	1,500
		Toronto, by boat	......	350
	19.	Elmira, N. Y	12	720
		Local points, Saturday half holiday	25	1,500
		Ohio and Indiana points	22	1,320
		Chautauqua (Newton party)	5	300
		Brantford, Ont., G. T. R. employees	31	1,860
		Toronto, Ont., Anglo-Saxon Society, by boat	......	1,050
	20.	Ohio points	20	1,200
		N. Y. C., West Shore, Erie and R., W. & O. railway	60	3,600
		Buffalo	60	3,600
	21.	Buffalo Street Railway Employees	5	300
	22.	Buffalo Street Railway Employees	5	300
		Buffalo Newsdealers and Stationers	5	300
		Syracuse, N. Y	6	360
		Chautauqua (Newton party)	9	540
	23.	Ohio and Indiana	24	1,440
		Buffalo Street Railway Employees	5	300

1899.		No. of cars.	Persons.
Aug. 24.	Lockport, N. Y., English Lutheran Church	8	480
	Buffalo Street Railway Employees.	5	300
25.	Wabash excursion	40	2,400
	Nickel Plate & Erie	40	2,400
	M. C. & Lake Shore	28	1,680
	Buffalo, N. Y	30	1,800
	Toronto, by boat	......	80
26.	Rochester and Syracuse, N. Y.	10	600
	Buffalo and local points	30	1,800
	Toronto, by boat	......	290
27.	Cleveland, Ohio	12	720
	Erie, Pa.	10	600
	Syracuse, Rochester and Lockport	30	1,800
	Main line, Erie R. R.	12	720
	Buffalo, N. Y.	60	3,600
29.	Main line D., L. & W.	10	600
	Main line W. N. Y. & P.	10	600
	Pennsylvania R. R.	10	600
	Ingersol, Ont.	6	360
	Buffalo, N. Y.	20	1,200
30.	Warsaw, N. Y.	10	600
	Main line Lehigh Valley R. R.	9	540
	Toronto, Ont.	20	1,200
	Avon and Attica, N. Y	10	600
	Hornellsville, N. Y.	14	840
31.	American Bar Association	5	300
	Main line P. B. & L. E.	10	600
	Main line L. S. & M. S.	10	600
	Toronto, Ont.	10	600

1899.		No. of cars.	Persons.
Sept.	1. Pittsburgh, Pa.	10	600
	Toronto, Ont. (annual fair)	20	1,200
	2. Wabash & Nickel Plate Line	36	2,160
	Cleveland, Ohio	20	1,200
	Toronto, Ont. (annual fair)	30	1,800
	Local points	20	1,200
	Toronto by boat	......	300
	3. Illinois points via Wabash	6	360
	New York city via Erie R. R.	5	300
	Akron, Ohio, and other points	205	12,300
	4. Carbondale, Pa.	12	720
	Toronto, by boat	......	100
	All other sources	250	15,000
	5. Points via Wabash R. R.	18	1,080
	Cleveland, Ohio	10	600
	Points via W. N. Y. & P. R. R.	18	1,080
	Other sources	30	1,800
	6. Points via Nickel Plate Line	10	600
	Toronto, by boat	......	500
	Other sources	30	1,800
	7. Toronto, by boat	......	700
	7. All sources	12	720
	9. Toronto, boat (annual fair)	35	2,100
	Other sources	40	2,400
	10. Points via Lehigh Valley	12	720
	Syracuse, Rochester, etc., via N. Y. C.	14	840
	Buffalo, N. Y.	15	900
	11. All sources	10	600
	12. English tourists (Cook excursion).	1	60

1899.	No. of cars.	Persons.
National ass'n Master Bakers....	5	300
13. Hartford, Ct., (Putnam Phalanx).	3	180
14. All sources	12	720
15. Lynchburg, Va., (Union Sunday school)	16	960
Detroit, Mich., (Odd Fellows)....	3	180
16. All sources	10	600
17. Buffalo, N. Y..................	20	1,200
20. Wheeling, W. Va..............	8	480
Bradford, Pa..................	8	480
21. Wheeling, W. Va..............	8	480
22. Detroit, Mich., Odd Fellows......	20	1,200
23. Detroit, Mich., Odd Fellows......	6	360
Syracuse, N. Y..................	10	600
25. New England Railway Association	6	360
	6,515	403,016

Recapitulation.

	Cars.	Persons
October, 1898	231	13,860
January, 1899	10	600
February, 1899	157	8,920
May, 1899	210	12,700
June, 1899	497	30,983
July, 1899	1,947	121,815
August, 1899	2,459	152,298
September, 1899	1,004	61,840
	6,515	403,016

GOAT ISLAND.

"The most interesting spot in all America."
Capt. Basil Hall

By PETER A. PORTER.

1900.

I have endeavored, in this article, to bring together a number of the opinions that have been expressed about Goat Island, in its various aspects. These expressions are mainly those of persons to whom the world has given a hearing, because of their abilities and prominence in their respective spheres. And joined to, and interwoven with these expressions, I have added such a chronology of the Island as I have been able to collect.

NIAGARA.

Author Unknown.

Great Fall, all hail:
Canst thou unveil
The secrets of thy birth;
Unfold the page
Of each dark age,
And tell the tales of earth?

When I was born
The stars of morn
Together sang — 'twas day:
The sun unrolled
His garb of gold
And took his upward way.

He mounted high
The eastern sky
And then looked down on earth;
And she was there,
Young, fresh, and fair,
And I, and all, had birth.

The word of power
Was spoke that hour:
Dark chaos felt the shock;
Forth sprung the light,
Burst day from night,
Up leaped the living rock.

Back fell the sea
The land was free,
And mountain, hill, and plain
Stood forth to view,
In emerald hue,—
Then sang the stars amain.

And I — oh thou:
Who taught me how
To hymn thy wondrous love
Deign to be near
And calm my fear.
O Holy one above.

I caught the word
Creation heard,
And by thy power arose;
His goodness gave
The swelling wave
That ever onward flows.

By his command
The rainbow spanned
My forehead and his will
Evoked the cloud
My feet to shroud,
And taught my voice to trill.

And who is he
That questions me?
From whom hast thou thy form,
Thy life, thy soul?
My waters roll
Through day, night, sunshine, storm.

In grateful praise
To him, I raise
A never ceasing song
To that dread one,
To whom stars, sun,
Earth, ocean, all belong.

Thou too adore
Him ever more
Who gave thou all thou hast;
Let time gone by
In darkness die
Deep buried in the past.

And be thy mind
To him inclined
Who made earth, heaven and thee—
Thy every thought
To worship wrought.—
This lesson learn of me.

Goat Island Group. From Upper Terrace. Canada.

OLD TERRAPIN TOWER.

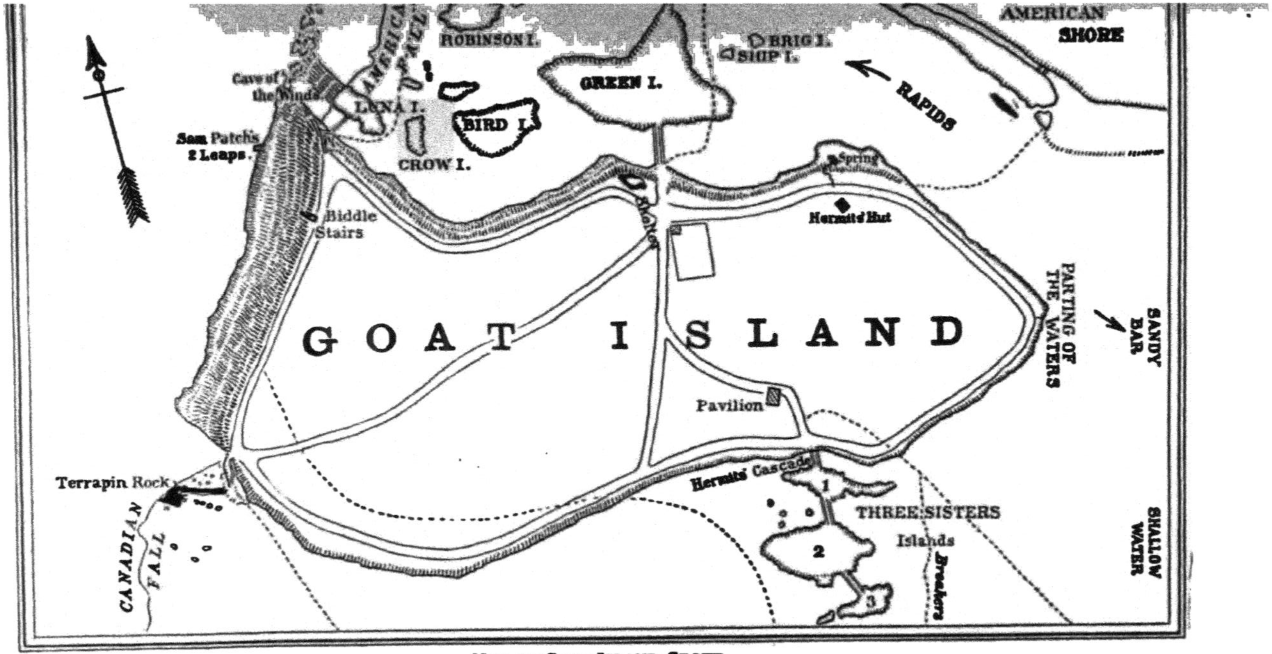

MAP OF GOAT ISLAND GROUP.

GOAT ISLAND.

Goat Island, as the words are ordinarily used, means the group of islands and islets situated between the American and Canadian rapids, at the verge of and just above the Falls of Niagara.

This group consists of Goat Island, which is half a mile long and a quarter of a mile broad, running to a point at its eastern end, comprising 70 acres; and 16 other islands or masses of rock, varying in size from an average of 400 feet to 10 feet in diameter.

Five of these islands and the Terrapin rocks are connected with Goat Island by bridges. Many years ago the two small islands above Green island were also thus accessible. As Goat island divides the Falls themselves, so it divides with them the interest of visitors; for it is *the* one spot at Niagara. If only one point here were to be visited, that one spot, beyond all question, should be Goat Island.

From it, with the one exception of the grand general view to be obtained from the Canada shore, are to be seen all the best views of Niagara, including both falls, both rapids, the gorge and the rainbows. And of Niagara, the Terrapin rocks, accessible only from Goat Island, are the scenic, as they are the geographical center, its very epitome. To Goat Island have been applied numerous epithets, among them the Temple of Nature, the Sacred Isle, the Fairy Isle, the Enchanted Isle, the Isle of Beauty, the Shrine of the Deity, and less poetic, but perhaps most truthful of all, the words quoted on the title page, "the most *interesting spot in all* America."

"It is interesting to consider that many of the trees now standing on Goat Island looked down on the first recorded visit of a white man to the Falls, and have remained the only living witnesses of those important scenes in the drama of European conquest in America, which were enacted at this all-important portage in the great water route to the heart of the continent. The savage chiefs and conquering generals, the tribes and armies that moved along this well-known track from Ontario, and launched their vessels on the river above Goat Island, are gone, but the trees that shadowed the flashing stream still remain to make the past real and brings vividly to memory our wonderful progress."

The Island embraces over two-thirds of the acreage, and by reason of its location is by far the most important part, of the New York State Reservation at Niagara.

"It is a paradise; I do not believe there is a spot in the world which within the same space comprises so much grandeur and beauty." This expression by a Boston divine, 70 years ago, is but a condensation of what many others since then have verbally expressed, in longer, but certainly in no more forcible, words.

The purchase of this property by the Empire State in 1885, was the tangible fulfillment of the following opinion, uttered half a century before, that "Niagara does not belong to Canada or America. Such spots should be deemed the property of civilized mankind; and nothing should be allowed to weaken their efficacy on the tastes, the morals and the enjoyments of all men."

It is a group, or speaking collectively, a spot, wondrous in many aspects; wondrous from its location, wondrous from its geology, wondrous from its botany, wondrous from its scenery, and famous, if not wondrous, from its history.

ITS GEOLOGY.

During the last 75 years geologists have written a great deal about Niagara, and from it speculatists have deduced theories as to the antiquity of the earth, trying to prove

> "That He who made it, and revealed its date
> To Moses, was mistaken in its age."

In early geological days this entire section was covered by the salt waters of the Devonian seas, which is proved by the shells of the Conularia Niagarensis, found in the shale underlying Goat Island and along the gorge; this shale having once been the muddy bottom of these seas, and this shell being found only in salt water.

At a later geological period, on top of what is now this shale, at the bottom of a warm ocean, still covering all this land, grew a vast, thick and solid bed of coral, of which ancient life the Niagara limestone of today is a monument.

Subsequently these two ancient and contiguous sea bottoms, then solid stone, were uplifted and by the configuration of the earth hereabouts the original Niagara river was formed. In general terms its course was similar to that of the present river (though its volume was not as great) as far north as the Whirlpool, from whence it ran, in a broadening channel, to St. Davids, westerly from its present outlet; and prior to the coming of the ice age it had cut this channel back certainly to the Whirlpool, and perhaps even farther south.

Next came the glacial period, when this part of the country was enveloped with a covering of ice, (working down from the northeast) similar to that now covering Greenland, though having a depth of perhaps a mile or more. This ice age, as approximately determined, lasted 50,000 years and closed about 200,000 years ago.

This ice sheet as it moved forward and southward broke off all the projecting points of rock, and scraped all the rocks themselves bare. Its presence and power are attested by the scratchings and markings on the smoothed surfaces of the top layer of rock wherever it is laid bare today, as far south as the Ohio river, and is apparent on Goat Island. This ice sheet brought down in its course not only boulders from the far north and northeast, but its own vast accumulations and scrapings and ebrasions, which we call "drift;" it being of a marine derivation; and with this drift the ice sheet filled up (and with its enormous weight pressed compactly) all valleys, gorges and indentations of the earth in its course, among them the old outlet or bed of the Niagara river from St. Davids to the Whirlpool.

The sectional view of Goats Island's rocky substrata shows what enormous grinding force must have been exerted on the top rock above the present western end of Goat Island, (for of course there was no gorge west of the Island then), so much of the limestone having been gouged out by the ice. In this excavated cavity, drift was deposited by the ice. Many of the boulders brought here in the ice age, carried perhaps hundreds of miles, have been collected in this section and used in the con struction of the handsome stone bridges that have been built on the Reservation, on the main shore opposite Goat island.

On the recession of the ice sheet a second Niagara river came into existence.

The weight of this vast ice sheet had canted or tilted the land to the northeast, so that at its recession the waters of the present three great northern lakes flowed east by the Ottawa and later, as the land rose, by the Trent valley. As this second Niagara river drained only the Lake Erie basin, and as Lake Erie was very much smaller than at present, it worked at first in a small channel, was of small volume and had but small rock cutting power to take up the work or erosive process of the earlier Niagara river, which had drained only this same Lake Erie basin.

This is the period, again referred to, when the present channel to the south and west of Goat Island (the Canadian Channel) was made.

It should be noted that the land to the northeast is even yet rising, or slowly regaining its former level. This bears on our subject in that in time, in the upper lake region the present slight slope to the southeast will be entirely overcome, and then the waters of the three great upper lakes will find their discharge to the westward, and the Niagara river will again drain only the Lake Erie basin and as a result will enormously decrease in volume.

If when this time comes the two falls shall have eaten their way back past Goat Island they will have left it an elevated and isolated Island, or more probably a promontory, whose little forest will be perched on a rocky base over 200 feet above the rapids, below the falls. The Island itself will be narrower than at present on account of the action of the elements.

If, however, when that time shall come the American Fall shall not have receded far (and judging from its recession during the last 200 years, it is improbable that it will have), its channel, by the great lessening of the flow of the river will become dry and Goat Island, and the American channel, between it and the main shore, will become once more a part of the American mainland, and there will be but one small fall in the Canadian channel.

The second Niagara river gradually merged itself into a vast fresh water lake, formed by the melting ice and heavy rainfalls, and covering all the Lake Erie basin, and gradually rose in level until it stood fully 100 feet above the present rocky bed of Goat Island.

Its northern boundary was the escarpment or ridge whose lowest point was just above the present village of Lewiston, which point is 32 feet above the present level of Lake Erie. Here the rising waters first broke over the dam and here Niagara Falls were born.

From here they cut their way back to the Whirlpool, for the waters found it easier to cut a new channel back through the soft rock from this point in the embankment than to scour out the old drift filled channel (which was at the very bottom of the lake) from the Whirlpool to St. Davids.

The flow of the lake set towards the falls and brought down from the Erie basin fluviate deposits in large amounts during the succeeding years, depositing them all along the bottom of the lake. It is of these fluviate deposits, consisting of sand, and loam (excepting a comparatively small layer of drift next to the top rock) that the soil of Goat Island is formed.

This Goat Island soil, more than any surface in this section is the geologists' paradise. While some lands and forests near here may not have been cultivated by man, the western end of Goat Island is an absolutely unique piece of virgin forest.

Most of the time it has been, in general terms, inaccessible to man; and since accessible by bridges, no cutting of the trees, no clearing of the land nor cultivation thereof, no pasturing of cattle, in fact no disturbance of the soil, has been permitted.

Here then is the original drift, with the subsequent over lying alluvial deposits and accumulations, undisturbed by man. And when, as in this case, in this undisturbed fluviate deposit are found fresh water shells, it proves that the Niagara river to-day flows through what was once the bottom of a vast fresh water lake that covered all this section.

As the falls cut their way back to the Whirlpool, so their height diminished and the level of this fresh water lake fell until finally there came a time when the land of what is now Goat Island, rose above the waters. That this lake existed at a comparatively recent geological period is proven by the fact that these shells now found on Goat Island are identical in species with those found inhabiting the Niagara river and Lake Ontario to-day. According to the most accurate calculation, the concensus of geological opinion is that 35,000 years have elapsed since the falls were at Lewiston, which is seven miles away; and that the fluvial deposits on the Island began as soon as the river rose over the moraine at the foot of Lake Erie, can scarcely be doubted.

That in 35,000 years there is no specific difference between the ancient shells found in the soil of Goat Island, and their existing representatives and progeny in this locality is wonderful indeed.

As geologists differ by thousands of years as to how long it took the falls to cut their way from Lewiston ridge to their present location it would be impossible to say when in the history of this section the waters had so far drained off, that the muddy deposits overlying the rocky bed of what is now Goat Island, first appeared above the slowly receding waters of the lake, unless we adopt some length of time for this work as a basis.

But it is not so difficult, by noting the elevation of the land, the trend of the rocks and the depth of the overlying "drift," to locate approximately where the falls were when this occurred. At that time, judging from the present levels of the land, the falls must have been at a point nearly a mile north of the present location of the Horseshoe Fall. And if we accept, as above, one foot a year as a fair average estimate of the recession of Niagara from Lewiston Heights in the more recent geological time, say since the Christian era, it must have been between four and five thousand years ago that the soil of Goat Island, then a part of the mainland, first appeared; and probably it is nearly as long since it became an island.

In speaking of the recession of Niagara, I refer to the recession of the Horseshoe Falls, for they recede several hundred times as fast as the American Falls; for in the time that the Horseshoe has receded from Prospect Point, at the lower or northern edge of the American Falls, across the width of these American Falls and across the width of Goat Island to their present position, the American Fall has receded but a very few feet.

Hence on these deductions, Goat Island has existed as an island from about the time of the Flood, or from about 2300 B. C.

This proves the statement that "In a scientific sense the island is of trifling antiquity, in fact it would be difficult to point out in the western world any considerable tract of land more recent in its origin."

As the Canadian Fall is lower in level than the American Fall, and as the main body of water and deepest channel appertain to this Canadian Fall, it is certain that the channel of the second Niagara river, which of course, after the lake was drained off, was at the lowest level of this old lake bed, was practically identical with the Canadian channel of the river just above the falls today; that is to the south and west of Goat Island.

Then Goat Island was a part of the American mainland, and the rocky bed of the river between the Island and the shore, where to-day are the American rapids, was also part of the mainland and covered with soil like that on Goat Island.

Then came a time, perhaps some hundreds of years afterwards, when, in the steady rerising of the land at the northeast towards the elevation that it had before it was depressed by the ice, the outlet of the three upper lakes to the east was cut off; and the waters seeking a new outlet found it by what is now the St. Clair river into Lake Erie.

By this means the volume of the Niagara river was suddenly and enormously increased. This permanently raised the level of the river, and part of this increased volume of water poured over the lowest point of the mainland near where Goat Island is to-day, this point being in the present channel of the American rapids and along the American shore up stream, and this rush of waters cut and swept away the soil down to the rock, leaving and thus forming Goat Island.

Probably at the same time and in the same manner were cut off and formed the small islands that now lie on both sides of Goat Island, though they were at the first larger and being joined together, fewer in number than at present.

Certainly up to the time of the cutting of the channel of the American Fall, the river shore of what is now Goat Island extended very much farther up stream, and probably after the Island itself was formed its upper end extended much farther eastward; for at its eastern end, now called "the parting of the waters," a sandy bar extends some hundreds of yards up stream. On this bar and south of it the depth of water is to-day less than three feet, and in the winter its whole length is covered with ice that lodges there. This entire bar was no doubt at one time covered with soil and was a part of Goat Island, the land being gradually washed away by the water, aided in its work by frost and ice.

One author says " One of the early chronicles states that the island contained 250 acres of land," but I have been unable to find that chronicle.

ITS BOTANY AND FOREST BEAUTY.

" The groves were God's first temples."

Sir Joseph Hooker, the noted English Botanist, has said that he found on Goat Island a greater variety of vegetation within a given space than he had found elsewhere in Europe or east of the Sierras in America, and Dr. Asa Gray, the greatest of American Botanists, confirms that statement.

The man today most familiar with the botany of Goat Island is David F. Day, who at the request of the Reservation Commissioners recently prepared a list of the Flora of the islands and Reservation. From his report to them and from his other writings, I quote:

"The vegetation of the island is that which might be expected to luxuriate upon a deep calcarious soil, enriched with an abundance of organic matter."

"The Flora of Goat Island presents few plants which may be called uncommon in Western New York."

"Goat Island is very rich in the number of its species."

"Its vernal beauty is attributable, not merely to its variety of plants, conspicuous in flower, but also to the extraordinary abundance in which they are produced. Yet it seems likely that there was a time, probably not long ago, when other species of plants of great beauty, were common upon the island, but which are not now to be found there. It is hardly possible that several orchidaceous plants and our three native lilies did not once embellish its woods and grassy places. Within a little while the harebell has gone and the Grass of Parnassus is fast going. This is undoubtedly due to careless flower gatherers, who have plucked and pulled without stint or reason. The same fate awaits others that do so much to beautify the island, unless the wholesale spoiliation is soon arrested."

Mr. Day then suggests that pains be taken to re-establish on the Island the attractive plants which it has lost, stating that the success of the effort would be entirely certain and thereby the pleasure of a visit to the Island would be greatly enhanced to many visitors. And he rightly adds "it would surely be a step and not an unimportant one in restoring the island to the state in which nature left it."

No doubt many of the seeds from which started the first foliage and forest, as well as many succeeding species were planted by the river at its inception and in subsequent decreasing levels.

In another article Mr. Day says: "The tourist who takes enjoyment in the shadows of a forest, almost unchanged from its natural condition, in the stateliness and symmetry of individual trees planted by the hand of nature herself; in the beauty and fragrance of many species of flowers growing without cultivation and in countless numbers; in the ever varying forms and hues of foliage and in the constantly shifting panorama of the animated creation so near the scenes of human activity and occupation and yet so free from their usual effects, will find on the islands which hang upon the brink of the great Cataract, an abundant gratification of his tastes and an exhaustless field for study."

"A calcarious soil enriched with a abundance of organic matter like that of Goat Island would necessarily be one of great fertility. For the growth and sustentation of a forest and of such plants as prefer the woods to the openings it would far excel the deep and exhaustless alluvians of the prairie states."

"It would be difficult to find within another territory so restricted in its limits so great a diversity of trees and shrubs and still more difficult to find in so small an area such examples of arboreal symmetry and perfection as the island has to exhibit."

"The island received its Flora from the mainland, in fact the botanist is unable to point out single instance of tree, shrub or herb, now growing upon the island not also to be found upon the mainland. But the distinguishing characteristic of its flora is not the possession of any plant elsewhere unknown, but the abundance of individuals and species, which the island displays." "There are to be found in Western New York about 170 species of trees and shrubs. Goat Island and the immediate vicinity of the river near the falls can show of these no less than 140." There are represented on the island four maples, three species of thorn, two species of ash, and six species, distributed in five genera, of the cone-bearing family. The one species of basswood belonging to the vicinity is also there.

Mr. Day's catalogue of plants, in his report to the Reservation Commissioners, gives 909 species of plants to be found on the Reservation, of which 758 are native and 151 are foreign. Margaret Fuller Ossoli wrote: "The beautiful wood on Goat Island is full of flowers, many of the fairest love to do homage there. The wake robin and the May apple are in bloom, the former white, pink, green, purple, copying the rainbow of the falls, and fit to make it garland for its presiding Deity when he walks the land, for they are of imperial size and shaped like stones for a diadem. Of the May apple I did not raise one green tent without finding a flower beneath."

Frederick Law Olmstead wrote: "I have followed the Appallachain chain almost from end to end, and travelled on horseback 'in search of the picturesque,' over 4,000 miles of the most promising parts of the continent without finding elsewhere the same quality of forest beauty which was once abundant about the falls and which is still to be observed on those parts of Goat Island where the original growth of trees and shrubs has not been disturbed, and where from caving banks trees are not now exposed to excessive dryness at the root.

"All these distinctive qualities, the great variety of the indigenous perennials and annuals, the rare beauty of the old woods, and the exceeding loveliness of the rock foliage I believe to be a direct effect of the falls and as much a part of its majesty as the mist cloud and the rainbow. They are all as it appears to me to be explained by the circumstance that at two periods of the year, when the Northern American forest elsewhere is liable to suffer actual constitutional depression, that of Niagara is assured against ills and thus retains youthful luxuriance to an unusual age.

"First the masses of ice which every winter are piled to a great height below the Falls and the great rushing body of ice cold water coming from the northern lakes in the spring, prevent at Niagara the hardship under which trees elsewhere often suffer through sudden checks to premature growth. And second, when droughts elsewhere occur, as they do every few years, of such severity that trees in full foliage droop and dwindle and even sometimes cast their leaves, the atmosphere at Niagara is more or less moistened by the constantly evaporating spray of the Falls, and in certain situations bathed by drifting clouds of spray."

In 1785, years before the island was bridged, St. John de Crevecoeur in a long letter describing Niagara wrote: "You then come to an island covered with trees and shrubs, whose foliage and situation have a very happy effect amidst the turbulent scenes around."

And nowhere else is to be found a more beautiful piece of virgin forest, where nature protected it from man's encroachment by its insular position; where a rich alluvial soil furnished the trees with food, and nature's bounty provided them with drink from the ever-present spray. And, lastly, luckily when man acquired occupation and possession, the Island and forest became the property of those by whom its soil was not disturbed, but was left as nature herself for hundreds of years had preserved it. Truly we can say with Longfellow:

> "This is the forest primeval. The murmuring pines and the hemlocks,
> Bearded with moss, and in garments green, indistinct in the twilight,
> Stand like Druids of eld, with voices sad and prophetic."

ITS HISTORY.

Indian Ownership, 1600-1764.

In taking up its history chronologically, let us start with Goat Island, in the very early pre-Columbian days, when this section was inhabited or certainly visited by those unknown Indians to whom we refer as Aborigines.

We do not know the name of the tribe that inhabited this section prior to about 1600, but at that time the Neuter nation dwelt on both sides of the Niagara river. In 1651 the Senecas, the nearest neighbors of the Neuters on the east, and themselves the westernmost tribe of the Iroquois, suddenly attacked the Neuters and annihilated them; and by reason of the conquest claimed their lands. And this claim was recognized as valid by the other Indian tribes, and therefore later on by the white man. In this way Goat Island passed into the hands of the Senecas, who held it for over 100 years. To the Senecas, as well as to the Neuters and the Aborigines, Goat Island was a sacred spot. To them it was the abode of the Great Spirit of Niagara. In the spray they saw the manifestation of their Deity, in the thunder of the cataract they heard his voice—

> "And the poor Indian whose untutored mind
> Sees God in clouds and hears him in the wind."

believed that he could sometimes even see, in the ever shifting clouds of mist, the outlined figure of Him whom he worshiped. The Island's use to the Aborigines appears to have been as a burial ground, and tradition says that in its soil rest the remains of many an Indian warrior, interred there hundreds of years ago; over whose mounds to-day stand trees of great age. Here, says the same untraceable tradition, was interred the body, when recovered, of the "fairest maiden of the tribe," who was annually sent over the Falls, in a white canoe decked with flowers, as the noblest possible sacrifice to the Great Spirit.

There is no written nor published record, that I know of, of any Indian burial taking place on the Island. Hennepin makes no mention of this use of it, as he would in all probability have done had the Senecas, or even had their immediate predecessors, the Neuters, buried their warriors here. But he says "the island is inaccessible." Hence we can only assume that these graves long antedate his visit, and are the graves of Aborigines.

In 1834, the skeleton of a young female that had been dug up on Goat Island shortly before, was in the Museum of the Boston Medical College. This may possibly have been the skeleton of that heroine of the "Legend of the white canoe" who was the last "fairest maiden" to be sacrificed to Niagara's Deity. It was found interred in a sitting posture; and it is said that "the graves on the island were in a sandy spot, each body in a separate grave, always in a sitting or squatting posture, and without ornaments." Can this position of burying their dead be any aid in tracing the tribe or stock to which the Aborigines about Niagara belonged? It has been further advanced as possible that these Indian burials on the Island took place when the Island was a part of the mainland, but this seems to me to be improbable.

Goat Island, practically as it is to-day, has existed for many hundred years, and its insular position, so difficult of access, added to its sacred character as the home of Deity, must have been one of the main reasons for its selection by the Indians as their warriors' burying ground.

Tradition tells us that the Indians of long ago made annual pilgrimages to Niagara, often coming great distances, to offer to the Great Spirit sacrifices of the spoils of the chase, of war, and of the crops. Further, the chiefs and warriors, invoking blessings for the future, used to cast into its waters offerings of their weapons and adornments. We must assume that at least these offerings were made from Goat Island, as no "brave" would have been considered worthy of the name who could not reach the insular abode of the Great Spirit, from thence to offer up his invocation.

While there are references to Niagara Falls, though not by name, in works published from 1604 on; in *Champlain*, in the

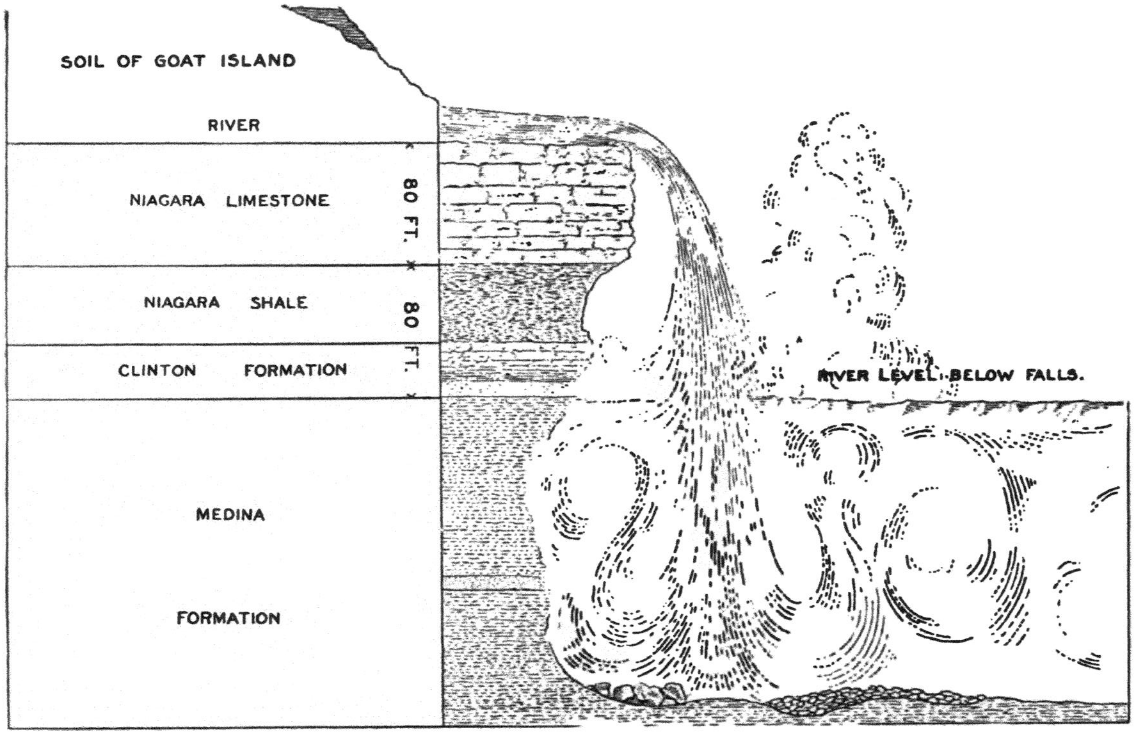

PROF. SHALER'S SECTIONAL VIEW OF NIAGARA.

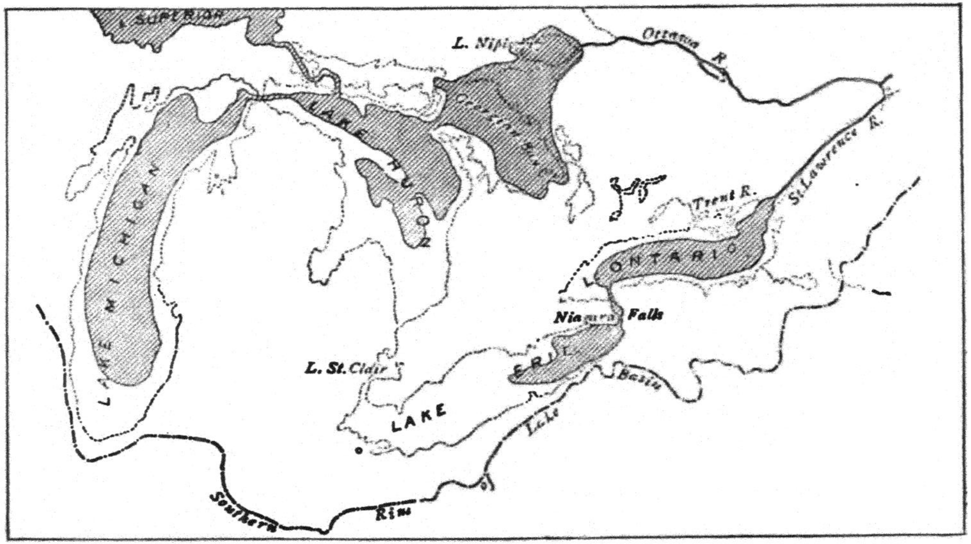

Lake Region, after the melting of the Great Glacier. Modern Lakes in Light broken lines. Ancient Lakes, shaded.

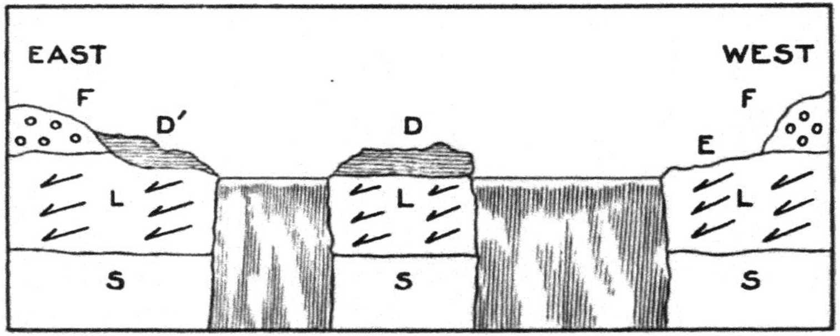

L. Limestone 80 feet thick. S. Shale 80 feet thick.
D. Fresh water Strata Goat Island 20 feet thick.
D′. Same formation. American shore.
E. Ledge bare Limestone Canada Shore. F. Ancient Drift.

SIR CHARLES LYELL'S SECTION OF NIAGARA.

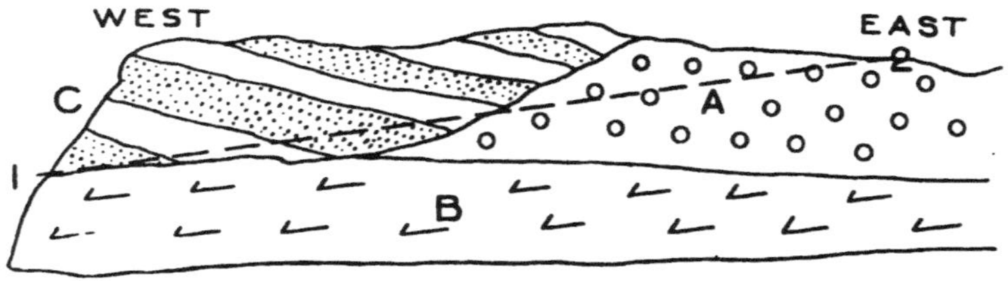

A. Upper thin bedded Niagara Limestone.
1, 2. Present surface of Niagara River, at the Rapids.
B. Massive compact Niagara Limestone.
C. Alluvial drift and fresh water beds of Gravel, Sand and Loam, with Shells.

SIR CHARLES LYELL'S SECTION OF GOAT ISLAND, 2500 FEET LONG.

PATH ON GOAT ISLAND.

First known Picture of Niagara. 1697.

Jesuit Relations, in *De Creuxius*, etc., I know of no reference to Goat Island until Hennepin, who first saw it in December, 1678, mentions it, saying of Niagara: "Its fall is composed of two sheets of water and a cascade with an island sloping down," and in the English edition of his works, he tells of "This wonderful downfall with an isle sloping along the middle of it."

And in the same work, when he again saw Niagara on his return from the West, he says: "After it has run thus violently for six leagues it meets with a small sloping island about half a quarter of a league long and near 300 feet broad, as well as one can guess by the eye, for it is impossible to come at it in a canoe of bark, the waters run with that force. The isle is full of cedar and firr, but the land of it lies no higher than that on the bank of the river. It seems to be all level even as far as the two great cascades that make the main fall. The two sides of the channel which are made by the isle, and run on both sides of it, overflow almost the very surface of the earth of said isle, as well as the land that lies on the banks of the river to the east and west, as it runs south and north. But we must observe that at the end of the isle on the side of the two great falls there is a sloping rock which reaches as far as the Great Gulph into which the said waters fall; and yet the rock is not at all wetted by the two cascades which fall on both sides, because the two torrents which are made by the isle throw themselves with a prodigious force, one towards the east and the other towards the west, from off the end of the isle where the Great Fall is."

La Hontan, who saw Niagara in 1687, when he accompanied De Nonville in the expedition to build Fort Niagara, wrote of the Island: "Towards the middle of the water-fall of Niagara we descry an island that leans toward the precipice as if it were ready to fall."

These remarks of Hennepin and La Hontan show that 200 years ago the upper portion of the western end of Goat Island projected out over the gorge, and, as the softer shale at the base of the cliff above the debris slope had then crumbled away, it must have given to this end of the island that sloping or about-to-fall appearance mentioned.

All of this overhanging cliff has, since 1790, tumbled into the gorge below.

In speaking of the beasts that try to cross the river just above it La Hontan calls it "that unfortunate island." He published no view of Niagara. He was a soldier and possible sites for forts interested him more than wonderful scenery.

For seventy years after Hennepin published his, the first known picture of Niagara Falls, and therefore of Goat Island, numerous pictures of them appeared, mostly in geographies and books of travel, published in many languages and in several countries of Europe. All of these pictures, while varying in details, were based mainly on Hennepin's; all showing Goat Island as extending far up stream; but some of them represented it as very narrow at the cliff and throughout its length, while others broadened it even more than Hennepin did.

Between 1719, when Joncaire established his cabin or warehouse at Lewiston, with French attendants, and 1725, when the French built and garrisoned their second Fort Niagara, some of these men may have and probably did visit the Island; indeed there is no one to whom we can, with more probability of being correct, ascribe the honor of having been the first white man to set foot on Goat Island than to Joncaire. He was an adopted

THE LEGEND OF THE WHITE CANOE.

AN EARLY PRINT OF NIAGARA. 1751.

AN EARLY PRINT OF NIAGARA. 1783.

ERIE'S SKETCH OF NIAGARA. 1763.

child of the Senecas, and the man to whom Charlevoix refers as speaking " with all the good sense of a Frenchman and with all the eloquence of an Iroquois."

As the garrison at Fort Niagara, from 1725 to 1759 was usually a large one, it is more than probable that a number of these adventurous French officers and soldiers were at various times piloted to the Island in the canoes of the Senecas, who lived in this section and who were the firm friends of the French. In January, 1751, there appeared in London, in the Gentlemen's Magazine, a picture of Niagara Falls and a letter from the Swedish Naturalist Peter Kalm, who had visited the Falls the year before.

This picture, without the ladders on the Goat Island cliff, was a fair sample of the pictures of Niagara up to that time, and is reproduced herewith. In the letter, Kalm tells of two Indians who, twelve years before (that is in 1738), had gone in a canoe on the river above the falls, but having some brandy with them, became intoxicated, and lying down to sleep in the canoe, were carried down stream so far that the noise of the falls awakened them. By great effort they reached Goat Island, but their canoe seems to have been carried over the falls. After some time, two or three days probably, being nearly starved, and seeing no other possible way of escape they made ladders of the long vines that grew on the Island, and fastening the ends at the bank above, let them down the cliff and descended by them to the waters edge below. Here they tried to swim across the river, but the waves repeatedly beat them back, bruised, onto the Island's base. Discouraged, they ascended their ladder and finally attracted, by their cries, the attention of two Indians on the main shore. These, seeing the situation, hastened to report it to the commandant at Fort Niagara.

"He caused four poles to be shod with sharp irons. As the waters that ran by the Island were then shallow, two Indians took upon them to walk thereto by the help of these poles, to save the other poor creatures, or perish in the attempt. They took leave of their friends as if they were going to death. Each had two poles in his hands to set to the bottom of the stream to keep them steady. So they went and got to the Island, and having given poles to the two poor Indians there, they all returned safely to the main shore. Those two Indians who in this above mentioned manner were first brought to this Island are still alive. They were nine days on the Island."

"Now, since the road to this island has been found, the Indians go there often to kill deer, which have tried to cross the river above the falls and were driven upon this island by the stream." But, Kalm adds, "If the king of France were to give me all Canada, I would not venture to go to this island; and were you to see it, Sir, I am sure you would have the same sentiment." Kalm also in this letter, makes the first mention I find anywhere of small islands adjacent to Goat Island, saying, "On the west side of this island are several small islands or rocks of no consequence."

Another account of evidently this same story, tells how the rescuers were provided by the blacksmith at Fort Niagara, with long stilts shod with iron points, on these they walked to the Island, carrying two extra pairs of stilts, and all four Indians "stilted" back to safety. While the inventor of this last story avoided the incongruity of having men walk on foot across a channel where the water now at least is ten or twelve feet deep, his stilt story is almost as absurd.

Later on a traveler heard the story in this way: "By making long bark ropes and carrying them a considerable distance up the stream, they succeeded in floating one end against the Island by which means they were enabled to rescue the poor wretches from certain death." The inventor of this story evidently did not know that the current would carry the end of the rope away from, not towards, Goat Island. In 1759 the English captured Fort Niagara and secured complete control of all this section. In 1763 the Senecas planned and executed the Devils Hole massacre, from which only one man of the English escort escaped, John Stedman by name. Amid a shower of bullets and arrows he spurred his horse and dashed in safety to Fort Schlosser, nearly five miles away. He subsequently claimed that the Senecas, marvelling at his escape, and believing the Great Spirit had given him a charmed life, gave him all the land between the Niagara river and the line of his flight, some five thousand acres in all. The Senecas do not appear to have paid any attention to his claim, although during his lifetime Stedman seems to have occupied unmolested, such lands in his claimed grant as he chose, but only a small part thereof. When his descendants set up their claim, under this Seneca grant, they could produce no deed nor proof of one. They claimed that Stedman gave the deed to Sir William Johnson for safe keeping, and that it was destroyed when Sir William's residence, Johnson Hall, was burned.

They kept up the fight until about 1823, when the State of New York, after their claim had been declared worthless, ejected them from such lands as they occupied under the claim.

In 1764, at the great treaty held at Fort Niagara, between Great Britain and nearly all the Indian tribes of North America, Sir William Johnson obtained for England from the Senecas all the land along the Niagara river, four miles wide, averaging two miles in width on each side thereof, from Lake Ontario to Lake Erie. The diplomatic Senecas specially excepted from this grant all the islands in the river.

Only the year before that nation had attacked the English, in the Devils Hole Massacre, and had then been obliged to sue to Sir William Johnson for peace and reconciliation. And even at this great treaty gathering they had not kept their promise to him of being present, and had come to it only after he had arrived at the fort and finding them unrepresented, had sent a special messenger to them and threatened to send Bradstreet's army to punish them if they did not at once appear and fulfil their former promises. These they had just fulfilled, and now they begged Sir William Johnson personally to accept from them all the islands in the Niagara river "as a token of their regard for him, and in remembrance of the trouble they had from time to time given him."

Johnson's influence with the Indians was unbounded, he had been married to a sister of the great Mohawk warrior Brant, he was England's Indian agent, and so far as dealing with Indians of all tribes was concerned, he was the most influential white man that ever trod the continent of North America. Such a man's friendship was worth having at any time, especially to the Senecas at that time, even if paid for by the gift of many islands, Goat Island included.

Sir William Johnson accepted the proffered gift, fearing a loss of influence with the Senecas if he refused. But the English military law of that period forbade officers to accept presents, and certainly in cases of gifts of land, which could not be kept secret, the law was obeyed. So Sir William at once presented all these islands to the English Crown.

And thus in 1764, this wondrous, though as yet unnamed Island, passed from the possession of the Senecas and into the possession of the Crown of England.

SOVEREIGN OWNERSHIP.

1764—1816.

In 1764 there came to Fort Niagara, in Bradstreet's army, in the British service, a man destined in after years to be a conspicuous figure in Colonial history, Israel Putnam. He was lieutenant-colonel of a Connecticut regiment, and tradition says that during the month that Bradstreet's army lay encamped at that fort he visited Goat Island on a wager; being the first white man to set foot thereon. A long rope was fastened to a boat, its other end being secured on the shore, and it was paid out as the boat was swiftly paddled, by its Indian guides, to the Island. The boat and its occupants were later hauled back to the mainland. The story in itself is by no means improbable, for it is easily possibly to-day to go to Goat Island by boat, starting well up stream and keeping over the bar that extends far easterly from the Island, and it has been very frequently done during the past 100 years. Stedman, referred to later on, is reported to have gone to the Island on horseback, and by swimming his horse out to the sandy bar well up stream and letting the animal walk to the Island on the bar, on which the water is always shallow, it might easily be accomplished. It is much more than probable, however, that white men had been on the Island before 1764.

In 1768, an English officer, Lieut. Wm. Pierie, then stationed at that same fort, made, from the Canadian side, a sketch of Niagara Falls, which was engraved and issued the next year. While containing inaccuracies, this view of the Falls stood forth to the world as the first picture of them ever published that had the merit of approximate truthfulness of delineation, and at the same time any artistic pretentions.

Prior to 1770 John Stedman, before referred to, as claiming under a deed from the Senecas all the land on the American side near Niagara Falls, had construed this claim so as to include Goat Island, and had cleared a portion of the upper end thereof and raised thereupon a fine crop of turnips. In the fall of that year he placed on the Island a number of animals, among them a male goat. His expressed object in putting these animals there was to get them out of the reach of the bears and wolves which then prowled, practically unmolested, about his home on the main shore, some two miles further up stream. That winter was a very severe one. Why he left the animals uncared for is unknown, but by spring all but the goat were dead.

His tenacity of life gave his name to his Island prison, and Goat Island it has been called ever since. Whether the goat died on the Island is not known. So thoroughly has this name become attached to the Island that it would seem impossible now to change it, were it so desired, which it is to be hoped it will not be. In 1819, when the Commissioners under the treaty of Ghent were engaged in determining the boundary line between the United States and Canada, Gen. Porter, one of the Commissioners, and also an owner of Goat Island, proposed to call it "Iris Island," and it was so designated in the minutes of, and on the maps published by, the Commissioners.

But the traveling public of the world would have none of it; Goat Island it was; Goat Island it should remain. So they called it; so they continued to call it; and so it is known even until to-day, in literature and in cartography; and that is why the title of this pamphlet reads, not "Iris Island," but "Goat Island."

At the close of the Revolution, in 1783, by the treaty of Paris, England relinquished all claim over her American colonies, and their lands. Thus Goat Island passed into the possession of the State of New York. That treaty provided that the line of division between Canada and the United States should run "along the middle of the communication [between Lake Erie and Lake Ontario] into Lake Erie."

Under this wording the State of New York most naturally claimed Goat Island, and subsequently the Commissioners, under the treaty of Ghent, fixed the following boundary line at this point, which is still in force: "Thence [from a point in Lake Ontario opposite the mouth of the Niagara river] to and up the middle of the said river to the Great Falls; thence up the Falls through the point of the Horseshoe, keeping to the west of Iris or Goat Island, and of the group of small islands at its head," thus fully sustaining New York's contention. It was not until a year and a half after the signing of the treaty of Ghent, which was signed March 24, 1814, that the State of New York parted with the title to Goat Island; and not until 1822 that the Commissioners under said treaty signed their decision and thus fixed our northern boundary line.

It is also certain, with the large English garrison at Fort Niagara from 1759 until after the Revolution, and even until 1796 (until which date England held Fort Niagara) that many adventurous Englishmen visited Goat Island, and of this we have more substantial proof than we have of the earlier visits of Frenchmen.

Isaac Weld, who visited the Falls in 1796, says, "The Commodore of the King's vessels on Lake Erie, who had been employed on that lake for upwards of thirty years, informed me that when he first came into the country [that would be in 1776], it was a common practice for young men to go to the island in the middle of the Falls; that after dining there they used frequently to dare each other to walk into the river towards certain large rocks in the midst of the rapids not far from the edge of the Falls; and sometimes to proceed through the water even beyond these rocks. No such rocks are to be seen at present; and were a man to advance two yards into the river from the island, he would be inevitably swept away by the torrent."

Chataubriand, who saw the Falls in 1790, says, "Between the two Falls there is an island, hollow underneath, and which hangs with all its trees over the chaos of the waves," thus proving Hennepin's statement of the island "sloping down."

P. Campbell, in 1793, relates a curious story about the Island having been so "overrun with rattlesnakes that it was dangerous for a person to walk on it until a parcel of swine were put on it and which nearly rooted them out."

The title to Goat Island was not involved in the dispute, at the commencement of this century, between Massachusetts and New York regarding the ownership of the western part of the latter state.

Judge Augustus Porter first visited Goat Island in 1805, going by canoe. He found at its upper end the clearing of a few acres made many years before by Stedman.

He also found carved on the trees thereon the dates, 1769, 1770, 1779, 1783; which is pretty substantial proof of visits to the Island having been made by Englishmen as before claimed.

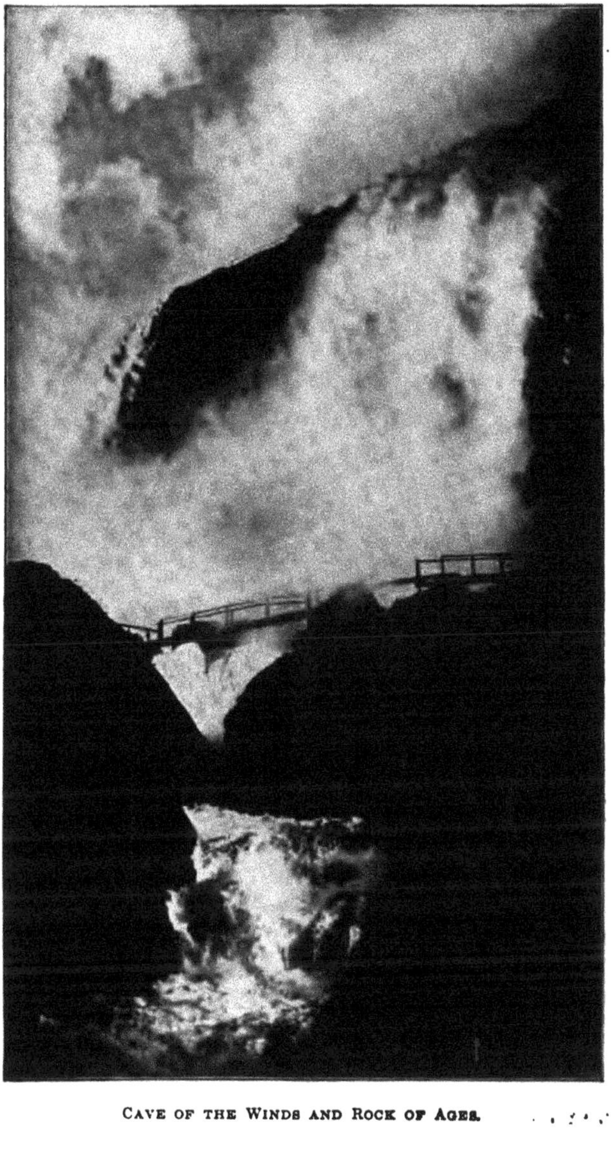

Cave of the Winds and Rock of Ages.

Of course, since the Island was bridged, thousands and thousands have visited it; so that an early date now readable on any tree thereon, may have been carved by a visitor of much more recent years.

In 1811 Augustus Porter, in behalf of his brother and himself, applied to the State of New York for the purchase of the Island. His petition read as follows:

"To the Honorable the Legislature of the State of New York, in Senate and Assembly convening; the petition of the subscriber humbly showeth, that your petitioner is an inhabitant of the town of Cambria, in the County of Niagara. That his place of residence is surrounded by a large body of unsettled lands, which are likely to remain so for a long time, which afford a shelter for wolves and other wild animals, owing to which the raising of sheep is rendered extremely difficult. That, in the Niagara river, directly opposite to the residence of your petitioner there is a small island owned by the people of the State, called Goat island, containing as your petitioner believes, about 100 acres, where sheep might be with great safety kept. Your petitioner therefore prays that your honorable body will pass a law authorizing the commissioners of the land office to sell to your petitioner this said island at a fair price, to be determined by appraisal, or in such other way as your honorable body in your wisdom may deem proper, and your petitioner will ever pray.

AUGUSTUS PORTER."

February 23, 1811.

The petition was referred to the Surveyor General, who reported as follows: "The surveyor general, on the petition respectfully reports, that the petitioner is settled on the shore of the Niaraga river opposite to an island of about 100 acres called Goat island, which he is desirous of obtaining for the purpose of keeping sheep free from wolves and other wild animals, which on account of the country it is difficult to do. This island is about 7 chains from the east shore, with its lower end butted on the precipice over which the Niagara river falls at the great Cateract. On account of the great velocity of the current which descends to the island and sweeps its sides, the passage to and from it is difficult and considered so dangerous that few have attempted it. The petitioner, however, thinks that by means of projections from the shore he can lessen the difficulty and danger of the passage, and is willing for that privilege he prays for, to pay the Senate a reasonable addition to what he appraised as its fair value. From the circumstances stated it must be evident that the value of the island must very materially depend on its being an appendage to the estate on the shore directly opposite it. Should the Legislature judge proper to authorize a grant of it to the petitioner, it ought to be with the proviso that the Indian title to it be first extinguished.

Respectfully,

SIMEON DE WITT."

February 22, 1811.

It would appear from the dates that the Surveyor General had made out a not unfavorable report on the petition, the day before the latter was signed.

The Legislature declined to authorize the sale however, stating as its reason that it expected to use the Island itself, erecting thereon in the near future either a State prison or a State arsenal.

Judge Porter still kept on raising sheep, and still wanted Goat Island, and he finally outwitted the State, and obtained it. In 1814 he found out that Samuel Sherwood, a prominent lawyer, owned an instrument called a "float," given to him by the State of New York, in consideration of a failure of title to some lands he had purchased of it. This "float" authorized the bearer to locate 200 acres on any of the unsold or unappropriated lands of the Commonwealth. For himself and his brother, Augustus Porter bought this instrument from Sherwood, and with it duly assigned and attested, he started east. As soon as the stagecoach could land him in Albany, he hastened to the office of the Land Commissioners, and stepping up to the desk laid down the "float," remarking, perhaps in a tone of exultation, "There, damn it, I want Goat Island;" stating at the same time that he located a sufficient acreage of the float to cover that and the adjacent islands.

He got them, but necessary formalities took nearly two years. In October, 1815, the necessary survey was completed, and it was only a few weeks before that the State extinguished the Indian title to the islands, and could give a good title to them. This cession from the Senecas was dated at Buffalo September 12th, 1815, and under it these Indians reserved the right of "hunting, fishing, and fowling in and upon the waters of the Niagara river and of encamping on the said islands for that purpose," which rights, in law, did they care to exercise them, the Senecas still possess. The compensation paid by the State of New York to the Senecas for the cession of all the islands in the Niagara river within the jurisdiction of the United States (which included Goat Island) was $1,000 in cash and $1,500 a year in perpetuity.

It was not until November 16th, 1816, that Daniel D. Tompkins, Governor of the State of New York, signed the "patent" or deed, transferring these islands to Augustus Porter, of which interesting document (now in the possession of the author) a copy is given in this pamphlet. Augustus Porter at once deeded a half interest in the Goat Island group to his brother, Gen. Peter B. Porter.

PRIVATE OWNERSHIP.

1816—1885.

The Porter brothers immediately made arrangements to get a bridge to Goat Island, and in the spring of 1817 a wooden structure (of which a reproduction is given) was erected, at a point some 50 rods up stream from the present bridge. When it was completed every visitor to Niagara was glad to pay toll in order to get on to the Island, and by the end of the year 1817 it was evident that Goat Island was worth more as a pleasure resort than it ever could be worth as a sheep pasture.

So the proverbial idea of separating the sheep from the goats (in this case putting the sheep on the Island and leaving the goats on the mainland) was abandoned. The small island above this first bridge, shown in the engraving, if it ever existed, has long since been washed away.

So bold was this enterprise of bridging the rapids considered, that years afterwards Margaret Fuller Ossoli suggested that the Great Spirit of Niagara "had punished General Porter's temerity with deafness, which must have come upon him when he sunk the first stone into the rapids."

The heavy masses of ice coming down the river in the early months of 1818 struck against the unprotected piers of the bridge with such force as to carry them away. Promptly with the coming of spring, 1818, the Porter brothers erected a second but a more substantial wooden bridge. They selected a site further down stream and built it from the mainland to Bath, or as it is now called Green Island, and from that island they built another bridge to Goat Island. These were built on the sites of the present bridges, their builders correctly assuming that by reason

Second Bridge to Goat Island. 1818.

Third Bridge to Goat Island. 1855.

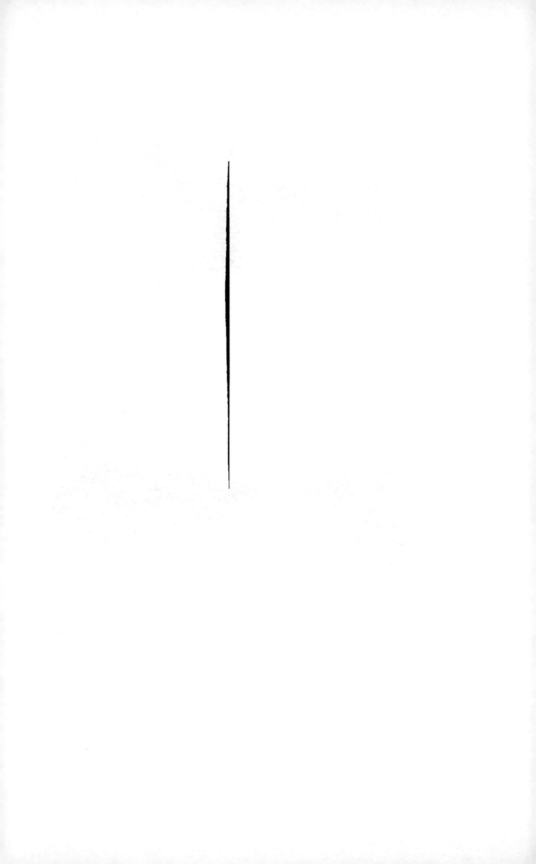

SHIP AND BRIG ISLANDS AND "LOVER'S BRIDGE."

First Bridge to Goat Island. 1817.

CAVE OF THE WINDS.

BIDDLE STAIRS FROM ABOVE.

BIDDLE STAIRS FROM BELO'

M1011.

of the descent of the river over the rocks, in the space between the destroyed and the new structure, the huge cakes of ice would be so broken up that comparatively little damage would be done to the new piers.

These two bridges (a cut of the one leading from Green Island to the main shore is given) with ordinary repairs stood till 1855, when they were replaced by the iron structures that to-day afford access to the Island.

In reply to the oft asked question how were these bridges built, let me answer; two giant trees about 80 feet long were felled in the vicinity, and hewed square on two opposite sides. A level platform, protected on the river side by cribbing, was built on the main shore. The two logs, parallel and some 8 feet apart, were laid on rollers, and with their shore ends heavily weighted with stone, were pushed out over the rapids. On each log a man walked out to the end, carrying with him a sharp iron pointed staff. A crevice in the rocky bed of the river having been found under the end of each of these logs, the staff was driven down into it, and to it the end of the log was firmly lashed. Plank were then nailed on these logs, and on this bridge stones were dragged out and laid in a pier, around these staves and under the end of either log, until a rocky foundation supported both timbers. Each succeeding span was then built in a like manner. While the bridge was in process of construction, Red Jacket, the famous Seneca, was on the bank an interested spectator. As the first span was successfully completed, and the erection of the bridge thus assured, some one asked him what he thought of it. Rising majestically, and drawing his blanket close about him, he muttered: "Damn Yankee," and stalked away.

Thus Goat Island was accessible to the public; and in 1818,

on the completion of the bridge, was made the first road around it. On the western and southern sides of the island it was built out beyond the upper edge of the land of to-day; for since that date some four rods in width on the western side and nearly 10 roads in width on the western half of the southern side of the Island have been washed away.

Here on the "Island of Iris, at the Falls of Niagara, Friday, the 4th day of June, 1819" (so read the minutes), when their survey had reached the mouth of the Niagara river, met Gen. Peter B. Porter, commissioner on the part of the United States of America, and John Ogilby, commissioner on the part of his Brittanic Majesty under the treaty of Ghent, with their secretary and attendants, in regular session. Among other things accomplished at this session, they resolved "that on the arrival of the surveyors, who were daily expected from Lake Ontario, where they had been engaged in completing some unfinished business of last year, they proceed to the survey of the Niagara river and its islands and on the completion thereof continue the survey of the [boundary] line between the United States and Canada."

Among the illustrious visitors to the island in 1825 came the Marquis of Lafayette, then the guest of the United States; who after a delightful walk of two hours left the Island, which appeared to him "like an aerial garden sustained by clouds and surrounded by thunder," regretting "that its distance from France would not permit him to purchase it as it would make a delightful residence."

Lafayette's secretary, M. La Vasseur, added to his account of the visit "The surrounding currents of water offer an incalculable moving power for machinery, which might be easily applied to all sorts of manufactories."

The owners of the Island were then power users and power developers, but were opposed to any such uses of this Island. They did develop power and erect mills on the main shore; and the one mill (a paper mill) whose erection was later permitted on one of the smaller islands, was allowed solely to enable one of the sons of Augustus Porter to start in business.

About 1826 a few deer (which had been plentiful in the vicinity) were placed on the Island, but the visitors of that day took such a delight in chasing them that, in their fright the animals, one by one, fled into the river and were carried over the falls. The great attraction on the Canadian side at this time was Table Rock, a projecting ledge just at the edge of the Horseshoe Fall, and, as an offset to that, in 1827 a bridge was built from Goat Island out to what is now known as Terrapin Rock. It was about 300 feet long, and the end of the bridge projected about 10 feet beyond the edge of the falls, forming an absolutely unique and dangerous point of observation. The heavy timbers of the bridge projected out some feet beyond the end of the bridge itself.

The next attraction built on the Island was the Biddle Stairs, enabling people to reach the slope below the island. They were erected in 1829, at the suggestion of Nicholas Biddle, of United States Bank fame, and he contributed a part of the expense of their erection. These stairs, after a period of 60 years of uninterrupted use, still afford the only means of descent to the debris slope below and to the Cave of the Winds.

Soon after their erection in the same year, there appeared at Niagara that man whose name is yet a synonym of high jumping, Sam Patch. The cliff of Goat Island appealed to him and his entreaties gained him permission to erect on the slope below the

Island and north of the Biddle Stairs, a platform from which he made, successfully, two leaps, 95 feet high, into the deep waters below. The platform from which he jumped was supported by (and also reached by) two enormous ladders whose lower ends rested upon the huge rocks at the waters edge, the ladders themselves leaning far out over the waters. Their upper ends were fastened by ropes to the top of the rocky slope on which the lower end of the Biddle Stairs rest. Midway of their length they were also fastened to the bank by ropes. Guy ropes, extending respectively up and down stream, kept the ladders from swaying sideways.

In the same year there came to Niagara Capt. Basil Hall, of the Royal British Navy; an extensive traveller and a voluminous writer. He admired and criticized Niagara; wrote learnedly and entertainingly about the pressure of the atmosphere behind the sheet of water, and left in his works his approbation of the decision of the owners of the Island to retain it in its natural state; and also took credit that his expressed views in favor of this course "may have contributed in some degree to the salvation of the most interesting spot in all America."

The same summer there appeared at Niagara that remarkable stranger, Francis Abbott, whose name will always be associated with this locality, as "The Hermit of the Falls." Young, learned, cultivated, and versed in the arts, he sought solitude and communion with nature. English relatives supplied him with ample money for his simple needs. Intending on his arrival to spend a week here, he passed the remaining year and a half of his life close to the great Cataract. He wanted to build a cabin on the first Sister Island which he proposed to reach by means of a drawbridge, but to this the owners of the Island could not consent. Obtaining permission to occupy an unused hut that stood on the north easterly side of Goat Island, he lived there for a year in solitude, save for his dog and his cat; preparing his own meals, writing much, but promptly destroying everything that he wrote, playing often on his flute and guitar; at all hours, but chiefly at night, when he would meet no human being, walking about the Island. He bathed daily, the year around, in the river, usually in the pool below the little fall between Goat Island and the first Sister Island, which thus has received the name of the "Hermit's Cascade." On the timbers that projected out beyond the edge of the bridge at Terrapin Rock, and which extended out even over the gulf, he would venture, walking rapidly right out to the end, and then turning quickly and fearlessly, retrace his steps. From the ends of these timbers he would hang by his hands, his body suspended in mid-air over the abyss, exhibiting absolute fearlessness and strength of will.

The increasing number of visitors induced him to leave the Island, and to occupy a hut on the mainland. Here he lived for six months, and one morning was drowned while bathing near the foot of the American fall. He is buried in the cemetery at Niagara Falls; and his life remains as a wonderful example of the all-pervading influence that Nature at Niagara can exert on an over-sensitive soul.

In the winter of this same year, a remarkable one in the Island's history, it is stated that the cold was so intense, and the ice in the river and in the rapids above so thick, that persons were able to cross to Goat Island without using the bridges; a remarkable fact, if true, and a condition which Nature has never vouchsafed us since; although during the intervening seventy years there have been some remarkably cold periods, notably in recent years, in 1874 and 1896. In the latter year, save for one wide break, over the deepest channel, a solid mass of ice accumulated, below the bridge to Green Island, and between the main shore and the smaller islands and Goat Island, on which many persons walked daily for nearly a week. And one man drove one afternoon from Bath Island down almost to the edge of the American Fall.

In 1833 was built of the stones of this immediate vicinity the Terrapin Tower, close to the edge or brink of the Horse Shoe Fall and quite a distance out from the Island. This tower was the one objective point of all visitors, the Mecca of all pilgrims. Of rude architectural design and construction, it stood for over forty years, a unique and not inharmonious adjunct to the great Cataract.

As the old Terrapin bridge was replaced with the present structure a few years afterwards, and as elderly visitors of to-day regret the disappearance of the old tower, a landmark of a past generation, I reproduce an old engraving of them as they were in 1834.

Familiar as the trip to-day is to many visitors, the first entrance of the Cave of the Winds, or Æolus's Cave, as it was first called, on July 15, 1834, marked an epoch at Niagara. For several years before that date visitors had penetrated a few feet behind the sheet of water below Table Rock on the Canadian side, but the passage behind the small sheet of water that flowed between Goat and Luna islands, and out beyond amidst the waters dashing and plunging in the sunlight, and the journey from rock to rock, and over rushing torrents, in front of this fall and back to Goat Island, was a new trip, with new sensations and new views. The trip is an experience which has been extolled by all who have ever enjoyed it, and it is a trip whose attractiveness has not been dimmed, but has increased, as the years have gone by; for the rushing, eddying spray and the sheets of water driven with great force against the face of the cliff have year by year eaten into the rocky back of the cave, making it larger and more wonderful with each succeeding summer.

On March 29, 1848, "for that day only," persons walked in the bed of the rocky channel of the American rapids between Goat Island and the mainland and from Goat Island out in the bed of the main channel towards Canada. But the river was not ice bound; its flow was diminished, not entirely cut off, its supply at Lake Erie having been temporarily blocked. Lake Erie was then full of floating ice, crowding to its outlet, the source of the Niagara river. During the previous afternoon a strong north-east wind had driven the ice back into the lake. During the night the wind veered suddenly and blew a gale from the west. This forced the ice floe sharply, in a mass, into the narrow channel or source of the river, quickly blocked it up, and the still advancing ice sealed up this source with a temporary barrier, pushed some feet into the air. It did not take long for the water north of this barrier to drain off, and in the morning, the Niagara river, as men knew it, "was not." The American Falls were dry. The Canadian Falls were a mere shadow of their former selves, a few threads or streams of water only falling over the edge. People, fearful every moment of an onrush of water from up stream, walked in the channels, where, up to that time," the foot of man had never trod." and where it has never trod since.

The roar of Niagara was reduced to a moan; the spray, and therefore, the rainbows disappeared. All day this phenomenon lasted, but by night the sun's rays and the pressure of Lake Erie's waters had made inroads on the icy dam, and during the night the barrier was swept away. By the next morning the river again rushed by in its might, and its roar once more proclaimed that Niagara had resumed its sway.

In 1860 two visitors of special note came to Niagara; Blondin, the man of iron nerve, and Albert Edward, heir apparent to the British Throne. The former wanted to stretch a rope from Goat Island's southwestern end to the Canadian shore opposite, and balance pole in hand to cross the gorge, where the column of spray might envelop him in its folds and shut him out of the view of the thousands who would throng the banks to see him risk his life. But Goat Island's owners refused to be parties to such an exhibition, and Blondin stretched his rope across the gorge about half a mile below, and there, in the presence of the Prince of Wales on one occasion, and in the presence of multitudes of people on others, several times crossed the gorge from side to side in safety.

New scenes of great beauty were opened up to visitors by the erection of the bridges to the Three Sister Islands in 1869; but the one point of vantage, the grand old Terrapin Tower, was needlessly torn down in 1873 in order that it might not prove an adverse attraction to the interests of a company which had bought and were about to fence in the last spot of land on the American shore, from which a near view of the Falls could be obtained; a point which so long as it remained in the possessin of the owners of Goat Island had been left free to the world. In 1877 the idea of the great hydraulic tunnel had been matured by Thomas Evershed. His plan and proposition was to have the outlet of this tunnel at the base of the slope directly under Goat Island, extending the tunnel eastwards under the Island and then under the bed of the river; placing the mills on the main shore and connecting their wheelpits with the main tunnel by lateral tunnels.

The passage in 1879, by the Legislature of the State of New York, of the preliminary act for the establishment of the State Reservation at Niagara precluded the adoption of that route, and necessitated the change thereof to its present location, a change that resulted financially to the benefit of the gigantic enterprise.

The next year Leonard Henkle advanced the idea of generating an electric current at Niagara that should supply New York city and intermediate points with light and power. A balance wheel, 100 feet in diameter, was to be fastened on, and parallel to, the face of the Goat Island cliff; and the induction coils, composed of miles and miles of wire were to be strung across the gorge between Goat Island and the Canadian shore. No progress was made in carrying this scheme into operation and the establishment of the New York Reservation has rendered its consummation, if ever feasible, impossible.

In 1885 an international sentiment in favor of State ownership of the land immediately surrounding the Falls, and rapids, and their restoration to a state of nature, and preservation for all time, free to mankind, took tangible form in the purchase by the State of New York, under its power of eminent domain, of 118 acres of land, including Goat Island, and a tract of land along the river on the American shore, Goat Island being the main feature of the reservation.

This land was bought under appraisal, $525,000 being paid for the Goat Island group; and on July 15, 1885, all the property so purchased became free forever to the world.

So after a family ownership of nearly 70 years the direct heirs of the original purchasers of this property from the State, ceded it back to it. Save for the one desecration of Bath Island, allowed, as stated before, purely for family reasons, the property was returned to the State in its original and natural condition. On all the other islands the owners had preserved the original forest beauty.

Since 1885 the plan has been to consistently restore, on the Reservation, the natural scenery. On Green Island all traces of the old mill have been removed.

And thus the islands remain, as nature intended them to be, and as they are destined to exist for all time, for "a thing of beauty is a joy forever."

ITS SCENERY.

"To him who in the love of Nature holds
Communion with her visible forms, she speaks
A various language."

The scenery of Goat Island is of a two-fold nature; that *on* the island and that *from* the Island. The scenery from the Island is the scenery of Niagara Falls, and I know of no reasonable way of describing that scenery, other than to quote the expressed thoughts of the master minds who have recorded their impressions of the great cataract. But to thus quote sufficiently, to even partially treat of the subject, would be to fill an entire volume. And so confining myself strictly to my subject, I feel constrained thus to leave out any material description of the scenery, *from* the island.

"The walk about Goat Island at Niagara Falls is probably unsurpassed in the world for wonder and beauty," wrote Charles Dudley Warner, and the judgment of the world agrees with him. And possibly, especially to that large number of persons who prefer the scenery of the rapids to that of the falls themselves, there is no more wondrous view about Niagara than that from the Terrepin Rocks, where the visitor, looking up the Canadian channel, sees before him naught but the upper line of the rapids meeting the sky.

It is of this view that the Duke of Argyle wrote, "The river Niagara above the falls, runs in a channel very broad, and very little depressed below the level of the country. But there is a deep declivity in the bed of the stream for a considerable distance above the precipice, and this constitutes what are called the rapids. The consequence is that when we stand at any point near the edge of the falls, and look up the course of the stream, the foaming waters of the rapids constitute the sky line. No indication of land is visible; nothing to express the fact that we are looking at a river. The crests of the breakers, the leaping and the rushing of the waters are still seen against the clouds, as they are seen on the ocean when the ship from which we look is in the trough of the sea. It is impossible to resist the effect on the imagination. It is as if the fountains of the great deep were being broken up, and that a new deluge were coming on the world. The impression is rather increased than diminished by the perspective of the low wooded banks on either shore, running down to a vanishing point, and seeming to be lost in the advancing waters. An apparently shoreless sea, tumbling towards one is a very grand and a very awful sight. Forgetting, then, what one knows, and giving oneself to what one only sees, I do not know that there is anything in nature more majestic than the view of the rapids above the Falls of Niagara."

To many others the view of the rapids, as one stands on and looks up stream from the bridge leading to Green Island, is the most beautiful at Niagara. Let me quote Margaret Fuller's description of these views: "At last, slowly and thoughtfully I walked down to the bridge leading to Goat Island, and when I stood upon this frail support, and saw a quarter of a mile of tumbling, rushing rapids, and heard their everlasting roar, my emotions overpowered me, a choking sensation rose to my throat, a thrill rushed through my veins, 'my blood ran rippling to my fingers' ends.' This was the climax of the effect which the falls produced upon me—neither the American nor the British fall moved me as did these rapids. For the magnificence, the sublimity of the latter I was prepared by descriptions and by paintings. When I arrived in sight of them I merely felt, 'Ah, yes, here is the fall, just as I have seen it in picture.' When I arrived at the Terrapin bridge, I expected to be overwhelmed, to retire trembling from this giddy eminence, and gaze with unlimited wonder and awe upon the immense mass rolling on and on, but, somehow or other, I thought only of comparing the effect on my mind with what I had read and heard. I looked for a short time, and then with almost a feeling of disappointment, turned to go to the other points of view to see if I was not mistaken in not feeling any surpassing emotion at this sight. But from the foot of Biddle's stairs, and the middle of the river, and from below the table rock, it was still 'barren, barren all.' And, provoked with my stupidity in feeling most moved in the wrong place, I turned away to the hotel, determined to set off for Buffalo that afternoon. But the stage did not go, and, after nightfall, as there was a splendid moon, I went down to the bridge and leaned over the parapet, where the boiling rapids came down in their might. It was grand, and it was also gorgeous, the yellow rays of the moon made the broken waves

American Rapids above Goat Island Bridge.

The Goat Island Home of the Hermit of Niagara.

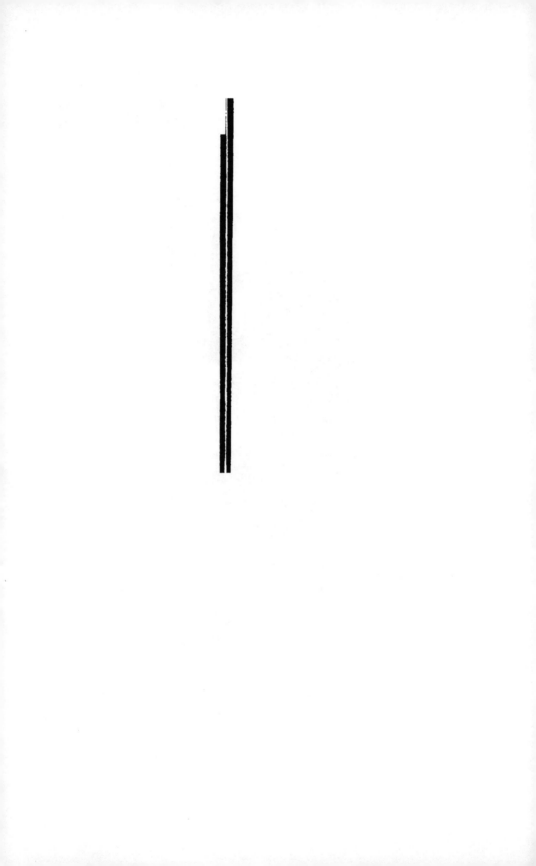

Terrapin Tower. First Bridge to it. 1829.

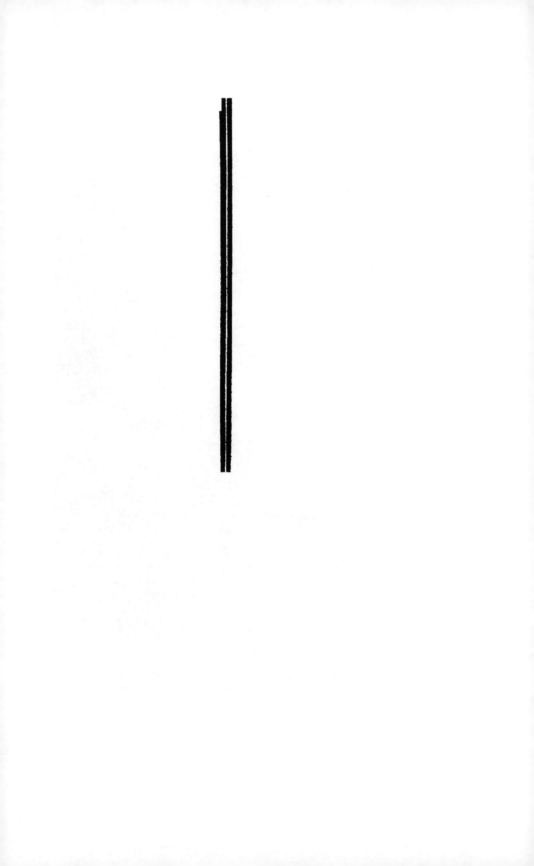

"The foaming waters of the Rapids constitute the sky line."

appear like auburn tresses twining around the black rocks. But they did not inspire me as before. I felt a foreboding of a mightier emotion rise up and swallow all others, and I passed on to the Terrapin bridge. Everything was changed, the misty apparition had taken off its many-colored crown which it had worn all day, and a bow of silvery white spanned its summit. The moonlight gave a poetical indefiniteness to the distant parts of the waters, and while the rapids were glancing in her beams, the river below the falls was black as night, save where the reflection of the sky gave it the appearance of a shield of blued steel. No gaping tourists loitered, eyeing with their glasses, or sketching on cards the hoary locks of the ancient river god. All tended to harmonize with the natural grandeur of the scene. I gazed long. I saw how here mutability and unchangeableness were united. I surveyed the conspiring waters rushing against the rocky ledge to overthrow it at one mad plunge, till, like toppling ambition, o'erleaping themselves, they fall on t'other side, expanding into foam ere they reach the deep channel where they creep submissively away. Then rose in my breast a genuine admiration, and a humble adoration of the being who was the architect of this and of all. Happy were the first discoverers of Niagara, those who could come unawares upon this view and upon that, whose feelings were entirely their own."

The scenery on the Island is its forest scenery, and by reason of its numerous flora and their abundance is, wonderfully attractive at all seasons; in the spring, when the natural forest blooms in its vernal foliage, and when the profusion of wild flowers carpet the ground; in the summer, when amidst the shaded walks and retreats on the little islands, fanned by the ever-stirring breezes created by the rapids, one wanders entranced;

in the fall, when the gorgeous coloring of the leaves, changed by the frost into all the colors of the rainbow, delight and dazzle the eye; in winter, when the glorious ice scenery covers every tree and twig, and Nature

"Wasteful decks the branches bare,
With icy diamonds rich and rare."

"Not one in 500, we are persuaded, knows anything about the apocalypse which is vouchsafed to him who in these glorious winter nights seeks the isle, not of Patmos, but of the Goat," wrote David Gray, and were one to have his choice of seeing Niagara but once, it would be hard to decide whether it should be in winter or summer, but probably in winter.

The scenery of Goat Island by moonlight, at any season, once seen is never to be forgotten. One might paraphrase and say

"If you would see this Isle aright,
Go visit it by pale moonlight."

It were useless to attempt a description of it. From the Terrapin Rocks and from Luna Island, the Lunar Bow is to be seen best in its glorious indistinctness, and it is to these points

"That many a Lunar belle goes forth,
To meet a Lunar beau."

And from the Terrapin Rocks and Luna Island each morning, when the sun is not obscured, one gazes entranced into the rising clouds of spray, from which the bow of promise, like

"An arch of glory springs,
Sparkling as the chain of rings,
Round the neck of virgins hung."

And when, on a bright afternoon, one stands among the rocks at the base of and in front of the Luna Island Fall, he is the centre of a complete rainbow circle.

Ice Scenery on Goat Island, from Terrapin Rocks.

Byron's description of Velino may properly be applied to Niagara:

> " A matchless Cataract
> Horribly beautiful! but on the verge,
> From side to side, beneath the glittering morn
> An Iris sits, amidst the infernal surge
> Like hope upon a deathbed, and unworn
> Its steady dyes, while all around is torn
> By the distracted waters, bears serene
> Its brilliant hues, with all their beams unshorn
> Resembling, midst the torture of the scene
> Love watching madness with unalterable mien."

Another likens the Island to " Love in the clasp of madness," while Tom Moore, who gazed at it from across the gorge in 1804, makes the Spirit say:

> " There amidst the island's sedge
> Just above the Cataract's edge
> Where the foot of living man
> Never trod since time began,"

which was poetic, but not founded on fact.

And still another wrote of

> " The isle that linked in wild Niagara's firm embrace,
> Still wears the smile of summer on its face."

ITS OWNERS.

The ownership of the islands may be summarized as follows:

The Aborigines	—1600
The Neuters	1600-1651
The Senecas	1651-1764
Sir William Johnson	1764
The English Crown	1764-1783
State of New York	1783-1816
The Porters	1816-1885
State of New York	1885-1900

ITS LITERATURE.

Much has been written about Niagara by thousands. Its description has been attempted by many who are well known in the literature of the world; and by many more who are unknown. The shortest, perhaps the most eloquent, probably the most suggestive, certainly the most non-descriptive description of Niagara ever penned was that by Fanny Kemble, whose journal tells of her approach to the brink of the abyss and closes with the words,

> "I saw Niagara,
> O God! who can describe that sight."

But while much has thus been written, a great deal of prose that is worth reading and a very little poetry that is worth remembering, it is of Niagara as a whole, as a unit, in its generality, in its comprehensiveness; treating the water, the Falls, the rapids, the gorge, the sky line of the river as seen from the brink of the Horseshoe, the spray, the rainbow, and the islands are component parts of one absorbing whole, that almost all writers treated it.

Some of them specially mention Goat Island; others, and they are in the vast majority, refer to it only as an incident. Neither Goat Island nor even Niagara Falls have ever elicited a strong poem from any poet of the first rank.

Some men, like Doré, have pictured Niagara without ever having seen it; some men, like Brainard, have written poetic effusions about it without ever having gazed upon it; but no important description of Niagara has ever been penned by one who has never gazed upon it and who has not known the sensation occasioned by the first view thereof; and certainly no one has ever written anything about Goat Island who has not visited it, studied it in all its varied aspects, and been held enthralled by its spell.

ICE SCENERY. CAVE OF THE WINDS. WINTER 1896.

ITS VISITORS.

Perhaps no one spot in the world has been visited during the last four score years by so many people, of both sexes, of so many varied occupations and of so many nationalities, as Goat Island.

Lovers of nature and of its unique and glorious scenery, travellers and tourists, scientists and artists, writers of prose and of poetry, divines and lawyers are numbered among its admirers and students.

Potentates and princes, rulers and statesmen, warriors and diplomats, adventurers and mountebanks and the leaders in every branch of science, knowledge and art have trod its paths.

And from its associations many of these have drawn inspirations that led them to higher and nobler aims. But in antithesis, from its edges men and women have leaped to self-destruction, while others have profaned its sanctity by availing themselves of the chances afforded by its solitude for murder.

ITS PROPOSED USES.

Many are the uses to which the ingenuity of man has, during the past 90 years, desired to turn the Island.

It was desired originally for a sheep pen.

The State Legislature designed to use it for a State prison or a State arsenal.

Lafayette as well as many others would have liked to have it for a residence park.

P. T. Barnum wanted to buy it for a circus ground.

Cornelius Vanderbilt, Sr., tried to buy it for use as a pleasure ground in connection with his railroads.

Jim Fiske wanted it for use as a picnic ground and as a terminal of the Erie railroad.

And among the many propositions which were made to its owners for its use were, as the site of a mammoth hotel, as a race track, as a botanical garden, as a rifle range, and as a site for a collection of manufacturies to be located along the shores of the Island and the power to be furnished by running tall piers out into the river and thus collecting the waters; and again by cutting a canal through the center of the Island from east to west and locating the factories along its banks.

DeWitt Clinton in 1810, noted its value for hydraulic works, and that use was suggested oftener than any other until the establishment of the State Reservation in 1885. And ever since then, plans have been urged with this object in view; some men seeming to be unable to realize (when they think they see a dollar for themselves) that the State's purchase was for the sole purpose of forever retaining the natural scenery, which private owners had happily preserved.

ADDENDA.

To give, even partially, reproductions of the best views *from* the Island would be to add so many illustrations of the scenery at Niagara, as to too greatly enlarge the bulk of this article. Hence, practically no views of the many sided modern scenery as seen from Goat Island have been reproduced.

In 1889 a hurricane blew down many trees on Goat Island, among them the Botanic "Monarch of the Isle," a cross section

THE BOTANIC MONARCH OF THE ISLE.

of whose trunk may be seen at the Niagara Falls Public Library. On it is inscribed: "I grew on Goat Island, and for over 400 years, stood sentinel over its Indian graves. I was a sturdy sapling when Columbus landed at San Salvador. I was 150 years old when the first white man gazed upon Niagara. I saw and knew this first white man, but cannot reveal his name. I was over 200 years old when La Salle and Hennepin visited Niagara. I was blown down in 1889, the oldest and largest tree within the sound of Niagara's roar."

On Luna Island is an embedded rock, whose top projects above the surface, and on this many years ago a cunning hand carved the words, still decipherable,

"All is change
Eternal progress
No death."

Who carved them no one knows, and where he lies entombed is a mystery; but here, in full view of thousands of annual visitors, stands his epitaph, and the ceaseless roar of Niagara sings his everlasting requiem.

In regard to all of Nature's handiwork, there are always men who think that certain parts of it would have been more effectively and better done if they could only have been consulted about it, and the case of Goat Island is no exception.

Perhaps one of the least objectionably worded of such criticisms on Goat Island, which is conceded to be one of the loveliest and grandest spots on earth, was written less than 40 years ago, in these words:

"It would be considered rather presumptious in any one to think of improving upon Niagara, but I cannot help thinking that the effect would be increased immensely if the island which divides the cataract into the Horse Shoe and the American Falls and the rock which juts up in the latter and subdivides it unequally, were moved or did not exist; then the river, in one grand front of over 1,000 yards, would make the leap en masse."

Fortunately the idea is now impracticable, and Goat Island exists because such is the will of the Creator.

Goat Island and Niagara, for they are synonomous terms, once seen can never be forgotten, nor will the influences derived from a leisurely visit to them ever be entirely lost.

Their impression on an appreciative mind was beautifully expressed many years ago, in the following poetic prose:

"Niagara, when once we become acquainted with it, is capable of exercising a strange power of fascination over the mind; and the imaginative individual should not be surpised, if he find mere water, earth and air, changing in its conceptions, into a creature of life. No wonder that the savages adored it, and peopled it with invisible beings, and imagined it the abode of the Great Spirit. With me it will always remain a vision of beauty, closely associated with that glory, with which, in my notion, I shadow and imagine the Supreme. I loved it as a fellow; I left it with regret. Its form still lingers before my eyes, its rushing voices still hymn in my ears. And often still, sleeping or waking am I, in heart, among the cedars of Iris Island."

SEVENTEENTH ANNUAL REPORT

OF THE

COMMISSIONERS

OF THE

State Reservation at Niagara.

FROM OCTOBER 1, 1899, TO SEPTEMBER 30, 1900.

TRANSMITTED TO THE LEGISLATURE MARCH 10, 1901.

ALBANY:
JAMES B. LYON, STATE PRINTER.
1901

Prospect Point

SEVENTEENTH ANNUAL REPORT

OF THE

COMMISSIONERS

OF THE

State Reservation at Niagara.

FROM OCTOBER 1, 1899, TO SEPTEMBER 30, 1900.

TRANSMITTED TO THE LEGISLATURE MARCH 13, 1901.

ALBANY:
JAMES B. LYON, STATE PRINTER.
1901.

STATE OF NEW YORK

No. 58.

IN ASSEMBLY,

MARCH 13, 1901.

SEVENTEENTH ANNUAL REPORT

OF THE

Commissioners of the State Reservation at Niagara.

To the Honorable the Speaker of the Assembly:

Sir: I herewith transmit for presentation to the Legislature Seventeenth Annual Report of the Commissioners of the State Reservation at Niagara, for the fiscal year ended September 30, 1900.

Respectfully yours,

ANDREW H. GREEN,

President.

SEVENTEENTH ANNUAL REPORT

OF THE

COMMISSIONERS

OF THE

STATE RESERVATION AT NIAGARA.

FOR THE FISCAL YEAR FROM

OCTOBER 1, 1899, TO SEPTEMBER 30, 1900.

Commissioners.

ANDREW H. GREEN, President.

GEORGE RAINES, CHARLES M. DOW,

THOMAS P. KINGSFORD, ALEXANDER J. PORTER.

Treasurer and Secretary.

PETER A. PORTER, Jr.

Superintendent.

THOMAS V. WELCH.

REPORT

To the Honorable the Legislature of the State of New York:

The Commissioners of the State Reservation at Niagara, as required by law, submit their report for the fiscal year begun October, 1899, and ended September 30, 1900, this being their seventeenth annual report.

The work of the year has been mainly in continuance of the system of improvement and restoration which the Commissioners have had in view, and in accordance with a previously adopted plan.

The last Legislature, appreciating well the importance of providing for the great number of people who visit Niagara, made liberal appropriations for improvements intended for the accommodation and safety of the public.

The two new bridges connecting the mainland with Goat Island are well under way, but the progress of the work has been somewhat impeded by unforeseen engineering difficulties arising from the difficult natural conditions. The Shelter and Administration building will be completed by the first of April. The improvement work on Goat and Luna Islands has been completed and meets with general favor. The further work of restoration contemplated for the year was completed in due time.

Since the abolishing of the toll charges and exactions on the Reservation, the number of visitors has annually increased, and owing to the prospects of an unprecedented number of visitors at the Falls the coming summer, the Commissioners ask your honorable body for an addition of $20,000 to the ordinary maintenance allowance. It will be necessary to use this sum in affording the visitors comfort, safety and a reasonable amount of freedom from annoyance. The unusual number of visitors will add materially to the receipts on the Inclined Railway, and the Superintendent estimates that as much or more than the $20,000 additional appropriation asked will be repaid into the State Treasury through the added income from carrying passengers at the Inclined Railway.

By provision in the act incorporating the Niagara Falls Power Company, and the acts amendatory thereto, that company is required to furnish electricity for light, heat and power, also water to be used for any purpose, to the State free of charge. The Commissioners have not heretofore availed themselves of this agreement owing in part to the expense of conducting the electricity and water from the power house to the Reservation property. We ask your honorable body for an appropriation of $7,100 to be used in adequately equipping the Reservation with electric conduits, wire and lamps. We have now but 20 lights on the Reservation, and for this light we are paying $600 per annum. The Reservation should be provided with at least 200 lights. The hydraulic appliances for sprinkling are grossly inadequate, and we ask for an appropriation of $7,500 to properly equip the grounds with a system of mains for water supply.

We ask $5,000 for filling and grading at the terminals of the two bridges connecting the mainland with Goat Island; the bridges being raised above the former bridge level will necessitate this filling.

A portion of the upper part of Goat Island was cleared of woods over a century ago. The Commissioners have directed the planting of forest trees in that section, thus restoring Goat Island to its primitive and natural condition so far as possible. For this we ask an appropriation of $1,500.

The following is our estimate and detailed statement of the work necessary to be done, and of the expenses of maintaining the Reservation for the fiscal year ending September 30, 1901.

Statement of Needed Improvements.

For extraordinary expenditures for the year 1901, care and maintenance by reason of Pan-American Exposition	$20,000 00
For filling and grading and putting grounds in order adjacent to stone arch bridges in course of construction from mainland to Goat Island	5,000 00
For electrical installations and lighting grounds and buildings	7,100 00
For system of water pipes for grounds and buildings.	7,500 00
For completion of and furnishing Shelter building	3,000 00
For bridge to Horseshoe Falls and Terrapin Point	1,000 00
For grading and planting	1,500 00
For paying tax, local assessment Nos. 21 and 22	2,165 15
	$47,265 15

The sum of $20,000 for extraordinary expenditures for the year 1901 is needed for extra police for preserving order, extra labor for keeping the Reservation grounds and buildings clean and orderly, and for extra guards for safety during the Pan-American Exposition, when ten times the usual number of visitors are expected.

The sum of $5,000 for filling and grading adjacent to the stone arch bridges in course of construction is needed to put the grounds adjacent to the bridges in order again, and to restore the slopes and shrubbery after the bridgework shall have been completed. The cost of constructing the bridges and the expenditures for engineering inspection, and care for the structures, and the safety of the public during the progress of the work will exhaust the appropriation for the bridges.

The sum of $7,100 for electrical installation and lighting grounds and buildings (to supplant the present inadequate lighting by 20 arc lamps at $50 each per annum).

Under the law the Commissioners of the Reservation are entitled to receive electricity from the Niagara Falls Power Company for lighting and power within the Reservation gratis.

The Commissioners now propose to install an adequate system, and abolish the present cost of lighting.

The sum of $7,500 for system of water pipes for grounds and buildings is needed to supersede the inadequate system of surface water pipes now in use.

Under the law the Commissioners of the State Reservation at Niagara are entitled to receive water for the Reservation grounds and buildings gratis from the Niagara Falls Power Company. It is now proposed to install an adequate water system for the buildings, and for the lawns and roads, within the Reservation.

The sum of $3,000 for completion of and furnishing Shelter building is needed for changes and completing rooms not covered by contract, and for furnishing the Shelter building for public use.

The sum of $1,000 for bridge to Horseshoe Falls and Terrapin

Building Centers for Stone Arch Bridge. Green Island to Goat Island.

Point is needed to replace the present bridge at that point, which is old, dilapidated and inadequate. The State Engineer, upon inspection, recommends the renewal of this structure.

The sum of $1,500 for grading and planting is needed for planting the upper end of Goat Island (now almost entirely denuded of growth) with native trees and shrubs.

The sum of $2,165.15 is for local assessments for paving Buffalo avenue, adjacent to the Reservation.

A statement of the local assessments has been forwarded to the Comptroller.

Ordinary Maintenance.

Salaries, office and traveling expenses	$4,600 00
Reservation police and caretakers	5,400 00
Labor	10,000 00
Material, tools, etc	5,000 00
	$25,000 00

Receipts.

In view of the large crowds which are expected to visit the Pan-American Exposition, and therefore the Reservation in 1901, any estimate of the receipts accruing to the State from the Reservation for the present fiscal year would be a matter of conjecture, but it would seem safe to assume that the income from the Inclined Railway will be very largely increased.

Estimated Receipts.

Inclined Railway	$20,000 00
Cave of the Winds	1,200 00
Steamboat landing	500 00
Carriage service	100 00
	$21,800 00

The Superintendent in his report has made a detailed statement of the work of the Commission for the last year, his report and that of the Treasurer and Secretary being appended to this report.

Respectfully submitted,

ANDREW H. GREEN,

President.

GEO. RAINES,

CHAS. M. DOW,

THOMAS P. KINGSFORD,

ALEXANDER J. PORTER,

Commissioners of the State Reservation at Niagara.

REPORT OF THE TREASURER AND SECRETARY

OF THE

State Reservation at Niagara

For the Fiscal Year Ending September 30, 1900.

Report of the Treasurer and Secretary.

To the Commissioners of the State Reservation at Niagara:

Gentlemen.—I respectfully submit the following report for the fiscal year begun October 1, 1899 and ended September 30, 1900.

As it embraces the last fiscal year of the century, and completes the fifteenth year of Free Niagara, it is the natural milestone in the progress of the advanced policy of the Empire State, in exercising its sovereign power for the acquisition of the lands at and adjacent to this world-famed spot, making them free to all mankind, and then proceeding to restore them, as far as possible, to their original, natural state; and a retrospect of the results thus far achieved may, therefore, be appropriately referred to.

The agitation for the sovereign ownership of Niagara which began in 1877 and ended July 15, 1885, on which date the State of New York entered into possession of the lands on the American side, was not an entirely new conception. Yet the idea, as then suggested, assumed such a tangible and business-like form, and appealed so strongly to all lovers of nature and of its unique and beautiful scenery, that it was at once encouraged by the influence of numerous prominent men, in many paths of life, not only in our own State but throughout the nation, and by reason of the world-wide love of Niagara, received the hearty endorsement of men, of great eminence, from across the ocean.

In 1834 Reverends Andrew Reed and Thomas Mattheson, as a deputation from the Presbyterian churches of Scotland, visited the Presbyterian churches of America, and in their journeys came to, and fully appreciated Niagara. In the history of their travels, in addition to expressing their admiration thereof, they left on record the following statement, which stands as the earliest public suggestion of an International Reservation at this place: "Niagara does not belong to Canada nor America. Such spots should be deemed the property of civilized mankind; and nothing should be allowed to weaken their efficacy on the tastes, the morals and the enjoyment of all men."

Fifty years later their suggestion became a reality; and what in 1834 was the far-sighted individual opinion of two eminent divines, became in 1885, by the action of the State of New York, supplemented in 1888 by the action of the Province of Ontario, an accomplished fact; sanctioned by the hearty endorsement of leading men everywhere and enacted by the votes of the elected representatives of the people, its fulfillment has been applauded and valued more and more each year by humanity in general.

For the permanent enjoyment of the annually increasing number of visitors to Niagara, occasioned by the increase in population and the constantly added facilities for travel, certain expenditures for additional improvements, whereby many beautiful points of view, now either unreached or reached only by such exertions as practically preclude access thereto by large numbers of people, on account of age or infirmity, may be made accessible, are urgently demanded.

During the decade and a half of Free Niagara's existence, all but two or three of the scores of unsightly structures that marred the harmony of the surroundings of the Niagara of old have been removed, and in their stead, restored banks, grassy slopes, graceful shrubbery and beautiful natural environments greet the eye. This gratifying result has been effected almost entirely on the main shore. For the Goat Island group, save for the mill on Bath, or as it is now called, Green Island (perhaps the most unsightly structure ever erected within the sound of the great cataract and of which every vestige has been happily swept away) needed no restoration. Goat Island and the adjacent islands were covered with an original bit of virgin forest, which, fortunately, has been conscientiously preserved in its primeval state, all undespoiled by man's so-called progress and improvements.

The experimental stage of Free Niagara, determining whether the State's action in reference thereto would be appreciated by the people, and the Falls visited by them in increased numbers, has long since been passed, and is answered by a hearty affirmative. In recording this gratifying verdict at the commencement of the Twentieth century the people strongly recommend a continuance and even an expansion of that liberal policy on the part of the State which has made the Reservation the beautiful spot it is to-day.

Of the certainly not less than five, and probably more, bridges which have in the past, and should in the future, afford access from Goat Island to the adjoining islands, two have already been reconstructed in enduring stone, and the remaining ones should be similarly rebuilt in the near future. One of these two stone bridges thus referred to, connects Goat Island with the first Sister Island. The last Legislature granted an appropriation of $3,000 for repairs to the existing bridges that connect the first with the second and the second with the third Sister Island.

Regarding another of these bridges, namely, the one from Goat

Island to the Terrapin Rocks, the recommendation is renewed for a substantial bridge at this point. Special stress may be laid on this appropriation, as the Terrapin Rocks are the geographical and scenic center of Niagara, a point which hardly any visitor to Goat Island, who is physically able to make the trip up and down the steps, fails to visit.

An elevator and new staircase by which easy access should be afforded to the debris slope below Goat Island has heretofore been recommended and is more urgently needed each year. Below the bank here is the Cave of the Winds which affords not only the most distinctly unique trip at Niagara, by means of which are to be obtained otherwise unobtainable views of little known and marvelous scenery, but the only place where one may pass completely around and behind the falling sheet of water. At the end of this debris slope next to the Horseshoe Falls, may be obtained new and almost unknown views, views which Professor Tyndall regarded as beyond compare the finest at the Falls, and at this point he penetrated farther than almost any visitor has ever done before around the curve of the Horseshoe, toward what he called the "Ultima Thule" of Niagara. The only access to this but comparatively little visited debris slope is by the winding stairway called the Biddle Stairs, in honor of Nicholas Biddle of United States Bank fame, who is credited with having suggested to the owners of Goat Island this means of accomplishing the descent, and is further said to have offered to share the expense of their construction, although this philanthropic part of his suggestion was declined.

Having been in use for nearly three-quarters of a century, they should be replaced by a more substantial, safe, and more modern means of descent.

TOWER HOTEL

The details of the work that has been accomplished on the Reservation during the fiscal year is fully and concisely set forth in the report of the Superintendent which accompanies this report. One piece of work, however, completed during the year demands special notice. The last Legislature appropriated $5,000 for extraordinary expenditures for the year 1900. The most of it was expended in replacing the wooden stairway at the American Falls on Goat Island with a handsome stone stairway protected by an indispensable iron guard rail. From Goat Island this stairway affords access to Luna Island, from which and from the stairway itself are to be obtained the most imposing views of the American Falls, of the Cave of the Winds Fall and of the gorge below.

The upper end of Goat Island, where is the open space cleared some hundred and twenty-five years ago, and on which it is recorded that John Stedman in early days raised a fine crop of turnips, has been ordered planted with appropriate and indigenous trees.

The liberal appropriation made by the last Legislature for the erection of a bridge between the mainland and Goat Island resulted in letting a contract for the erection of a steel concrete and stone bridge to replace the iron trussed bridge, which, during the last half century has afforded access to so many millions of visitors to Goat Island.

The engineering difficulties were numerous, but by reason of the vast improvements of modern science and the aid of the old bridge, were far less than those encountered when the first bridge was constructed on this site in 1818.

In spite of the dangers of the work of construction, thanks to the special care and precautions employed by the contractors, no accident has occurred during the building of the bridge that was attributable to the hazardous nature of the work, or its location in and over the rushing rapids of Niagara. While the progress of this bridgework has been much slower than expected, the work will be prosecuted with all possible speed and the bridge will be completed and available for use in the early part of the summer 1901, practically in time to accommodate the vast crowds which are expected at Niagara as a result of the Pan-American Exposition at Buffalo, in anticipation of which event the appropriation for this bridge was granted by the Legislature.

On the main shore in addition to the detail work fully set forth by the Superintendent in his report, two features deserve special notice. The granting of a revocable license to the International Traction Company (which controls largely the trolley lines on both sides of the river at Niagara, and also two bridges across the gorge) to lay a track on the Riverway between Falls street and the north end of the Reservation, has already been reported, and thus far the grant has worked to the advantage of the Reservation and the accommodation of visitors. With the completion of the company's projected terminal station, facing the Reservation, this grant will continue to result in decided advantages of access to the patrons of Free Niagara.

The inadequacy of the small building on Green Island used as an office for the Commissioners, and the insufficiency of the Superintendent's office located in the old Inclined Railway building, led to the recommendation for the erection of an Administration building, where should be located all the offices of the Commissioners and officials, and which would also furnish a much needed Shelter building for the accommodation of visitors, which building should be located in Prospect Park. The Legislative appropriation of 1898 for this purpose proved inadequate to erect the beautiful Gothic structure as originally designed by the State Architect. Subsequently, through his co-operation, certain changes were made reducing the height but not the ground floor area of the building and still preserving its classic beauties. The structure commenced last summer is now well advanced and is expected to be ready for occupation by April 1, 1901.

During the past year, in the interest of good order on the Reservation, it was decided to abolish all the hack stands theretofore allowed on the State property. This action was induced by reason of the many complaints that were made of petty annoyances and solicitations.

On February 1, 1900, Mr. Richard F. Rankine, who had so efficiently filled the office of secretary and treasurer to the Commissioners, resigned that office, and Mr. Peter A. Porter, Jr., was elected to fill the position. Thanks are due to Mr. Edward A. Bond, State Engineer, for his personal interest in the work of the Commission and his repeated attendance at the meetings and the suggestions and aid thus given. Also to Mr. G. L. Heins, State Architect, for his valuable co-operation and assistance in matters relating to his department.

It is again recommended that a topographical survey of the Falls be authorized and provided for. The value of the survey made ten years ago, and furnishing in connection with several prior surveyors a minute and accurate record, not only of the recession of the Falls themselves, but of the changes, erosive and creative, in the lines of the banks, renders this recurring ten-year period, and the commencement of the twentieth century a suitable time for this important work.

The present stone structure, built many years ago, which covers the upper end of the Inclined Railway is not only incommodious and unsightly, but is decidedly detrimental in that it obstructs what would otherwise be an uninterrupted view of the Falls from a very large portion of Prospect Park. With the facilities afforded by the new Shelter building, it will become merely the protection for the upper end of the Inclined Railway, and it is urged that an appropriation be granted for the purpose of demolishing this building and substituting in its place a suitable, permanent structure, located below the level of the land, whose roof shall be raised so little, if any, above the surface as to offer no obstruction to the view. At the same time such a building specially designed for the purpose, by its lower level would increase the water power, slightly lessen the length of the incline, and thus increase the facilities for handling the large crowds which frequently congregate at this point, all anxious to descend to the slope below.

In view of the Pan-American Exhibition to be held at Buffalo from May 1st to November 1st, 1901, and toward which the State has appropriated $300,000 for a suitable building and exhibit, and the reasonable and certain deduction, that of the several millions of people who are expected to visit it, a large majority will visit Niagara, and therefore the Reservation (for as an attraction Niagara Falls will be fully as great as the Exposition itself) a very much larger maintenance appropriation is desired, to the end that the comfort of and facilities for these extra visitors may be provided and their safety conserved, a most important consideration at a spot like Niagara, where amid such vast crushes of humanity at points of special scenic attraction, beauty and danger often go hand in hand. The needed increase in the force of caretakers and policemen will also tend to the preservation of the property of the Reservation, especially from the wanton destruction of small trees and shrubs, many of them recently planted by the order of the Commissioners, and will aid in and be required for the preservation of that good order which, especially among such throngs of visitors, it is the State's imperative duty to maintain for the benefit of all, and notably in behalf of the youthful, the aged and the infirm.

Respectfully submitted,

PETER A. PORTER, Jr.,

Treasurer and Secretary.

The Commissioners of the State Reservation at Niagara in account with Peter A. Porter, Jr., Treasurer, for the fiscal year begun October 1, 1899, and ending September 30, 1900.

1899.			
Oct. 1.	Balance on hand this date		$27 51
	Maintenance Receipts.		
Nov. 3.	Quarterly advance from State Comptroller	$6,250 00	
1900.			
Jan. 21.	Quarterly advance from State Comptroller	6,250 00	
April 23.	Quarterly advance from State Comptroller	6,250 00	
July 18.	Quarterly advance from State Comptroller	6,250 00	
			25,000 00

Special appropriation as per Chapter 569, Laws of 1899:

1899.				
Oct.	16.	Payment by State Comptroller on account	$485 00	
Dec.	28.	Payment by State Comptroller on account	921 78	
1900.				
June	12.	Payment by State Comptroller on account	16 85	
July	25.	Payment by State Comptroller on account	6 75	
Sept.	20.	Payment by State Comptroller on account	21 75	
	27.	Payment by State Comptroller on account	10 00	
				$1,462 13

Special appropriation as per Chapter 420, Laws of 1900:

1900.				
Aug.	31.	Payment by State Comptroller on account	$453 11	
Sept.	20.	Payment by State Comptroller on account	596 14	
	27.	Payment by State Comptroller on account	413 46	
	30.	Payment by State Comptroller on account	7,803 00	
	30.	Payment by State Comptroller on account	9,554 00	
				18,819 71

Special appropriation as per Chapter 419, Laws of 1900:

1900.			
July 13.	Payment by State Comptroller on account	$349 75	
25.	Payment by State Comptroller on account	331 43	
Aug. 7.	Payment by State Comptroller on account	1,498 29	
Sept. 20.	Payment by State Comptroller on account	1,983 49	
27.	Payment by State Comptroller on account	61 25	
			$4,224 21

Receipts of Reservation.

1899.		
Nov. 1.	Draft on Bank of Niagara for October receipts	$308 10
Dec. 1.	Draft on Bank of Niagara for November receipts	78 95
1900.		
Jan. 1.	Draft on Bank of Niagara for December receipts	35 55
Feb. 1.	Draft on Bank of Niagara for January receipts	70 20
March 1.	Draft on Bank of Niagara for February receipts	348 50
April 1.	Draft on Bank of Niagara for March receipts	251 70
May 1.	Draft on Bank of Niagara for April receipts	72 10

1900.			
June 1.	Draft on Bank of Niagara for May receipts	$230 80	
July 1.	Draft on Bank of Niagara for June receipts	645 30	
Aug. 1.	Draft on Bank of Niagara for July receipts	1,612 45	
Sept. 1.	Draft on Bank of Niagara for August receipts	2,809 70	
30.	Draft on Bank of Niagara for September receipts	2,076 45	
			$8,539 80
1900.			
March 31.	Interest on balances in Manufacturers and Traders' Bank	$6 05	
June 30.	Interest on balances in Manufacturers and Traders' Bank	8 07	
			14 12
	Total		$58,087 48

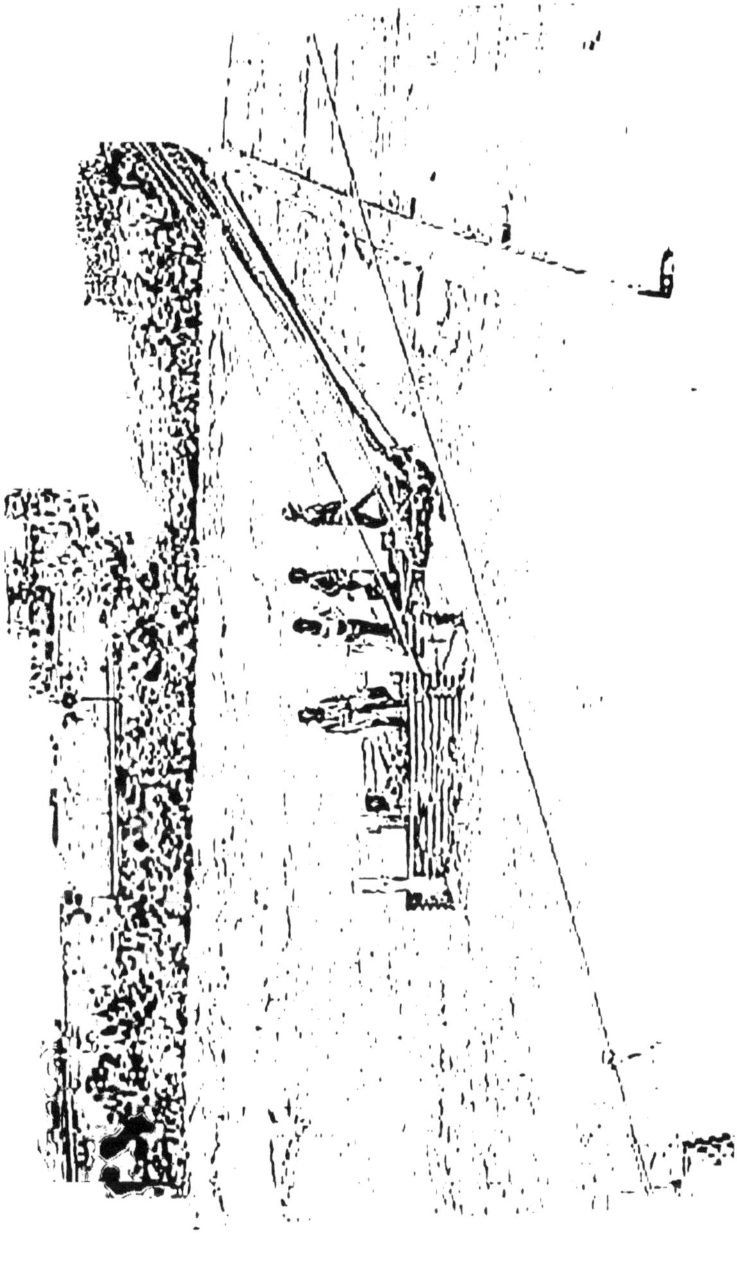

[redacted]

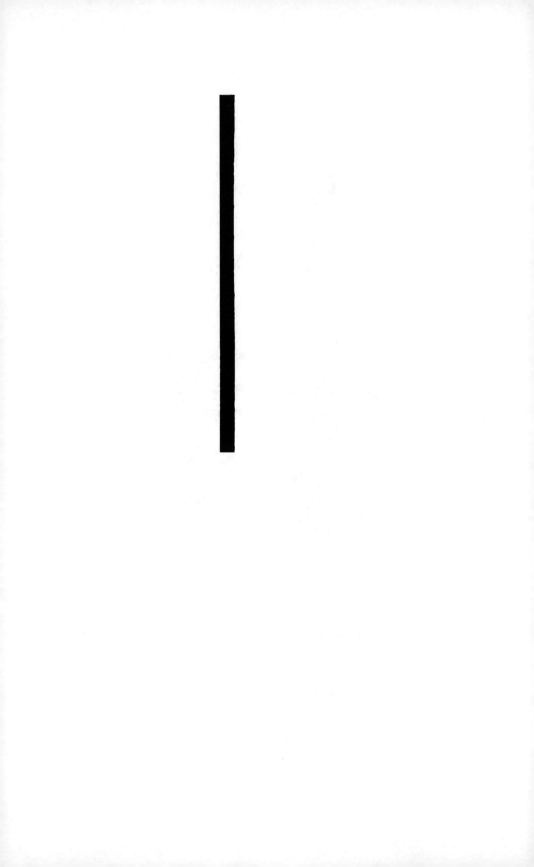

Expenditures.

Abstract CXVIII.

Date.	No. of Voucher.	Name.	Amount.
1899.			
Nov. 4.	1752..	R. F. Rankine, Secretary and Treasurer, office expenses	$7 28
	1753..	R. F. Rankine, Secretary and Treasurer, salary October, 1899	83 33
6.	1754..	Pay roll, October, 1899	1,579 33
7.	1755..	Thomas V. Welch, Commissioner's and office expenses	45 67
17.	1756..	Globe Ticket Co., Inclined Railway	10 50
	1757..	Hardwicke & Co., water pipes	2 52
	1758..	Walker & Paterson, tools, buildings, Inclined Railway	10 47
	1759..	Walker & Paterson, tools, buildings, Inclined Railway	15 55
	1760..	James Davy, stationery	11 00
	1761..	N. F. Hydraulic Power and Mfg. Co., electric lighting	50 00
	1762..	N. F. Hydraulic Power and Mfg. Co., electric lighting	50 00
	1763..	Andrew H. Green, Commissioner's expenses	41 70

Abstract CXIX.

Date.	No. of Voucher.	Name.	Amount.	
1899.				
Nov. 17.	1764..	J. H. Cook & Co., buildings.	$30 10	
	1765..	P. J. Davy, buildings......	47 13	
	1766..	T. P. Kingsford, Commissioner's expenses.........	152 10	
Dec. 2.	1767..	Pay-roll, November, 1899...	1,324 03	
	1768..	Thos. V. Welch, Superintendent's office expenses.	18 24	
	1769..	R. F. Rankine, Secretary and Treasurer, salary November, 1899..........	83 33	
				$3,562 28
Dec. 9.	1770..	P. C. Flynn & Son, buildings, iron railing.........	$70 12	
	1771..	Peterson & Grant, buildings	7 00	
	1772..	P. J. Davy, water pipes....	53 05	
	1773..	Hardwicke & Co., water pipes	55 97	
	1774..	Gazette Publishing Co., Superintendent's office expenses	2 25	
	1775..	Walker & Paterson, tools, buildings	21 62	
	1776..	N. F. Hydraulic Power and Mfg. Co., electric lighting.	50 00	
14.	1777..	A. J. Porter, Commissioner's expenses........	24 50	

Date.	No. of Voucher.	Name.	Amount.	
1899.				
Dec. 22.	1778..	Alexander Henschel, President's office expenses....	$70 00	
30.	1779..	Pay-roll, December, 1899...	1,392 55	
	1780..	Thos. V. Welch, Superintendent's office expenses..	49 05	
				$1,796 11

Abstract CXX.

1900.				
Jan. 8.	1781..	R. F. Rankine, Secretary and Treasurer, traveling expenses	$28 37	
	1782..	R. F. Rankine, Secretary and Treasurer, salary, December, 1899............	83 33	
				111 70

Abstract CXXI.

Feb. 2.	1783..	R. F. Rankine, Secretary and Treasurer, salary January, 1900	$83 33	
	1784..	R. F. Rankine, Secretary and Treasurer, office expenses	3 59	
	1785..	Pay-roll, January, 1900.....	1,198 29	
	1786..	Pay-roll, January, 1900, supplemental	475 25	
3.	1787..	Gazette Publishing Co., Superintendent's office expenses	2 25	

Date.	No. of Voucher.	Name.	Amount.
1900.			
Feb. 3.	1788..	James Davy, Superintendent's office expenses.....	$6 85
	1789..	Thos. E. McGarigle, bridges.	30 55
	1790..	Alvah Bushnell Co., Superintendent's office expenses	6 00
	1791..	Welch Coal & Wood Co., coal	60 50
	1792..	J. H. Cook & Co., bridges, buildings	34 69
	1793..	J. H. Cook & Co. buildings.	44 79
	1794..	N. F. Hy. Power & Mfg. Co., electric lighting.....	50 00
	1795..	N. F. Hy. Power & Mfg. Co., electric lighting.....	50 00
	1796..	Walker & Paterson, buildings	13 68
	1797..	Walker & Paterson, tools, buildings	11 21
	1798..	Thos. V. Welch, Superintendent's office expenses..	41 50
8.	1799..	Charles M. Dow, Commissioner's expenses	398 14
13.	1800..	National Press Intelligence Co., Secretary and Treasurer, office expenses.....	1 10
17.	1801..	Peter A. Porter, Jr., Secretary and Treasurer, office expenses	26 38

Date.		No. of Voucher.	Name.	Amount.
1900.				
Feb.	17.	1802	A. J. Porter, Commissioner's expenses	$22 89
	21.	1803	Pay-roll, February, 1900, Prospect Park	191 50
	28.	1804	Peter A. Porter, Jr., Secretary and Treasurer, salary, February, 1900	83 33
March	1.	1805	Pay-roll, February, 1900	1,007 91
	2.	1806	Thos. V. Welch, Superintendent's office expenses	35 10
	5.	1807	Milton E. Johnson & Co., Secretary and Treasurer, office expenses	14 00
	7.	1808	Pay-roll February, 1900, supplemental	294 50
	29.	1809	A. J. Porter, Commissioner's expenses	19 43
April	2.	1810	Peter A. Porter, Jr., Secretary and Treasurer, salary, March, 1900	83 33
		1811	Thos. V. Welch, Superintendent's office expenses	47 51
		1812	Pay-roll, March, 1900	1,486 17
	6.	1813	Joseph McDonald, coal	11 95
		1814	Grout Electrical Co., Inclined Ry	7 60
		1815	Elizabeth Read, ice	44 16
		1816	N. F. Hydraulic Power and Mfg. Co., electric lighting	50 00

Date.	No. of Voucher.	Name.	Amount.	
1900.				
April 6.	1817.	Gazette Publishing Co., Superintendent's office expenses	$15 00	
	1818.	Gazette Publishing Co., Superintendent's office expenses	6 25	
	1819.	N. F. Hy. Power & Mfg. Co., electric lighting	50 00	
	1820.	Walker & Paterson, tools, buildings	18 72	
	1821.	Walker & Paterson, tools, buildings	9 05	
	1822.	W. A. Shepard, bridges, buildings	17 99	
	1823.	Arthur C. Kugel, roads	6 30	
				$6,060 79

Abstract CXXII.

April 27.	1824.	A. J. Porter, Commissioner's expenses	$26 95	
30.	1825.	Peter A. Porter, Jr., Secretary and Treasurer, salary, April, 1900	83 33	
May 2.	1826.	Thos. V. Welch, Superintendent's, office expenses.	49 90	
	1827.	Pay-roll, April, 1900	2,040 40	
				2,200 58

Abstract CXXIII.

Date.		No. of Voucher.	Name.	Amount.	
1900.					
May	22.	1828	George Raines, Commissioner's expenses	$37 62	
	23.	1829	P. C. Flynn & Son, buildings	191 02	
		1830	F. W. Oliver Co., tools	75	
		1831	Brophy & Co., seed	3 50	
		1832	Welch Coal and Wood Co., coal	76 25	
		1833	J. H. Cook & Co., buildings	8 39	
	23.	1834	William F. Wall, Inclined Railway	231 07	
		1835	Walker & Paterson, buildings, tools, Inclined Railway and water pipes	78 11	
		1836	F. W. Oliver & Co., tools	5 65	
		1837	N. F. Hydraulic Power and Mfg. Co., electric lighting	50 00	
	31.	1838	Peter A. Porter, Jr., Secretary and Treasurer, salary May, 1900	83 33	
June	1.	1839	Pay-roll, May, 1900	2,202 50	
	4.	1840	Thos. V. Welch, Superintendent's office expenses	32 58	
	5.	1841	Peter A. Porter, Jr., Secretary and Treasurer, office expenses	10 86	
					$3,011 63

Abstract CXXIV.

Date.	No. of Voucher.	Name.	Amount.	
1900.				
June 7.	1842.	Andrew H. Green, Commissioner's expenses	$31 65	
13.	1843.	A. J. Porter, Commissioner's expenses	19 90	
15.	1844.	Thos. V. Welch, Superintendent's office expenses.	25 95	
18.	1845.	Peter A. Porter, Jr., Secretary and Treasurer, traveling expenses	25 00	
20.	1846.	Thos. V. Welch, Superintendent's office expenses.	24 50	
	1847.	A. J. Porter, Commissioner's expenses	25 00	
30.	1848.	Peter A. Porter, Jr., Secretary and Treasurer, salary, June, 1900	83 33	
July 3.	1849.	Pay-roll, June, 1900	1,772 62	
				$2,007 95

Abstract CXXV.

Date.	No. of Voucher.	Name.	Amount.
July 9.	1850.	National Press Intelligence Co., Secretary and Treasurer's office expenses	$5 00
19.	1851.	Elderfield-Hartshorn Co., tools, buildings	54 28
	1852.	Niagara Electric Equipment Co., inclined railway	1 28
	1853.	P. J. Davy, buildings	23 10

Building Coffer Dam in Rapids

Date.	No. of Voucher.	Name.	Amount.
1900.			
July 31.	1854.	Peter A. Porter, Jr., Secretary and Treasurer, salary, July, 1900	$83 33
Aug. 2.	1855.	West Disinfecting Co., buildings	48 00
	1856.	N. F. Hy. Power and Mfg. Co., electric lighting	50 00
	1857.	N. F. Hy. Power and Mfg. Co., electric lighting	50 00
	1858.	F. W. Oliver Co. tools	2 70
	1859.	Niagara Electric Equipment Co., inclined railway	2 89
	1860.	Elderfield-Hartshorn Co., water pipes tools, buildings	30 22
	1861.	J. H. Cook & Co., buildings	8 90
	1862.	Charlotte Haeberle, inclined railway	11 00
	1863.	Thos. McKnight, roads	100 00
	1864.	Thos. V. Welch, Superintendent's office expenses	42 27
	1865.	Pay-roll, July, 1900	1,888 61
21.	1866.	Pay-roll, August, 1900	339 00
	1867.	Thos. V. Welch, Superintendent's office expenses	49 46
31.	1868.	Peter A. Porter, Jr., Secretary and Treasurer, salary, August, 1900	83 33

Date.	No. of Voucher.	Name.	Amount.
1900.			
Sept. 4.	1869	Pay-roll, August, 1900	$1,200 87
	1870	Peter A. Porter, Jr., Secretary and Treasurer, office expenses	41 68
	1871	Elderfield-Hartshorn Co., tools	24 84
	1872	Elderfield-Hartshorn Co., water pipes, tools, buildings	72 39
	1873	James B. Lyon, extra annual reports	45 00
	1874	W. A. Shepard, bridges	71 03
	1875	Gazette Publishing Co., Superintendent's expenses	7 25
	1876	N. F. Hydraulic Power and Mfg. Co., electric lighting	50 00
	1877	N. F. Hydraulic Power and Mfg. Co., electric lighting	50 00
	1878	Globe Ticket Co., inclined railway	51 50
	1879	P. C. Flynn & Son, buildings	285 21
14.	1880	Pay-roll, September, 1900	45 00
30.	1881	Peter A. Porter, Secretary and Treasurer, September, 1900	83 37
	1882	Niagara Electric Equipment Co., inclined railway	1 50

Date.	No. of Voucher.	Name.	Amount.	
1900.				
Sept. 30.	1883.	Cataract Ice Co., ice	$87 60	
	1884.	P. C. Flynn & Son, signs	24 75	
	1885.	Elderfield-Hartshorn Co., tools, water pipes, buildings	33 00	
	1886.	N. F. Hydraulic Power and Mfg. Co., electric lighting	50 00	
	1887.	Dean & Hoffman, roads, stairways	38 45	
	1888.	Whitmore & Beaser, inclined railway	17 50	
	1889.	Globe Ticket Co., inclined railway	6 50	
	1890.	Thos. V. Welch, Superintendent's office expenses	49 49	
	1891.	Peter A. Porter, Jr., Secretary and Treasurer, office expenses	15 98	
	1892.	Fred Batchelor, seed	9 00	
	1893.	Pay-roll, September, 1900	1,003 37	
				$6,238 65

Payments out of $30,000, *as per chapter* 569, *Laws of* 1899.

Abstract II. Series K.

1899.				
Oct. 16.	8.	Pay-roll, September, 1899	$399 25	
	9.	Fred Batchelor, grass seed	9 00	
	10.	William S. Egerton, planting, roads	75 00	
	11.	Franklin Pfletcher, grass sod	1 75	
				485 00

Abstract III. Series K.

Date.	No. of Voucher.	Name.	Amount.	
1899.				
Dec. 22.	12.	William E. Egerton, planting	$70 30	
	13.	Elmanger & Barry, vines	2 88	
	14.	Pay-roll, November, planting	645 08	
	15.	Fred Batchelor, grass sod	6 00	
	16.	Charlotte Haeberle, planting	4 50	
	17.	Pay-roll, December, planting	193 02	
				$921 78

Abstract IV. Series K.

Date.	No. of Voucher.	Name.	Amount.	
1900.				
June 20.		The Engineering Record, shelter building	$6 60	
	19.	William Pool, shelter building	3 50	
	20.	Gazette Publishing Co., shelter building	6 75	
				16 85

Abstract V. Series K.

Date.	No. of Voucher.	Name.	Amount.
July 28.	21.	Journal Publishing Co., shelter building	$6 75
Sept. 22.	22.	Walter Jones, shelter building	17 50
	23.	Cataract Publishing Co., shelter building	2 50
	24.	William Pool, shelter building	1 75

Date.	No. of Voucher.	Name.	Amount.	
1900.				
Sept. 27.	25	Walter Jones, shelter building	$10 00	
				$38 50

Payments out of $120,000, *as per chapters* 419 *and* 420, *Laws of* 1900.

Date.	No. of Voucher.	Name.	Amount.
Aug. 31.	1	J. E. Boutrite, arch bridge to Goat Island	$150 00
	2	H. A. Reid, arch bridge to Goat Island	125 00
	3	Edward A. Bond, arch bridge to Goat Island	48 36
	4	D. S. Hollenbeck, arch bridge to Goat Island	18 40
	5	D. D. Waldo, arch bridge to Goat Island	90 08
	6	William P. Judson, arch bridge to Goat Island	21 27
Sept. 22.	7	Engineering News Publishing Co., arch bridge to Goat Island	57 60
	8	The Engineering Record, arch bridge to Goat Island	36 80
	9	J. E. Boutrite, arch bridge to Goat Island	17 95
	10	Edward A. Bond, arch bridge to Goat Island	30 04
	11	Walter Jones, arch bridge to Goat Island	56 25

Date.	No. of Voucher.	Name.	Amount.	
1900.				
Sept. 22.	12	..Pay-roll, August, 1900, arch bridge to Goat Island....	$397 50	
27.	13	..Walter Jones, arch bridge to Goat Island..........	120 00	
	14	..D. S. Hollenbeck, arch bridge to Goat Island...	81 00	
	15	..D. D. Waldo, arch bridge to Goat Island	156 96	
	16	..Pay-roll, September, 1900, arch bridge to Goat Island	55 50	
30.	17	..W. H. Keepers & Co., first estimate, arch bridge to Goat Island	7,803 00	
	18	..W. H. Keepers & Co., second estimate, arch bridge to Goat Island	9,554 00	
				$18,819 71

Payments out of $5,000, *as per chapter* 419, *Laws of* 1900.

Abstract I. Series N.

July 12.	1	..Pay-roll June, 1900, Briddle stairs (iron railing, bridges)	$349 75	
18.	2	..Dobbie Foundry & Machine Co., iron railings........	307 64	
	3	..Arthur C. Kugel, iron railing	7 13	
	4	..P. C. Flynn & Co., iron railing	16 66	

Date.	No. of Voucher.	Name.	Amount.
1900.			
Aug. 7.	5	Pay-roll July, 1900 (iron railing, stone stairway)	$702 73
	6	R. D. Young, stone stairway, Goat Island	140 75
	7	Arthur C. Kugel, stone stairway, Goat Island	140 00
	8	F. J. Munson, stone stairway, Goat Island	158 17
	9	Queenston Quarry Co., stone stairway, Goat Island	356 64
Sept. 22.	10.	J. H. Cook & Co., stone stairway, Goat Island	15 76
	11	Dobbie Foundry and Machine Co., stone stairway, Goat Island	134 95
	12	Walter Jones, stone stairway, Goat Island	21 25
	13	Elderfield-Hartshorn Co., stone stairway, Goat Island	48 56
	14	William S. Egerton, stone stairway, Goat Island	41 75
	15	R. D. Young, stone stairway, Goat Island	249 50
	16	F. J. Munson, stone stairway, Goat Island	467 10
	17	Pay-roll August, 1900, stone stairway, Goat Island	1,004 62

Date.	No. of Voucher.	Name.	Amount.	
1900.				
Sept. 27.	18.	Pay-roll September, 1900, stone stairway, Goat Island	$53 25	
	19.	Walter Jones, stone stairway, Goat Island	8 00	
				$4,224 21

Remittances to the State Treasurer:

1899.			
Nov. 1.	Draft for October receipts	$308 10	
Dec. 1.	Draft for November receipts	78 95	
1900.			
Jan. 2.	Draft for December receipts	35 55	
Feb. 2.	Draft for January receipts	70 20	
March 2.	Draft for February receipts	348 50	
April 2.	Draft for March receipts	251 70	
May 1.	Draft for April receipts	72 10	
June 4.	Draft for May receipts	230 80	
July 2.	Draft for June receipts	645 30	
Aug. 2.	Draft for July receipts	1,612 45	
Sept. 3.	Draft for August receipts	2,809 70	
30.	Draft for September receipts	2,076 45	
			8,539 80
1900.			
April 6.	Check for interest on balances in Manufacturers and Traders' Bank, Buffalo, N. Y.	$6 05	
July 2.	Check for interest on balances in Manufacturers and Traders' Bank, Buffalo, N. Y.	8 07	
			14 12
1900.			
Sept. 30.	Cash balance on hand		37 82
	Total		$58,087 48

Temporary Bridge. Mainland to Green Island.

Classification of Accounts.

Maintenance.

Secretary and treasurer, office expenses	$124 77
Secretary and treasurer, salary	1,000 00
Salaries, superintendent and clerk	3,066 65
Police	4,500 00
Inclined railway	2,324 83
Prospect Park	3,356 84
Goat Island	2,075 72
Roads	1,591 02
Walks	1,540 73
Commissioner's expenses	829 43
Superintendent's office expenses	539 62
Water pipes	189 34
Tools	137 90
Electric lighting	650 00
President's office expenses	70 00
Secretary and treasurer, traveling expenses	53 37
Bridges	172 76
Compost	475 25
Coal	148 70
Ice	131 76
Seed	12 50
Buildings	1,896 05
Extra annual reports	45 00
Signs	24 75
Stairways	32 70
	$24,989 69

Improvements under chapter 569, *Laws of* 1899.

Series K.

Grading	$399 25	
Grass sod	16 75	
Planting	962 90	
Roads	25 00	
Vines	2 88	
Shelter building	53 35	
		$1,462 13

Improvements under chapters 419 *and* 420, *Laws of* 1900.

Series L.

Arch bridge, mainland to Goat Island............		18,819 71

Improvements under chapter 419, *Laws of* 1900.

Series N.

Iron railing	$669 88	
Biddle stairs	49 50	
Bridges	37 00	
Stone Stairway, Goat Island...........	3,467 83	
		4,224 21
Total remittances to State Treasurer.............		8,553 92
Cash balance in hand September 30, 1900..........		37 82
Total		$58,087 48

PETER A. PORTER, Jr.,

Treasurer.

We, the undersigned, hereby certify that we have examined the foregoing report of the treasurer for the fiscal year ended September 30, 1900, the vouchers and other papers, and we find the report and accompanying documents correct, and that the treasurer has properly accounted for all moneys received and disbursed by him during the fiscal year ended September 30, 1900.

ALEXANDER J. PORTER,
GEORGE RAINES,
Commissioners of the State Reservation at Niagara.

Attest:

CHARLES S. DURKEE,
Professional Accountant and Auditor, Niagara Falls, N. Y.

Report of the Superintendent

OF THE

STATE RESERVATION AT NIAGARA

FOR THE

Fiscal Year Ending September 30, 1900.

Report of the Superintendent.

To the Commissioners of the State Reservation at Niagara:

Gentlemen.—The most important improvements made during the fiscal year ending September 30, 1900, have been the work upon the stone arched bridges from the mainland to Goat Island, and the work upon the new shelter building in Prospect Park. The construction of an iron guard railing, on Luna Island, and the construction of a stone stairway, and observation points, at the American Falls on Goat Island

STONE ARCHED BRIDGES, MAINLAND TO GOAT ISLAND.

June 30th, work of constructing temporary bridges and of removing portions of the whipple truss iron bridges, was commenced. The railings and sidewalks were removed from the iron bridge from the mainland to Green Island. The iron bridge from Green Island to Goat Island was entirely removed.

All the space desired for materials, and every facility possible for the advancement of the work, have been accorded to the contractors. Although a considerable force of men has been employed the work has not progressed as rapidly as expected. No attempt was made to work double shifts during the summer months when the work could be done to advantage. That the contractors came without an adequate plant and without experienced overseers for the work has been evident from the outset.

At the close of the fiscal year the piers and abutments for the bridge from the mainland to Green Island are not in place, and the arches for the bridge from Green Island to Goat Island are unfinished.

Before the winter sets in the bridge from Green Island to Goat Island will probably be completed, excepting the sidewalks and the piers and abutments will be in place for the bridge from the mainland to Green Island.

The iron bridge, from the mainland to Green Island will probably be allowed to remain during the winter for the accommodation of the public.

SHELTER BUILDING, PROSPECT PARK.

The new shelter building was located in Prospect Park between Falls street and Niagara street. Work was begun in August and has progressed steadily since that time. Some delay has been experienced in obtaining the stone for the structure.

The walls are approaching completion, and if the building is enclosed before the winter sets in it can be completed within the contract time.

STONE STAIRWAY, GOAT ISLAND.

The wooden stairway at the American Falls on Goat Island has been removed and a substantial stone stairway constructed.

There are 57 steps of Queenston stone, 10 feet wide, 14-inch tread, and 6-inch rise, laid in Portland cement, on stone foundations 4 feet deep, so as to be below the frost line. There are four flights of stairs slightly winding, and three spacious landings, for rest and observation. An iron guard railing has been

American Falls, from Goat Island.

erected on the side towards the high bank, of the river and a rustic embankment constructed on the other side. The margins of the stairway have been graded and covered with sod and will be planted with vines and trailing plants the coming spring.

OBSERVATION POINTS.

The walk and look-out at the foot of the stone stairway on Goat Island have been raised from 1 to 6 feet, and an iron guard railing constructed. Observation points have also been constructed at the top of the stairway on each side of the entrance, and the iron guard railing adjusted to the new arrangement.

The curbing and paving at the carriage turn-out near the top of the stairway have been taken up and are being replaced at a point about 50 feet further distant from the entrance to the stairway, and the observation points for visitors at the American Falls on Goat Island.

IRON GUARD RAILING, LUNA ISLAND.

The old guard railing on Luna Island has been removed and a new iron railing of the pattern designed by Mr. Calvert Vaux, substituted. The standards are bolted to stone flags, laid in concrete. There are no straight lines in the railing. The new railing is safer and more sightly than the one removed.

PAVING BUFFALO AVENUE.

Buffalo avenue, between Mill slip and First street, adjacent to the Reservation, has been paved with asphalt by order of the city of Niagara Falls. Buffalo avenue, between Seventh street and the eastern terminus of the Reservation, has also been paved with macadam by direction of the city authorities.

The paving of Buffalo avenue made it necessary to raise the grade of First street within the Reservation about 2 feet.

Filling and grading were also required back of the curb on the Green below First street.

Advantage was taken of the paving of Buffalo avenue to obtain material for filling the low places along the riverway and on the terrace and on Willow Island.

Some filling and grading has also been done on the uneven open space adjacent to the Loop driveway at Port Day.

SUSPENSION BRIDGES TO SISTER ISLAND.

Several examinations of the Suspension bridges to the Sister Islands have been made by the State Engineer. Plans and specifications for the repairs of the bridges were received. After a subsequent examination requested by the Board, the plans and specifications were returned to the State Engineer for revision and no further action has been taken

It may be desirable to abandon the idea of repairing the present bridges and to apply to the Legislature for an appropriation for stone arch bridges to the Sister Islands, in keeping with the bridge to the first island already constructed.

PUBLIC CARRIAGE STANDS.

The public carriage stands on the reservation were abolished May 26th in accordance with the resolution adopted by the Board. A copy of the resolution was mailed to each driver of a licensed carriage. The drivers have complied with the requirements almost without exception.

The licensing of improper persons by the city government necessitated the abolition of the carriage stands.

During the year two licensed carriage drivers have been excluded from the Reservation for violation of the ordinances.

RESERVATION CARRIAGE SERVICE.

During the progress of the work upon the arched bridges to Goat Island the Reservation carriage service has been operated by placing horses and carriages on Goat Island.

The Reservation Carriage Service Company has been notified to deliver passengers only on the river side of the Riverway and to use the Reservation carriages for hire only within the Reservation.

The Reservation carriage service should be operated without combination of any kind with any other local business enterprise.

THE BIDDLE STAIRS.

Repairs have been made upon the Biddle Stairs and the steps leading to them. The structures are so old and unsuitable that the question of their public use after the present season, is respectfully submitted to the Board for consideration.

COVERING CRIBWORK GOAT ISLAND.

A large number of the sumacs that were rapidly spreading over the upper end of Goat Island have been removed. The timber docking and cribwork along the southern shore of Goat Island have been covered by the sumacs removed and by brush collected in the thicket on Goat Island so as to be invisible, restoring the natural appearance of the shore.

The banks adjacent to the cribwork have been planted with Virginia creepers and willows.

ELECTRIC RAILWAY ON THE RIVERWAY.

The work of laying a single track for the electric cars on the riverway was commenced November 13, 1899. The cars are run on the riverway only a portion of one block between Falls avenue and the entrance to the International Steel Arch Bridge.

The cars have been in operation nearly a year under regulations formulated by the Board.

Connecting the railway systems at the foot of Falls street and at the Steel Arch Bridge, has been a great convenience to visitors to the Falls. There is not the entire freedom for vehicles formerly enjoyed in that part of the riverway, but aside from that the Reservation has suffered no injury. The operation of the cars as a rule, has been satisfactory, proving the wisdom of the State in retaining the ownership of the track and the absolute control of the operation of the cars within the distance traveled inside the Reservation.

ELECTRIC RAILWAY STATION.

The site for the railway structure proposed to be built by the Electric Railway Company on the riverway has been cleared and sketches for the building have been submitted, but the structure has not yet been erected.

The station should be dignified and substantial with architectural character worthy of so prominent a location.

GRADING AND PLANTING.

Planting has been done in Prospect Park on the Riverway on Green Island and at the entrance to Goat Island. A list of the trees, shrubs and vines planted is herewith submitted.

A large number of dead trees and tree stumps in Prospect Park have been grubbed out and removed.

The riprapping of the river shore at Fourth street with large stones has been completed and the ground made by filling, graded, sown with grass seed and planted with trees and shrubs.

The riprapping, grading and planting at the new stone arch bridge to the First Sister Island has been completed.

A large quantity of compost for the lawns which had been collected in the gravel pit on Goat Island during the summer and turned and mixed from time to time was spread upon the lawns during the winter months.

The fountain in Prospect Park has been taken down, the basin removed, and the locality filled and graded.

ROADS.

A heavy covering of gravel has been placed on the Riverway the entire distance from Falls avenue to First street, and from the entrance to the Steel Arch Bridge at Niagara street. Extensive repairs have been made to the roads on Goat Island, and the carriage turn-out at the Inclined Railway has been graveled.

The margin of the Riverway between the Steel Arch Bridge and Niagara street has been filled and covered with sod, and a low section of Prospect Park where the water lodged in the spring and fall has been filled.

The carriage turn-out at the Three Sister Islands should be enlarged and the walks through it changed before another season.

The artificial banks beside the road in Prospect Park have been graded to a natural slope and the road widened at the Niagara street entrance.

The head of Green Island, where erosion took place during the winter, has been riprapped with large stone.

The shrubs in the nursery have been transplanted and rearranged. Cuttings of Virginia creepers, bitter sweet and other native plants have been planted in the nursery for propagation. A screen of evergreens has been planted on Goat Island to shut out the view of the lumber yard from the driveway.

A bay in the pond in Prospect Park, in which the water was stagnant, has been filled.

WALKS.

A gravel walk, eight feet wide, has been constructed on the Riverway, from the entrance to the Steel Arch Bridge to the northern terminus of the Reservation at Niagara street.

Gravel walks in Prospect Park, on the Riverway and on Goat Island have been repaired.

EXAMINATION OF STRUCTURES BY STATE ENGINEER.

The superintendent communicated with the State Engineer and Surveyor, as directed by the Board, concerning the examination of the bridges and structures on the Reservation.

December 27th to 31st, inclusive, a careful examination was made by C. T. Middlebank, C. E., of the State Engineer's Department. The report made upon such examination by W. R. Davis, chief bridge designer, is herewith submitted.

The report substantially recommended the removal of the iron bridges, the Biddle Stairs and the building covering the Inclined Railway, and substitution of more substantial structures.

THE INCLINED RAILWAY.

The tracks and machinery of the Inclined Railway have been examined and repaired. After being in use only three months, the manilla cable of the Inclined Railway was found to be defective. It was removed, an old cable substituted and a new cable was ordered and received and was attached to the cars. The defective cable has been returned with a claim for an allowance on the new cable.

Many of the timbers in the roadbed of the Inclined Railway should be renewed. The frame structure covering the Inclined Railway should also be renewed.

An elevator shaft in the locality would be less obtrusive than the Inclined Railway structure and would afford less lodgment for the ice in the winter season. The enormous accumulation of frozen spray on the present structure is a great element of danger and expense every winter. In making changes in the locality it would be well to consider the advisability of substituting an elevator for the present Inclined Railway.

STEAMBOAT LANDING.

During the year the steamboat landing has been leased to the Maid of the Mist Steamboat Company, which has kept the landing in repair.

Only one boat has been in operation.

BRIDGES.

The roadway of the Boulder Bridge on the Riverway has been covered with a heavy coating of cement to prevent the soakage of water through the masonry of the arch.

The foundations of the retaining walls at the rustic stone bridge, at the weir below First street, have been repaired.

GUARD RAILINGS.

The iron guard railings in Prospect Park and along the high bank on Goat Island and Luna Island and at the Stone Stairway have been painted.

WATER PIPES.

The water pipes conveying the supply for the roads and buildings on Goat Island have been removed from the iron bridges to the temporary bridges and arrangements made for supplying the bridge contractors with water for their work.

The water pipes on the surface of the ground, the standpipes for sprinkling the roads and the drinking fountains have been disconnected for the winter season.

BUILDINGS.

The roof of the terminal station at the foot of the Inclined Railway has been repaired in places where it was damaged by the ice during the winter.

April 16th the gable on the northern side of the lower terminal station was torn off by the ice. The damage has been repaired. The structure covering the Inclined Railway has been repainted. The cottage on Goat Island has been reshingled. The slat sides and ends have been removed from the Shelter Building at the Three Sister Islands and the structure repainted.

The Inclined Railway building and the pavilion in Prospect Park have been repainted. The pavilion building should be reshingled the coming spring.

The gate house and summer house in Prospect Park, the shelter building at the Horse Shoe Falls, the police booth on Green Island, the structures at the Cave of the Winds building and the park seats and tables have been repainted.

The park seats and tables have been stored in the open pavilion in Prospect Park and on Goat Island for the winter season.

VISITORS.

The number of persons who visited the Reservation during the fiscal year is estimated at 750,000. Of this number 456,249 came in "excursion" parties. A statement of the excursions to the Reservation during the year is hereto appended.

EMPLOYEES.

The number of regularly classified employees is nine, to wit: 1 superintendent, 1 treasurer and secretary, 1 clerk, 2 police and caretakers, Goat Island; 3 police and caretakers, Prospect Park; 1 Inclined Railway police and caretaker.

A statement of the number of laborers and teamsters employed each month is herewith submitted.

STATEMENT OF NUMBER OF EMPLOYEES.

Maintenance Rolls.

1899.	Foreman.	Asst. foremen.	Teamsters.	Laborers.
October	1	3	1	17
November	....	3	....	18
December	1	3	1	13
January	1	3	2	18
February	1	3	2	19
March	1	2	1	14
April	1	3	2	38
May	1	3	1	31
June	1	2	1	24
July	1	2	1	35
August	....	1	1	28
September	....	1	....	6

Improvement Rolls.

Chapter 569 of the Laws of 1899.

1899.	Foreman.	Asst. foreman.	Teamsters.	Laborers.
November	1	....	2	19
December	....	....	1	23

Chapter 419, Laws of 1900.

June	1	1	1	6
July	1	2	3	17
August	1	3	3	20
September	1	2	1	18

Chapter 420, Laws of 1900.

August	....	....	....	8
September	....	....	....	9

Detailed statements of the receipts and expenditures of the superintendent, the amount of the pay-rolls for each month and the classification of the pay-rolls and accounts are hereto appended.

Respectfully submitted,

THOMAS V. WELCH,

Superintendent.

EXCURSIONS.

OCTOBER, 1899, TO SEPTEMBER, 1900, INCLUSIVE.

1899.		Cars.	Persons.
Oct.	6. Special party from Philadelphia..	5	300
	Boston, Mass., Knights of Columbus..........................	1	25
	8. Cleveland and Erie, Pa...........	50	3,000
	11. National Druggists' Association..	10	600
	12. Pennsylvania Superintendents of Poor........................	6	360
	Special Mexican party...........	1	60
	15. Buffalo, N. Y., and Cleveland, O...	30	1,800
	20. Baltimore and Washington.......	8	480
1900.			
Jan.	14. Buffalo, N. Y....................	5	300
Feb.	4. Buffalo, N. Y....................	10	600
	7. Buffalo, N. Y....................	6	360
	11. Buffalo, Rochester and Lockport, N. Y........................	30	1,800
	18. Buffalo, Rochester and Lockport, N. Y........................	20	1,200
March	4. Buffalo, Rochester and Lockport, N. Y........................	15	900
	11. Buffalo, Rochester and Lockport, N. Y........................	15	900
	18. Buffalo, N. Y..................	10	600
	25. Buffalo, N. Y..................	12	720

1900.		No. of cars.	Persons.
May	5. Lockport, N. Y., Arbor Day.......	10	600
	11. Buffalo, Central High School.....	4	240
	14. Norwegians from Minnesota to Europe......................	10	600
	16. Rochester, N. Y., the Eagles......	4	240
	24. Queen's birthday excursion.......	30	1,800
June	1. Detroit, Mich., Coal Dealers......	2	120
	3. Detroit, Mich., Coal Dealers......	2	120
	Sunday excursions	20	1,200
	4. Canadian Editorial Ass'n........	2	120
	Butchers' Convention	10	600
	5. American Pattern Makers........	3	130
	Canadian Press Ass'n............	2	120
	N. Y. State Butchers' Ass'n......	3	180
	9. Union Sunday Schools, Hamilton, Ont.........................	2	120
	Dunkirk, N. Y..................	12	720
	City officials, Allegheny, Pa......	2	120
	10. Sunday excursions (all sources)...	65	3,900
	12. Bradford, Pa.	8	480
	City officials, Toronto, Ont., and friends.....................	3	180
	13. Buffalo, N. Y., Knights of Honor..	2	120
	Pennsylvania R. R. officials and newspaper men	3	180
	Toronto, by boat	......	200
	14. Pennsylvania R. R. officials.......	6	360
	Strathroy, Ont.	7	420
	Stratford, Ont.	4	240

1900.			No. of cars.	Persons.
June	15.	Buffalo, N. Y., Wholesale and Retail Coal Dealers	3	180
	16.	Hamilton, Ont., Tailors' Union	7	420
		Buffalo Coal Dealers	14	840
		Delegates to Philadelphia Convention	10	600
		Toronto, Ont., by boat	......	100
	17.	Carbondale, Pa.	10	600
		Western N. Y. & Pa.	10	600
		Cleveland, Ohio	10	600
		Rochester and Buffalo, N. Y.	141	8,420
		Niagara on the Lake, Canadian militia	......	2,000
		Susquehanna	9	540
	18.	Topeka, Kan. (Turnverein)	1	60
		Milwaukee, Wis. (Turnverein)	1	60
	19.	Teachers and Scholars, Public Schools, Cleveland, O.	6	360
		Toronto, Ont., by boat	......	75
	20.	Toronto, Ont., by boat	......	200
	21.	Epworth League, Rochester, N. Y.	5	300
		McCaul, M. E. Church, Toronto, Ont		240
		Epworth League, by boat, Toronto, Ont	......	400
		Niagara on the Lake Ontario, Canadian militia	......	500
	22.	Toronto, Ont., by boat	......	100
	23.	Hornellsville, N. Y.	9	540
		Saturday half-holiday excursion	10	600

1900.			No. of cars.	Persons.
June	23.	English Paper Manufacturers....	1	60
		Hamilton, Ont., brewers.........	8	480
		Orangeville, Ont., Foresters......	9	540
		Mooneyville, Ont., by boat.......	......	950
	24.	Sunday excursions...............	167	10,000
	26.	Carlton M. E. Church, Toronto, Ont., by boat	......	900
	27.	Baptist Church, Toronto, Ont., by boat	......	275
	28.	Buffalo, N. Y., Miss Remington's School	2	120
		Delegates to National Saengerbund	1	60
		Homeopathists, en-route to Paris.	2	120
		Lockport, N. Y., Thursday Club..	1	60
		Toronto, Ont., by boat...........	......	525
	30.	Delaware, Lackawanna & Western	16	960
		Half-holiday excursions..........	20	1,200
		Maine, H. H. Delegates to Kansas Convention	1	60
July	1.	Rochester, N. Y.................	12	720
		Cleveland, O	10	600
		Buffalo, N. Y...................	60	3,600
	2.	Dominion Day excursions from Canada	140	8,400
		Toronto, by boat	......	4,000
	3.	Railroad employees, T. H. & B. Line	8	480
		Toronto, by boat	......	125
	4.	From all sources................	250	15,000

1900.			No. of cars.	Persons.
July	5.	Presbyterian Sunday School, Toronto, by boat	......	554
	6.	Methodist Sunday School, Toronto, by boat	......	750
	7.	National Confectioners' Convention	2	120
		Saturday half-holiday excursions..	40	2,400
		Employees, Mathew Harris, Hamilton, Ont	10	600
		Toronto, Ont., Foresters, by boat.	......	400
	8.	Sunday excursions, all sources....	85	5,100
	9.	Kansas, Neb., Delegates returning.	2	120
		Chautauqua, N. Y.	4	240
		National Whist Players Convention	6	360
	10.	Employees, International Traction Co	12	720
		Toronto, Farmers' Institute, by boat	......	900
	11.	Washington, Ohio, Grocers.......	7	420
		American Knife Mfrs	1	60
		Buffalo, N. Y., Feohsein Bowling Club	1	60
		Toronto, Ont., by boat	......	100
	12.	Carbondale, Pa	6	360
		Rochester, N. Y.	2	120
		Buffalo Traction Co.'s employees.	15	900
		Welland, Ont., Orangemen	28	1,680
		Toronto, by boat	......	500
	13.	Grand Trunk R. R., employees...	84	5,040

1900.			No. of cars.	Persons.
July	13.	Buffalo Traction Co., employees...	9	540
		Main line, Erie R. R.............	16	960
		Lehigh Valley R. R.............	8	480
	14.	Cincinnati, Ohio	17	1,020
		Chautauqua, N. Y...............	7	420
		Saturday half holiday excursions.	40	2,400
		Toronto Brantford, Waterloo, Ont.	36	2,160
		Toronto, by boat, Plumbers......	......	3,000
	15.	Pittsburg, Pa....................	18	1,080
		Sunday excursions, all sources....	90	5,400
	16.	Guelph, Ont....................	8	480
		Buffalo, N. Y. Traction Co.'s employees	17	1,020
	17.	National American Paper Box Makers	2	120
		Pittsburg, Pa....................	10	600
		Detroit, Mich., Glass Blowers....	4	240
		St. Mary Magdalene Church, Toronto, Ont......................	6	360
	18.	Hamilton, Ont., Grocers' Union...	117	7,020
		Toronto, Ont., by boat...........	......	5,000
	19.	Bradford, Pa....................	7	420
		Jamestown, N. Y...............	10	600
		Buffalo, N. Y., Traction Co.'s employees	14	840
		Daughters of Rebecca............	......	200
		Toronto, Ont., by boat...........	......	4,000
	20.	Cleveland, O....................	8	480
		Baltimore and Washington......	12	720
		Toronto, Ont., by boat............	......	500

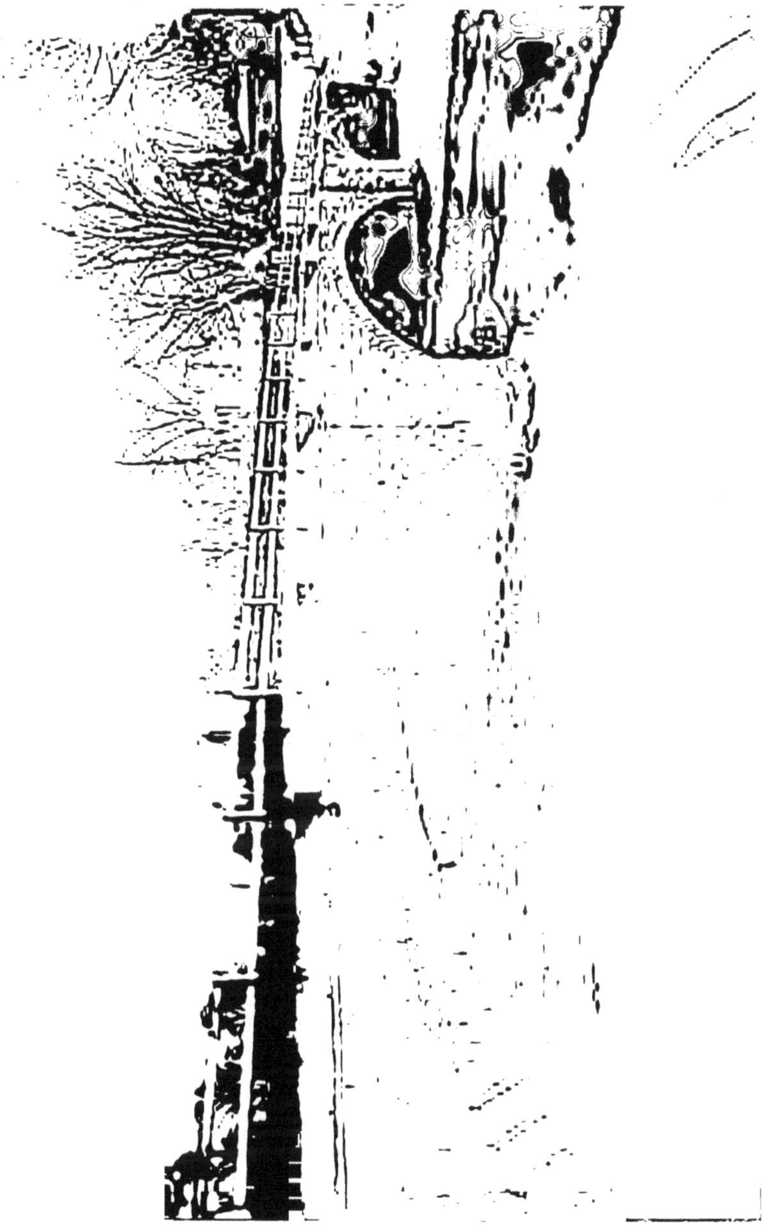

1900.			No. of cars.	Persons.
July	21.	Cleveland and Detroit...........	16	960
		Saturday half-holiday excursions..	45	2,700
		Lancaster, N. Y.................	2	120
		Hamilton, Ont., Clothiers' Union..	6	360
		Grand Trunk Railroad employees.	60	3,600
	22.	Pittsburg, Pa....................	10	600
		Wheeling, W. Va................	10	600
		Cleveland, O....................	12	720
		All other sources...............	120	7,200
	23.	Western N. Y. and Pennsylvania excursion	10	600
		Other sources....................	10	600
		Toronto, Ont., by boat...........	......	500
	24.	Toronto, Ont., by boat, Odd Fellows	......	1,100
	25.	R. W. & O., main line...........	18	1,080
		St. Thomas, Ont., Grocerymen....	17	1,020
		Toronto, Ont., Foresters, by boat.	......	1,200
	26.	Olean, N. Y.....................	9	540
		St. Thomas, Ont.................	17	1,020
		Toronto, Ont., by boat...........	......	700
	27.	Buffalo, N. Y., Public Schools Nos. 3, 42 and 44....................	9	540
		Washington and Baltimore......	20	1,200
		Ohio and Indiana................	30	1,800
		Toronto, Ont., Baptist Church....	4	240
		Hamilton, Ont..................	18	1,080
	28.	Batavia, N. Y...................	60	3,600
		Pittsburg, Pa....................	30	1,800
		General half holiday excursion....	65	3,900

1900.			No. of cars.	Persons.
July	28.	Hamilton, Ont	30	1,800
		Toronto, Ont., by boat	......	2,000
	29.	Buffalo and Rochester, N. Y	90	5,400
		South Bend, Ind. and Cleveland, O.	30	1,800
		Pittsburg, Pa	10	600
		Phi Gamma Delta Fraternity	3	180
	30.	Guelph, Ont., Foresters	34	2.040
		Toronto, Ont., by boat	......	900
	31.	Jamestown, N. Y	9	540
		Buffalo, N. Y	10	600
		Toronto, Ont., by boat	......	500
Aug.	1.	Toronto, Ont., by boat	......	700
		Owen Sound, Ont., by boat	......	600
	2.	German Lutheran Church (city)	9	540
		Lockport, N. Y., Congregational Church	5	300
		Toronto, Ont., by boat	......	1,000
	3.	Washington and Baltimore	14	840
		Philadelphia and Reading	20	1,200
		M. C. and Lehigh Valley railroads	140	8,400
		Lake Erie and Western Railroad	40	2,400
		Cincinnati and Columbus, Ohio	40	2,400
		Buffalo, N. Y	20	1,200
	4.	Chautauqua, N. Y	9	540
		Hamilton, Ont	10	600
		Port Jervis, Ont	18	1,080
		Saturday half holiday excursion	30	1,800
	5.	Paris, Ont., Wholesale Grocers	5	300
		Cleveland, Ohio	10	600

1900.			No. of cars.	Persons.
Aug.	5.	Buffalo, Lockport and Rochester, N. Y.	167	10,020
		R., W. & O. main line	8	480
	6.	Chautauqua, N. Y.	8	480
		Woodstock, Ont., Foresters	7	420
	7.	Western points	10	600
		Buffalo, N. Y.	15	900
		Hamilton, Ont., Odd Fellows	10	600
		Toronto, Ont., by boat	......	7,000
	8.	Niagara County Veterans	15	900
		National Ass'n Window Trimmers.	3	180
		Big Four excursion	20	1,200
		Chautauqua, N. Y.	9	540
		Firemen's Convention	74	4,440
		Toronto, Ont., by boat	......	520
	9.	Buffalo, Rochester and Pittsburg.	10	600
		Canadian Brotherhood Locomotive Engineers	5	300
		Main line W. N. Y. & Pa.	9	540
		Main line C., H. & D.	18	1,080
		Wabash line	16	960
		Other sources	20	1,200
		Toronto, Ont., by boat	......	1,000
	10.	Michigan Southern R. R.	30	1,800
		Buffalo, N. Y.	40	2,400
		Grand Rapids, Mich	18	1,080
		Hamilton, Ont	10	600
		Toronto, Ont., by boat	......	800
	11.	Main line, R., W. & O.	16	960
		St George's, Ont	25	1,500

1900.			No. of cars.	Persons.
Aug.	11.	Stony Creek, Ont.	20	1,200
		Saturday half-holiday excursion	60	3,600
		Toronto, by boat	......	1,500
	12.	Main line N. Y. C. and Buffalo	115	6,900
		Cincinnati, Ohio	23	1,380
		Cleveland, Ohio	14	820
	13.	Woodstock and Hamilton, Ont	34	2,040
		Lakewood, N. Y	8	480
		Toronto, Ont., by boat	......	500
	14.	Chautauqua, N. Y	8	480
		Syracuse and Auburn	10	600
		Pittsburg, Pa	12	720
		Cleveland, Ohio	10	600
		Buffalo, N. Y	30	1,800
		Toronto, Ont., by boat	......	500
	15.	Western points	75	4,500
		Elmira, N. Y	10	600
		Cleveland, Ohio	10	600
		Toronto, Ont., Hibernians	34	2,040
		Canadian Pacific and boat	20	1,200
	16.	Lockport, N. Y., Lutheran Church	4	240
		Buffalo and Rochester	40	2,400
		N. Y. C. & H. R. R. R	30	1,800
		W. N. Y. & Pa	18	1,080
		Pennsylvania R. R	20	1,200
		D., L. & W	20	1,200
		Clover Leaf line	20	1,200
	17.	All sources	202	12,120
		Orangeville, Ont., by boat	......	6,000
	18.	T., H. & B. R. R	27	1,620

1900.			No. of cars.	Persons.
Aug.	18.	Chautauqua, N. Y	8	480
		Saturday half-holiday excursion	40	2,400
	19.	Jamestown, N. Y	8	480
		N. Y. C., W. S. and C. & B	100	6,000
		Nickel Plate line	20	1,200
	20.	Wisconsin Press Ass'n	2	120
		Kingston, Ont., by boat	......	700
	21.	Brandford and Simcoe, Ont	20	1,200
		National Florists' Convention	2	120
		Chautauqua, N. Y	7	420
		Youngstown	8	480
	22.	L. S. & M. S. R. R	9	540
		Chautauqua, N. Y	4	240
		Buffalo, N. Y	15	900
		Toronto, Ont., by boat	......	200
	23.	Black Rock Union School	4	240
		Albion, Brockport and Rochester	16	960
		Main line, Nickel Plate	12	720
		Cleveland and Buffalo	12	720
		St. Thomas, Ont	34	2,040
	24.	Philadelphia and Washington	20	1,200
		Brooklyn, N. Y., G. A. R	9	540
		Grand Island Evangelical Church	3	180
		Toronto, Ont., by boat	......	200
	25.	Brooklyn, N. Y., G. A. R	20	1,200
		Maine and New Hampshire G. A. R	9	540
		Hamilton, Ont	16	960
		Saturday half-holiday excursion	25	1,500
		Toronto, Ont., by boat	......	200

1900.			No. of cars.	Persons.
Aug.	26.	Knights of Pythias..............	4	240
		Sunday excursions	80	4,800
		St. Catherines, Ont..............	8	480
	27.	Erie, Pa.......................	10	600
	27.	Jamestown, N. Y...............	4	240
	28.	Main line Pa. R. R...............	12	720
		Chautauqua, N. Y. Society of Friends	5	300
		Pennsylvania (special excursion).	40	2,400
		Elmira, N. Y....................	14	840
		Jamestown, N. Y................	6	360
	29.	Toronto, Ont. (annual fair)......	30	1,800
		Toronto by boat................	......	1,300
	30.	Allegany Valley	12	720
		Toronto, Ont....................	15	900
		Lancaster, Pa...................	14	840
		Toronto by boat................	......	300
	31.	Knights of Pythias..............	5	300
		Philadelphia and Baltimore......	20	1,200
		All other sources	50	3,000
Sep.	1.	G. A. R. en route to Philadelphia..	8	480
		New York city	10	600
		Saturday half-holiday excursion..	75	4,500
		Toronto, by boat	......	900
	2.	Baltimore, Washington and Philadelphia	20	1,200
		Lake Erie and Western R. R.....	15	900
		Buffalo, Rochester, etc..........	85	5,100
	3.	New York, Mozart Verein	4	240
		Rochester, Syracuse and Lockport	65	3,900

1900.			No. of cars.	Persons.
Sept.	3.	R. W. & O. & L. S. & M. S........	30	1,800
		Buffalo, N. Y..................	20	1,200
		Scranton, Pa.	10	600
		Hornellsville, N. Y..............	10	600
		Toronto, Ont., by boat...........	......	3,000
	4.	Cleveland, O.	5	300
		W. N. Y. & P. Line..............	9	540
		Chicago, Ill.	18	1,080
		Toronto, Ont., by boat..........	......	900
	5.	Mansfield and Toledo, O.........	14	840
		Pittsburg, Pa.	12	720
		Philadelphia, Washington and Baltimore	20	1,200
		Toronto, Ont.	25	1,500
		Nickel Plate R. R...............	10	600
		California (Special Party)	2	120
		Hamilton, Ont.	20	1,200
		Toronto, Ont., by boat...........	......	1,600
	6.	Canton, O.	14	840
		Toronto, Ont., by boat...........	......	5,000
	7.	G. A. R. returning..............		60
		Toronto, Ont., by boat...........	......	1,500
	8.	Nickel Plate R. R...............	20	1,200
		Detroit, Mich., Mail Carriers.....	9	540
		Saturday half-holiday excursions.	10	600
		Toronto, Ont., cars and boat.....	10	1,500
	9.	M. S. & L. S....................	21	1,260
		Sunday excursions	45	2,700
	10.	Toronto, Ont., by boat..........	......	900
	11.	Fire Underwriters Convention...	5	300

1900.			No. of cars.	Persons.
Sept.	11.	Laundry Men's Association......	5	300
	12.	National Association, Embalmers	5	300
		Braumeisters' Convention	8	480
		Simcoe, Ont.	12	720
	12.	Buffalo, N. Y., C. M. B. A. Society.	4	240
	13.	Insurance agents	3	180
		Toronto, Ont., by boat..........	......	300
	14.	Baltimore, Washington and Philadelphia	16	960
		Toronto, Ont., by boat...........	......	250
	16.	Nickel Plate R. R...............	10	600
		L. S. & M. S. R. R...............	10	600
		Rochester, Lockport and Buffalo.	15	900
	17.	Toronto, Ont., by boat...........	......	200
	19.	Street Railroad Association	4	240
		Toronto, Ont., by boat..........	......	200
	20.	Carbondale, Pa.	6	360
		Scranton, Pa.	6	360
		Hornellsville, N. Y..............	5	300
	21.	Washington and Baltimore......	8	480
		Toronto, Ont., by boat..........	......	350
	22.	Port Jarvis	[illegible]	240
	23.	Rochester, Syracuse and Buffalo..	32	1,920
	24.	Stratford and Galt, Ont., Methodist churches	10	600
		St. Catherines, Ont.............	3	180
	25.	Conshohocton, Pa., Firemen.....	1	60
		Emporium, Pa.	5	300
	26.	New York (Special Party).......	2	120
		St. Catherines, Ont.............	3	180

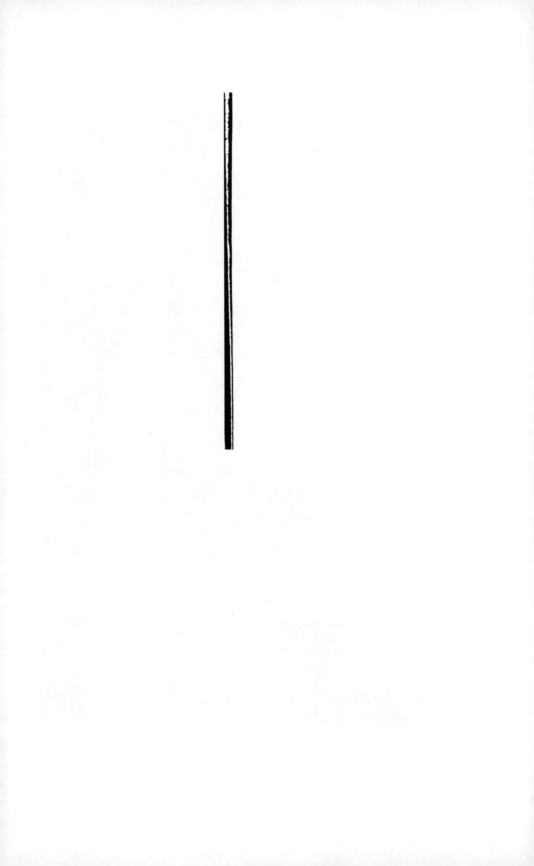

1900.		No. of cars.	Persons.
Sept.	26. Toronto, Ont., by boat...........	......	600
	27. North Carolina Press Association.	2	120
	Washington, Baltimore and Philadelphia....................	9	540
	St. Catherines, Ont............	4	240
	Toronto, Ont., by boat..........	......	250
	28. Newfane, N. Y., Basket Factory employees..................	2	120
	30. Buffalo, N. Y..................	20	1,200
	Rochester, N. Y................	6	360
	St. Catherines, Ont.............	5	300
		6,389	456,249

RECAPITULATION.

1899.		
October................................	111	6,625
November..............................	......	....
December..............................	......	....
1900.		
January................................	5	300
February..............................	66	3,960
March.................................	52	3,120
April..................................	......	....
May....................................	58	3,480
June...................................	636	44,325
July....................................	2,055	150,229
August................................	2,559	176,540
September............................	847	67,670
	6,389	456,249

The Story of Niagara

BY

PROF. C. H. HITCHCOCK.

Reprinted from the *American Antiquarian*, January, 1901.

THE STORY OF NIAGARA.

BY C. H. HITCHCOCK, LL. D.

[Reprinted from the American Antiquarian, January, 1901.]

There is, so far as is known, no better locality than the region of the Niagara gorge to afford an insight into the length of time that has elapsed since the disappearance of the ice-sheet from the northern United States. At first it was thought to be a very simple problem. Given a gorge seven miles long, known to be lengthening annually by a definite number of feet and inches, we can ascertain the number of years required by dividing the sum total of feet by the annual recession. This result is believed to represent the period that has elapsed since the ice age, because the Niagara River commenced its existence in consequence of the damming of the ancient Erie (Erigan) River. The original channel having been blocked, Lake Erie accumulated, and its lowest line of exit was by way of the Niagara River. Sir Charles Lyell in 1841 estimated this period to have been about 36,000 years. He supposed the annual wear to be one foot. The latest figures indicate the amount of recession to be four times as great. Hence, if there were nothing more to be said, our conclusion must be that the time needed was only one-fourth of 36,000.

But this estimate must be modified by a careful consideration of various factors. It must be conceded that the Niagara River began to flow only after the obliteration of the Erigan Channel. Then the peculiar elbow at the whirlpool indicates a second blockade, and consequent divergence of the river toward Lewiston; for in none of the numerous canons of the West do we find any such arrangement. The gorge varies greatly in width and depth, as if the volume of the river had been diminished at times. Again, the work may have been interrupted by another glacial blockade, or modified by the presence of lakes not now existent. The story of the gorge, then, shows much complication. It will be our task to state what the principal events in the history of this region have been, so as to be able to judge of the value of computations based upon the work performed.

There is no lack of literature pertaining to this subject. The facts are of very easy apprehension; hence a multitude of authors, attracted by the sublimity of the roaring cataract, have placed their opinions upon record. Several hundred titles of these writings might be enumerated—of travellers, historians, poets, artists, statesmen, generals, divines, besides professional physicists and geologists. Among those worthy of special mention are the early pictures by Father Hennepin, as seen in 1678; by Peter Kalm in 1751; by Capt. Basil Hall in 1827 with a camera lucida; by famous artists, Church and Bierstadt, and finally by the photographer with the best of modern appliances. Bird's-eye views, with the geology, or estimates of age, came from Robert Bakewell, Jr., in 1830; Sir Charles Lyell, 1841; Prof. James Hall, 1843; Dr. Julius Pohlman, 1888, and others The two more recent authors, who have surpassed all others by their discernment and brilliant suggestions, are Drs. J. W. Spencer and G. K. Gilbert in their writings during the past twenty years. Stations for the accurate measurement of the recession were established by James Hall in 1842; verified and observed by the United States Lake Survey in 1875, R. S. Woodward in 1886, and A. S. Kibbe in 1890. An altogether different class of papers relate to the use of the power derived from the falling water, and its transmission to great distances.

Niagara River is a part of the St. Lawrence system of drainage, commencing with the five great lakes, Superior, Michigan, Huron, Erie, and Ontario, having an area of 94,650 square miles, of which 87,410 have their outlet at Buffalo. This discharge proves to be about 275,000 cubic feet per second, or 7,000 tons, equal to the latent power every day of 200,000 tons of coal. The St. Lawrence River rises in Lake Ontario, flowing irregularly to the ocean. The entire system is a very modern river, geologically speaking, full of rapids, pools and still water. Had it existed for a long time, both rapids and lakes would have disappeared by erosion and drainage, save as an overflow prevails because of submergence.

The Niagara River is a strait, connecting Lakes Erie and Ontario. Its depth is very constant, but varies with storms. A westerly gale may crowd water from Lake Erie; while a northerly wind may push it in a contrary direction. For this reason the falls were once almost dry for several hours.

The following statement of the various physical features is needed for the discussion of the history and age of the gorge. They are compiled from the maps of the United States Lake Survey and the United States Geological Survey.

	Miles or feet.	Altitude above sea.	Descent in feet.	Width.
Lake Erie	....	573	......	1 2-5 miles.
Buffalo to head of rapids	22.50 miles.	563	10	
Rapids to head of falls	.75 mile.	517	46	1,060 feet.
American fall	....	520	165	3,010 feet.
Canadian fall	....	517	160	average 1,300 ft.
Great gorge to railroad bridge	11,750 ft. 2.25 miles.	340	17	750 feet.
Whirlpool Rapids	3 900 ft. .74 mile.	295	45	1,200 feet.
Eddy Basin	1,500 ft. .28 mile.	285	10	1,800 feet.
Whirlpool Basin	1,900 feet.	282	3	1,300 ft. to 1.25 m.
Whirlpool to edge of Escarpment at St. David's	3 miles.	......	....	1,300 feet.
Cove section	3,750 feet.	......	....	1,300-1,700 ft.
Foster's Island to Catholic College	5,754 feet.	......	....	1,100-1,300 ft.
Catholic College to mouth of gorge	8,468 feet.	......	....	1,100-1,700 ft.
Whirlpool to Lewiston	3,465 miles.	249	33	2,400 feet.
Lewiston to Lake Ontario	7 miles.	247	2	
Lake Erie to Lake Ontario	37,165 miles.	......	326	
Niagara Plateau	....	600	..... .	

EARLY GEOLOGICAL HISTORY.

In the study of the geological history of the vicinity of Niagara, three great periods may be designated, each characterized by peculiar conditions. First, the laying down of the crystalline foundations; second, the deposition of the Silurian and later sediments upon the Archean; third, erosion and the accumulation of earth, sands, and clays. The first witnessed the action of fire; the second indicated the presence of the ocean, and the third demanded the movement of enormous glaciers with the formation of great lakes.

The nearest locality of the Archean crystallines outcropping at the surface, is in Canada, more than a hundred miles to the north. The rocks are of the granite order—derived from the primeval crust. The southern border of this Archean coincided with a shore line from the Thousand Islands to Georgian Bay on Lake Huron, against which the sands, limestones and clays of the Lower Silurian terranes were slowly deposited in definite order. A boring 1,600 feet deep at the edge of Lake Ontario, through these lower deposits, might reach the crystalline schists, which must also extend below Niagara. A restoration of the conditions prevailing here all through Silurian time would show bands of sediment added one by one to the crystalline tract—each age indicating a growing area of land, while the shore line moved southerly. At Lewiston the red clays and sands of the Medina terrane prevail, and they constitute the visible foundation of the plateau. The sands represent an ancient sea beach which supported worms, mollusca and trilobites, besides marine plants. The waves pushed shells upwards, and as the water receded there would be left the lines made by the upper limit of the surf, and rill marks produced by the rush of the current

The Driveway. Goat Island.

over the valves of the brachiopods. Every vestige of wave action is perfectly preserved in the solid rock. In another period limestone was produced. Evidently this was a coral growth, for their remains are abundant. Stated definitely, the following is the order of the terranes: first, the Medina group, consisting of soft red shales and sandstones below, a gray sandstone, and another series of red sandstone and shales; in all a thousand feet in thickness. Above the Medina is the Clinton group of shales and limestone, followed by the Niagara shale and limestone. At Lewiston above the water-level the Medina is 210 feet thick; the Clinton, 29; the Niagara shale, 82, and the limestone, 20. The last becomes thicker in passing southerly, reaching 65 feet, of which 55 are exposed at the falls. This variation is due to a greater erosion at the north. The shale was a calcareous mud containing organic matter, which still emits an oily smell when freshly broken.

After the deposition of the layers named, there ensued the Salina period, when large bodies of salt accumulated from the evaporation of ocean water. This rock lies to the south of the gorge, and before reaching Buffalo the Lower Helderburg limestone succeeds; so that the whole of the Upper Silurian is displayed along the Niagara River, with a thickness of about 900 feet. These rocks constitute a plateau, extending as far east as Rochester, and north to Georgian Bay; the northern edge has a steep slope, which runs parallel to Lake Ontario and then turns northwesterly and reaches the promonotory between the two sections of Lake Huron.* The strata underlying this plateau slope southerly about twelve feet to the mile. The altitude of the plain is about six hundred feet above the sea.

* See plate 1.

The Genessee and the Niagara are the most important streams that have sawed through it. The smaller streams, like the Tonawanda, Chippewa, and the Grand of Ontario, flow along depressions in this plateau, and its general mass may be compared to a dam, causing the water from the south to accumulate as Lake Erie.

After the deposition of the Upper Silurian terranes the work of continent-building shifted to the south: thick flags, grits and shale mainly constituting the Devonian system, followed by carboniferous sediments supporting an immense vegetation over the region of the Appalachian coal fields, and eventually rising into mountains at the close of the Paleozoic. Thus originated a grand system of drainage, arising in West Virginia, southwestern Pennsylvania, eastern Ohio, and western New York, tributary to the St. Lawrence. The Monongahela and Youghiogheney, rising in West Virginia, flowed northerly down the Ohio to the great bend in Beaver County, and then continuing in the same direction in the valley of the Big Beaver River, the Conoquenessing, the Shenango, and over the present divide by Conneaut Lake into Lake Erie. The rocky floor slopes from Pittsburg northerly. The drainage of the upper Allegheny River was also to the north, the reverse of that now prevailing. A part of it joined the stream just described, flowing from Oil City up French Creek, thence into the Conneaut. Another river began to the southeast of Warren, Pa., ascended the present Conewango Creek, and at Falconer, N. Y., received the drainage of the upper Allegheny from Potter County, Pa., receiving also the outlet from Chautauqua Lake, and reached Lake Erie at Dunkirk. Of course the present streams discharging into Lake Erie from Buffalo around to Lake St. Clair flowed as at present. As the result, there was a very large river made up from these tributaries, which followed the main channel of Lake Erie, leaving it at the mouth of the Grand River of Ontario, pursuing a northerly and northeasterly course till it reached Lake Ontario at Hamilton. Plate I. is drawn to represent the drainage system of this ancient Erigan River. Like the modern Niagara, it was compelled to cut its way through the plateau. Prof. J. W. Spencer was the first to point out the existence of this great river across the Ontario peninsula, and he clearly proved its course through the Dundas Valley in a cañon much exceeding the dimensions of the modern Niagara, now covered by drift.* This system of drainage doubtless existed through all Mesozoic and Tertiary time. Coeval with the flowage of the Erigan River, was the excavation of the basin of Lake Ontario.

It was stated above that the Lower Silurian rocks, which are calcareous, occupy the region between the escarpment and the crystallines of Ontario. The ordinary subaerial agents of erosion must have been occupied with the removal of these limestones during the whole time from their deposition to the present. Because these rocks were more easily removed than the overlying sandstone, the drainage was made to move easterly from the very first. In a similar manner the whole of the Appalachian Valley from the St. Lawrence to Alabama has been excavated; eight and ten thousand feet of strata having disappeared. Perhaps it may be possible to explain the excavation of Lake Ontario below the sea level by the constant flowage of water containing the limestone in soluton. In this way rock can be excavated below the natural fall from a higher to a lower level, and thus save

* American Philosophical Society Proceedings, Vol. XIX. 1881.

the necessity of a belief in a change of altitude to explain deep seated excavations in soluble materials.

ADVENT OF THE ICE AGE.

Near the close of the Tertiary Age the continent rose thousands of feet, judging from the submerged fiords upon both coasts. It may have been that this elevation was great enough to allow Lake Ontario to flow to the Atlantic. The amount was sufficient to bring on intense cold, so as to cover the land with glaciers. The latest view, hardly as yet elaborated, represents the succession of three or more glacial epochs, the first culminating in a Cordilleran ice cap, with its maximum in British Columbia. Glaciers moved across the Saskatchewan plains and down the upper Missouri Valley. It is not proved that it extended as far as to the great lakes. With a change in the climate, the Cordilleran cap disappeared, followed by a temperate epoch; and then a second ice cap accumulated on the northwest shores of Hudson's Bay. Though upon a low foundation this ice grew to mountainous altitudes, known as the Kewatin mass, which moved southerly into the Mississippi Valley and southeasterly over the great lakes. As southeast courses of striæ are common in the Ontario peninsula, it is probable that this incursion brought material to fill up the Erigan Valley; and thus in the temperate period following, the drainage of that water-course accumulated to give us Lake Erie. As the discharge by the way of Grand River and the Dundas Valley had been impeded by the glacial debris, the water was compelled to seek an outlet through the Niagara depression, and hence the birth of the Niagara River. No one can tell how extensive were the previously existing excavations made by the pre-glacial Tonawanda and Chippewa Rivers. Whatever they were, the new river took possession of the valley and pushed northerly past the whirlpool and the St. David's channel to a confluence with Lake Ontario at its extreme northerly end. This early Niagara carried no less water than it does now—perhaps more—and its sources may have been confined to the Erigan Basin. These were capable of eroding the greater Dundas Valley; and why not the earlier Niagara as well? The gorge, as we lose sight of it at the whirlpool, equals in its dimensions those of any part of the chasm: while the St. David's terminus much exceeds in size the outlet at Lewiston. If the entire Erigan drainage were not availabl, the upper great lakes must have been in existence and flowed through Lake St. Clair, as at present. The third glacial epoch will be mentioned later.

Reality of the St. David's Channel.

The following considerations may aid us in the belief in the reality of the St. David's Channel: (1) The direct route is the natural course of the river, and for 1,000 feet the water still flows as in the former time. (2) The bend of the river at the whirlpool is phenomenal, nothing like it has yet been described among the numerous canyons of the Cordilleras. (3) The character of the material in the gorge is such as implies glacial transportation. At the end of the whirlpool about a hundred feet thickness of sand and clay are well exposed to view, underlying red till filling the whole gorge nearly forty feet higher than the flat surface of the limestone just at the outlet. This, till at the trestle over Bowman's Creek, holds considerably clay over a very stony mass. The creek clings to the west wall, not following the middle of the gorge, as commonly represented upon the Lake Survey maps. About one and a half miles northwesterly wells have been sunk a hundred feet in earth without reaching rock. Half a mile farther, two railroads have cut deeply through sand; and at the edge of the plateau are several cuttings and pits of gravel, certainly fifty feet deep. This modified drift may be a mile wide. It resembles an esker; and must represent the melting of a glacial tongue pushing from the main sheet up a notch in the edge of the plateau. The width of the gap in the rocky escarpment is about one and one-third miles. (4) The west wall of the canyon has been glaciated, for fully a half mile northerly from a point opposite the outlet. The smoothing and striation is upon both the Niagara and Clinton limestones, with a southeast direction, corresponding to the course of the gorge itself, but differing fully seventy degrees from that observed upon the surface of the plateau. The same course has been noted at Thorold, and is quite common near the Dundas Valley; so that it is not necessarily a local deflection produced by the crowding of the ice through the St. David's Channel; it may result from the movement that blocked the Erigan gorge, though it seems better to refer it to the later epoch. (5) The depth of the Niagara River in the whirlpool and just below, in the lower gorge, are best explained by supposing the existence of an earlier St. David's Channel. The water is 150 feet deep in the whirlpool, is very shallow at the outlet, making a shelf there, while the normal great depth is renewed below. Deep excavations are made by a cataract; shallow ones, by rapids. When the river had worn back nearly to the whirlpool, there would be only a narrow rim separating the cataract from the ponded water above. The pressure of the water above would knock the wall down laterally, and thus there would have been no pounding of the lower layers by the cataract. If the wearing action had been effected by falling water continuously around the great bend, the depth would have been uniform, whereas now there is a shallow river. Hence, it is evident that the whirlpool pit existed before the origination of the lower gorge.

Granting the existence of St. David's Channel the following events seem to have been necessitated: (1) A Niagara River, fully equal in dimensions to the present stream, flowed from the whirlpool to the edge of the escarpment near St. David's, and thence to Lake Ontario. A cataract commenced at the cliff and cut its way back the entire distance of three miles. As intimated above, we can understand that sufficient water came down the Erigan river to accomplish this result; and, therefore, it is not necessary to say that the upper great lakes contributed their share, as now, to swell the stream; yet they may have done so. (2) Frigid conditions brought on a third ice sheet [Labrador] which pressed against the edge of the bluff, broadened the mouth of the gorge, especially eroding the western wall because of the movement in a southwesterly direction, and filled the channel with till and modified drift, to the height of 300 feet. (3) The return of temperate conditions melted the ice and restored Niagara River, but it had not sufficient volume to overflow the barrier of till. Hence, it turned its direction to the northeast, following a slight depression in the plateau to the edge of the escarpment at Lewiston. After the stream had cut its way back to what remained of the original channel, the whirlpool was set in motion.

Erosion of the Lower Gorge.

There must have been some erosion in pre-glacial times along the lower part of this section, as is evidenced by the tributary canyons at the Devil's Hole and Smeaton Creek. What it amounted to in the main stream, no one can tell. The top of the cliff at Lewiston is about 575 feet, and the fall 325 feet. It would not have been well defined at first, because of the greater altitude of Lake Ontario and the character of the rock. The fall being over the Niagara limestone and shale directly into the level of the lake, would present a cataract 100 feet high, and could not act energetically upon the Clinton limestone beneath. Hence this becomes an upper fall, receding more rapidly than a second one, which would make its appearance as soon as the lake disappeared. For similar reasons this would wear away only the Clinton limestone and shale of 30 feet thickness. Thus there would be two cataracts, separated by a considerable interval. There remains the Medina sandstones and shales, which would have their own cataract, and so Niagara commenced the excavation of the lower gorge by a triple division of its energy.

This view is confirmed by a study of the Falls of the Genessee, 70 miles further east, at Rochester. The lower fall there, is over the Medina terrane. One-fourth of a mile higher, is the second fall of 25 feet over the Clinton limestone. A mile and a half above is the greater fall of 110 feet over the Niagara limestone and shale, precisely as at Niagara. The strata dip to the south in a similar way, but the limestone is only 70 feet thick, as opposed to 140 feet farther west. The escarpment is less pronounced at Rochester.

Two very important conclusions follow: First, the upper falls, receding more rapidly at first and more slowly afterwards, must have reached the island at Wintergreen flat before the lower cataracts caught up with it. Second, the falls being worn out individually, the depth of the excavations must have been small. The profound depths of the upper great gorge are wholly

THE FOOTWAY. GOAT ISLAND.

M oll

wanting below the island. The width of the lower gorge is nearly as great as that above the railroad bridges, but the water is not very deep. Even if the whole volume of the present river was concerned in the erosion, the depth would be small.

Of the 330 feet exposed in the cliff at the mouth of the gorge, about 200 belong to the Medina terrane, and must have been the material over which the lower fall poured. This dips southerly, so that only 40 feet of it rise above the river at the whirlpool. While this fall receded slowly at first, its pace would be greatly quickened in passing southerly. The second fall must have receded uniformly, because the thickness of the Clinton is the same throughout. The upper limestone is twice as thick as the whirlpool as at Lewiston. So the upper fall could not recede so fast as at first, and the lower cataract gained constantly upon it.

Looking at the amount of water in the early Ontario, the Niagara must have fallen into it at a level 135 higher than it is now. This would have brought the level of the water to the top of the lower grey sandstone, or that just at the river's surface at the whirlpool. The lower fall, then, did not start so early as the others. But in a later epoch the lake receded so that its surface was over 80 feet lower than now. With the greater depth of fall the erosion of the lower rock would have proceeded more energetically. Perhaps the average recession has been the same as if there had been no variation in its rate. We do not consider in this sketch the excavation of the softer parts of the Medina terrane below Lewiston. Its presence neither accelerated nor retarded the excavation of the greater gorge.

A mile below the whirlpool there seems to have been an island in the midst of the channel analogous to Goat Island between the American and Canadian falls. Foster's flat shows a surface that was swept by the river before the excavation of the gorge. About fifteen feet below the level of the electric road at Wintergreen flat, is a rock terrace, 750 feet long, and 600 feet broad, which is believed to have been fashioned by the western part of the early Niagara; and at its northern end is a precipice over which the water fell into Foster's glen. The sloping area between the flat and the river, 2,500 feet long, and 750 feet wide, is covered by immense blocks of Niagara limestone. The explanation given by Dr. Gilbert* is essentially as follows. The river wore its way back to the lower limit of Foster's glen in a single broad cataract. Here, it was divided by a narrow strip analogous to Goat Island at the existing falls; the main channel being upon the American, instead of the Canadian side. The falls receded faster upon the eastern side; so that when the island had been passed, all the stream was captured by the American branch, and the Canadian fall ceased to exist. In this erosion the river wore out a peculiar notch pointing northeasterly to the Wintergreen platform. The island disappeared later because of the undermining of the shale and the consequent fall of the limestone. The published contours show the northeast extension of the rock below Wintergreen flat. The cataract did not excavate anything below the Clinton limestone upon this island. These and other details in the gorge are exhibited in Plate II.

History of the Great Lakes.

In continuing our study of the cañon, the next section to be described is that of the whirlpool rapids, where the gorge is very narrow. To appreciate the views entertained respecting the condition of Niagara River when this section was excavated, it will

* "Niagara Falls and Their History," monograph of National Geographic Society, 1895.

be necessary to take into account the variations in the size and drainage of the upper great lakes developed during the melting of the Labrador ice-sheet. At present these bodies of water discharge through Niagara. In the past, this has not always been the case; and whenever the drainage of any of these lakes has been diverted elsewhere, the volume of the Niagara has been just so much diminished. First, we endeavor to determine the position and succession of these lakes; secondly, to correlate these bodies of water with the making of the different sections of the gorge. Necessarily these conclusions at present must be regarded as only tentative.

It is premised that the whole of the region of the great lakes, from Ontario to Superior, was covered by the Labrador ice-sheet, moving southwesterly and receding northeasterly. The greatest movement would have been in the valleys, as that was in the region of least resistance. The valleys thus occupied were those of Lake Superior at Duluth and south of Keweenaw point; Lake Michigan at Green Bay and Chicago; Lake Huron near Saginaw, Lake St. Clair and Georgian Bay. The ice thawed at the ends of these glacial lobes; and as they receded considerabe water accumulated above the level of the several outlets, or what are now watersheds. Hence, quite early in the recession there were simultaneous lakes at Duluth, the water draining into the St. Croix River 468 feet above Lake Superior; at Chicago, only 58 feet higher than Lake Michigan, flowing into the Des Plaines River, and at Fort Wayne, Indiana, 225 feet higher than Lake Erie, emptying into the Wabash River. Each of these small lakes increased in size as fast as the ice to the northeast melted; until finally, without mentioning details, the waters of Lakes Superior, Michigan, Huron, and Erie were combined into one body, discharging at Chicago, because this was the lowest point in the whole shore line, the altitude being about 600 feet. This has received the name of Warren Lake.

Later in the history the outlet shifted to the east, and while the St. Lawrence was still blockaded the discharge was through the Mohawk River into the Hudson and thence to the sea. This has been called Algonquin Lake, continuous from Superior, across Huron and Erie to Ontario. When the water had fallen to 144 feet above Erie, a strait connected the two lower lakes. A beach and spit, formed at this time at Lundy's Lane, near Niagara, has been described by Spencer, who notes it also at Fort Hill, Ontario, and Ebenezer, New York. No epoch of prolonged rest seems to have been reached in the further process of shallowing, till Erie was ponded separately from Ontario by the Johnson Ridge (so named by Spencer), the highest divide crossing the Niagara gorge near Inspiration Point. This would represent the beginning of the present Niagara, discharging the waters of Erie into the Ontario basin, with an outlet still through the Mohawk. For a time the water rested at a level 300 feet below the Lundy beach, and 220 feet above the present Ontario, and has been named Lake Iroquois. It received the outflow from Lake Algonquin through the Georgian Bay of Huron, into Lake Simcoe and Trent River. The St. Clair outlet was probably closed while the Trent River connected Huron with Iroquois.

The ice front receded next so as to change the outlet of the upper lakes from the Trent to the Ottawa River, via Lake Nipissing, originating the sprawling Nipissing great lake, entirely disconnected with Lake Erie, which now had an independent existence with a Niagara much reduced in size. This epoch is regarded by F. B. Taylor as coeval with the highest extension of the sea to Lake Ontario in the Champlain time. Before this, the ice barrier to the east had disappeared and the St. Lawrence begun its independent existence, relieving the Mohawk.

The study of the deserted strands of these several lakes has led to a belief in their deformation. A breach encircling a lake, should certainly be formed at the same level; when carefully followed from point to point, it has been discovered to be unequally elevated above the water. The rise of the land has been invariably to the northeast; and the decline to the southwest. When the land to the northeast rises, the water will move to the southwest, and the old beach be elevated at one end and depressed at the other. This result may be illustrated by tipping a bowl containing water. The liquid will rise towards the edge that has been depressed and fall away from the raised part. If a lake had its outlet on the southwest side, the canting opposite may have been great enough to cause it to discharge in the opposite direction. Such a deformation may have been the cause of the shifting of the outlet of Lake Warren from Chicago to the Mohawk; if the canting should be in the opposite direction, the place of discharge might also be changed. Dr. Gilbert presents considerations to show that the northeast uplifting of the rims of the lake barriers is still in progress. Should this continue a few thousand years from now henceforth, Niagara would cease to run, the wharves of Buffalo would need extension, and the outlet for all the great lakes, except Ontario, would be at Chicago. Lake Erie is now only nine feet lower than Lake Michigan; so that but little change of level would be required to effect this change.

A brief statement of the altitudes of the Iroquois beach around Lake Ontario may render the preceding paragraph better understood. This beach is now much lower at the west end. Prof. Spencer finds it to stand at 362 feet above the sea at Hamilton, Ontario, and to rise to 632 feet at Trenton, Ontario, at the east end of Ontario. Dr. Gilbert finds this beach to be 370 feet at St. David's, Ontario, and traces it to 657 feet at Adams, New York. The rise of this beach near the west end of Ontario is 1.60 foot per mile. Near Oneida Lake the uplift is 3.5 feet per mile, and near Watertown, New York, it is 5 feet per mile. The total rise from Hamilton to Trenton is 270 feet; from Hamilton to Adams, 295 feet. Other beaches have been found beneath the Iroquois, which pass under the lake in going west. Thus near Watertown, there is one 200 feet lower; at Oswego, one 185 feet lower than the Iroquois beach. Continuing this lower beach westward, it should be found 78 feet below the surface of Ontario, off the mouth of the Niagara River, which stream would have extended four miles further north, cutting a channel through shales. The existence of the conditions requisite for the formation of this beach affords us the evidence of the former greater fall of Niagara over the Lewiston escarpment, mentioned above.

It is interesting to note the most western limit of the sea in the Ontario basin. Ramsay mentions the occurrence of marine shells at Kingston, Ontario; and the cetacean cited by J. W. Dawson at Smith's Falls cave from an altitude of 440 feet. Taylor states that the top of this deposit is about thirty feet higher, which would be 223 feet above Ontario. The beach, 200 feet below the Iroquois level at Watertown, must be 210 feet above Ontario. Is it possible that this is of marine origin?

The near approach of the levels to each other proves the former connection of the ocean with Lake Ontario in the Champlain period. It extended likewise above Pembroke, in the valley of the Ottawa. This marine shore-line is traced down the St. Lawrence to the gulf, culminating near the Saguenay, at about 800 feet above the sea; so that the differential elevation indicated by the Lake Iroquois beds was continued through most of Lower Canada.

From the statements now made of the history of the great lakes, there seems to have been two former possible outlets for the waters above St. Clair—by the Trent river and Lake Nipissing; and it has been suggested by Dr. Gilbert that, when the discharge was by the former route, the Niagara river cut its way through the shoals at Wintergreen flat; and that when the water flowed through the Nipissing channel, the gorge of the whirlpool rapids was excavated. Both those routes are shown distinctly upon the map, Plate I. It is not easy to show just what part of the lower gorge was excavated when the upper lakes discharged through the Trent outlet. At the present writing, it may be best to say that, because of the limited time during which this discharge was effective, no marked effect was produced.

It is different with the Nipissing outlet. Mr. Taylor has shown that this discharge must have operated for a very long period. The Nipissing beach is recognized upon all the upper lakes, being the best defined of any of the ancient strands. At the outlet at North Bay it is 120 feet above Lake Huron; 45 feet in Mackinaw Straits; 50 feet above Lake Superior at Sault Ste. Marie; and passes beneath the lake level 40 feet at Port Huron, 100 feet at Chicago, and 25 feet at Duluth. Fully eight-ninths of the water passed to the Ottawa River, leaving only one-ninth of the present volume for the Niagara. Consequently the gorge made by the Erie-Niagara must have been comparatively small. Its width at the top is 750 feet, and the depth of the river estimated at 35 feet. The descent is 45 feet, and the length three-fourths of a mile.

There is quite a contrast between these and the rapids above the cataract, which descend the same amount and have a similar length. The action is deliberate in the one, and impetuous in the other. The shallowness may result because of the want of erosive power in the smaller volume of water falling over the cliff, or may be due in part to the superior hardness of the rock at the base of the falls, as suggested by Professor James Hall.

Some authors, following Pohlman, believe this part of the gorge was made by a smaller stream in pre-glacial times, perhaps in connection with the St. David's Channel, and before the modern Niagara came into being. This seems inadmissible, because the St. David's Channel evidently required a river of large volume for its excavation, greater than the Erie-Niagara, and the whirlpool rapids' gorge had a later origin. These facts also enforce the conclusion that it is not likely that in any case a small gorge would have been enlarged to the full width by the later greater volume of water. Were this true, this narrow section should show evidences of increasing enlargement during the latest episode of excavation. All agree that the river which excavated this section was of comparatively small size, and, consequently, that the duration of this episode must have been correspondingly great.

There is a shoal ledge between the whirlpool and Eddy basin not unlike the one between the whirlpool and cove, which was accounted for by supposing the presence of the water removed a rim of rock rapidly, without the aid of the cataract. Mr. Taylor suggests a species of weathering for its presence here. It is presumed that the St. David's Niagara eroded the gorge back to this ledge, when the work suddenly stopped. If the cliff were exposed to the elements for a time, the falling of talus blocks would ensue, with some weathering. Or the ice known to have moved through the buried gorge may have impinged against this ledge and fractured it. It must be remembered that, according to our views, this ledge was exposed for a very long time—while the whole of the gorge between the whirlpool and Lewiston was being excavated; so that the cliff could hardly fail to have been operated upon in some way. When the cataract commenced above the whirlpool, it would be occupied first with removing débris, and not in channeling out the bottom. These suggestions may show why that ledge remained shoal.

The Upper Section.

Between the falls and the railroad bridges, a distance of two and one-fourth miles, the interpretation of the history is easy. The gorge is uniformly of great width and depth, 1,300 feet in width and with the ordinary depth of 160 feet for the river. At the very cataract the bottom of the river is 100 feet lower than the level of Lake Ontario. The water moves slowly, falling about seven feet to the mile. The rocks vary but little in this section. All agree that the conditions determining the formation of this gorge are practically the same with those now prevailing; and, consequently, the only change in the scenery has been the position of the cataract. Having the same volume as now, the levels of all the upper lakes have not varied appreciably. Without doubt, during this episode, the falls have been viewed by the aborigines of our continent as they carried their canoes and cargoes from lake to lake.

The increase in the volume of water is supposed to have been occasioned by the rising of the land at North Bay; damming the water and compelling it to wear its channel through Lake St. Clair to Lake Erie. When the rush began, it bared many acres of rock on both sides of the river, leaving various denuded tracts, as the flat two miles in length, south of the steel arch bridge, on the Canadian side, and a very narrow strip on the American side, reaching as low as to the whirlpool, give evidence. These flats are like the bed of Niagara above the rapids, if the water were removed and vegetation had gained a footing. Because of the presence of these old river beds where the Johnson ridge crosses the gorge, it seems to us that it furnishes no obstacle to the free passage of the water. This ridge is an anticlinal swelling in the strata, but had been broken down by early erosion, so that really no greater mass of limestone has been excavated here than elsewhere in this section.

Goat Island rose higher than the water on either side, and, therefore, dividing the stream into unequal portions. Its former extension northerly may be indicated by the promontory directly opposite the north end of the American fall, utilized as a pier for the landing of the Maid of the Mist, for there is a rescission in the cliff just behind. It would seem that the former northerly extent of the island crowded the Canadian fall so that it was compelled to excavate a place for itself, and cut into the wall back of the promontory.

The occurrence of fluviatile shells in various places along the banks illustrates another phase in this history: the presence of water, sometimes thirty feet above the present level. The highest locality observed is at the Bowen place, three miles above the falls. Next, they are well shown in the recent excavations for the new wheel-pit of the Cataract Construction Company. The opening shows two feet of clay and sand at the surface, overlying a thicker sandy mass containing *Cyclas*, *Goniobasis*, *Unio* and *Planorbis* in considerable abundance. At the base of the earthy material is a mass of till ten feet thick, resting upon the glaciated surface of the Niagara limestone. The locality upon Goat Island has been studied by Lyell and Hall. In a gravel pit now being excavated these same shells occur plentifully; some of the *Unios* and *Cyclas* still having their valves closed. This deposit is fully twenty-five feet thick, mostly of course gravel, readily correlated with corresponding beds in the villages upon both sides of the river. At Clifton sixteen species have been found, adding to the above the genera *Physa*, *Limnæa*, *Paludina*, *Amnicola*, *Margaritana* and *Pisidium*. Other localities are just below the railroad bridges on the west side, and on both sides of the whirlpool. Their surface altitudes vary from 566 to 575 feet above the sea. Professor Hall says that the pebbles upon Goat Island have been transported northerly, as they contain fragments of the Black Rock limestone at Buffalo. He thinks the action was fluviatile, the first condition having been that of a quiet lake, followed by quite a strong current. Hence, the correlation is with the lake covering this region after the deposition of the Lundy beach.

Recession of the Cataract.

Everyone has heard how the cataract recedes and the gorge elongates. The water pours over a cliff of hard limestone, from 140 to 170 feet deep, and pounds upon the soft rocks beneath.

As these give way, the higher, harder strata project like cornices, and when they have assumed considerable prominence, being unsupported they fall down. As seen from Horseshoe Fall, erosion is more effective in the center of the current. The stream rushes more and more furiously into the middle, and tends to abstract some water from the sides—perhaps leaving large surfaces of rock bare. The undermining of the cliffs cause these "table-rocks" to fall, eventually, and thus the width of the gorge is kept essentially uniform; and in the long run, the breadth keeps pace with the extreme points or erosion.

Only a few years' interval is required for noting these changes. Within the 30 years of my own observation, the middle of the Horseshoe Fall has become more pointed, and the amount of water at the margins runs more feebly, or else has developed bare surfaces. In illustration of the increase of the width of the gorge, I may quote from a telegram sent to the Press 25 years since, which may be duplicated every 25 years:

GREAT FALL OF ROCK AT NIAGARA FALLS.—Buffalo, September 5th.—Several hundred tons of rock, which projected from the side of the bank in Prospect Park, beneath the platform which overlooks the Maid of the Midst landing, and about midway from the top of the bank, fell this morning. No damage was done. The immense mass of débris fell with a crash which was heard in the village. It does not in any way affect the top of the bank in this vicinity.

The piles of talus upon both flanks of the gorge throughout its seven miles of extent, testify to the wide extent and constancy of this movement. The fragments at the base of the cataract will be broken quite rapidly, as nothing can resist the force of the fall. The depth of the water here is greater below the level of the stream than the height of the rock above. The limestone is from 60 to 80 feet thick; the shale, 82; the Clinton limestone,

30 feet. Beneath the water level there may be 200 feet of the Medina terrane.

By comparing the best views of the falls taken at different dates, it is very easy to note the fact of recession. Precision has been given to the rate by careful measurements, with the most delicate instruments, of the positions of every part of the cataracts at stated intervals. It is usual to compare the positions determined trigonometrically by Prof. Hall in 1842, and in 1890 by Engineer A. S. Kibbe. During this period of 48 years, about 220 feet in length of the limestone has been removed, or about 4½ feet each year. This means that the gorge is lengthened by this amount each year. Although the uppermost part of the fall is narrow, the gorge will broaden as fast as the apex recedes, as explained above; and therefore it is reasonable to use the rate of recession of the fall as equivalent to the average of elongation of the gorge. Here and there the gorge may widen or contract, according to variations in the volume of the water, its velocity, the width of the channels, the thickness of the limestones and other rocks, their comparative hardness, etc. It is not our purpose to present allowances for these varied conditions, as the approximate estimate derived from the rate of recession will indicate how the calculations should be made.

Estimates of Age in Years.

Let us assume the present rate of recession to be four feet annually. This is less than what has been observed, but some think that the time for the wearing action must have been greater, and the omission of the half foot may be nearer the true figure. There is a greater thickness of limestone at the Johnson ridge than elsewhere, in the first section of the gorge, from the falls to near the railroad bridges, and possibly more time would be required for its erosion. But it seems as if this excess of thickness was removed in pre-glacial times, because the flat produced by the ancient bed of the river, before the erosion of the rock had been effected, slopes uniformly, without regard to the presence of the ridge. Hence this excess of rock has not materially affected the rate of erosion here. The total length of this first section of the gorge is 11,750 feet. This measurement was made in 1875, and the gorge must now be over 100 feet longer, making the entire length 11,850 feet. Divide this by four, the number of feet the cataract recedes annually, and the quotient is 2,962; *i. e.* the beginning of the great cataract dates back to 1062 B. C., 300 years before the time of Romulus, or to the reign of King David at Jerusalem.

The next section is that of the whirlpool rapids, 3,900 feet. If the water of Niagara came from Lake Erie alone, under existing conditions, its volume would have been three-fourteenths of the present flow. This is less than the volume of the water pouring over the American fall, which recedes 0.64 of a foot annually. This stream may be conceived to wear away the rock at the rate of six inches annually. If so, it would have required 7,800 years for the formation of the gorge of the whirlpool rapids.

The Eddy basin seems to have been worn by the normal Niagara. With the same rate as has been given for the first section, only 375 years would be required to erode this distance of 1,500 feet. The whirlpool belongs to the St. David's epoch, and will be considered in another connection.

The Cove section, or the part from the whirlpool to the narrows at Foster's Island, is 3,750 feet long. No one imagines the rate of the recession for this section to be different from that first named, as the width of the gorge and the depth of the water are the same. This would have required 937 years for its excavation.

Professor Spencer has calculated the age of the lower section of the gorge from Lewiston to Foster's Island to be 17,200 years, upon the assumption that the conditions would be those of the American falls. But measurements of the breadth below the Cove section give an average of width of the top of the gorge, fully equal to that of the upper gorge as far as to the Catholic College, 5,754 feet, while the other part, 8,448 feet, may for convenience be comparable to the American falls. Calculations upon these bases give 1,438 years for the upper part, and 5,406 years for the lower, a total of 6,844 years.

The section from the beginning of the whirlpool to the edge of the escarpment, near St. David's, is about three miles in length. Its erosion was effected before the last ice-epoch, according to our interpretation, and some might say pre-glacial. It would appear that the volume of the water passing through was fully equal to that of the present epoch, and may have been derived from the same sources. At the rate of recession adopted for the greater gorge, the time must have been at least 3,960 years. This does not account for the greater width of the mouth of the gorge, which may have been either waterworn, or eroded by the ice-sheet.

The sum of the figures given above amounts to 18,918 years for the present gorge, to which might be added the figure for the St. David's channel—making in all 22,878 years.

Professor Spencer's estimate of the age, including the whirlpool, but not the St. David's channel, sums up 31,000 years, to which he adds a thousand years for the duration of the river before the advent of the falls. Sir Charles Lyell gave the figure of 36,000; Professor Warren Upham, 10,000, and F. B. Taylor would estimate it to lie between 36,000 and 50,000.

Other estimates of the age of parts of the gorge are based upon the alleged time required for the warping of the basins. Involving the correctness of suggested correlation, they cannot have the value of the direct estimates, but are useful by way of comparison. That presented by Dr. Gilbert of the age of the Nipissing outlet well illustrates the method employed. First of all came the determination of the fact of tilting, and its rate. Three pairs of stations were selected, one between Milwaukee, Wis., and Port Austin, Mich., connecting Lakes Michigan and Huron; another between Cleveland, Ohio, and Port Colborne on Lake Erie, and the third between Charlotte and Sacket Harbor, N. Y., on Lake Ontario. Bench marks were established, and accurate observations recorded of the height attained by the water each day for three months, in the summer of 1896. Care was taken to eliminate from the observed figures the levels affected by tides, seiches, storms, and the seasons. At some of the stations, observations had been made for many years, which were used for comparison. The general conclusion reached, was that there is a rise in the land from southwest to northeast, at the rate of 0.42 feet per century. Assuming this rate to have been uniform and secular, Dr. Gilbert estimates that the time when the drainage was shifted from North Bay to Port Huron must have been about 10,000 years. This he would make to correspond with the date of the position of the cataract at the head of the whirlpool rapids. Professor Spencer's figure, for the same date, is 7,800 as deduced from differential elevations. For the total length of time since the Algonquin Lake existed, which corresponds to the age of Niagara, Professor Spencer obtains the figure of 28,000 years.

PRINCIPAL EVENTS IN THE NIAGARA DISTRICT.

It may be useful to summarize the various episodes in the geological history of the region of Niagara, as now delineated. In the not distant future, geologists may be able to give them value in centuries or years.

1. Laying down of the crystalline Archean floor, upon which the Cambrian and Lower Silurian terranes were deposited in order, before the dry land reached as far south as Lake Ontario.

2. Deposition of the Upper Silurian members of the series: (a) Medina clays and sands resting upon the raised Lower Silurian land, and extending from Lake Ontario to some distance south of the falls. (b) Clinton shale and limestone. (c) Niagara shale and limestone, both formed below the ocean's surface; coral reefs extensive. (d) Onondaga, or Salina, salt group.

3. Land gradually rising and attaining a considerable altitude at the close of the Paleozoic era. Subaerial denudation continued through later periods, during which time the basin of Ontario takes shape.

4. Age of Erigan River. Drainage of country from the Appalachian coal fields northerly, discharging through the Dundas Valley into Lake Ontario.

5. First ice age, that of the Cordilleras. Not certain that its influence was felt in the Niagara region.

6. Second ice age, the Keewatin. Blocking up of Erigan River. Ice moves southerly; its melting gives birth to Lake Erie.

7. Temperate epoch. Wearing of the gorge through the St. David's Channel, including the whirlpool, 3,960 years.

8. Accumulation of ice upon the Labrador highlands and its movement southwesterly, as far south as to the Ohio River, filling the St. David's Channel. Terminal moraines left as it recedes; one near Buffalo, and the next from Lockport across to Toronto.

9. Warren Lake, discharging at Chicago, occupying parts of Lakes Michigan, Huron and Erie.

10. Algonquin Lake was composed of Lakes Superior, Michigan and Huron, discharging into the Ontario basin through Lake Erie, or by way of Balsam Lake and Trent River, and thence through the Mohawk into Hudson River. Erie was tributary. The waters sank to the level of Lundy Lake, after which the Niagara River began its flow; accumulation of deposits with fluviatile shells.

11. Iroquois Lake, as expanse of Lake Ontario succeeded, discharging through the Mohawk. Beginning of the erosion of the lower gorge. River divided by the island below Wintergreen flat. Time of erosion 6,844 years.

12. Erosion of the Cove section requiring a river as large as the present Niagara. Breaking of the barrier at outlet of whirlpool. Time, 937 years.

13. Period of erosion of the gorge of the whirlpool rapids, when the water of Lake Erie alone was concerned. Outlet of upper lakes perhaps through Lake Nipissing. Time, 7,800 years.

14. Restoration of full volume of Niagara River. Upper great gorge excavated from the railroad bridges to the existing cataract. Time, 2,962 years.

Time since the water fell over the Niagara escarpment, 18,918 years. Between this epoch and the shifting of the

Labrador ice-sheet from the Niagara River valley there intervened Lakes Warren and Algonquin (events 9 and 10), probably not of long duration. Nothing about Niagara gives us any clue to the duration of the Labrador ice-sheet. The time of the wearing of the whirlpool basin, 475 years, is not included in the estimate of the age of the general gorge.

EIGHTEENTH ANNUAL REPORT

OF THE

From

COMMISSIONERS

OF THE

State Reservation at Niagara

FROM OCTOBER 1, 1900
TO SEPTEMBER 30, 1901

With appendix: Guide to the geology of the falls etc.
(Bulletin of N. Y. State museum, no. 45)

TRANSMITTED TO THE LEGISLATURE FEBRUARY 24, 1902

ALBANY
J. B. LYON COMPANY, STATE PRINTERS
1902

President McKinley at Niagara September 6, 1901.
(Frontispiece)

[illegible] Party at Niagara July 27, 1901.

EIGHTEENTH ANNUAL REPORT

OF THE

COMMISSIONERS

OF THE

State Reservation at Niagara

FROM OCTOBER 1, 1900,
TO SEPTEMBER 30, 1901.

TRANSMITTED TO THE LEGISLATURE FEBRUARY 24, 1902.

ALBANY
J. B. LYON COMPANY, STATE PRINTERS
1902

[redacted]

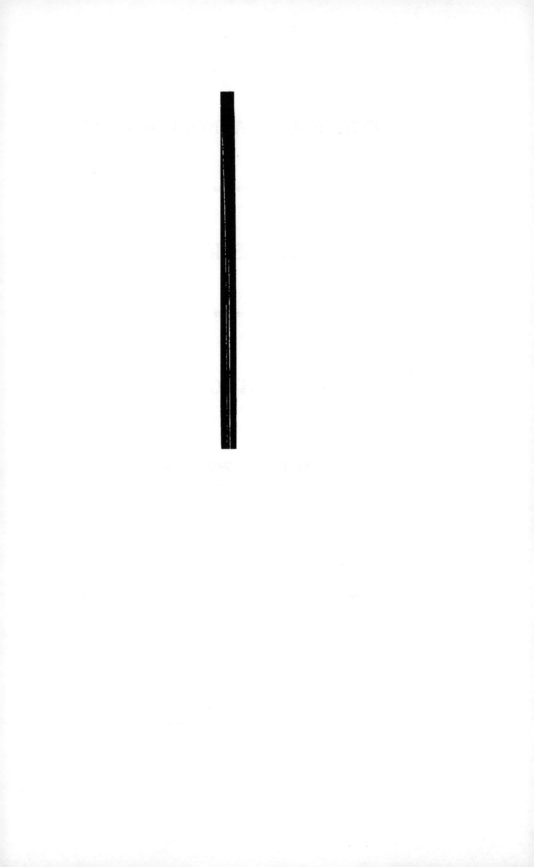

STATE OF NEW YORK.

No. 45.

IN ASSEMBLY,

FEBRUARY 24, 1902.

EIGHTEENTH ANNUAL REPORT

OF THE

Commissioners of the State Reservation at Niagara.

To the Honorable the Speaker of the Assembly:

Sir.—I herewith transmit for presentation to the Legislature the Eighteenth Annual Report of the Commissioners of the State Reservation at Niagara for the fiscal year ended September 30, 1901.

Respectfully yours,

ANDREW H. GREEN,

President.

OF THE

COMMISSIONERS

OF THE

State Reservation at Niagara.

FOR THE FISCAL YEAR FROM

OCTOCER 1, 1900, TO SEPTEMBER 30, 1901.

Commissioners.

ANDREW H. GREEN, President.

GEORGE RAINES, CHARLES M. DOW,
THOMAS P. KINGSFORD, ALEXANDER J. PORTER.

Treasurer and Secretary.

PETER A. PORTER, JR.

Superintendent.

THOMAS V. WELCH.

REPORT.

To the Honorable the Legislature of the State of New York:

The Commissioners of the State Reservation at Niagara, as required by law, submit their report for the fiscal year begun October 1, 1900, and ended September 30, 1901, this being their eighteenth annual report.

The two stone arched bridges connecting the mainland with Goat Island, which were being built at the date of the last report, are now in use by the public. The approaches have been filled, graded and planted. The building of these permanent stone bridges was the most important work of the year. It is also the greatest work of construction accomplished since the establishment of the State Reservation at Niagara.

The new Shelter building was completed July 4, 1901, and has been partly furnished. It affords many conveniences and accommodations for the public, as well as offices for the administration of the affairs of the Reservation.

Photographs of the stone arch bridges and of the Shelter building are herewith submitted.

A new bridge from Goat Island to Terrapin Point at the Horseshoeshoe Falls has been constructed; new cables, anchors, needle beams and braces have been attached to the suspension bridges to the Three Sister Islands, and the woodwork of the bridges renovated.

The Cave of the Winds building on Goat Island has been moved to a more suitable location. The frame structure in Prospect Park, adjacent to the new Shelter building, has been taken down.

On September 27th the new bridges from the mainland to Goat Island, the new Shelter building and other improvements were inspected by Governor Odell, Senator Ellsworth, President pro tem. of the Senate, Senator Higgins, Chairman of the Finance Committee of the Senate, Assemblyman Allds, Chairman of the Ways and Means Committee of the Assembly, and other State officers.

The improvements and needs of the Reservation were explained in detail to the Governor and his party, who expressed deep interest in the work of the Commission.

On September 6th the Reservation was visited by President McKinley and party three hours before the assassination of the President at the Pan-American Exposition at Buffalo.

It is estimated that the Pan-American Exposition at Buffalo brought to the State Reservation three millions of visitors. No disorder or accident occurred within the State Reservation.

The following is an estimate and detailed statement of the work necessary to be done, and the expenses of maintaining the Reservation for the fiscal year ending September 30, 1903:

Statement of Needed Improvements.

For completing stone arch bridges	$4,454 14
For safety brakes for Inclined Railway	1,000 00
For alteration of Inclined Railway building	910 13

These improvements may be made by the reappropriation of $6,364.27 of the unexpended balance of the $20,000 appropriated for extraordinary expenditures for care and maintenance during the year 1901, by reason of the Pan-American Exposition. If this reappropriation is made, no new appropriation will be required for improvements during the coming year. Owing to the work upon the stone arched bridges and the great numbers of visitors from the Pan-American Exposition, the sum of $6,500 appropriated last year for the improvement of the water supply has not yet been expended.

Plans and estimates have been obtained, and the work of installing a system of water mains, as far as the appropriation allows, will be done the coming spring.

One thousand dollars appropriated last year for grading and tree planting has not yet been expended.

A plan and estimate for a lighting system has been obtained. The estimated cost is over $18,000, which is more than twice the amount of the appropriation of $7,000 made last year. As soon as the weather will permit in the spring, the work will be done as far as the appropriation of $7,000 will allow. When the current of the Niagara Falls Power Company is brought to the Reservation for use without charge, as provided by law, the question of operating the Inclined Railway by electricity will be submitted for consideration.

Ordinary Maintenance.

The appropriation made for ordinary maintenance during the current year is $25,000. The estimate of receipts for the past fiscal year, made in the last report, was $21,800. The actual receipts by the Superintendent, paid into the State treasury, were $21,710.40 for the fiscal year, and for the calendar year 1901 $22,635.10, or within $2,364.90 of the amount appropriated for ordinary maintenance of the Reservation. The large increase in the amount of receipts was due to the crowds attracted to the Reservation by the Pan-American Exposition at Buffalo.

The following is our estimate of the amounts required for ordinary maintenance for the fiscal year ending September 30, 1903:

For salaries and traveling expenses	$5,600 00
For Reservation police and caretakers	8,100 00
For labor	13,800 00
For materials, tools, etc	7,500 00
	$35,000 00

The appropriation of $25,000 made each year for ordinary maintenance has been found to be insufficient, and each year we have been obliged to dispense with necessary labor toward the close of the fiscal year, when greatly needed to keep the grounds in order. The amount will be more inadequate in the future, because the care of the new shelter building requires additional labor. The new stone arch bridges to the islands are not provided with gates, thus requiring the service of watchmen day and night, and the increasing amount of territory under cultivation requires an additional amount for care and mainetenance. For these reasons $35,000 is asked for care and maintenance, instead of the $25,000 usually appropriated.

Estimated Receipts.

The following is our estimate of the receipts for the fiscal year ending September 30, 1902:

Inclined Railway	$8,000 00
Cave of the Winds	1,200 00
Steamboat landing	500 00
Carriage service	100 00
	$9,800 00

The Superintendent in his report has made a detailed statement of the work of the Commission for the past year. The report of the Superintendent and that of the Treasurer and Secretary are hereto appended, also the contracts and specifications for the stone arch bridges, and three chapters of Bulletin of the New York State Museum, Frederick J. H. Merrill, Director, No. 45, vol. 9, April, 1901, Guide to the Geology and Paleontology of Niagara Falls and Vicinity, by Amadeus W. Grabau S. D.

Respectfully submitted,

ANDREW H. GREEN,

President.

GEO. RAINES,

THOMAS P. KINGSFORD,

CHAS. M. DOW,

ALEXANDER J. PORTER,

Commissioners of the State Reservation at Niagara.

Report of the Treasurer

For the Year Beginning October 1, 1900, and Ending September 30, 1901.

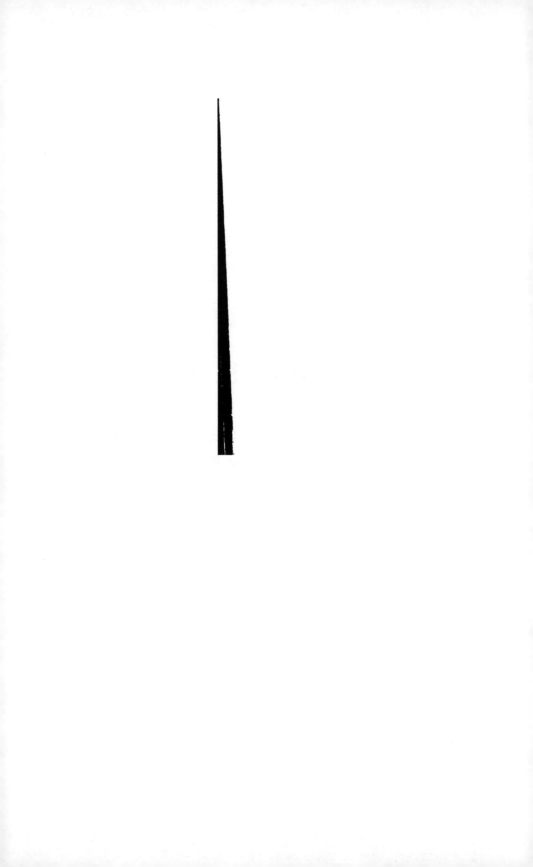

Report of the Treasurer.

THE COMMISSIONERS OF THE STATE RESERVATION AT NIAGARA, In Account with PETER A. PORTER, JR., Treasurer, for the Fiscal Year Begun October 1, 1900, and Ended September 30, 1901.

1900.				
Oct.	1.	Balance on hand this date		$37 82
		MAINTENANCE RECEIPTS.		
Oct.	18.	Quarterly advance from State Comptroller	$6,250 00	
1901.				
Jan.	24.	Quarterly advance from State Comptroller	6,250 00	
April	23.	Quarterly advance from State Comptroller	6,250 00	
July	17.	Quarterly advance from State Comptroller	6,250 00	
				25,000 00
		Special appropriation as per Chap. 569, *Laws of* 1899 ($30,000):		
1900.				
Oct.	31.	Payment by State Comptroller on account	$4,910 50	
Dec.	14.	Payment by State Comptroller on account	4,145 45	
	31.	Payment by State Comptroller on account	5 25	

1901.			
Jan. 21.	Payment by State Comptroller on account	$765 00	
Feb. 7.	Payment by State Comptroller on account	3,504 55	
April 19.	Payment by State Comptroller on account	2,569 42	
May 22.	Payment by State Comptroller on account	1,071 87	
Sept. 11.	Payment by State Comptroller on account	8,151 87	
			$25,123 91

Special appropriation as per Chap. 420, *Laws of* 1900 ($120,000):

1900.		
Oct. 11.	Payment by State Comptroller on account	$264 75
22.	Payment by State Comptroller on account	8,953 78
31.	Payment by State Comptroller on account	162 82
Nov. 12.	Payment by State Comptroller on account	252 00
21.	Payment by State Comptroller on account	18,224 00
26.	Payment by State Comptroller on account	1,583 23
Dec. 17.	Payment by State Comptroller on account	1,054 99

1900.			
Dec.	24.	Payment by State Comptroller on account	$7,803 00
	31.	Payment by State Comptroller on account	302 60
1901.			
Jan.	18.	Payment by State Comptroller on account	652 39
Feb.	15.	Payment by State Comptroller on account	472 54
March	14.	Payment by State Comptroller on account	435 08
	28.	Payment by State Comptroller on account	13,149 69
April	13.	Payment by State Comptroller on account	431 47
	19.	Payment by State Comptroller on account	867 00
May	16.	Payment by State Comptroller on account	4,060 45
June	10.	Payment by State Comptroller on account	3,323 84
	15.	Payment by State Comptroller on account	7,599 00
July	10.	Payment by State Comptroller on account	142 50
	11.	Payment by State Comptroller on account	858 65
	25.	Payment by State Comptroller on account	10,132 00

1901.			
July 29.	Payment by State Comptroller on account................	$43 03	
Aug. 16.	Payment by State Comptroller on account................	715 32	
31.	Payment by State Comptroller on account................	1,734 00	
Sept. 13.	Payment by State Comptroller on account................	327 02	
30.	Payment by State Comptroller on account................	54 59	
			$83,599 74

Special appropriation as per Chap. 420, *Laws of* 1900 ($3,000):

1901.			
May 16.	Payment by State Comptroller on account................	$2,800 00	
June 15.	Payment by State Comptroller on account................	4 75	
July 17.	Payment by State Comptroller on account................	193 74	
			2,998 49

Special appropriation as per Chap. 419, *Laws of* 1900 ($5,000):

1900.			
Oct. 11.	Payment by State Comptroller on account................	$548 25	
31.	Payment by State Comptroller on account................	86 15	
Nov. 12.	Payment by State Comptroller on account................	141 39	
			775 79

Special appropriation as per Chap. 645, *Laws of* 1901 ($44,665.15):

1901.				
June	13.	Payment by State Comptroller on account................	$3,011 06	
	15.	Payment by State Comptroller on account................	194 85	
July	10.	Payment by State Comptroller on account................	2,272 85	
	11.	Payment by State Comptroller on account................	2,228 06	
	17.	Payment by State Comptroller on account................	887 15	
	29.	Payment by State Comptroller on account................	900 00	
Aug.	16.	Payment by State Comptroller on account................	2,673 74	
	31.	Payment by State Comptroller on account................	425 90	
Sept.	13.	Payment by State Comptroller on account................	2,228 82	
	17.	Payment by State Comptroller on account................	381 34	
				$15,203 77

RECEIPTS OF RESERVATION.

1900.			
Nov.	1.	Receipts for month of October.	$327 90
Dec.	1.	Receipts for month of November	41 30

1901.				
Jan.	1.	Receipts for month of December	$56 75	
Feb.	1.	Receipts for month of January..	41 80	
March	1.	Receipts for month of February	529 00	
April	1.	Receipts for month of March...	191 20	
May	1.	Receipts for month of April....	88 15	
June	1.	Receipts for month of May.....	405 70	
July	1.	Receipts for month of June....	2,450 40	
Aug.	1.	Receipts for month of July.....	4,843 25	
Sept.	1.	Receipts for month of August..	6,635 55	
	30.	Receipts for month of September	6,099 40	
				$21,710 40
		Total		$174,449 92

EXPENDITURES.

Maintenance.

Abstract CXXVI.

Date.		Voucher.	Name.	Amount.
1900.				
Oct.	22.	1894.	Charles M. Dow, Commissioner's expenses......	$190 65
		1895.	Thomas P. Kingsford, Commissioner's expenses.....	194 72
	31.	1896.	Peter A. Porter, Jr., Secretary and Treasurer office expenses	13 12
		1897.	Peter A. Porter, Jr., Secretary and Treasurer salary October, 1900	83 33

Date.	Voucher.	Name.	Amount.
1900.			
Nov. 2.	1898.	Pay-roll October, 1900	$1,882 10
	1899.	Thomas V. Welch, Superintendent's office expenses	30 95
12.	1900.	Pay-roll, supplemental, October, 1900	10 50
	1901.	A. J. Porter, Commissioner's expenses	3 40
20.	1902.	William R. Davies, bridges	13 35
22.	1903.	Thomas V. Welch, Superintendent's office expenses	22 90
	1904.	Niagara Falls Hydraulic Power & Mfg. Co., electric lighting	50 00
	1905.	Gazette Publishing Co., Superintendent's office expenses	3 50
	1906.	James Davy, Superintendent's office expenses	30 93
	1907.	P. J. Davy, plumbing	48 75
	1908.	P. C. Flynn & Son, iron railing, buildings	95 24
	1909.	Elderfield-Hartshorn Co., tools, buildings	19 13
Dec. 3.	1910.	Peter A. Porter, Jr., Secretary and Treasurer, salary November, 1900	83 33
	1911.	Pay-roll, November, 1900	1,447 37

Date.	Voucher.	Name.	Amount.	
1900.				
Dec. 13.	1912..	A. J. Porter, Commissioner's expenses	$6 05	
	1913..	Thomas V. Welch, Superintendent's office expenses.	42 86	
				$4,272 18

Abstract CXXVII.

18.	1914..	Peter A. Porter, Jr., Secretary and Treasurer, traveling expenses	36 45
	1915..	William S. Humbert, roads.	3 94
	1916..	Niagara Sand Co., roads...	9 00
	1917..	Elderfield-Hartshorn Co., tools, buildings	12 03
	1918..	Fred Batchelor, seed	9 00
	1919..	Charles S. Durkee, Superintendent's office expenses.	15 00
	1920..	Thomas E. McGarigle, Inclined Railway	16 42
	1921..	Gazette Publishing Co., Superintendent's office expenses	4 00
	1922..	Niagara Falls Hydraulic Power & Mfg. Co., electric lighting	50 00
	1923..	Dean & Hoffman, roads...	42 25
	1924..	Thomas V. Welch, Superintendent's office expenses.	45 40

Date.		Voucher.	Name.	Amount.	
1900.					
Dec.	20.	1925..	William F. Wall, Inclined Railway	$64 11	
		1926..	Alexander Herschel, President's office expenses....	65 60	
	24.	1927..	Alexander J. Porter, Commissioner's expenses	17 55	
		1928..	Alexander J. Porter, Commissioner's expenses	23 50	
	27.	1929..	George Raines, Commissioner's expenses	42 60	
	31.	1930..	Peter A. Porter, Jr., Secretary and Treasurer, salary December, 1900.........	83 33	
1901.					
Jan.	2.	1931..	Pay-roll, December, 1900..	1,425 25	
					$1,965 43

Abstract CXXVIII.

Date.		Voucher.	Name.	Amount.
1901.				
Jan.	9.	1932..	A. J. Porter, Commissioner's expenses...........	$7 55
		1933..	Thomas V. Welch, Superintendent's office expenses.	27 86
	26.	1934..	Peter A. Porter, Jr., Secretary and Treasurer, office expenses	10 54
	31.	1935..	Peter A. Porter, Jr., Secretary and Treasurer, salary, January, 1901	83 33

Date.		Voucher.	Name.	Amount.
1901.				
Feb.	2.	1936.	Pay-roll, January, 1901	$1,417 00
	4.	1937.	Thomas V. Welch, Superintendent's office expenses	49 63
	8.	1938.	Addison Johnson, Superintendent's office expenses	7 35
		1939.	P. J. Davy, buildings	29 80
		1940.	J. H. Cook & Co., buildings	16 73
		1941.	Elderfield-Hartshorn Co., buildings, Inclined Railway	8 96
	15.	1942.	A. J. Porter, Commissioner's expenses	19 20
		1943.	Thomas V. Welch, Superintendent's office expenses	31 07
	28.	1944.	Peter A. Porter, Jr., Secretary and Treasurer, salary, February, 1901	83 33
		1945.	Peter A. Porter, Jr., Secretary and Treasurer, traveling expenses	21 85
March	2.	1946.	A. J. Porter, Commissioner's expenses	10 10
		1947.	Pay-roll, February, 1901	1,426 75
	14.	1948.	Niagara Falls Hydraulic Power & Mfg. Co., electric lighting	50 00
		1949.	Niagara Falls Hydraulic Power & Mfg. Co., electric lighting	50 00

Date.	Voucher.	Name.	Amount.	
1901.				
March 14.	1950..	Niagara Falls Hydraulic Power & Mfg. Co., electric lighting	$50 00	
	1951..	W. W. Read, ice	67 48	
	1952..	Thomas E. McGarigle, Inclined Railway	4 62	
	1953..	William S. Humbert, roads, bridges	17 64	
	1954..	Welch Coal & Wood Co., coal	83 90	
	1955..	Elderfield-Hartshorn Co., tools	4 78	
	1956..	Globe Ticket Co., Inclined Railway	45 00	
	1957..	Elderfield-Hartshorn Co., tools, Inclined Railway	10 83	
				$3,635 30

Abstract CXXIX.

March 27.	1958..	Thomas V. Welch, Superintendent's office expenses	$49 79
	1959..	George Raines, Commissioner's expenses	26 06
28.	1960..	Peter A. Porter, Jr., Secretary and Treasurer, traveling expenses	20 00
	1961..	A. J. Porter, Commissioner's expenses	7 90

Date.		Voucher.	Name.	Amount.	
1901.					
April	1.	1962	Charles M. Dow, Commissioner's expenses	$168 07	
		1963	Peter A. Porter, Jr., Secretary and Treasurer, salary, March, 1901	83 33	
	2.	1964	Pay-roll, March, 1901	1,588 75	
					$1,943 90

Abstract CXXX.

April	8.	1965	P. J. Davy, buildings	$18 11
		1966	Domenick Penders, roads	144 30
	13.	1967	Thomas P. Kingsford, Commissioner's expenses	162 32
May	1.	1968	Peter A. Porter, Jr., Secretary and Treasurer, salary, April, 1901	83 33
	2.	1969	Pay-roll, April, 1901	2,677 12
		1970	Pay-roll, April, 1901, supplemental	18 00
	3.	1971	Thomas V. Welch, Superintendent's office expenses	34 06
	9.	1872	A. J. Porter, Commissioner's expenses	9 80
	13.	1973	Peter A. Porter, Jr., Secretary and Treasurer, office expenses	10 00
	18.	1974	P. C. Flynn & Son, buildings	112 88

Date.	Voucher.	Name.	Amount.
1901.			
May 18.	1975	P. C. Flynn & Co., buildings and bridges	$205 68
	1976	Welch Coal and Wood Co., coal	41 25
	1977	Charlotte Haeberle, buildings	48 86
	1978	J. H. Cook & Co., tools	5 87
	1979	Dobbie Foundry and Machine Co., Inclined Railway	41 69
	1980	Elderfield-Hartshorn Co., Inclined Railway	7 38
	1981	Elderfield-Hartshorn Co., buildings, tools, water pipes	179 11
	1982	Niagara Falls Hydraulic Power & Mfg. Co., electric lighting	50 00
	1983	Niagara Falls Hydraulic Power & Mfg. Co., electric lighting	50 00
	1984	Fred Batchelor, seed	10 00
31.	1985	Peter A. Porter, Jr., Secretary and Treasurer, salary, May, 1901	83 33
June 3.	1986	Pay-roll, May, 1901	1,388 12
	1987	J. H. Cook & Co., buildings	102 62

Date.	Voucher.	Name.	Amount.	
1901.				
June 3.	1988..	Niagara Falls Hydraulic Power & Mfg. Co., electric lighting	$50 00	
	1989..	Gazette Publishing Co., Superintendent's office expenses	7 50	
	1990..	P. J. Davy, plumbing	82 28	
	1991..	William Young, carting	9 69	
6.	1992..	Thomas V. Welch, Superintendent's office expenses	52 99	
	1993..	Thomas V. Welch, Superintendent's office expenses	55 96	
	1994..	Howard H. Baker & Co., Inclined Railway	5 50	
13.	1995..	Milton C. Johnson & Co., Secretary and Treasurer, office expenses	14 00	
July 1.	1996..	Peter A. Porter, Jr., Secretary and Treasurer, salary, June, 1901	83 33	
	1997..	Pay-roll, June, 1901	1,117 75	
				$6,962 83

Abstract CXXXI.

July 18.	1998..	Elderfield-Hartshorn Co., water pipes, tools, buildings	$78 32
	1999..	P. C. Flynn & Son, buildings	176 50

Date.	Voucher.	Name.	Amount.
1901.			
July 18.	2000	Gazette Publishing Co., Superintendent's office expenses	$12 50
	2001	Rush Electric Co., Inclined Railway	7 00
	2002	Gazette Publishing Co., Superintendent's office expenses	7 50
22.	2003	Peter A. Porter, Jr., Secretary and Treasurer, office expenses	18 45
31.	2004	Peter A. Porter, Jr., Secretary and Treasurer, salary, July, 1901	83 33
Aug. 2.	2005	Pay-roll, July, 1901	1,713 25
5.	2006	Gazette Publishing Co., Secretary and Treasurer, office expenses	7 75
7.	2007	James Davy, Superintendent's office expenses, buildings	65 42
	2008	Frank E. Giraux, roads	24 00
	2009	W. A. Shepard & Co., walks	250 00
	2010	Dean & Hoffman, walks	66 70
	2011	F. W. Oliver Co., tools	10 67
	2012	W. S. Humbert, walks	27 65
	2013	Elderfield-Hartshorn Co., tools, buildings	47 44

Date.	Voucher.	Name.	Amount.
1901.			
Aug. 7.	2014	P. C. Flynn & Son, signs, flagpole	$66 25
	2015	Elderfield–Hartshorn Co., tools, buildings, pipes, railings, Inclined Railway.	81 92
	2016	Thomas V. Welch, Superintendent's office expenses, roads	46 41
9.	2017	Peter A. Porter, Jr., Secretary and Treasurer, office expenses	28 50
	2018	Dobbie Foundry and Machine Co., drainage	22 13
	2019	Welch Coal and Wood Co., coal	2 75
	2020	Niagara Falls Hydraulic Power & Mfg. Co., electric lighting	3 83
	2021	Niagara Falls Hydraulic Power & Mfg. Co., electric lighting	50 00
	2022	Niagara Falls Hydraulic Power & Mfg. Co., electric lighting	50 00
	2023	J. H. Cook & Co., roads	5 81
	2024	Charlotte Haeberle, buildings	49 87
	2025	Charlotte Haeberle, buildings	23 61

Date.	Voucher.	Name.	Amount.	
1901.				
Aug. 9.	2026	Globe Ticket Co., Inclined Railway	$66 25	
10.	2027	Pay-roll, July, 1901, supplemental	147 25	
13.	2028	Pay-roll, August, 1901, supplemental	141 00	
16.	2029	Thomas V. Welch, Superintendent's office expenses.	50 25	
Sept. 4.	2030	Peter A. Porter, Jr., Secretary and Treasurer, salary, August, 1901	83 33	
	2031	A. J. Porter, Commissioner's expenses	2 50	
	2032	Pay-roll, August, 1901	1,774 12	
				$5,292 26

Abstract CXXXII.

Sept. 30.	2033	Peter A. Porter, Jr., Secretary and Treasurer, salary, September, 1901	$83 37	
	2034	Pay-roll, September, 1901	805 00	
				888 37

Payments out of $30,000, *as per chapter* 569, *Laws of* 1899.

Abstract VI—Series K.

1900.				
Oct. 31.	26	W. A. Shepard & Co., Shelter building, first estimate	$4,910 50	
Dec. 14.	27	W. A. Shepard & Co., Shelter building, second estimate.	4,145 45	
				9,055 95

Abstract VII—Series K.

Date.		Voucher.	Name.	Amount.	
1900.					
Dec.	31.	28	Gazette Publishing Co., Shelter building	$5 25	

Abstract VIII—Series K.

Date.		Voucher.	Name.	Amount.	
1901.					
Jan.	21.	29	Breen Bros., Shelter building.	$765 00	
Feb.	7.	30	W. A. Shepard & Co., Shelter building, third estimate...	3,504 55	
					$4,269 55

Abstract IX—Series K.

Date.		Voucher.	Name.	Amount.	
April	19.	31	W. A. Shepard & Co., Shelter building, fourth estimate..	$2,569 42	
May	22.	32	William Lansing, Shelter building	199 87	
		33	Burleson Hardware Co., Shelter building	872 00	
					3,641 29

Abstract X—Series K.

Date.		Voucher.	Name.	Amount.	
Sept.	11.	34	F. W. Oliver Co., Shelter building	$275 00	
		35	Burleson Hardware Co., Shelter building	198 00	
		36	W. A. Shepard & Co., Shelter building, final estimate...	6,070 87	
		37	Breen Bros., Shelter building.	1,608 00	
					8,151 87

Payments out of $120,000, as per chapter 420, Laws of 1900.

Abstract II—Series L.

Date.		Voucher.	Name.	Amount.
1900.				
Oct.	12.	19..	Pay-roll, September, 1900, arch bridge to Goat Island.	$264 75
	31.	20..	Edward A. Bond, arch bridge to Goat Island..........	46 09
		21..	D. S. Hollenbeck, arch bridge to Goat Island..........	20 00
		22..	D. S. Hollenbeck, arch bridge to Goat Island..........	19 29
		23..	D. S. Hollenbeck, arch bridge to Goat Island..........	78 28
		24..	D. D. Waldo, arch bridge to Goat Island............	13 50
		25..	D. D. Waldo, arch bridge to Goat Island............	37 25
		26..	D. D. Waldo, arch bridge to Goat Island............	162 82
		27..	David R. Lee, arch bridge to Goat Island............	69 37
		28..	W. H. Keepers & Co., third estimate, arch bridge to Goat Island............	8,670 00
Nov.	12.	29..	Pay-roll, October, 1900, arch bridge to Goat Island.....	252 00
	21.	30..	W. H. Keepers & Co., fourth estimate, arch bridge to Goat Island............	18,224 00

Date.		Voucher.	Name.	Amount.	
1900.					
Nov.	26.	31	D. D. Waldo, arch bridge to Goat Island	$259 55	
		32	David R. Lee, arch bridge to Goat Island	125 57	
		33	D. S. Hollenbeck, arch bridge to Goat Island	118 51	
		34	E. N. Allendorf, arch bridge to Goat Island	28 43	
		35	David R. Lee, arch bridge to Goat Island	10 22	
		36	R. S. Buck, arch bridge to Goat Island	515 95	
		37	Walter Jones, arch bridge to Goat Island	255 00	
		38	Walter Jones, arch bridge to Goat Island	270 00	
					$29,440 58

Abstract III—Series L.

Date.		Voucher.	Name.	Amount.
Dec.	18.	39	Pay-roll, November, 1900, arch bridge to Goat Island	$225 00
		40	E. N. Allendorf, arch bridge to Goat Island	129 43
		41	D. S. Hollenbeck, arch bridge to Goat Island	115 71
		42	D. D. Waldo, arch bridge to to Goat Island	185 20
		43	F. W. Hamilton, arch bridge to Goat Island	101 10

Date.	Voucher.	Name.	Amount.	
1900.				
Dec. 18.	44	Walter Jones, arch bridge to Goat Island	$275 00	
	45	Edward A. Bond, arch bridge to Goat Island	23 55	
24.	46	W. H. Keepers & Co., fifth estimate, arch bridge to Goat Island	7,803 00	
31.	47	R. S. Buck, arch bridge to Goat Island	290 74	
	48	E. N. Allendorf, arch bridge to Goat Island	11 86	
				$9,160 59

Abstract IV—Series L.

Date.	Voucher.	Name.	Amount.
1901.			
Jan. 18.	49	Pay-roll, December, 1900, arch bridge to Goat Island.	$181 50
	50	Fred W. Hamilton, arch bridge to Goat Island	68 37
	51	D. S. Hollenbeck, arch bridge to Goat Island	119 57
	52	Walter Jones, arch bridge to Goat Island	90 00
	53	D. D. Waldo, arch bridge to Goat Island	192 95
Feb. 15.	54	Pay-roll, January, 1901, arch bridge to Goat Island	186 00
	55	D. D. Waldo, arch bridge to Goat Island	172 97

Date.	Voucher.	Name.	Amount.	
1901.				
Feb. 15.	56	D. S. Hollenbeck, arch bridge to Goat Island	$113 57	
March 14.	57	D. S. Hollenbeck, arch bridge bridge to Goat Island	96 00	
	58	Edward A. Bond, arch bridge to Goat Island	25 15	
	59	D. D. Waldo, arch bridge to Goat Island	145 93	
	60	Pay-roll, February, 1901, arch bridge to Goat Island	168 00	
				$1,560 01
		Abstract V—Series L.		
28.	61	W. H. Keepers & Co., sixth estimate, arch bridge to Goat Island	$10,404 00	
	62	W. H. Keepers & Co., seventh estimate, arch bridge to Goat Island	1,734 00	
	63	W. H. Keepers & Co., eighth estimate, arch bridge to Goat Island	884 00	
	64	R. S. Buck, arch bridge to Goat Island	127 69	
				13,149 69
		Abstract VI—Series L.		
April 13.	65	D. D. Waldo, arch bridge to Goat Island	$184 40	
	66	D. S. Hollenbeck, arch bridge to Goat Island	107 57	

Date.		Voucher.	Name.	Amount.
1901.				
April	13.	67..	Pay-roll, March, 1901, arch bridge to Goat Island....	$139 50
	19.	68..	W. H. Keepers & Co., ninth estimate, arch bridge to Goat Island.............	867 00
May	16.	69..	W. H. Keepers & Co., tenth estimate, arch bridge to Goat Island.............	3,468 00
		70..	Pay-roll, April, 1901, arch bridge to Goat Island....	135 00
		71..	F. W. Hamilton, arch bridge to Goat Island...........	95 22
		72..	D. S. Hollenbeck, arch bridge to Goat Island...........	103 71
		73..	D. D. Waldo, arch bridge to Goat Island.............	163 07
		74..	H. S. Ball, arch bridge to Goat Island.............	95 45
June	10.	75..	W. H. Keepers & Co., duty, arch bridge to Goat Island.	2,615 96
		76..	Pay-roll, May, 1901, arch bridge to Goat Island....	139 50
		77..	D. S. Hollenbeck, arch bridge to Goat Island...........	113 57
		78..	H. S. Ball, arch bridge to Goat Island.............	118 86
		79..	F. W. Hamilton, arch bridge to Goat Island...........	118 86

Date.	Voucher.	Name.	Amount.	
1901.				
June 10.	80	D. D. Waldo, arch bridge to Goat Island	$217 09	
15.	81	W. H. Keepers & Co., eleventh estimate, arch bridge to Goat Island	7,599 00	
				$16,281 76

Abstract VII—Series L.

July 10.	82	Pay-roll, June, 1901, arch bridge to Goat Island	$142 50
11.	83	Edward A. Bond, arch bridge to Goat Island	23 26
	84	D. D. Waldo, arch bridge to Goat Island	121 57
	85	D. D. Waldo, arch bridge to Goat Island	200 40
	86	Thad. Wilson, arch bridge to Goat Island	80 16
	87	F. W. Hamilton, arch bridge to Goat Island	128 57
	88	A. B. Williams, arch bridge to Goat Island	25 47
	89	J. B. Barrett, arch bridge to Goat Island	99 23
	90	F. V. Searls, arch bridge to Goat Island	64 28
	91	D. S. Hollenbeck, arch bridge to Goat Island	115 71

Date.		Voucher.	Name.	Amount.	
1901.					
June	25.	92..	W. H. Keepers & Co., twelfth estimate, arch bridge to Goat Island.............	$10,132 00	
	29.	93..	George McDonald, arch bridge to Goat Island....	23 31	
		94..	R. S. Greenman, arch bridge to Goat Island...........	19 72	
Aug.	16.	95..	R. S. Buck, arch bridge to Goat Island.............	203 37	
		96..	D. D. Waldo, arch bridge to Goat Island.............	187 80	
		97..	D. S. Hollenbeck, arch bridge to Goat Island...........	119 57	
		98..	Thad. Wilson, arch bridge to Goat Island.............	123 64	
		99..	F. W. Hamilton, arch bridge to Goat Island...........	80 94	
	31.	100..	W. H. Keepers & Co., thirteenth estimate, arch bridge to Goat Island...........	1,734 00	
					$13,625 50

Abstract VIII—Series L.

Sept.	13.	101..	Thad. Wilson, arch bridge to Goat Island.............	$52 40
		102..	D. D. Waldo, arch bridge to Goat Island.............	167 05
		103..	D. S. Hollenbeck, arch bridge to Goat Island...........	107 57

Date.	Voucher.	Name.	Amount.	
1901.				
Sept. 30.	104	D. D. Waldo, arch bridge to Goat Island	$44 07	
	105	D. S. Hollenbeck, arch bridge to Goat Island	10 52	
				$381 61

Payments out of $3,000, *as per chapter* 420, *Laws of* 1900.

Abstract I—Series M.

May 16.	1	John A. Roebling's & Sons, repairs to Sister Island bridges	$2,800 00	
June 15.	2	Stowell & Cunningham, repairs to Sister Island bridges	4 75	
				2,804 75

Abstract II—Series M.

July 17.	3	P. C. Flynn & Son, repairs to Sister Island bridges	$143 24	
	4	F. W. Oliver Co., repairs to Sister Island bridges	50 50	
				193 74

Payments out of $5,000, *as per chapter* 419, *Laws of* 1900.

Abstract II—Series N.

1900.				
Oct. 12.	20	Pay-roll, September, 1900, stone stairway, Goat Island	$548 25	
31.	21	F. J. Munson, stone stairway, Goat Island	31 50	

Date.	Voucher.	Name.	Amount.	
1900.				
Oct. 31.	22..	W. S. Egerton, stone stairway, Goat Island.........	$44 00	
	23..	R. D. Young, stone stairway, Goat Island.............	10 65	
Nov. 12.	24..	Pay-roll, October, 1900, stone stairway, Goat Island	141 39	
				$775 79

Payments out of $44,665.15, as per chapter 645, Laws of 1901.

Abstract I—Series O.

Date.	Voucher.	Name.	Amount.
1901.			
June 13.	1..	Wm. F. Wall, Inclined Railway, Ext. Ex............	$230 31
	2..	C. T. Middlebrook, Inclined Railway, Ext. Ex........	72 00
	3..	L. B. Ackley, grass sod, Ext. Ex....................	338 00
	4..	Wm. A. Shepard & Co., completing and furnishing Shelter building..........	90 15
	5..	Wm. A. Shepard & Co., completing and furnishing Shelter building..........	74 60
	6..	Wm. A. Shepard & Co., completing and furnishing Shelter building..........	581 00
	7..	Wm. A. Shepard & Co., completing and furnishing Shelter building..........	69 75

Date.	Voucher.	Name.	Amount.	
1901.				
June 13.	8..	Pay-roll, May, 1901, Extraordinary expenditures......	$1,555 25	
15.	9..	Baker Art Gallery, folding map and guide, Ext. Ex..	194 85	
				$3,205 91

Abstract II—Series O.

July 10.	10..	Pay-roll, June, 1901, Extraordinary expenditures....	$2,272 85
11.	11..	City of Niagara Falls, local paving assessment........	2,004 78
	12..	L. B. Ackley, grass sod, Ext. Ex....................	223 28
17.	13..	Schumacher & Moyer, completing and furnishing Shelter building..........	70 60
	14..	Howard H. Baker & Co., completing and furnishing Shelter building..........	15 00
	15..	Breen Brothers, completing and furnishing Shelter building................	718 00
17.	16..	Breen Brothers, completing and furnishing Shelter building................	83 55
29.	17..	David Phillips, bridge to Terrapin Point..............	900 00
Aug. 16.	18..	Metal Stamping Co., completing and furnishing Shelter building................	21 63

Date.	Voucher.	Name.	Amount.
1901.			
Aug. 16.	19..	Clark Furnishing Co., completing and furnishing Shelter building.........	$21 50
	20..	F. W. Oliver Co., completing and furnishing Shelter building.........	13 40
	21..	A. Cutler & Son, completing and furnishing Shelter building................	225 90
	22..	Schumacher & Moyer, completing and furnishing Shelter building.........	14 25
	23..	Burleson Hardware Co., completing and furnishing Shelter building.........	7 00
	24..	Burleson Hardware Co., completing and furnishing Shelter building..........	8 00
	25..	Wm. A. Shepard & Co., completing and furnishing Shelter building..........	42 15
	26..	Wm. A. Shepard & Co., completing and furnishing Shelter building.........	8 60
	27..	Pay-roll, July, 1901, improving ground near arch bridges.................	555 12
	28..	Pay-roll, July, 1901, extraordinary expenditures.....	1,790 92

Date.		Voucher.	Name.	Amount.	
1901.					
Aug.	16.	29	Salem G. LeValley, police badges, Ext. Ex.	$6 20	
		30	The Courier Co., folding map and guide, Ext. Ex.	150 60	
		31	Dobbie Foundry and Machine Co., bridge to Terrapin Point	34 37	
	31.	32	Howard Iron Works, park seats, Ext. Ex.	200 00	
					$9,387 70

Abstract III—Series O.

Date.		Voucher.	Name.	Amount.
Sept.	13.	33	Pay-roll, August, 1901, extraordinary expenditures	$2,026 25
		34	Williams Lansing, completing and furnishing Shelter building	202 57
	17.	35	Pay-roll, September, 1901, supplemental, Ext. Ex.	26 13
		36	Globe Ticket Co., Inclined Railway, Ext. Ex.	57 50
		37	Gazette Publishing Co., printing, Ext. Ex.	9 75
		38	Gazette Publishing Co., printing, Ext. Ex.	7 50
		39	Fred J. Allen, Inclined Railway, Ext. Ex.	18 25
		40	Wm. A. Shepard & Co., completing and furnishing Shelter building	201 21

Date.		Voucher. Name.	Amount.	
1901.				
Sept.	17.	41..F. W. Oliver & Co., completing and furnishing Shelter building	$55 50	
		42..Schumacher & Moyer, completing and furnishing Shelter building	5 50	
				$2,610 16

REMITTANCE TO STATE TREASURER.

1900.				
Nov.	5.	Draft for October receipts........	$327 90	
Dec.	1.	Draft for November receipts......	41 30	
1901.				
Jan.	2.	Draft for December receipts......	56 75	
Feb.	2.	Draft for January receipts........	41 80	
March	5.	Draft for February receipts.......	529 00	
April	2.	Draft for March receipts.........	191 20	
May	3.	Draft for April receipts...........	88 15	
June	3.	Draft for May receipts...........	405 70	
July	1.	Draft for June receipts...........	2,450 40	
Aug.	1.	Draft for July receipts...........	4,843 25	
Sept.	3.	Draft for August receipts........	6,635 55	
	30.	Draft for September receipts......	6,099 40	
				21,710 40
		Cash balance on hand....................		77 55
		Total		$174,449 92

CLASSIFICATION OF ACCOUNTS.

Maintenance.

Secretary and Treasurer, office expenses......	$102 36	
Secretary and Treasurer, salary.............	1,000 00	
Salaries, Superintendent and Clerk..........	3,300 00	
Police	4,350 00	
Inclined Railway.........................	2,296 65	
Prospect Park............................	2,578 24	
Goat Island..............................	1,645 49	
Roads	2,323 56	
Walks	2,184 09	
Buildings	1,922 78	
Commissioner's expenses..................	891 97	
Superintendent's office expenses............	653 43	
Water pipes..............................	203 23	
Tools	133 66	
Electric lighting	503 83	
President's office expenses................	65 60	
Secretary and Treasurer, traveling expenses..	78 30	
Bridges	197 60	
Coal	127 90	
Ice	67 48	
Seed	19 00	
Signs	48 25	
Plumbing.................................	131 03	
Iron railings	81 62	
Cartage	9 69	
Flag pole	18 00	
Railings	4 38	
Drainage	22 13	
		$24,960 27

Improvements under chapter 569, Laws of 1899—Series K.

Shelter building		$25,123 91

Improvements under chapter 420, Laws of 1900—Series L.

Arch bridge, mainland to Goat Island		83,599 74

Improvements under chapter 420, Laws of 1900—Series M.

Repairing Sister Island bridges		2,998 49

Improvements under chapter 419, Laws of 1900—Series N.

Stone stairway, Goat Island		775 79

Improvements under chapter 645, Laws of 1901—Series O.

Inclined Railway, Ext. Ex.	$488 31	
Grass sod, Ext. Ex.	561 28	
Prospect Park, Ext. Ex.	2,640 37	
Goat Island, Ext. Ex.	1,331 86	
Folding map and guide, Ext. Ex.	345 45	
Roads, Ext. Ex.	734 12	
Walks, Ext. Ex.	1,194 50	
Buildings, Ext. Ex.	366 87	
Detectives, Ext. Ex.	1,293 43	
Police badges, Ext. Ex.	6 20	
Park seats, Ext. Ex.	200 00	
Printing, Ext. Ex.	17 25	
Improving ground near arch bridges	555 12	
Completing and furnishing Shelter building	2,529 86	
Bridge to Terrapin Point	934 37	
Local paving assessment	2,004 78	
		15,203 77
Remittances to State Treasurer		21,710 40
Cash on hand September 30, 1901		77 55
Total		$174,449 92

PETER A. PORTER, Jr.
Treasurer.

We, the undersigned, hereby certify that we have examined the foregoing report of the Treasurer for the fiscal year ended September 30, 1901, the vouchers and other papers, and we find the report and accompanying documents correct, and that the Treasurer has properly accounted for all moneys received and disbursed by him during the fiscal year ended September 30, 1901.

THOMAS P. KINGSFORD,

CHAS. M. DOW,

Commissioners of the State Reservation at Niagara.

Attest:

C. S. DURKEE,

Professional Accountant and Auditor and member of the National Association of Public Accountants, office and post-office address Room N, Arcade Building, Niagara Falls, N. Y.

Correct:

EDWARD H. PERRY.

Report of the Superintendent

OF THE

State Reservation at Niagara

FOR THE

Fiscal Year Ending September 30, 1901.

Bridge to Green Island. State Reservation at Niagara. Erected 1901.

Report of the Superintendent.

To the Commissioners of the State Reservation at Niagara:

Gentlemen.—The most important improvements made during the fiscal year ending September 30, 1901, have been the completion of the two stone arch bridges between the mainland and Goat Island; the completion of the Shelter building in Prospect Park; the construction of a new bridge from Goat Island to Terrapin Point, at the Horseshoe Falls; the renovation of the two suspension bridges to the Three Sister Islands, and the grading and filling adjacent to the new stone arch bridges.

Stone Arched Bridges, Mainland to Goat Island.

At the close of the last fiscal year the stone arched bridge from Green Island to Goat Island was completed, excepting the parapet walls and the sidewalks, while the piers only for the bridge from the mainland to Green Island were in place. The temporary wooden foot bridge between the mainland and Green Island was almost entirely swept away by the high water and floating ice during the winter months.

April 1, 1901, work upon the bridges was resumed, and the parapet walls of the bridge from Green Island to Goat Island were quickly completed. A new wooden foot bridge was built from the mainland to Green Island. Considerable delay was caused by the uncertainty concerning the floating ice from Lake Erie, which would endanger the centers for the arches placed in the rapids between the mainland and Green Island. For the first time in many years, there was no run of ice from Lake Erie. As soon as it was considered certain that the ice would not interfere with the construction of the centers, work was commenced on the bridge from the mainland to Green Island. The last concrete arch was completed on June 29th, the bridge was opened to the public July 14th, the stone work was completed July 26th, and the sidewalks on August 24th.

The approaches to the bridges have been filled, graded and planted, the roads and walks readjusted to meet the new conditions, the water pipes to Goat Island laid in the roadbed, drains and catch basins constructed, and all traces of the work of construction removed.

Photographs of the new bridges are herewith submitted, and the contract and supplemental contract for the construction of the bridges are hereto appended.

Suspension Bridges to Three Sister Islands.

The repairs of the two suspension bridges to the Three Sister Islands were completed on May 9th by the John A. Roebling's Sons Company of New York. New cables, anchors, metal needle beams and braces were provided.

The woodwork of the bridges was repaired and the bridges painted; the stairs at the approaches to the bridges were removed and inclined walks of rustic stone and gravel substituted, which are safe and convenient and more natural in appearance.

The suspension bridges bore the heavy traffic of the past season without any apparent strain.

Bridge to Terrapin Point.

The wooden bridge from Goat Island to Terrapin Point was removed during the spring months, and a new bridge of steel and iron, with plank floor, constructed.

The bridge to Luna Island has been repainted, and the retaining walls at the rustic bridge over the overflow on the riverway repaired.

The Shelter Building.

The new Shelter building on the riverway, in Prospect Park, was completed and opened to the public on July 4th. It contains spacious rooms for the accommodation of visitors, the offices of administration, toilet rooms and storerooms. The windows in the basement story have been enlarged and the two large end rooms finished, the completion of which was not included in the original contract.

A large amount of filling, grading and sodding has been done around the Shelter building, gravel walks constructed, and a stone walk, ten feet wide, laid from the Shelter building to the riverway. The old open frame pavilion, immediately adjacent to the new Shelter building in Prospect Park, has been removed.

Cave of the Winds Building.

The Cave of the Winds building on Goat Island has been moved eastward across the road to a less conspicuous place, where it does not obstruct the view of the Falls and the river. The location formerly occupied has been filled, iron guard railings, stairways and gates constructed, and repairs made upon the Biddle stairs leading to the Cave of the Winds.

The Cave of the Winds building has been repainted and a veranda constructed for the accommodation and shelter of visitors.

One of the buildings used for drying clothes was destroyed by fire. The prompt and energetic action of the city fire department prevented the flames from spreading. The building was rebuilt by the lessee of the Cave of the Winds.

Inclined Railway Building.

The frame structure covering the Inclined Railway is old and should be renewed. Many of the sills and other timbers have been renewed during the past year. The weight of snow and ice upon the structure during the winter months is very great. In April, the gable on the south side was torn off by sliding ice. The damage was repaired, the building painted, the flume and the machinery in the wheelpit repaired, the cars renovated, a new cable attached to the cars, a new ticket office provided, and everything made ready for the great multitude of visitors from the Pan-American Exposition in Buffalo.

Roads.

The Paving of Buffalo avenue adjacent to the Reservation made it necessary to widen and raise the grade of First street, between Buffalo avenue and the riverway.

The road around the head of Goat Island from the entrance to the Three Sister Islands was much worn and in need of repairs. An opportunity occurred for obtaining a large quantity of gravel from an excavation, for building purposes, at a cost of 10 cents per load. With the gravel so obtained, the roads on Goat Island, Prospect Park and on the riverway were extensively repaired and put in order for the heavy traffic of the Pan-American Exposition year.

Walks.

The wooden stairs at the right at the entrance to Goat Island have been removed and a winding, inclined, gravel walk constructed to the top of the bank.

A winding, inclined walk has also been constructed on the left of the entrance to Goat Island, and the flat, artificial spaces on either side of the road, at the approach to Goat Island, filled and graded, and large boulders placed at the approaches to the new stone arched bridges. The graded banks were covered with sod and planted with shrubs and vines.

The upper half of the stairway at the Horseshoe Falls on Goat Island was removed and an inclined, gravel walk with rustic stone embankments constructed, making the descent more safe and natural. The large amount of filling done on Green Island at the approaches to the new stone arched bridges made it necessary to rebuild the gravel walks on either side of the road, the entire width of the island. Repairs were also made upon the gravel walks along the riverway.

Grading and Filling.

The completion of the new stone arched bridges, in August, found the grounds at Bridge street and on Green Island in great disorder and greatly damaged by the work of construction. The damaged portions of the grounds were repaired and seeded down, the approaches to the bridges covered with sod and planted with shrubs and vines; the roadway on Green Island was filled to the grade of the new bridges and the surface of the island cultivated and seeded down. The paving of Buffalo avenue adjacent to the Reservation at Port Day necessitated a large amount of filling and grading. The river shore at " The Wing " below Fourth street was riprapped with large stones and filled and graded.

Two landslides in Prospect Park and on Goat Island necessitated the filling of the cavities and the removal of the iron guard railing further back from the edge of the high bank.

Planting.

A number of trees have been planted in Prospect Park to take the place of old trees which have died out. Vines and drooping shrubs have been planted at the new stone stairway on Goat Island, at the new walks, at the approaches to Goat Island, at the approaches to the new stone arched bridges, and at the rustic bridges on Riverway. Vines have also been planted about the walls of the new Shelter building. The stock in the nursery has been cultivated, the plantations of shrubbery pruned, and a large amount of compost collected for spreading upon the lawns during the winter season.

The Ponds.

In order to insure an adequate supply of water for the operation of the Inclined Railway during the Pan-American Exposition, the water was shut off from the pond in Prospect Park, used as a reservoir, the growth of eel-grass removed, and the pond deepened. A wing of the pond in which the water was stagnant was filled, covered with sod. The basin of the fountain in Prospect Park was removed and the site filled and covered with sod. The eel-grass was also removed from the pond at the loop at Port Day, the work requiring a flat boat and a crew of four men more than a month for its performance.

The Electric Railway in the Riverway.

The station of the Electric Railway Company, on the east side of the riverway, has been completed. A double track has been laid in the riverway from Niagara street to the entrance to the steel arch

bridge, and a second track in the riverway from the entrance to the steel arch bridge to the electric railway station, the street having been widened for that purpose, according to the agreement with the Electric Railway Company.

Before the completion of the railway station, the attention of the railway company was called by the Superintendent to the unnecessary blocking of the sidewalks by cars, the dangerous speed of cars at times, and the usage of the riverway by the company as a terminus and standing place for cars, in violation of the agreement. Since the completion of the railway station, some of the causes for complaint have been remedied, but there is still a tendency on the part of the company to have cars stand upon the riverway longer than is necessary. During the Pan-American Exposition as many as fifteen thousand visitors came from Buffalo and returned by the electric railway in one day. Aside from the frequent overcrowding of the cars, the service was excellent and a great convenience and accommodation to the public. Notwithstanding the tracks in the riverway, visitors on arrival were unloaded at the corner of Falls and Prospect streets, the headquarters of many confidence men and gamblers, where the lack of order and protection to visitors gave rise to many serious causes for complaint. Visitors should be brought by the railway company to the railway station and to the State Reservation as agreed, otherwise the object of the Commissioners in granting a revocable license to the railway company in the riverway will be defeated, in which case it seems questionable whether the railway company should be allowed to use the tracks in the riverway, which are owned by the State.

Reservation Carriage Service.

Until the completion of the stone arched bridges, the Reservation carriage service was operated by placing horses and carriages on Goat Island.

The multitude of visitors from the Pan-American Exposition made it necessary for the lessee to procure a large additional number of carriages. Even with such additional facilities, it was difficult to transport the number of visitors to the Reservation. The service was operated during the year without accident or complaint.

Licensed Carriage Drivers.

During the year no public carriage stands have been allowed upon the Reservation. Very little difficulty is experienced with the majority of the carriage drivers, especially those who own their carriages and are disposed to be courteous and accommodating to visitors. But many improper persons are licensed as carriage drivers by the city government, which gives rise to constant cause for complaint. Several of these have been discovered acting in collusion with gamblers and confidence men, brought here by the Pan-American Exposition. During the year eight licensed carriage drivers have been excluded from the Reservation for violation of the ordinances.

The Cave of the Winds.

Notwithstanding the change in the location of the Cave of the Winds building during the Pan-American year, the Cave of the Winds was visited by the greatest number of people of any year since it has been operated as a point of interest. The general question whether the State should any longer allow visitors to be encouraged to pass through the Cave of the Winds is respectfully submitted to the Commissioners for consideration.

This is entirely aside from the matter of the Biddle stairs, forming the mode of access to the Cave, which are old and unsuitable. Many arguments can be adduced against submitting visitors to this tiresome and hazardous undertaking. The removal of the dressing rooms, dry house and clothes yard, and the discontinuance of the business, would be in keeping with the work of the State, in restoring the grounds about the Falls, so far as possible, to a natural condition.

Steamboat Landing.

During the year the steamboat landing has been kept in repair by the Maid of the Mist Steamboat Company. Two boats were in operation during the season, which, because of the Pan-American Exposition, was greatly advantageous to the steamboat company.

Inspection by Governor Odell and Party.

On September 27th the State Reservation was visited and the new bridges, Shelter building and other improvements inspected by Governor Odell, Senator Ellsworth, Presiment *pro tem.* of the Senate; Senator Higgins, Chairman of the Finance Committee of the Senate; Assemblyman Allds, Chairman of the Ways and Means Committee of the Assembly, and other State officers.

President McKinley and Party.

On September 6th the State Reservation was visited by President McKinley and party, three hours previous to the assassination of the President at the Pan-American Exposition in Buffalo.

The President expressed deep interest in the restoration of tne grounds about the Falls, and said that the State of New York was doing a good work at Niagara.

The Pan-American Exposition Visitors.

The Pan-American Exposition brought a great number of visitors to the Reservation, the estimated attendance reaching as high as fifty thousand visitors in one day. On several occasions it was necessary to discontinue the sale of tickets at the Inclined Railway because of the inability to carry the number of persons seeking transportation. The increased number of visitors necessitated the employment of a large additional number of laborers and caretakers. Extra precautions were taken to insure the safety of all bridges, stairways, railings and observation points in the Reservation, and to prevent disorder or accident of any kind.

Notwithstanding the work upon the bridges over the rapids, the many dangerous places in the Reservation, and the unprecedented number of visitors, no disorder or accident of any kind occurred.

Gamblers and Confidence Men.

The Pan-American Exposition brought to the Reservation a large number of gamblers and confidence men. The method of operation of the confidence men was to engage a promising stranger in conversation and induce him to accompany them to some resort up town where "souvenirs of Niagara" or "tickets to points of interest" were to be obtained gratis. On their arrival at the place, confederates appeared and a trap was sprung to induce the stranger to take part in "three card monte" and similar games. In many cases unwary strangers were fleeced in this manner.

The resorts and the operations carried on in them were well known to the police department of the city and to the city authorities. In many cases, in which the injured party reported the facts to the Superintendent of the Reservation, complaint was made to the city police headquarters, where the victims were informed that nothing could be done for them. In one case the confidence men had secured a draft for $200, which they forwarded for collection through the regular channels, endorsed by a local "politician" and proprietor of a resort for gamblers and confidence men, but payment was stopped by advice of the Reservation officers. In another case an aged Western farmer barely escaped robbery of a draft for $2,000, the proceeds of the sale of his farm and the earnings of his lifetime.

In still another case, a stranger notified a Reservation officer that he had been approached by a confidence man and invited to take part in a well-known confidence game. When the officer ordered the offender to leave the Reservation, he informed the officer that he was a friend of the alderman of the First ward of the city, whom he named, and for that reason expected to escape molestation. Among the swindlers who infested the grounds were four women, well known to the police officers of the city.

The frequency of these complaints caused the Superintendent to place in the hands of the Reservation officers notices for distribution to visitors, of which the following is a copy:

CAUTION TO VISITORS!

BEWARE OF "CONFIDENCE MEN" AND PICKPOCKETS!

DO NOT BE ENTICED ANYWHERE BY STRANGERS, WHO MAY BE "SHARPERS!"

PLEASE REPORT ANY ATTEMPT TO SWINDLE YOU AT ONCE TO THE RESERVATION OFFICERS.

T. V. WELCH,

Superintendent State Reservation at Niagara.

After the notice to the public of their presence and the exposure of their operations the confidence men disappeared, but came again from time to time during the season, seemingly upon the suggestion of persons here who controlled their operations in this locality.

The Pan-American Exposition also brought to the border of the Reservation, on the eastern side of the riverway, a number of fakirs' immoral and disgusting exhibitions, and gambling places evidently organized and controlled in the same manner as the operations of the confidence men. Buildings on the eastern border of the Reservation were rented, and several vacant lots adjacent to the Reservation were leased and temporary structures erected on them by an "Amusement Company" licensed by the city of Niagara Falls. These "places of amusement" were called "The Midway," "Congress Museum," "Saints and Sinners," "Streets of Asia," "Pan-American Beauties," "Amusement Parlor" and other names, the names being changed from time to time.

In these places degrading and immoral exhibitions were licensed by the city authorities, and attempts to revoke the licenses of the most brutal and vicious were defeated by an element in the common council, in harmony with the existing state of affairs. Several of these so-called "places of amusement" were, in reality, gambling places of the worst character, into which unsuspecting strangers were enticed by swarms of "Cappers," systematically swindled, and in some cases violently robbed of their money. Daily complaints were made at the office of the Reservation Superintendent by the victims, in many cases left penniless long distances from their homes. In some cases the whole or part of the money was recovered, the city police acting as go-betweens in the transactions, but no steps were taken by the city police department to punish the offenders or to close the unlawful resorts until the season was over.

In one case, in which the Reservation Superintendent brought to the mayor a stranger from Nova Scotia who had been swindled out of $100, all the money in his possession, in one of these gambling places said to be owned by a city alderman, the mayor sent for the chief of police and ordered him to recover the money and to forthwith close the place where the offence was committed. The place was not closed and was in daily operation thereafter, showing that the chief of police did not obey the order of the mayor, or the mayor and chief of police were not acting in good faith. Complaints were made by persons swindled to the mayor, the police justice, the chief of police and to the patrolmen, but no action was taken, and in some cases the victims complaining were threatened with arrest.

Visitors seated in Prospect Park were obliged to see the exhibition of "Dancing Women," and to witness the operations of professional "Cappers" for gambling places of the worst character, and their enjoyment of the Reservation was disturbed, even on Sunday, by the noises and outcries made to attract the attention of those passing by.

Under such circumstances, although the gambling places were not within the Reservation, but under the control of the city authorities, the Superintendent of the Reservation placed an officer at the boundary of the Reservation, in front of each resort, with instructions to hand to each stranger entering the resorts a notice of warning, of which the following is a copy:

CAUTION TO VISITORS!

BEWARE OF SO-CALLED "PLACES OF AMUSEMENT" WHERE GAMBLING GAMES AND OTHER SWINDLES ARE CONDUCTED!

DO NOT BE ENTICED BY GAMBLERS OR "CAPPERS" TO TAKE PART IN ANY GAMBLING GAME!

T. V. WELCH,
Superintendent State Reservation at Niagara.

As directed at the meeting of the Commissioners held July 2d, the Superintendent transmitted to the mayor copies of the resolution of the Board and the report of the Superintendent relating to the unlawful exhibitions, licensed by the city, adjacent to the Reservation. So far as known, no action was taken by the mayor upon the protest of the Board. The gambling places remained open until the season was over, when the last of them, "The Streets of Asia," which was known to be about to close its doors, was closed with a great show of authority by a city police sergeant and a squad of policemen in a manner that under the circumstances was farcical in the extreme and which only added to the already great disrepute of the police department of the city.

Only a small number of the cases of swindling which occurred were reported to the Reservation officers.

The evils existing in the neighborhood of the Cataract before the establishment of the State Reservation were more annoying to visitors, but were never so degrading and objectionable as these licensed by the city authorities in the immediate vicinity of the Falls during the past year. Aside from the injuries to individuals and to public morals, they operated to destroy the quiet and peaceful enjoyment of the scenery of the Cataract, which was the object of the State in establishing the Reservation.

The primary responsibility for the present discreditable condition of affairs rests with the citizens of the city of Niagara Falls, who elected the mayor of the city. The direct responsibility rests upon the mayor, who is charged with the enforcement of the laws. In order to fix the responsibility, the city charter was amended so as to confer upon the mayor the absolute power of appointment of the city boards. The mayor appointed the board of police commissioners, of which he is the president. The responsibility for the present scandalous attempt to nullify and defeat the labor and expenditure of the State of New York in establishing the State Reservation at Niagara therefore rests upon the mayor of the city of Niagara Falls, and I respectfully submit the matter to the consideration of the Commissioners for such action as may be deemed proper.

Visitors.

The number of persons who visited the Reservation during the year is estimated at three millions. During the Pan-American Exposition, from May 20th until November 1st, the number of visitors ranged from ten thousand to fifty thousand each day. The new stone arched bridges were completed in time to accommodate the great multitudes of visitors in July, August, September and October. The care for the property of the State, the spirit of order and refinement pervading the moving masses of humanity, is a gatifying memory of the multitudes attracted to Niagara by the Pan-American Exposition.

Employees.

The number of regularly classified employees is ten—1 superintendent, 1 treasurer and secretary, 1 clerk, 1 police superintendent, 2 police and caretakers on the Islands, 3 police and caretakers in Prospect Park, 1 Inclined Railway operator.

During the Pan-American Exposition 10 additional police were employed, and a large number of additional laborers and teamsters were engaged in keeping the grounds, roads, walks, bridges and stairways in order. Of the $20,000 appropriated for extraordinary expenditure during the Pan-American Exposition, $6,364.27 remains unexpended. A statement of the number of laborers and teamsters employed each month is herewith submitted.

Statement of Number of Employees.

1900.	Foreman.	Ass't foreman.	Teamsters.	Laborers.	Maintenance.
October	1	3	2	27	
November	1	-	2	13	
December	1		-	13	
1901.					
January	1	-		13	
February	1	2		13	
March	1	2	1	14	
April	1	2	6	54	
May	....	......	......	20	
June	....	......	......	12	
July	....	-		26	
August	....	2		32	
September	....	......	......	3	
		Chapter 419, Laws 1900.			
1900.					
October	....	-	-		8 Carriage turnouts (Goat Island).

In the woods on Goat Island.

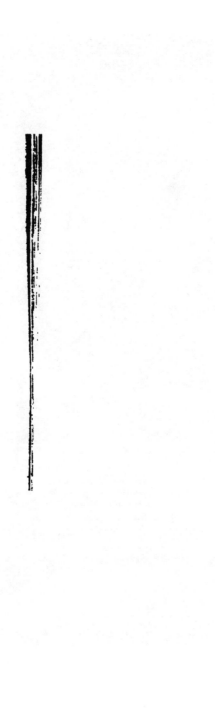

		Chapter 420, Laws 1900.			
1900.	Foremen.	Ass't. Foremen.	Teamsters.	Laborers.	Maintenance.
October	....	......	......	5	
November	....	......	......	5	
December	....	......	......	5	
1901.					
January	....	......	......	4	Arch bridges—Mainland to Goat Island.
February	....	......	......	4	
March	.....	......	......	3	
April	....	......	......	3	
May	....	......	......	3	
June	....	......	......	3	
		Chapter 645, Laws 1901.			
1901.					
May	1	4	5	29	
June	1	3	3	57	Care and maintenance Pan-American.
July	1	-	2	27	
August	1	2	3	30	
September	1	4	2	49	
		Chapter 645, Laws 1901.			
1901.					
July	....	2	2	10	Grading, filling and putting in order grounds adjacent to stone arched bridges.

Water Supply.

Plans and estimates for a water supply for the Reservation grounds and buildings have been obtained, but owing to the work upon the stone arched bridges, and the great multitudes of visitors from the Pan-American Exposition, the work of installing a system of water mains has been deferred until the coming spring.

Electric Lighting.

A plan and estimate for a lighting system for the Reservation grounds has been obtained. The estimated cost of the system is over $18,000.

This work may be done during the coming spring if an adequate appropriation is obtained.

Receipts.

The receipts by the Superintendent during the fiscal year were as follows:

From the Inclined Railway	$19,660 40
From lease of the Cave of the Winds	1,200 00
From lease of the Steamboat landing	750 00
From lease of Reservation carriage service	100 00
Total	$21,710 40

The amount received has been paid into the treasury of the State of New York.

Detailed statements of the receipts and expenditures by the Superintendent, the amount of the pay-rolls for each month, and the classification of the pay-rolls and accounts are hereto appended.

Respectfully submitted,

THOMAS V. WELCH,

Superintendent.

Contracts and Specifications

FOR

Stone Arched Bridges from Mainland to Goat Island,
State Reservation at Niagara.

Erected 1901.

Contracts and Specifications.

By Authority of Chapters 419 and 420 of the Laws of 1900, State of New York.

THIS AGREEMENT, made this twenty-third day of June, in the year nineteen hundred, by and between William H. Keepers, James H. Wynkoop and John M. Braly, composing the firm of W. H. Keepers and Company, of the city of New York, in the State of New York, party of the first part, and hereinafter called the contractor, and the People of the State of New York, by Andrew H. Green, George Raines, Charles M. Dow, Thomas P. Kingsford and Alexander J. Porter, composing the Commissioners of the State Reservation at Niagara, of the second part.

WITNESSETH, That the said contractor has agreed, and by these presents does agree, for and in consideration of the covenants and payments to be made as hereinafter provided for, to furnish all of the materials, appliances, tools and labor required and build and complete, in the most substantial and workmanlike manner, a concrete-steel arch bridge with stone facing, gravel roadway and granitoid sidewalks, extending across the Niagara River, from the mainland to Green Island, in the Niagara State Reservation, and a concrete-steel arch bridge with stone facing, gravel roadway and granitoid sidewalks across the Niagara River, from Green Island to Goat Island, in the Niagara State Reservation. Each of said bridges to consist of three spans, with radius of arch to be approved by State Engineer: Each bridge to have a twenty-foot roadway and two ten-foot sidewalks. The bridges generally to be in accordance with the plans exhibited at the office of Andrew H. Green June 16, 1900, which have been marked for identification and are hereby referred to and made a part of this agreement.

The said contractor agrees to execute the above-described work for the sum of one hundred and two thousand and seventy dollars ($102,070) lawful money, to be paid in the manner hereinafter described; and for the further consideration of the right to use the existing bridges, abutments and piers or such materials contained therein as may be suitable, in the execution of said work; and the right to dispose of any part of said existing bridges or said material therein contained which shall not be used in the execution of said work for their own use and benefit.

The said work shall be done in conformity with the plans and specifications which are hereby referred to and made a part of this agreement, which plans and specifications shall be approved by the State Engineer and Surveyor.

The contractor further agrees to conform to the provisions of chapters 415 of the Laws of 1897, 444 of the Laws of 1897, and 567 of the Laws of 1899, relative to the Labor Law and the assignment and subletting of contracts.

The contractor further agrees to comply with the laws providing that no laborer, workman or mechanic in the employ of the contractor, or of any subcontractor, or any other person doing or contracting to do a whole or a part of the work contemplated by these specifications and this contract, shall be permitted or required to work more than eight hours in any one calendar day, except in cases of extraordinary emergency, caused by fire, flood or danger to life or property.

It is further provided that each such laborer, workman or mechanic employed by such contractor, subcontractor or other person on, about or upon such work shall receive not less than the prevailing rate of wages for a day's work in the same trade or occupation in the locality within the State where such labor is performed, or where such public works on, about or in connection with which such labor is performed in its final or completed form is to be situated, erected or used. It is further provided that any contract shall be void and of no effect unless the person or corporation making or performing the same shall comply with the provisions of section 3 of the "Labor Law," as amended by chapter 567 of the Laws of 1899, and chapter 298 of the Laws of 1900.

The contractor further agrees that he will promptly begin the work herein embraced and so prosecute the same that it shall be entirely completed on or before January 1, 1901.

It is further mutually agreed that if at any time during the prosecution of the work the contractor has, in the judgment of the State Engineer, failed to provide such men, labor, tools, appliances or materials as will ensure the completion of the said work on or before the said first day of January, nineteen hundred and one, and after ten days' notice in writing from the State Engineer (a copy of which notice shall also at the same time be sent by mail to the bondsmen of said contractor) still fails to provide such men, labor, tools, appliances or materials, then the said Commissioners may order and supply such men, labor, tools, appliances or materials at the contractor's expense and provide in their discretion for the payment of the same from any moneys due or to become due under this contract, and in case such expense shall exceed the amount due or to become due the contractor, on the final completion of the work herein embraced, then it is expressly understood and agreed that the said contractor shall be personally liable for such excess.

It is further mutually agreed that no change shall be made in the plans or specifications, except by means of a supplemental agreement, to be in writing and signed by the State Engineer, the contractor and each of the said Commissioners.

It is further mutually agreed that upon the completion of the work and before the payment of the final estimate, the contractor shall furnish a guarantee for a term of five (5) years, to be of form approved by the said Commissioners.

On the faithful performance of the work herein embraced as set forth in the foregoing contract and the accompanying specifications and plans, which are a part thereof, the said Commissioners of the State Reservation above named hereby agree to use and dispose of said existing bridges and the materials therein contained as hereinbefore fully set forth, and agree to pay to the contractor the sum of one hundred and two thousand and seventy dollars ($102,070) in the following manner, to wit:

On or before the fifteenth day of each month the proportionate value, to be determined by the State Engineer, of all work done up to the first of that month, reserving fifteen per cent. (15%) thereof until the final completion, the balance to be paid within thirty days of the final completion of the work and its acceptance by the said Commissioners and the State Engineer; provided, however, that the contractor shall have given the said Commissioners such reasonable assurance as they may require that all bills connected with this work which might constitute a claim against the said Commissioners have been paid; and provided, also, that the contractor shall, if required, execute a bond with sufficient sureties and in form and amount

Hermit's Cascade. Goat Island.

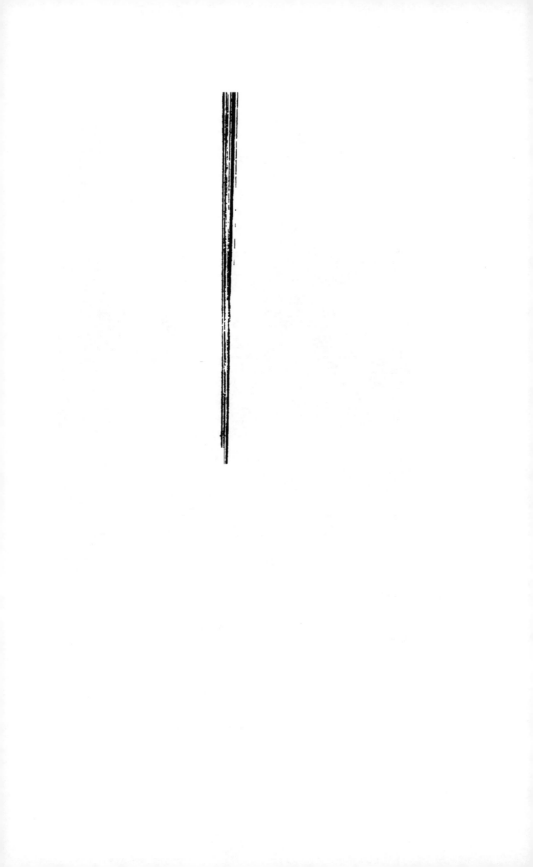

satisfactory to said Commissioners for the protection of the State from such claims.

Witness our hands and seal on the day and year first above written.

W. H KEEPERS & CO.,
Party of the First Part,
By John M. Braly.

The People of the State of New York by

ANDREW H. GREEN,
THOMAS P. KINGSFORD,
CHARLES M. DOW,
GEORGE RAINES,
ALEXANDER J. PORTER,
Commissioners of the State Reservation at Niagara.

Approved June 23, 1900.

Edward A. Bond,
State Engineer and Surveyor.

State of Massachusetts, }
County of Worcester, } *ss.:*

On this 7th day of July, 1900, before me personally appeared Andrew H. Green, one of the Commissioners of the State Reservation at Niagara, to me known to be the person described in and who executed the foregoing instrument and acknowledged that he executed the same as such Commissioner.

[Notary Seal.] WILLIAM W. MACOMBER,
Notary Public.

STATE OF NEW YORK, }
County of Monroe, } ss.:

On this 30th day of June, 1900, before me personally appeared George Raines, one of the Commissioners of the State Reservation at Niagara, to me known to be the person described in and who executed the foregoing instrument and acknowledged that he executed the same as such Commissioner.

[Notary Seal.] EDWARD R. FOREMAN,
Notary Public.

STATE OF NEW YORK, }
County of Chautauqua, } ss.:

On this 27th day of June, 1900, before me personally appeared Charles M. Dow, one of the Commissioners of the State Reservation at Niagara, to me known to be the person described in and who executed the foregoing instrument and acknowledged that he executed the same as such Commissioner.

[Notary Seal.] MABLE S. HARMON,
Notary Public.

STATE OF NEW YORK, }
County of Oswego, } ss.:

On this 2d day of July, 1900, before me personally appeared Thomas P. Kingsford, one of the Commissioners of the State Reservation at Niagara, to me known to be the person described in and who executed the foregoing instrument and acknowledged that he executed the same as such Commissioner.

[Notary Seal.] CHARLES A. BENTLEY,
Notary Public.

STATE OF NEW YORK, }
County of Niagara, } ss.:

On this 26th day of June, 1900, before me personally appeared Alexander J. Porter, one of the Commissioners of the State Reservation at Niagara, to me known to be the person described in and who executed the foregoing instrument and acknowledged that he executed the same as such Commissioner.

[Notary Seal.] F. J. COE,
Notary Public.

STATE OF NEW YORK, }
County of New York, } ss.:

On this 23d day of June, 1900, before me personally appeared John M. Braly, of the firm of W. H. Keepers & Co., described in and who executed the foregoing instrument and who acknowledged that he executed the same as a member and on behalf of the said firm of W. H. Keepers & Co.

[Notary Seal.] HENRY S. GREENBERGH,
Notary Public, New York County, No. 75.

GUARANTEE.—We the undersigned members of the firm of W. H. Keepers & Co., contractor for the construction of the within described bridges, hereby guarantee that the said bridges shall be free from any change of form, or any failure of the whole or any part thereof due to defects or imperfections of material or of workmanship, for a period of five (5) years from the final completion of said bridges and their acceptance by the State Engineer as herein provided. Said guarantee being given in consideration of the price of $102,070 herein named.

WILLIAM H. KEEPERS,
JAS. H. WYNKOOP,
JOHN M. BRALY,

Parties of the First Part.

Approved:

ANDREW H. GREEN,
THOMAS P. KINGSFORD,
CHAS. M. DOW,
GEO. RAINES,
ALEXANDER J. PORTER,

Commissioners of the State Reservation at Niagara.

COPY OF BOND.

(Original filed with contract in Comptroller's office August 4, 1900.)

Know all men by these presents, That we, William H. Keepers, James H. Wynkoop and John M. Braly, of the city of New York, in the county of New York, are held and firmly bound unto the People of the State of New York, in the sum of ($15,000) fifteen thousand dollars, to be paid to the said People: For which payment well and truly to be made, we bind ourselves, our and each of our heirs, executors and administrators, jointly and severally, firmly by these presents.

Sealed with our seals. Dated this thirty-first day of July, in the year nineteen hundred.

The condition of this obligation is such, That if the said William H. Keepers, James H. Wynkoop and John M. Braly, who have executed the five-year guarantee attached to a contract dated June 23, 1900, between said William H. Keepers, James H. Wynkoop, John M. Braly and the Commissioners of the State Reservation at Niagara, which said guarantee is attached to said contract and has been approved by said Commissioners shall in all respects well and faithfully execute and perform conditions of said guarantee then this obligation to be void, otherwise to remain in full force and virtue.

WILLIAM H. KEEPERS,
JAMES H. WYNKOOP,
JOHN M. BRALY.

STATE OF NEW YORK, }
New York County, } *ss.:*

On this 1st day of August, 1900, before me appeared James H. Wynkoop and John M. Braly, of New York city, who severally acknowledged that they executed the within instrument, and I certify that I know the persons who made the said acknowledgment to be the individuals described in and who executed the said instrument.

[Seal.] HENRY S. GREENBERGH,

Notary Public, New York County, No. 75.

STATE OF NEW YORK, }
Albany County, } ss.:

On this 31st day of July, 1900, before me appeared William H. Keepers, of New York city, who acknowledged that he executed the within instrument, and I certify that I know him to be the individual described in and who executed the said instrument.

[Seal.] JOHN J. ALLEN,
Notary Public, Rensselaer County.

Certificate filed in Albany county.

STATE OF NEW YORK, }
Albany County, } ss.:

William H. Keepers, of New York city, one of the persons named in the within bond, being duly and severally sworn, says that he is worth the sum of fifteen thousand dollars over and above all debts and liabilities entered into or incurred.

WILLIAM H. KEEPERS.

Subscribed to and sworn before me
this 31st day of July, 1900.

[Seal.] JOHN J. ALLEN,
Notary Public, Rensselaer County.

Certificate filed in Albany county.

STATE OF NEW YORK, }
New York County, } ss.:

James H. Wynkoop and John M. Braly, both of the city of New York, who have executed the within bond, being duly and severally sworn, each for himself, says that he is worth the sum set opposite his name over and above all debts and liabilities entered into or incurred, to wit:

James H. Wynkoop, fifteen thousand dollars.

John M. Braly, seven thousand five hundred dollars.

JAS. H. WYNKOOP,
JOHN M. BRALY.

Subscribed to and sworn before me
this 1st day of August, 1900.

[Seal.] HENRY S. GREENBERGH,
Notary Public, New York County, No. 75.

COPY OF BOND.

(The original is attached to contract filed in Comptroller's office August 4, 1900.)

Revenue stamps for $2.42 attached to original.

BOND.

Know all men by these presents, That we, William H. Keepers, James H. Wynkoop and John M. Braly, composing the firm of W. H. Keepers & Company of the city of New York, in the State of New York (hereinafter called the "Principal"), and the United States Fidelity and Guaranty Company, a corporation organized under the Laws of Maryland (hereinafter called the "Surety"), are held and firmly bound unto the People of the State of New York in the full and just sum of fifty-one thousand dollars ($51,000), good and lawful money of the United States of America, to the payment of which said sum of money, well and truly to be made and done, the said Principal binds himself, his heirs, executors and administrators, and the said Surety binds itself, its successors and assigns, jointly and severally, firmly by these presents: Signed, sealed and dated this second day of August, A. D. 1900.

Whereas, Said Principal has entered into a certain written contract bearing date on the 23d day of June, 1900, with the People of the State of New York, by Andrew H. Green, George Raines, Charles M. Dow, Thomas P. Kingsford and Alexander J. Porter, composing the Commissioners of the State Reservation at Niagara, for the construction of two three-span each concrete-steel arch bridges with stone facing, twenty-foot gravel roadway and two ten-foot granitoid sidewalks extending across the Niagara River, from the mainland to Green Island and from Green Island to Goat Island, in the Niagara State Reservation, for the sum of one hundred and two thousand and seventy dollars ($102,070), and the right to use the existing bridges, abutments and piers or materials therein.

Now, therefore, the condition of the foregoing obligation is such that if the said Principal shall well, truly and faithfully comply with and perform all of the terms, covenants and conditions of said contract on his part to be kept and performed, according to the tenor of said contract, and shall protect the said State of New York against, and pay any excess of cost as provided in said contract, and all amounts, damages, costs and judgments which may be recovered against the said State or its officers or agents or which the said State of New York may be called upon to pay to any person or corporation by reason of any damages arising or growing out of the doing of said work, *or the repair or maintenance thereof, or the manner of doing the same, or the neglect of the said Principal, or his agents or servants, or the improper performance of the said work by the said Principal, or his agents or servants, and well and truly pay or cause to be paid in full the wages stipulated and agreed

---

*Albany, N. Y., July 31, 1900—Words "or the repair or maintenance thereof" in 5th line are erased by consent of both parties before the execution of this bond by the Surety Co.—WM. PIERSON JUDSON, Deputy State Engineer of New York.

Stone Stairway, Goat Island. State Reservation at Niagara.
Erected 1900.

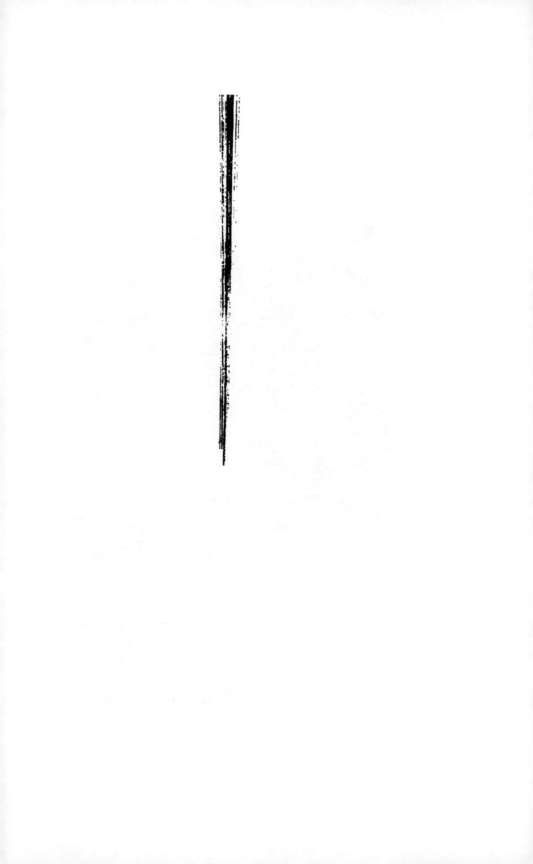

to be paid to each and every workman, laborer or employee employed by the said Principal or his agent or agents in the execution of said work described in the said contract, then this obligation shall be null and void, otherwise to remain in full force and virtue.

And the said Surety hereby stipulates and agrees that no change, extension, alteration or addition to the terms of this said contract or specifications accompanying the same, shall in anywise affect their obligation on this bond.

In testimony whereof, the said Principal has hereunto set his hand and seal, and the said Surety has caused this instrument to be signed by its second vice-president and its assistant secretary, and its corporate seal to be hereunto affixed the day and year first above written.

WILLIAM H. KEEPERS, [L. S.]
JAS. H. WYNKOOP, [L. S.]
JOHN M. BRALY, [L. S.]
Principals.

Signed, sealed and delivered
in presence of
WM. PIERSON JUDSON.

THE UNITED STATES FIDELITY AND GUARANTY CO.,
Surety.

(Corporate seal of Surety Co.) By EDWD. J. SIMMSON,
President.

RICD. D. LAW.

STATE OF NEW YORK, }
County of New York, } *ss.:*

On this 23d day of June, 1900, before me personally came William H. Keepers, James H. Wynkoop and John M. Braly, to me known to be the persons described in and who executed the foregoing instrument, and they severally acknowledged that they executed the same.

[Notary Seal.] HENRY S. GREENBERGH,

Notary Public, New York County, No. 75.

(Acknowledgment By Surety Company.)

(Statement sheet and oath of officers of Surety Company attached to original on file with Comptroller, attached to contract.)

SPECIFICATIONS

FOR

Concrete Steel Bridge at Niagara Falls, N. Y.

1. *Plans.*—The work shall be constructed complete in accordance with the general plans, sections and diagrams herewith submitted, and these specifications.

The specifications and drawings are intended to describe and provide for the complete work. They are intended to be cooperative, and what is called for by either is as binding as if called for by both.

The work herein described is to be completed in every detail, notwithstanding that every item necessarily involved is not particularly mentioned.

The contract price shall be based upon these specifications and drawings, which are hereby signed and made a part of the contract.

2. *Conditions of calculations.*—

	Pounds.
Modulus of elasticity of concrete	1,400,000
Modulus of elasticity of steel	28,000,000
Maximum stress per square inch on steel	10,000
Maximum compression per square inch on concrete	500
Maximum shear per square inch on concrete	100
Maximum tension	50

The above to be exclusive of temperature stresses.

The steel ribs under a stress not exceeding their elastic limit, must be capable of taking the entire bending moment of the arch without aid from the concrete and have a flange area of not less than one one hundred and fiftieth part of the total area of the arch at crown.

3. *Discrepancies.*—In the event of any discrepancies between the drawings and the figures written on them, the figures are to be taken as correct, and in case of any discrepancy between the drawings and specifications, the specifications are to be adhered to.

4. *Foundations.*—All foundations shall be shown on plans, and to conform to the dimensions marked thereon.

In preparing foundations, a temporary cofferdam shall be used and all loose boulders, shale, sand mud or other soft material shall be removed from the rock, and depressions in the rock filled and leveled up with concrete upon which the piers and abutments shall be built; the intention being to obtain a suitable and proper foundation regardless of whether it is found above or below the elevation shown on the plans.

The cofferdams shall be reasonably tight so that they can be pumped out, thus permitting the work to be done under proper conditions.

5. *Cement.*—The cement shall be a true Portland cement, made by calcining a proper mixture of calcareous and clayey earths; and if required, the contractor shall furnish a certified statement of the chemical composition of the cement, and the raw materials from which it is manufactured.

The fineness of the cement shall be such that at least 99 per cent. will pass through a sieve of 50 meshes per lineal inch, at least 95 per cent. will pass through a sieve of 100 meshes per lineal inch, and at least 70 per cent. will pass through a sieve of 200 meshes per lineal inch.

Samples for testing may be taken from each and every barrel delivered, unless otherwise specified. Tensile tests will be made on specimens prepared and maintained until tested at a temperature of not less than 60 degrees Fahrenheit. Each specimen will have an area of one square inch at the breaking section, and after being allowed to harden in moist air for 24 hours will be immersed and maintained under water until tested.

The sand used in preparing the test specimens shall be clean, sharp, crushed quartz, retained on a sieve of 30 meshes per lineal inch, and passing through a sieve of 20 meshes per lineal inch.

No more than 23 to 27 per cent. of water shall be used in preparing the test specimens of neat cement, and in the case of test specimens of one cement and three sand, no more than 11 or 12 per cent. of water by weight shall be used.

Specimens prepared from neat cement shall after seven days develop a tensile strength of not less than 450 pounds per square inch. Specimens prepared from a mixture of one part cement and three parts sand, parts by weight, shall after seven days develop a tensile strength of not less than 160 pounds per square inch, and not less than 220 pounds per square inch after 28 days. Specimens prepared from a mixture of one part cement and three parts sand, parts by weight, and immersed after 24 hours in water maintained at 176 degees Fahrenheit, shall not swell nor crack, and shall after seven days develop a tensile strength of not less than 160 pounds per square inch.

Cement mixed neat with about 27 per cent. of water or only sufficient to form a stiff paste shall after 30 minutes be appreciably indented by the end of a wire one-twelfth inch in diameter loaded to weigh one quarter pound. Cement made into thin pats on glass plates shall not crack, scale nor warp under the following treatment: Three pats shall be made and allowed to harden in moist air at from 60 to 70 degrees Fahrenheit; one of these will be subjected to water vapor at 176 degrees Fahrenheit for three hours, after which it shall be immersed in hot water for 48 hours; another shall be placed in water at from 60 to 70 degress Fahrenheit, and the third shall be left in moist air.

Cement for mortar shall not be used directly from any original package, but the contents of five packages shall first be mixed dry in order to obtain uniformity. This dry mixed cement shall then be measured by bulk as wanted and the specified proportions of dry sand shall also be measured by bulk.

All cement shall be kept housed and dry until wanted in the work.

10. *Stone facing.*—The stone facing is to cover the entire structure, including the piers and abutments below water, only excepting the intrados of the arches between the ring stones on each face, and that portion of the abutments buried in the banks.

Portland cement concrete.—Concrete shall be composed of crushed stone which shall be of approved kind and quality of rock which must be known, before crushing, to be free from soil or mud or dirt, or of gravel having irregular surface; the sizes of such stone or gravel to be as hereafter specified. With this crushed stone or gravel shall be mixed clean sharp sand as hereafter specified, which shall be washed if necessary to remove loam or silt, and American Portland cement of the quality above specified; the proportions of sand and cement to be as hereafter specified.

Sand.—Sand which is to be used for forming the concrete and mortar shall be of the best quality available and shall be the cleanest and sharpest found in the vicinity of the work.

Samples of sand.—The contractor shall inform the State Engineer as soon as the contract is awarded what sand is proposed to be used. Samples of this sand shall be obtained by the Engineer and will be examined and tested at the cement testing office of the State Engineer, and if found to contain an injurious amount of loam or silt or material that is friable or soluble, the contractor will be required to wash the sand before it is brought on the work. It will be the duty of the engineer in charge to see that the soil overlaying the sand bank (if the sand is thus obtained) is cleared away so that no soil shall slide or wash into the sand during its use, and special attention will be given during the progress of the work to see that dirty sand shall not be used in making concrete or mortar. If the sand is obtained by pumping from Lake Erie or the Niagara river, care will be taken that the sand is obtained from the locality free from sewerage and silt.

These materials shall be mixed by machine for all portions of the work except for minor parts requiring small quantities.

The machines used shall be of approved kind for which the proportions for each batch shall be exactly measured; consideration shall not be given to any continuous mixer in which the proportions depend upon shovelers, it being required that the proportions of the several materials and of water shall all be accurately measured for each batch.

The stone or gravel, sand and cement in specified proportions shall be mixed dry in the mixer and the due proportion of water shall then be added and the mixing continued until the product is so thoroughly mixed that every face of every particle of stone or of gravel is completely coated with the cement and sand.

Hand mixing.—Mixing by hand may be permitted on minor pieces of the work where the quantity is less than a charge of the mixer. It shall then be done in the following manner:

The stone or gravel shall be sprinkled while in the pile so that there shall be no dry dust, and the measured quantity for each batch shall then be spread on a plank bed in a layer not more than four inches thick; the measured quantity of mortar shall then be made on a separate mortar-bed where the specified proportions of cement and of dry sand shall be mixed dry by turning with shovels or by screening as being the most effective and easiest way of securing a perfect mixture which shall be of uniform color and without streaks which will be required before adding water to make the mortar, which shall be done gradually to avoid washing the cement from the sand. The quantity of mortar thus made at each batch shall be no greater than shall be used in 45 minutes after beginning to mix it. If this mortar has set at all before using, it shall not be retempered but shall be thrown away.

Voids.—Before beginning construction the engineer in charge shall determine the voids in the crushed stone or the gravel which is to form the aggregate of the concrete.

The voids shall be determined by filling a twelve-inch cubical box with well-shaken dry crushed stone or gravel, placing this on accurate scales and weighing the water required to cover the contents.

Proportion.—The proportion of mortar which is to form the matrix of the concrete shall then be varied slightly if necessary in order that it shall exceed the voids by not more than one to two per cent. of the total mass of the aggregate. The proportion shall be used until a change in the character of the aggregate may require a slight variation in the proportion of mortar; the relative propotions of sand and cement being as specified.

Two grades of concrete shall be used in the work for which the proportions shall be as follows, subject to the modification above described, the volume of the material being measured loose:

For the arches between skewbacks, one part Portland cement, two parts sand and four parts broken stone or gravel, which shall pass through a one and one-quarter inch ring, including the total product

of the crusher between one and one-quarter inch and one-quarter inch.

For the foundation abutments, piers and spandrells, one part Portland cement, three parts sand and six parts broken stone or gravel, which shall pass through a two-inch ring, including the total product of the crusher between two inches and one-quarter inch. In the last named portion of the structure, large stones containing not less than one and one-half cubic feet each may be placed in the body of the piers and abutments, provided that each stone is carefully set in mortar on its broadest bed, and so placed that no stone is nearer than eight inches to another stone or to the face of the work, and so that the concrete shall be properly rammed into intervening spaces. Each of these stones shall be washed perfectly clean.

Depositing.—In joining new concrete to other concrete that has already set, precaution shall be taken to secure a perfect bonding by sweeping and washing the work already in place, and by spreading over its surface a thin layer of mortar before the new concrete is placed.

Separate batches.—In any given layer the separate batches shall follow each other so closely that each one shall be placed and rammed before the preceding one has set so that there shall be no line of separation between them. When a machine mixer is used, the successive layers must also follow each other before the preceding layer has set, so that each day's work shall form a monolith. After the concrete has set it shall not be walked upon in less than 12 hours. *See* "Arches."

Ramming.—The operation of ramming shall be so conducted as to form a compact, dense, impervious, artificial stone whose specific gravity is close to the natural rock which was crushed to form the aggregates. The ramming must be so thorough as to perfectly compact the concrete and to fill the voids so that the mortar comes to the surface and that the concrete shall show a perfectly smooth face when the forms are removed.

7. *Concrete facing.*—If concrete facing is used it shall be composed of one part Portland cement and two and one-half parts of clean sharp sand and shall have a thickness of at least one inch and not more than two inches on the arch soffits, arch faces, abutments, piers, spandrells or other exposed surfaces. Facings and backings must be formed simultaneously in the same horizontal layer. In order to gauge the thickness of the facing correctly, a plate of thin metal with convenient handles shall be set on edge parallel to and two inches from the wall of the form; the facing material shall be deposited in the space between this plate and the form; the concrete of the backing shall then be deposited against the back of this plate, which may then be withdrawn and the whole mass thoroughly rammed so as to bond the facing and the backing and to efface the line of demarkation between them.

No piece of stone from the concrete shall be forced nearer to the exposed surfaces than one inch.

8. *Plastering.*—No plastering will be allowed on the exposed faces on the work, and any concrete which is porous must be removed and replaced at the expense of tl e contractor. The inside spaces of the spandrell walls covered by the filling may be plastered with mortar having the same composition as specified for facing.

9. *Protection of work from changes in temperature.*—All finished and unfinished work until thoroughly set shall be kept moist by sprinkling at short intervals in warm weather.

Concrete must be covered with canvas or otherwise protected from the sun, and in cold weather it must be covered in such a way as to retard freezing. During freezing weather the mortar and the concrete shall be mixed with brine, carefully made by dissolving salt to give about one per cent. of saturation for each degree of temperature below the freezing point.

The brine shall be made in barrels and shall be stirred from the bottom before using to insure uniformity, as shown by frequent observations of salometer. An excess of salt is injurious and a saturated solution which shows free salt lying in the bottom of the barrel shall never be used.

No concrete shall be made when the mercury falls below 20 degrees above zero Fahrenheit, and when the temperature is below freezing point the sand and water shall be heated to 200 degrees Fahrenheit by use of steam coils and otherwise, and special provision shall be made for covering the work at night to prevent freezing as far as possible.

The ring stones shall extend into the work alternately ten and eighteen inches, and the face stones of spandrells and piers shall bed not less than eight inches. Every fifth stone shall be a header and extend into the work at least eighteen inches.

For stone facing the ring stones, cornices and faces of spandrells, piers and abutments shall be of an approved quality of stone. The stone must be of a compact texture, free from loose seams, flaws, discolorations, or imperfections of any kind, and of such a character as will stand the action of the weather. The spandrell walls will be backed with concrete or rubble masonry, to the thickness required. The stone facings shall in all cases be securely bonded or clamped to the backing. All stone shall be rock faced with the exception of cornices and string courses, which shall be sawed or bushhammered. The ring stones shall be dressed to true radial lines and laid in Portland cement mortar, with one-quarter inch joints. All other stones shall be dressed to true beds and vertical joints. No joint shall exceed one-half inch in thickness, and shall be laid to break joints at least nine inches with the course below. All joints shall be cleaned, wet and neatly pointed.

The faces of the walls shall be laid in true lines, and to the dimensions given on plans, and the corners shall have a chisel draft one inch wide carried up to the springing lines of the arch or string course. All cornices, moldings, capitals, keystones, brackets, etc., shall be built into the work in the proper positions, and shall be of the form and dimensions shown on plans.

11. *Artificial stone.*—If molded railings, etc., are used, woven wire shall be molded into them in such manner as to strengthen all projectings, and they shall be of the designs shown on plans, and be molded in suitable molds. The mortar for at least one inch thick shall consist of one part Portland cement and two and one-half parts sand, and, when the size of the molding will admit, the interior may be composed of concrete of the same composition as specified for the arches. When pedestals, posts or panels carry lamp posts, a four-inch wrought iron pipe shall be built into the concrete from top to bottom, and at bottom shall be connected with a three-inch pipe extending under the sidewalk and connected with gas pipe or electric wire conduit. The pipes shall have no sharp bends, all changes in direction being made by gentle curves.

12. *Arches.*—The concrete for the arches shall be started simultaneously from both ends of the arch, and be built in longitudinal sections wide enough to inclose at least two steel ribs, and of sufficient width to constitute a day's work. The concrete shall be deposited in layers, each layer being well rammed in place before the previously deposited layer has had time to partially set. The work shall proceed continuously day and night if necessary to complete each longitudinal section. These sections while being built

shall be held in place by substantial timber forms, normal to the centering and parallel to each other, and these forms shall be removed when the section has set sufficiently to admit of it. The sections shall be connected as specified under "depositing," and also by steel clamps or rib connections built into the concrete.

13. *Drainage.*—Provisions for drainage shall be made at each pier as follows: A wrought iron pipe of sufficient diameter shall be built into the concrete, extending from the center of each space over pier to the soffit of the arch near springing line, and project one inch below soffit. The surface of concrete over piers shall be formed so that any water that may seep through fill above will be drained to the pipes. The line of drainage will be covered with a layer of broken stone, and the top of pipes will be provided with screens to prevent clogging.

14. *Steel ribs.*—Steel ribs shall be imbedded in the concrete of the arch. They shall be spaced at equal distances apart, and be of the number shown on plans. Each rib shall consist of two flat bars of the sizes marked on plans. The bars shall be in length of about 30 feet, thoroughly spliced together and extending into the abutments as shown. Through the center of each bar shall be driven a line of rivets eight inches c to c with heads projecting about seven-eighths inch from each face of bar, except through splice plates, where ordinary heads will be used. The bars shall be in pairs with their centers placed two inches within the inner and outer lines of the arch respectively as shown. All steel must be free from paint and oil, and all scale and rust must be removed before imbedding in the concrete.

The tensile strength, limit of elasticity and ductility shall be determined from a standard test piece cut from the finished material and turned or planed parallel. The area of cross-section shall not be less than one-half square inch; the elongation shall be measured after breaking on an original length of eight inches. Each melt shall be tested for tension and bending.

Test pieces from finished material prepared as above described shall have an ultimate strength of from 60,000 to 68,000 pounds per square inch, an elastic limit of not less than one-half of the ultimate, shall elongate not less than 20 per cent. in eight inches, and show a reduction of area at point of fracture of not less than 40 per cent. It must bend cold 180 degrees around a curve whose diameter is equal to the thickness of the piece tested without crack or flaw on convex side of bend. In tension tests the fracture must be entirely silky.

15. *Rivet steel.*—Test pieces from finished material prepared as above described shall have an ultimate strength of from 54,000 to 62,000 pounds per square inch, an elastic limit of not less than one-half the ultimate strength, shall elongate not less than 20 per cent. in eight inches and show a reduction at point of fracture of not less than 50 per cent. It must bend cold 180 degrees and close down on itself without fracture on convex side of bend. In tension tests the fracture must be entirely silky.

16. *Inspection.*—All raw and finished steel and iron and all workmanship thereon shall at all times and at all stages of the work or manufacture be subject to the inspection and acceptance or rejection of the State Engineer and Surveyor or of his authorized representative, who shall at all times while the work or manufacture is in progress have free access to all parts of the furnaces, mills or shops in which the work, or any part thereof, is in progress. The contractor shall freely furnish all desired facilities for inspecting and testing raw material, ingots and finished material at the furnaces and rolling mills, and shall facilitate examination of workmanship

in the shops and during erection. The contractor shall bear no part of the cost of making such tests and inspection, except the furnishing of required facilities and samples.

As soon as any order for rolled or cast steel or iron is placed by the contractor, a complete copy of the order, together with copies of accompanying diagrams or drawings, shall be furnished by the contractor to the State Engineer and Surveyor or his authorized inspector; such orders shall, in all cases, show the name of the furnaces or mills where the material is to be manufactured, with full description of the kind of material wanted. The contractor shall notify the inspector when the material is ready for inspection, and any delay in making such inspection shall be reported to the State Engineer and Surveyor, who will accept no material which has not been duly passed upon by his authorized representative.

The acceptance at any time of any material or work will not be a bar to its future rejection if subsequently found to be defective.

17. *Workmanship.*—The rivet holes for splice plates of abutting members shall be so accurately spaced that when the members are brought into position the holes shall be truly opposite before the rivets are driven. When members are connected by bolts, the holes must be reamed parallel and the bolts turned to a driving fit. Rivets must completely fill the holes, have full heads concentric with the rivets, and be machine driven when practicable.

18. *Centering.*—The contractor shall build an unyielding false work or centering. The lagging shall be dressed to a uniform size so that when laid it shall present a smooth surface, and this surface shall conform to the lines shown on the drawings. The center shall not be struck until at least 28 days after the completion of the arch, unless permission to remove same earlier is given by the State Engineer and Surveyor. Great care shall be used in lowering the

centers so not to throw undue strains upon the arches. The tendency of the centers to rise at the crown as they are loaded at the haunches must be provided for in the design, or, if not, the centers must be temporarily loaded at the crown, and the load be so regulated as to prevent distortion of the arch as the work progresses.

19. *Casing.*—When concrete facing is used, all piers, abutments and spandrell walls shall be built in timber forms. These forms shall be substantial and unyielding, of the proper dimensions for the work intended, and all parts in contact with exposed faces of concrete shall be finished to a perfectly smooth surface, by plastering or other means, so that no mark or imperfection shall be left on the work.

20. *Waterproofing.*—The waterproofing shall consist of a thin coat of mortar composed of one and one-half parts of sand and one part Portland cement.

After the completion of the arches and spandrells, and before any fill is put in, the top surface of arches, piers and abutments, and the lower six inches of the inner surface of the spandrell walls, shall be covered with a suitable waterproof material so as to effectually exclude water.

21. *Fill.*—The space between spandrell walls shall be filled with sand, earth, cinders, or other suitable material, and be thoroughly compacted by ramming, steam road roller, saturated with water, or other effective means, and finished to the proper grade to receive the curbing and pavements.

The roadway on top of the fill shall be composed of about twelve inches of bank gravel to generally conform with the roadways on the approaches and adjacent drives.

22. *Granitoid sidewalks.*—The spaces over which the sidewalks are to be laid shall be first covered with six inches of cinders well com-

pacted. On this shall be laid four inches of concrete, consisting of one part Portland cement, three parts sand and six parts broken stone or gravel—small size—well rammed. The flag divisions shall then be marked off to the desired size. On the surface of the concrete shall then be laid a wearing surface, one inch thick, composed of two parts Portland cement and three parts broken granite or other acceptable stone, in size from three-eights inch downward. It shall be well rammed and troweled to a perfect even surface and rolled with a toothed roller. This wearing surface must be spread on the concrete while the latter is still soft and adhesive, and neat connections must be made with cornices and curbs.

23. *Roadway*. (See paragraph No. 21.)

24. *Balustrades and handrailings.*—The balustrades shall be of the material and of the form and dimensions shown on plans, and shall be brought to true alignment and be firmly fastened to the outside of each sidewalk in the position shown. If an iron handrailing is used, it shall receive two coats of approved paint after erection. The contractor shall submit drawings of stone rail with openings and shall exhibit sample railings during the progress of the work if required.

25. *Erection.*—The contractor shall employ suitable labor for every kind of work, and all stone work shall be laid by competent masons. The contractor will furnish all staging, piling, cribbing, centering, casing and material of every description required in the erection of the work; also all plant, including dredges, engines, pumps, derricks, barges, mixing machines, pile drivers, conveyors or other appliances necessary for carrying on all parts of the work. The contractor shall assume all risks for loss or damage incurred by ice, floods, fire or other causes during the construction of the

work, and shall sufficiently watch and light the work at night during construction.

26. *Cleaning up.*—Upon completion of work, and before final acceptance thereof, the contractor shall remove all temporary work from the river and all rubbish from the streets as provided in paragraph 28.

27. *Maintaining public travel.*—During the progress of the work the contractor shall build and maintain a temporary foot bridge at least six feet wide. It shall be made strong and safe for the use of pedestrians at all hours.

28. *Removal of old bridge.*—The site of the proposed structure is occupied by an old bridge, which with its piers shall be removed by the contractor in a manner satisfactory to the commission. During the preparation for the work and during its progress and after its completion the contractor shall store and dispose of all materials, both new and old, in a manner satisfactory to the Superintendent of the Reservation.

29. *Work embraced by contract.*—The work embraced by contract will be for the structure complete from out to out of abutments or retaining walls, as per plans and specifications, and will embrace fill, pavement, sidewalks and balustrades complete for this length, unless otherwise mentioned.

The approaches to each of the bridges shall be earth fill with top dressing of about twelve inches of bank gravel. They shall connect with the bridges at the elevations shown on the drawings, and each approach shall slope away in such manner as will be in best harmony with the contour of the land adjoining, and of such easy grades as the engineer may approve.

Wing walls shall be built of the forms shown on the plans, and connecting with abutments. They shall be founded at depths well below frost lines and the sidewalk railing shall be extended over them, as shown on plans.

30. *Approaches.*—The approaches will commence where the work above mentioned ends, and if all or any part of them is included in contract, the same shall be specially mentioned, as provided in paragraph No. 29.

31. *Inspection.*—All materials furnished by the contractor and all workmanship shall be subject to the inspection and acceptance or rejection of the State Engineer and Surveyor.

32. *Name plates.*—Two neat tablets of bronze shall be let into the sidewalk face of the panels of the railing, one inscribed with the names of the officers of the Reservation and the other with the names of the contractors and the year of completion.

33. *Interpretation of plans and specifications.*—The decision of the State Engineer and Surveyor shall control as to the interpretation of the plans and specifications during the execution of the work thereunder.

34. *Expansion joints.*—Expansion joints shall be made in the spandrell walls of each arch above springing lines, or at such other points as may be noted or shown on drawings.

35. *Spandrells.*—The spandrell walls shall have a thickness of not less than 18 inches at any point, and a thickness at bottom of not less than four-tenths of the height of the wall, measured from the top of cornice.

36. *Estimates.*—Approximate estimates of work done shall be made on or about the last day of every month, and a valuation of the same in proportion to contract prices for the completed work will be made by the State Engineer and Surveyor, which sum shall be paid to the contractor in cash on or before the 15th day of the

following month, less a deduction of 15 per cent. upon said valuation, which shall be retained until the final completion of the work.

37. *Final payment.*—Upon the final completion of the work and its acceptance by the Commissioners and the State Engineer and the execution of a guaranty for five years to be of a form approved by the Commissioners, the contractor shall be promptly paid any balance of the contract price which shall then remain due and unpaid.

38. *Imported stone.*—In case it is found that rock which is more satisfactory to the State Engineer and to the contractor can be obtained in Canada, it is understood that the rough stone can be imported uncut, and that this bid is made with the understanding that such importation shall be free from duty.

THIS AGREEMENT, dated this nineteenth day of March in the year one thousand nine hundred and one, made between the firm of W. H. Keepers & Company, parties of the first part, and the people of the State of New York, by the Commissioners of the State Reservation at Niagara, parties of the second part, and supplemental to an agreement between the same parties dated June 23, 1900.

SUPPLEMENTARY AGREEMENT.

WITNESSETH, That in consideration of and pursuant to a provision contained in said agreement of June 23, 1900, providing for the making of an agreement supplementary to that agreement, the said parties do hereby agree as follows: This agreement and the agreement aforesaid of June 23, 1900, shall be read together, as an agreement between the said parties for the construction of two (2) bridges in said agreements mentioned and referred to, it being the intention of the said parties that this supplemental agreement shall affect and change said agreement of June 23, 1900, only so far as the changes therein and thereto are expressly herein provided for.

It is further mutually agreed that the plans which were approved by the State Engineer and Surveyor on December 3, 1900, shall be considered a part of the contract dated June 23, 1900, and of this contract, and that all unfinished parts and portions of the work covered by said plans approved December 3, 1900, shall be completed in accordance with said plans taken together with said agreements and the specifications forming a part thereof. It is further agreed that the plans approved December 3, 1900, provide for the following changes from the plans originally submitted:

(a) In the bridge between the mainland and Green Island the number of spans shall consist of one (1) center span of one hundred and ten (110) feet and two (2) side spans each one hundred and three (103) feet and six (6) inches.

(b) The piers of the bridges between the mainland and Green Island shall be thirteen (13) feet six (6) inches thick, instead of twelve (12) feet six (6) inches thick.

The piers of the bridges between Green Island and Goat Island shall be eight (8) feet thick, instead of six (6) feet eight (8) inches thick.

(c) The ice breakers of the downstream side of the piers of both bridges shall be omitted; and the facing of the upstream side of the breakers shall be of granite instead of limestone from the level of the foot course to the springing of the arch.

(d) The steel ribs shall be made three (3) inches in from the intrados and extrados of the arch, instead of two (2) inches as called for by the original plans.

(e) Each pair of steel ribs shall be connected together by three-quarter (¾) inch bolts at intervals of about two (2) feet.

(f) The contractors are to furnish and put in place a sufficient additional amount of concrete which will be required to decrease the allowable compression of the concrete to four hundred (400) pounds per square inch rather than five hundred (500) pounds per square inch, as provided for in the original contract. The amount and disposition of such additional concrete to accomplish this desired result to be in accordance with the plans aforesaid approved December 3, A. D. 1900.

It is further agreed by and between the said parties that the parties of the first part shall have and receive as full compensation for any and all claims on their part arising out of said changes, or any of them, the sum of five hundred dollars ($500) for and on account of extra thickness of piers of said bridges caused by said changes, and the sum of seven dollars ($7) per cubic yard for each and every yard additional concrete in such bridges caused by said changes; and that said contractor shall receive the sum of three hundred dollars ($300) in full compensation for extra richness of concrete due to direction of the State Engineer and Surveyor; and that said contractor shall receive additional compensation for all the bolts required by subdivision D hereinbefore contained at the rate of three (3) cents per pound for said bolts placed in the work completed.

It is further agreed between the parties hereto that the contractor shall during the progress of the work upon said bridges build and maintain a temporary foot bridge ten (10) feet in width which shall be strong and thoroughly safe for the use of pedestrians at all hours, which bridge shall be divided as to its floor space into two (2) equal parts by a strong, substantial hand railing extending lengthwise through the bridge, and that said contractor shall receive the sum of five hundred dollars ($500) in full compensation for building and maintaining said foot bridge, ten (10) feet in width, instead of six (6) feet in width, which last named width was specified in the original specifications.

And the parties of the second part hereby agree that the parties of the first part shall be paid and receive the aforesaid items of additional compensation in addition to the original contract price of $102,070, and that no deduction shall be made from said original contract price of $102,070 for or by reason of any of the changes hereinbefore specified.

It is further agreed by and between the parties hereto that the parties of the first part shall not later than July 1, 1901, so far complete the work of constructing the bridge between the mainland and Green Island, under and in accordance with the provision of said agreement of June 23, 1900, and of this agreement and of the original specifications, and in accordance with the plans approved December 3, 1900, that said bridge shall then be in such condition that teams and pedestrians shall be able to pass over it with comfort and safety not later than that day, and in case the parties of the first part shall fail to so far complete the construction of the work of said bridge between the mainland and Green Island by July 1, 1901, that teams and pedestrians are unable to pass over it with comfort and safety, then and in such case the parties of the first part shall become liable to the State of New York for and they do hereby promise and agree to pay as liquidated damages for such delay in the completion of such construction work the sum of seventy-five dollars ($75) for each day after the said 1st day of July, 1901, that said bridge shall be in such condition that teams and pedestrians are unable to pass over it with comfort and safety, it being further mutually agreed and understood between the parties hereto that as the liquidated damage clause next above hereinbefore contained is based on the supposition that the contractor will be able to resume operations in the Niagara River on or before April 1, 1901, and that in case the condition of the Niagara River as to volume of water or running ice shall be such as to prevent the contractor at that time from commencing and continuing operations connected with the springing of the arches in the said bridge between the mainland and Green Island, that all of the time after April 1, 1901, during which they shall be prevented from such causes from commencing and continuing said operations, shall be considered as a credit to the contractor and deduction shall be made therefor in determining the time at which the liquidated damage clause should begin to be enforced.

It being understood and agreed by and between the parties hereto that the liquidated damage as above described and provided for shall not apply in case of delays which will be due to acts of God or to the elements or unpreventable accidents beyond the power of the contractor to overcome.

It being left to the discretion of the State Engineer and Surveyor and said Commissioners to determine whether any delays to which the hereinbefore liquidated damage clause would apply have been caused by the condition of the Niagara River, running ice therein, acts of God, the elements, or unpreventable accidents beyond the power of the contractor to overcome.

It is hereby mutually understood and agreed by and between the parties hereto that the foregoing supplemental agreement extends and relates to all matters of difference now or heretofore existing between the contractor and the Commissioners, and this supplemental agreement is entered into for the purpose of adjusting and determining all such differences which have existed or now exist, and it is hereby agreed between the parties hereto that the same are now fully and finally adjusted by this supplemental agreement.

In witness whereof the parties to this agreement have hereunto set their hands this 19th day of March, A. D. 1901.

W. H. KEEPERS & CO.,

Contractor.

By W. H. KEEPERS.

ANDREW H. GREEN,

THOMAS P. KINGSFORD,

GEORGE RAINES,

ALEX. J. PORTER,

CHAS. M. DOW,

Commissioners of the State Reservation at Niagara.

Approved March 19, 1901.

EDWARD A. BOND,

State Engineer and Surveyor.

AGREEMENT

FOR

ADDITIONAL TRACKS IN THE RIVERWAY, STATE RESERVATION AT NIAGARA.

THIS AGREEMENT, Made the 2d day of July, 1901, between The Niagara Falls and Suspension Bridge Railway Company, party of the first part, and the Commissioners of the State Reservation at Niagara, party of the second part, Witnesseth:

Whereas, under and pursuant to an agreement made by and between the parties hereto, dated the 23d day of September, 1899, the party of the first part has constructed for the party of the second part, a single street railway track upon and along Riverway upon the State Reservation at Niagara, between Falls and Niagara streets, which track the party of the first part is using under a license therefore issued by the party of the second part, dated the 23d day of September, 1899; and

Whereas, the party of the first part desires the use of a double street railway track along a portion of said Riverway and also of turnouts to its passenger station fronting upon said Riverway, all as laid down upon the map hereto attached.

Now, for the purpose of changing or altering said single street railroad track to a double track with turnouts, as shown upon the said map attached, it is agreed as follows:

1. The party of the first part agrees to construct on and along Riverway upon the State Reservation at Niagara, between Falls and Niagara streets, street railroad tracks with turnouts in complete conformity to the maps, plans and specifications hereto attached and made a part hereof, and to pay all expenses connected therewith

incurred by the Commissioners, and to make all changes and alterations in the said existing track necessary to make the tracks as finally constructed and completed, conform to the map hereto attached.

2. Such work shall be commenced at once and prosecuted diligently to completion.

3. The work of construction shall be carried on under the direct supervision of said Commissioners and subject to such regulations and directions as they may from time to time make and give, and in giving such instructions and directions, the Superintendent of the Reservation shall be treated and obeyed as the agent and representative of the Commissioners.

4. The party of the first part shall keep an exact and itemized statement of all the cost and expense of such construction, including expenses incurred by the Commissioners in connection therewith for counsel, engineering or otherwise, and shall furnish to the Commissioners a detailed written statement thereof, and when the work is completed the Commissioners shall determine and certify the reasonable cost of such work, having power to determine what are proper expenses, payments and charges.

5. The said tracks and all work, material and constructions connected therewith within the limits of the State Reservation, shall at all times be and remain the property of the State, free from all lien, claim or charges of the party of the first part, which shall have no claim or charge whatever against the Commissioners or either of them, or the State, for the work done or payments made hereunder.

6. Upon the completion of said tracks, and when the same are ready for use, the said first party may use the same for street railroad purposes pursuant to and under the terms of a license granted

by the said Commissioners to the party of the first part, dated the 23d day of September, 1899, for the use of street railroad tracks in said Riverway. And it is further agreed that the terms of the paragraph of said license numbered one are hereby changed and amended so that the party of the first part shall not be called upon or required to make the first payment of the license fee of $1,000 per annum in advance for the use of said tracks, and the same shall not be due and payable until the expiration of such time as license fees at said rate computed from the 23d day of September, 1899, shall equal the amount expended or advanced by the party of the first part in the construction of said street railroad tracks in said Riverway under and pursuant to an agreement between the parties hereto made the 23d day of September, 1899, and the further amount expended or advanced by it under and pursuant to this agreement.

7. This agreement is made subject to the provisions of section 108 of the Railroad Law and shall not be of force until the Commissioners of the Land Office have given the consent therein required.

8. The construction of said work and the payment of the expenses connected therewith shall not constitute any claim or ground for claiming that said license shall not be revocable at the pleasure of the Commissioners.

9. The party of the first part, in consideration of said license, agrees to indemnify and save harmless the State of New York, the Commissioners of the State Reservation at Niagara and each of them, and their and each of their successors in office, against all costs, payments, expenses, suits and damages whatsoever, arising from or growing out of the improper condition of said tracks in said Riverway, or the use thereof, by the first party or any of its

allied roads named in the said license during the construction of said tracks and the continuance of said license.

In witness whereof, the parties hereto have executed this agreement in duplicate the day and year first above written.

NIAGARA FALLS AND SUSPENSION BRIDGE RAILWAY COMPANY,

By W. CARYL ELY, *President.*

GEO. RAINES.

T. P. KINGSFORD.

CHAS. M. DOW.

ALEX. J. PORTER.

Attest:

R. F. RANKINE,

Secretary.

[Seal.]

STATE OF NEW YORK, } *ss.:*
County of .... }

On this fourteenth day of August, in the year of our Lord one thousand nine hundred and one, before me personally came W. Caryl Ely of the city of Buffalo, N. Y., to me personally known, who being by me duly sworn, did depose and say that he resides in the city of Buffalo, N. Y., and is the president of the Niagara Falls and Suspension Bridge Railway Company; that he knows the corporate seal of said company; that the seal hereunto affixed is said corporate seal, and that it was hereunto affixed by order of the board of directors of the said company, and he signed his name thereto as president, by like order.

HUGH McDONALD,

Notary Public, Niagara County, N. Y.

Certificate filed in Erie county.

SPECIFICATIONS

FOR

Additional Tracks in the Riverway of the State Reservation at Niagara Falls.

The tracks shall be laid in conformity with the attached plan, and the grade shall conform to the street levels as now occupied. The track now in place shall be moved two (2) feet towards the river; the gutter on the river side of the driveway shall be moved back two feet and carefully relaid, the present location of the gutter to be replaced by macadam. The curb on the east side of driveway, commencing at a point three feet southerly of the first iron pole of overhead construction, shall be set back on curve so that when tangent is reached it will be two feet back of its present position, the whole of the straight line to be moved back two feet and care taken that the curb at Niagara street is properly recut and set. Any curbstone broken must be immediately replaced and no stone shall be less than thirty inches in length. All surface and overhead work must strictly conform to the tracks already in place, and the engineer in charge shall have full authority to cause the removal or correction of work improperly executed. All detail specifications of original work of every character shall apply to these additional tracks. Where tracks cross the sidewalk in front of the car barn, great care must be exercised to close up every opening excepting the space necessary for the flange of the wheel, and no guard-rail or part of frog shall show above the level of the sidewalk. All necessary changes shall be made in catch basins, or otherwise, to take care of drainage, and when work is completed the whole length of the Riverway disturbed shall be properly resurfaced and thoroughly rolled after being well wet down.

At a meeting of the Commissioners of the Land Office, held at the office of the Secretary of State, on Thursday the 27th day of February, 1902, at 9.30 o'clock A. M.

Present: John T. McDonough, Secretary of State; Nathan L. Miller, Comptroller; John P. Jaeckel, Treasurer; John C. Davies, Attorney-General; Edward A. Bond, State Engineer and Surveyor.

The Secretary of State in the chair.

Mr. F. W. Stevens, Attorney for the Commissioners of the State Reservation at Niagara, applied for consent of the Commissioners of the Land Office, pursuant to the provision of section 108 of the Railroad Law, as amended by chapter 710, Laws of 1899, to construct, without expense to the State, additional street railroad tracks upon and along Riverway, between Falls and Niagara streets, in the city of Niagara Falls.

Upon motion of the State Engineer and Surveyor, the following resolution was adopted:

Resolved, That the Commissioners of the Land Office do hereby consent, pursuant to section 108 of the Railroad Law, as amended by chapter 710, Laws of 1899, that the Commissioners of the State Reservation at Niagara may construct, without expense to the State, additional street railroad tracks upon and along that portion of Riverway, so called, between Falls and Niagara streets, in the city of Niagara Falls, as shown upon a map submitted and filed herewith, and that they may issue revocable licenses for the use of said tracks as provided in section 108 of the Railroad Law.

STATE OF NEW YORK, }
Office of the Secretary of State, } *ss.:*

I have compared the preceding extract from the minutes of the proceedings of the Commissioners of the Land Office with the proceedings of said Commissioners on file in this office, and I do hereby certify the same to be a correct transcript therefrom and of the whole thereof relating to the construction of street railroad tracks upon and along a portion of Riverway, so called, in the State Reservation at Niagara.

Witness my hand and the seal of office of the Secretary of [Seal.] State, at the city of Albany, this twenty-eighth day of February, one thousand nine hundred and two.

JOHN T. McDONOUGH,

Secretary of State.

APPENDIX.

BULLETIN

OF THE

New York State Museum.

FREDERICK J. H. MERRILL, Director.

No. 45. April, 1901.

GUIDE TO THE

GEOLOGY AND PALEONTOLOGY OF NIAGARA FALLS AND VICINITY

By AMADEUS W. GRABAU, S. D.,
Professor of Geology Rensselaer Polytechnic Institute, and Lecturer in Geology Tufts College.

PREFACE

With the support and cooperation of the Buffalo society of natural sciences and the department of paleontology of the state museum, Dr Grabau has prepared this guide to the geology and paleontology of Niagara falls and vicinity with the special purpose of affording to visitors to Buffalo during the season of the Pan-American exposition in 1901 a viaticum in their tours through this region renowned for its scenic features and classic in its geology. The ground has been the subject of a multitude of scientific treatises concerned now with the succession of events in the upbuilding of the rock strata along the canyon of the river; again with the nature of the organic remains inclosed in these strata; sometimes with the changes which the falls have undergone in historic times, but for the most part with the perplexing problems of the origin of the present drainage over the great escarpment and through the gorge, the raison d'être of the falls, the various changes in the course and work of the Niagara river since its birth and the significance of the present topography of the region. These scientific investigations began with the careful surveys instituted by the late Prof. James Hall, state geologist and paleontologist, during the years of his explorations in the 4th geologic district of this state from 1837-43, who, in addition to his record of the work done by this tremendous agent, derived from this region an important term in the New York series of rock formations, the Niagara group, and portrayed the organisms of the various strata which are so superbly exposed along its great channel. Lyell and Bigsby, Logan, Gilbert, Upham, Spencer, Leverett and Taylor are among the names of others who have contributed, from various points of view, facts and hypotheses relating to the geologic history of the river. In no one place however has the general purport of all these various studies been brought together so that the intelligent traveler or student can acquire them in convenient form. It is for this reason that Dr Grabau's work in bringing together in concise form the essence of these investigations, tempered and proved by his own review of them in the field, will not fail to prove serviceable to a large element of the public.

JOHN M. CLARKE
State paleontologist

Niagara Falls from Father Hennepin's view point.

INTRODUCTION —NIAGARA FALLS

AND

HOW TO SEE THEM[1]

The falls of Niagara have been known to the world for more than 200 years. Who the first white man was that saw the great cataracts is not known, but the first to leave a description was the French missionary, Father Louis Hennepin, who, in company with La Salle, visited the falls in 1678. He was the first white man to use the name, Niagara, for the river and the falls, a name which had been applied by the Neuter Indians, who occupied the territory on both sides of the river prior to the year 1651, when they were conquered by the Senecas, who after that occupied and possessed the territory.[2] In the native language the name is said to signify " the thunder of the waters ".

The first sight of the great cataracts must have made a powerful impression on Father Hennepin, unprepared as he was by previous descriptions save those given him by his Indian allies and guides. He speaks of the falls as " a vast and prodigious Cadence of Water which falls down after a surprizing and astonishing manner, insomuch that the Universe does not afford its Parallel ".[3] He considered the falls " above Six hundred foot high ", and adds that " the Waters which fall from this horrible Precipice, do foam and boyl after the most hideous manner imaginable, making an outrageous Noise, more terrible than that of Thunder, for when the

[1] Niagara falls are reached from Buffalo by train or electric cars, both of which run at frequent intervals. A direct line of railway runs from Rochester to the falls by way of Lockport. Direct railway connection with western cities is obtained by way of Suspension bridge, while from Toronto and other cities north of Lake Ontario the falls may be reached by train direct, or by boat to Lewiston or Queenston, and thence by train or electric road to Niagara. All electric cars on the New York side run to or past Prospect park, and most of them pass the railway stations. The railway stations are within walking distance of the falls.

[2] Porter, Peter A. Goat island. 16th an. rep't comr's state reservation, 1900.

[3] A new discovery of a vast country in America. 1698. p. 29. Reprinted in part in special report N. Y. state survey for 1879.

Wind blows out of the South, their dismal roaring may be heard more than Fifteen Leagues off."[1]

If today, from our vantage ground of precise knowledge, we smile on the extravagant estimates of Father Hennepin, it may be asked, who among us, that is capable of admiration and enthusiasm at the sight of nature's grand spectacles, would, on coming unprepared on these great cataracts, be able to form a calm and just estimate of hight and breadth and volume of sound?

Since the time of La Salle and Hennepin, the falls have been viewed by a constantly increasing number of visitors. For Americans of the present generation and for people of other lands as well, Niagara has become a sort of Mecca, to which they hope once in their life time to journey. With many this is a hope long deferred in realization, with most perhaps it is a dream never realized, but those who do go and see, come away with their conceptions of nature much enlarged and with memories which cling to the end of life. Fully to appreciate Niagara, one must give it more than a passing glance from the carriage of an impatient driver, who is anxious to have you " do " Niagara in as short time as possible, that he may secure another " fare ". Learn to linger at Niagara, and give nature time to impress you with her beauty and her majesty. " Time and close acquaintanceship," says Tyndall, " the gradual interweaving of mind and nature, must powerfully influence any final estimate of the scene ". And the growing impression which this incomparable scene produced on him, served to strengthen the desire " to see and know Niagara falls as far as it is possible for them to be seen and known".[2]

It is surprising that many of the visitors to the falls allow themselves to be hurried past some of its most beautiful spots and to bestow on others only casual attention, and then waste a wholly disproportionate amount of time in the museums and curio stores looking for souvenirs purporting to come from Niagara but generally manufactured elsewhere. Real and valuable Niagara souvenirs may be had for the trouble of picking them up, in the minerals, fossils and shells which abound in the vicinity of the falls.

[1] *Loc. cit.*

[2] Tyndall. Fragments of science, " Niagara ".

And, while one gathers these, one's knowledge of Niagara becomes broadened, and the perception grows that there are other lessons to be learned in this region, lessons of even more tremendous import than those taught by the cataracts.

The pedestrian has by far the best opportunity to see and enjoy nature as she is only to be seen and enjoyed at Niagara. The stately forest beauty of Goat island, unequaled in the estimation of those competent to judge, by that of any other wooded spot of similar size; the constantly changing views of gorge and falls and rapids which are obtained from nearly every path on the islands and the mainland on both sides of the river; the magnificence of the turbulent waters as they rush toward you, wave piling on wave, till it seems as if the frail-looking structure on which you stand must inevitably be carried away by them—none can enjoy these to their full extent while sitting in a carriage, though it move never so slowly, or while being compelled to listen to the descriptions and explanations of an unsympathetic and unappreciative driver. If you must ride, patronize the reservation carriages, which leave you wherever you wish to stop and take you on again at your own pleasure.[1]

Views from the New York side

The first view of the falls which the visitor on the New York side enjoys is generally from Prospect point, or some of the more elevated view points along the brink of the gorge in Prospect park (*see* frontispiece, pl. 1). While impressive, this view by no means reveals to its full extent the matchless grandeur of the cataracts, and in this respect the visitor on the Canadian side has the advantage. However, the views from Prospect point and Father Hennepin's view point should be obtained by every one, and it may be that some will find greater attraction in these than in the more comprehensive views obtained from the other side. While in Prospect park, it is well to descend to the foot of the inclined railway, and get the views

[1]These carriages are run at intervals of 15 minutes, starting from Prospect park, and making the circuit of Goat island. The fare is 15c for the round trip, and stop-overs at all places, and for any length of time on the same day, are allowed.

of the falls from below. The ride on the *Maid of the Mist* will be found an interesting and novel experience, besides affording views of the cataracts obtainable in no other way; but most people will defer this till they have seen more of the cataracts and rapids from above. In visiting the foot of the falls, an umbrella should be taken, while a waterproof cloak will be found of great advantage, for the visitor is apt to be drenched by the spray which will be blown on him unawares. Caution is necessary here, as everywhere at Niagara, to avoid accidents. In the talus heaps of limestone fragments, minerals and occasionally fossils may be found.

From Prospect point the visitor should next turn his attention to Goat island, "the most interesting spot in all America", as Capt. Basil Hall called it. The unpoetic name of this island is, as Mr Porter tells us[1], commemorative of the power of endurance of a male goat, which, in company with a number of other animals, had been left on this island uncared for during the severe winter of 1770-71, and proved the only survivor.

From the bridges which cross to Green, and thence to Goat island, memorable views of the rapids above the falls may be obtained; and the visitor will do well to pause, that he may become impressed by the magnificence of the spectacle. Perhaps he will feel as did Margaret Fuller, who said: "This was the climax of the effect which the falls produced upon me—neither the American nor British fall moved me as did these rapids." The naturalist will be interested to note that, in spite of the fearful rush of water, fresh-water mussels have found a lodging place among the more protected rocks, where they seem to thrive well. Along the shores of the islands, in places where other animals would find it hard to gain a foothold, numerous small gastropods may be found clinging to the slippery rock surfaces.

On Goat island, despite the so-called "improvements" for the convenience of visitors, nature still reigns supreme. The virgin character of the forest is almost undisturbed, as it was when the red man regarded this as the sacred abode of the Great Spirit of

[1] Porter. Goat island.

Plate 2

Luna Falls, and the limestone fragments at its base (Copyright by Underwood & Underwood, New York)

Niagara. The botanist will here find a greater variety of plants within a given space than in almost any other district.[1]

But it is in the wonderful views of the falls and the rapids and the gorge which can be obtained from this island, that its chief attraction lies. The various view points are easily found, and the stroller about Goat island would best come on them unawares. Mention may be made of the glimpses of the American falls obtained from the head of the stairway leading to Luna island, as well as from the island itself, and the panorama of rapids, falls and gorge from the Terrapin rocks at the edge of the Horseshoe falls. Every visitor is advised to descend the Biddle stairway and view the falls from below. No charge is made unless one wishes to enter the Cave of the winds, a most thrilling experience for a person of nerve and one unparalleled by any other which may legitimately be obtained at Niagara. But, even if one does not care to go behind the falls, a visit to the foot of the stairway, and a walk along the path at the base of the vertical cliff of limestone will well repay the exertion of the climb. Many noble views of the gorge and the falls may be obtained from the stairway, while from certain points below, impressive sights of the small central fall are to be had. Here too can be seen the undermining action of the spray, which removes the soft shale, leaving the limestone ledges projecting till in the course of time they fall for want of support. On the talus slopes at the foot of the cliff good specimens of minerals and occasional fossils may generally be obtained.

After leaving the Biddle stairway, and the Terrapin rocks, the visitor will proceed southward along the river bank to the bridge leading to the Three Sister islands. On the way the geologist will pause where a wood-road leads off to the left into the famous gravel pit of Goat island, since there the shell-bearing gravels are exposed.[2]

---

[1]A catalogue of the flowering and fern-like plants growing without cultivation in the vicinity of the falls of Niagara, by David F. Day, is published in the 14th annual report of the commissioners of the state reservation. In this a total of 909 species are recorded, a large proportion of which are credited to Goat island.

[2]These shells are described in chapter 5.

A small fall known as " The Hermit's cascade " lies between Goat island and the First Sister. In the pool at the foot of this fall Francis Abbot, the Hermit of Niagara, was wont to take his daily bath.

From the bridges and from the islands unsurpassed views of the upper rapids are obtained. These are particularly impressive when seen from the rocks of the Third Sister. Of these rapids as seen from the Terrapin rocks, the Duke of Argyle wrote:

When we stand at any point near the edge of the falls, and look up the course of the stream, the foaming waters of the rapids constitute the sky line. No indication of land is visible—nothing to express the fact that we are looking at a river. The crests of the breakers, the leaping and the rushing of the waters, are still seen against the clouds as they are seen in the ocean, when the ship from which we look is in the trough of the sea. It is impossible to resist the effect of the imagination. It is as if the fountains of the great deep were being broken up, and that a new deluge were coming on the world. The impression is rather increased than diminished by the perspective of the low wooded banks on either shore, running down to a vanishing point and seeming to be lost in the advancing waters. An apparently shoreless sea tumbling toward one is a very grand and a very awful sight. Forgetting, then, what one knows, and giving oneself to what one only sees, I do not know that there is anything in nature more majestic than the view of the rapids above the falls of the Niagara.

On returning to Goat island the visitor may take the reservation carriage for a return to Prospect park, or he may continue his walks around or across Goat island.

In front of Prospect park the electric cars may be taken to cross the river, the bridge-toll which every foot passenger has to pay, being included in the car fare.

Views from the Canadian side

The Canadian side is reached either by bridge or by the steamer *Maid of the Mist*.[1] Every visitor to the falls should obtain the views from the Canadian side, which are in many respects superior to any obtainable on the New York side. Several rustic arbors have been constructed along the brink of the gorge in Queen Victoria park, and here the visitor may tarry for hours and not weary of

[1] If the visitor plans to take the belt line ride—Niagara, Queenston, Lewiston—he will have opportunity to stop off in Queen Victoria park, and need not make a special crossing.

the sights he beholds. The remarkable vivid green of the water of the Horseshoe falls will excite the observer's interest, and question. Tyndall observes that, while the water of the falls as a whole "bends solidly over and falls in a continuous layer. . . close to the ledge over which the water rolls, foam is generated, the light falling upon which, and flashing back from it, is sifted in its passage to and fro, and changed from white to emerald-green."[1]

Near the edge of the Horseshoe falls are the remains of Table rock, formerly a projecting limestone shelf of considerable extent, and a favorite view point. Huge portions of this rock have fallen into the gorge at various times, the most extensive falls occurring in 1818 and 1850, with minor ones between and since. On one occasion some forty or fifty persons had barely left the rock before it fell. From the remaining portion of this rocky platform a good near view of the Horseshoe falls is obtained, though the visitor is apt to find himself in a drenching shower of spray at almost all times.

Beyond Table rock, in the upper end of the park, and on the Dufferin islands many attractive walks are to be met with. These are generally little visited and afford an opportunity for solitude and escape from the crowds of sightseers. Some of the best views of the rapids above the falls are to be obtained here. A wooded clay cliff bounds the park on the landward side, generally rising steeply to the upland plateau. Here on July 25, 1814, the memorable battle of Lundy's Lane was fought between the British and the Americans; "within sight of the falls, in the glory of the light of a full moon, the opposing armies engaged in hand-to-hand conflict, from sundown to midnight, when both sides, exhausted by their efforts, withdrew from the field ".[2]

At the head of the park, a road leads to the upland, where is situated the famous burning spring. The inflammable gas which here bubbles through the water of the spring is chiefly sulfureted hydrogen, but the quantity is such as to produce a flame of considerable magnitude, and it is asserted that the supply has not diminished for the hundred years or more that the spring has been known to exist.[3]

---

[1] *Loc. cit.*

[2] Porter.

[3] An admission fee is charged here.

The gorge below the falls

The gorge of the Niagara river should be seen from both sides. Here as elsewhere the pedestrian with abundant time has the best opportunity to see the numerous interesting and attractive features; but, since distances here are considerable, it is perhaps more advisable to avail one's self of the conveyances afforded.[1]

The best view of the gorge is afforded by going down the river on the Canadian side and returning by the gorge road. In this way the passenger on the cars gets nearest to the river, particularly if the right hand seats are selected. If the visitor however prefers to go down the river on the gorge road, and return by the Canadian line, let him choose the left side of the cars as nearest to the river in both cases.

After passing Clifton on the Canadian side, and the last of the bridges which here span the gorge, the observer begins to have a view of the whirlpool rapids, which even from this elevation have a threatening aspect. It was through these rapids and through the whirlpool below, that the first *Maid of the Mist* was safely navigated in 1861, having at the time three men on board—a feat which has never been repeated. Through this same stretch of rapids Capt. Webb made his fatal swim, paying for the foolhardy attempt with his life. After passing the rapids we reach the whirlpool, of which good views are afforded from many places along the top of the bank. After crossing several small ravines, that of Bowmans creek is reached. This ravine is a partial reexcavation of the old drift that filled St Davids channel.[2] From the upper end of the bridge which crosses the ravine, a path leads down to the water's edge, the ravine being one of singular attractiveness to the lover of wild woodland scenery. A short distance beyond the bridge is the Whirlpool station of the electric road. Here, from a little shelter built on the extreme point, fine views of the whirlpool and the river above and below it are obtained. The river here makes a right-angled bend, the whirlpool forming the swollen elbow. In the rocky point projecting from the

---

[1] The visitor will do well to purchase a belt line ticket, which entitles him to make the circuit in either direction and to stop at all important points. The Canadian scenic route will take him along the top of the bank, while the gorge road, on the New York side, takes him close to the edge of the water.

[2] *See* map, and chapter 1.

bank on the New York side the succession of strata is finely shown[1]; and from this point northward the New York bank exposes a nearly continuous section as far as the mouth of the gorge at Lewiston.

A short distance below the whirlpool we reach Foster's flats, or Niagara glen, as it is more appropriately called. This is visited by comparatively few tourists, though it is one of the most attractive spots along the gorge.[2] It marks the site of a former fall, and, besides its interest on that account deserves to be visited for its silvan beauty and its wild and picturesque scenery of frowning cliff, huge moss-covered boulders and dark cool dells, where rare flowers and ferns are among the attractions which delight the naturalist. Many good views of the river and the opposite banks may here be obtained, and the student of geology will find no end to instructive features eloquent of the time when the falling waters were dashed into spray on the boulders among which he now wanders. After leaving Niagara glen the visitor should stop at Queenston hights and obtain the view which is here afforded.[3] If possible the more comprehensive views from the summit of Brock's monument should be obtained.[4]

After descending and crossing to the New York side, one may return directly by the gorge road, leaving the inspection of the fossiliferous strata for another day, or one may, after a rest at the hotel, or on the river bank, spend some hours in studying the sections exposed along the New York Central railroad cut.[5]

The return journey by the gorge road is one of great interest, as it carries the visitor close to the rushing waters of the river. Walking along the roadbed is forbidden, and stops are made only at the regular stations.[6] The first of these is the Devil's hole, a cavern in the rock, of the type described in chapter 3 and supposed to have figured in Indian lore. The ravine of Bloody run, a small stream generally dry during the summer season, was the scene of a fearful massacre of the English soldiers by the Seneca Indians in 1763, the

[1] For a description of these, *see* chapter 3.
[2] *See* chapter 2.
[3] *See* chapter 1.
[4] An admission fee is charged here.
[5] Waggoner's hotel near the Lewiston suspension bridge makes a convenient stopping place, specially if one desires to visit the fossiliferous sections. The Cornell, at the ferry landing, opposite the Lewiston railroad station, is also recommended.
[6] In stopping off, be sure to obtain stop-over checks from the conductor.

whole party with the exception of two, with wagons and horses, being driven over the cliff by the savages, and dashed to pieces on the rocks below. Next above the Devil's hole is Ongiara[1] park, a picturesque wooded slope opposite the southern end of Foster's flats, and like parts of that region are dotted with enormous blocks of limestone, which have fallen from the bank above. A short distance above this we come to the whirlpool, where a stop of some time can profitably be made. But by far the most attractive place at which to stop is the whirlpool rapids. The water which here rushes through a narrow and comparatively shallow channel, makes a descent of nearly 50 feet in the space of less than a mile, and its turbulence and magnificence are indescribable. Seen at night by moonlight, or when illuminated by the light from a strong reflector, the spectacle is beyond portrayal. We may perhaps not inaptly apply Schiller's description of the Charybdis to these waters:

Und es wallet und siedet und brauset und zischt,
Wie wenn Wasser mit Feur sich mengt.
Bis zum Himmel spritzet der dampfende Gischt,
Und Well' auf Well' sich ohn' Ende drängt,
Und wie mit des fernen Donner's Getose,
Entstürzt es brüllend dem finstern Schosse.

Fossiliferous sections

These sections are to be seen on the cut of the New York Central and Hudson river railroad, Lewiston branch, and along cuts of the Rome, Watertown and Ogdensburg railroad at Lewiston hights. The former are approachable from Lewiston on the north or the Devil's hole station on the south. The approach from Lewiston is the more natural, as it will give the strata in ascending order. Waggoner's hotel makes a convenient starting point. Follow the car tracks southward to where a road leads off on the left. Entering this, a wood-road is found to lead off on the right, which when followed will bring you on the terrace formed by the quartzose sandstone bed, and on which the bridge towers stand. A quarry in the white sandstone by the roadside gives an opportunity to study this rock, which is practically barren of fossils. Beyond this the tracks of the New York Central railroad are reached, which, after traversing a short tunnel hewn out of the Medina sandstone, bring you to the sections in the gorge (plate 12). Care must be exercised in exploring

[1] One of the 40 ways of spelling Niagara.

these sections, as trains are frequent, and rockfalls from the cliffs are among the daily occurrences. With a little caution however the sections may be studied without danger. The total amount of walking necessary from Waggoner's hotel to the Devil's hole is about 3 miles. Near the upper end of the section, where the track enters a rock cutting, a steep path along the river bank leads to the top of the rocky plateau, and a short walk along the top of the bank will bring you to the Devil's hole station. One may also climb the bank in the quarry at the head of the section, and, passing along the top, reach the Devil's hole station by crossing the bridge over the rock-cut before mentioned. At the Devil's hole station[1] one may either take the surface car, which runs to Niagara falls at frequent intervals (5c fare), or, by paying the admission to the Devil's hole, descend to the gorge road and continue the journey to the falls. (A ticket or 50c fare is required here.)

If the sections are approached from the upper end, the Devil's hole station may be reached by the surface electrics[2] or the visitor may leave the cars of the gorge road at the lower Devil's hole station, and, paying the admission fee, ascend the banks by the stairs and paths. The path from the Devil's hole station to the sections leads close along the brink of the gorge. If the sections are visited in the forenoon, the investigator will find himself in the shadow of the cliffs, which is most grateful on a warm summer day.

The sections on the Rome, Watertown and Ogdensburg railroad are reached from Waggoner's hotel by paths leading up " the mountain " one of which begins on the New York Central tracks not far north of the tunnel.

Geologic nomenclature

Geologic time is divided into five great divisions, based on the progress of life during the continuance of each. These are:

5 Cenozoic time, or time of " modern life "
4 Mesozoic time, or time of " medieval life "
3 Paleozoic time, or time of " ancient life "
2 Proterozoic time, or time of " first life "
1 Azoic time, or time of " no life "

[1] Refreshments may be obtained here.

[2] These electrics run from near Prospect park to the Devil's hole and return, at short intervals.

Each of these time divisions is farther divided into great eras, those of Paleozoic time being given in the annexed table. Each era is in general divisible into three periods of time, the early, middle, and later, for which the prefixes *paleo* (or *eo*), *meso* and *neo* are used. The farther division of the periods is into epochs.

During the continuance of each great time division of the geologic history of the earth, more or less extensive rock systems were deposited, wherever the conditions were favorable. Thus the Paleozoic rock system is that deposited during Paleozoic time. That part of the Paleozoic rock system which was deposited during the Siluric era, is called the Siluric rock series, and similarly, the name of each of the other great *eras* is also applied to the rock series deposited during its continuance. In like manner each geologic *period* has its corresponding *group* of rocks deposited during its continuance. These rock groups and their farther subdivision into stages have, in New York, received local names, the name of the locality where the rocks are best developed being selected. The rocks formed during Proterozoic and Azoic time are generally spoken of as pre-Cambric.

The following table embodies the result of the latest studies.[1] The thicknesses are chiefly obtained from well records published by Prof. I. P. Bishop. The relations of these strata to each other in this region are shown in the north and south section from Canada to the New York-Pennsylvania line, presented in fig. 1.

Ever since the days of Lyell and Hall the life history of Niagara and the origin of the Great lakes has engaged the attention of geologists the world over. Among the names prominent in connection with studies of the geology of Niagara in one or more of its aspects, may be mentioned those of Bishop, Clarke, Claypole, Davis, Fairchild, Gilbert, Hall, Hitchcock, Lesley, Lyell, Newberry, Pohlman, Ringueberg, Shaler, Spencer, Tarr, Taylor, Upham and Wright, besides a host of others.[2]

---

[1] Clarke and Schuchert. Science. n. s. Dec. 15, 1899. 10:3. It will be found to differ in some respects from the table published in the author's Geology of Eighteen Mile creek, etc.

[2] In the field work I have had the efficient assistance of my friend Mr R. F. Morgan of Buffalo.

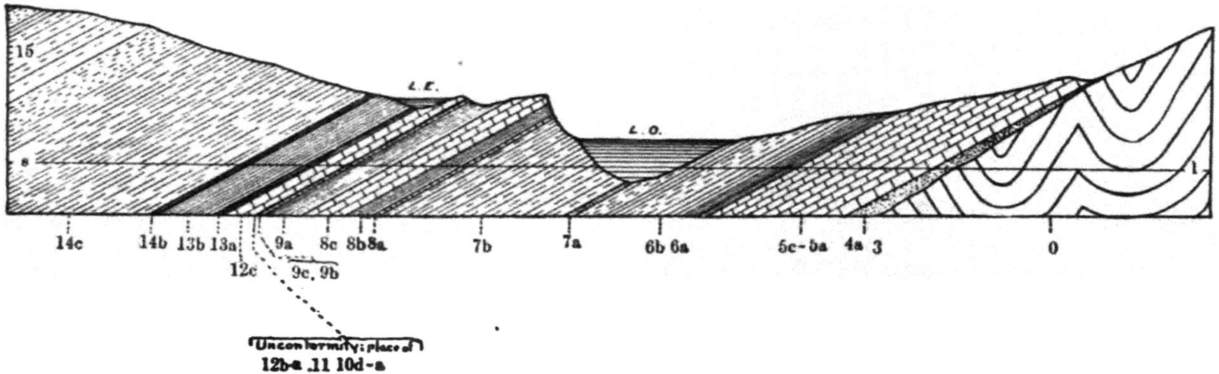

Fig. 1 Section from the Canadian highlands across western New York to the Pennsylvania line, showing the succession of strata. The numbering of the beds corresponds to that of the succeeding table. Owing to the great exaggeration of the vertical over the horizontal scale, the dip of the strata appears much too steep. s. l. = sea-level, L. E. = Lake Erie, L. O. = Lake Ontario. Horizontal scale 1 inch = 38 miles. Vertical scale, 1 inch = 1500 feet.

Table of Paleozoic subdivisions

Time scale	Formation scale	New York state rock equivalents	Thickness in the Niagara region
Paleozoic time	*Paleozoic system*		
5 Carbonic era	5 Carbonic series		
c Neocarbonic period	*c* Upper Carbonic group	Absent in New York	
b Mesocarbonic period	*b* Middle Carbonic group	"	
a Paleocarbonic period	*a* Lower Carbonic group	"	
4 Devonic era	4 Devonic series		
c Neodevonic period	*c* Upper Devonic group	*c* Upper Devonic group	
		15 Chautauquan subgroup	?
		15a Chemung beds (Catskill sandstone, local facies)	
		14 Senecan subgroup	
		14c Portage beds	
		3) Oneonta beds } local facies	
		2) Ithaca beds } local facies	938′ to 1541′
		1) Naples beds } local facies	
		14b Genesee shale	17′ to 25′
		14a Tully limestone	absent
b Mesodevonic period	*b* Middle Devonic group	*b* Middle Devonic group	
		13 Erian subgroup	
		13b Hamilton beds	76′
		13a Marcellus shale	211′
		12 Ulsterian subgroup	
		12c Onondaga limestone (including Corniferous)	108′ to 250′ ?
		12b Schoharie grit	absent
		12a Esopus grit	absent
a Paleodevonic period	*a* Lower Devonic group	*a* Lower Devonic group	
		11 Oriskanian subgroup	
		11a Oriskany beds	thin streaks of sandstone
		10 Helderbergian subgroup	
		10d Kingston beds (upper shaly)	absent
		10c Becraft limestone (upper Pentamerus)	absent

		10b New Scotland beds (Delthyris shaly)	absent
		10a Coeymans limestone (Lower Pentamerus)	absent
3 Siluric era	3 Siluric series		
c Neosiluric period	*c* Upper Siluric group	9 Cayugan group	
		9c Manlius limestone	7′ to 8′
		9b Rondout waterlime	60′
		9a Salina beds	386′
b Mesosiluric period	*b* Middle Siluric group	8 Niagaran group	
		8d Guelph dolomite	absent?
		8c Lockport limestone (Niagara limestone)	200′ to 247′
		8b Rochester shale (Niagara shale)	70′ to 80′
		8a Clinton beds	23′ to 40′
a Paleosiluric period	*a* Lower Siluric group	7 Oswegan group	
		7b Medina sandstone	1266′ ±
		7a Oswego sandstone, Oneida conglomerate or Shawangunk grit	75′ ±
2 Ordovicic era	2 Ordovicic series		
c Neordovicic period	*c* Upper Ordovicic group	6 Cincinnatian group	
		6c [Richmond beds] absent	....
		6b Lorraine beds } 6a Utica shale }	630′
b Mesordovicic period	*b* Middle Ordovicic group	5 Mohawkian group	
		5c Trenton limestone } 5b Black river limestone }	680′ to 720′
		5a Lowville limestone (=Birdseye limestone)	absent?
a Paleordovicic period	*a* Lower Ordovicic group	4 Canadian group	
		4b Chazy limestone	absent?
		4a Beekmantown limestone (=Calciferous sandstone) }	10′ to 110′
1 Cambric era	1 Cambric series		
c Neocambric period	*c* Upper Cambric group	3 Potsdamian group } 3a Potsdam sandstone and [illegible] }	(10′ to 110′)
b Mesocambric period	*b* Middle Cambric group	2 [Acadian]	absent
a Paleocambric period	*a* Lower Cambric group	1 Georgian group	absent
		1a Shales and limestones of Troy and Washington co. N. Y.	

Geologic map

A few words, descriptive of the accompanying geologic map may be added.

The topography is indicated chiefly by contour lines. These lines are 20 feet apart, and each connects the points which have the same elevation above sealevel. Thus wherever the 300 foot contour line occurs, every point along that line is supposed to be 300 feet above sealevel. The level of Lake Ontario is 247 feet above the sea; therefore the hight of any point above Lake Ontario can be calculated from the contours. Where the contours are close together, the slope of the country is steep; where far apart, it is gentle.

The various color patterns indicate what geologic formations would be shown on the surface of any given area, if the drift covering were removed. The beds of this region all dip gently southward; and, as we proceed northward, the lower beds rise from beneath the covering of the higher. Where steep cliffs occur, as in the gorge of the river or at Lewiston or Queenston, the lower beds crop out beneath the upper ones for only a very short space; hence they appear on the map as narrow color bands only. The character of the outcrops in the buried St Davids channel is only approximately delineated, to the extent indicated by well borings. It is probably much more irregular than is shown.

The outlines of the edges of the various beds from Lewiston eastward are taken from a map by G. K. Gilbert, the man who more than any other is identified with geologic studies at Niagara. The outcrops of the Onondaga and waterlime beds are taken from a map by Prof. I. P. Bishop. For the other outlines the author is responsible.

A few statistics[1]

Hight of American falls, Oct. 4, 1842	167.7	feet
" Horseshoe falls, "	[2]158.5	"
Mean total recession of American falls between 1842 and 1890	30.75	"

[1]Chiefly from the annual reports of the commissioners of the state reservation.

[2]The hights vary from 4 to 20 feet with the elevation of the water in the river below the falls.

Mean annual recession of American falls between 1842 and 1890	.64	feet
Mean total recession of Horseshoe falls between 1842 and 1890	104.51	"
Mean annual recession of Horseshoe falls between 1842 and 1890	2.18	
Length of crest line of American falls in 1842	1080	
in 1890	1060	
Horseshoe falls in 1842	2260	
in 1890	3010	
Total area of rock surface which has disappeared at the American falls between 1842 and 1890	32,900	sq. ft
or	.755	acres
Total area of rock surface which has disappeared at Horseshoe falls between 1842 and 1890	275,400	sq. ft
or	6.32	acres

These changes are graphically shown in the successive crest lines of the Horseshoe falls, given in fig. 19, p. 81.

Volume of water passing over the falls each minute[1]	22,440,000	cu. ft
or	1,402,500,000	pounds
or more than	7,000,000	tons

Depth of water in the channel of Niagara river below the falls[2]: (*see* fig. 18)

At foot of Horseshoe falls (center)	150 to 200 feet
Upper Great gorge, from the falls to the beginning of the Whirlpool rapids (from soundings)	160 to 190 "
Whirlpool rapids	35 "
Whirlpool	150 "
Outlet of whirlpool	50 "
Opposite Wintergreen flat	35 "
Below Foster's flats	70 "

[1]Blackwell, Am. jour. sci. 1844, 46 : 67.

[2]Estimated by Gilbert. Am. geol. 1896, 18 : 232-33 and elsewhere.

Width of Niagara gorge[1] (approximate):

Opposite the extreme west end of Goat island, and just in front of the Horseshoe falls	1250	feet
Opposite the center of the American fall	1700	"
Opposite inclined railway	1350	"
Between carriage and railroad bridges, narrowest point midway between the two	1000–1350	"
Just south of railroad bridges	950	"
Gorge of the whirlpool rapids	700–750	"
100 rods south of south side of whirlpool	1200	"
Same at water line	850	"
Inlet to whirlpool	1000	"
Same at water line	550	"
Outlet of whirlpool	900	"
Same at water line	450	"
South of Ongiara park	1300	"
Just south of Wintergreen flat	1600	"
River opposite Foster's flats (bottom)	300	"
Just south of Foster's flats (top)	1700	"
North of Devil's hole	1000	
At the tunnel on the New York Central railroad (plate 12)	1300	"
Average width below Lewiston	2000	"

[1]Chiefly after Taylor. Bul. geol. soc. Am. 9 : 61-65. Top width is given unless otherwise stated.

Plate 3

Chapter 1

PHYSICAL GEOGRAPHY OF THE NIAGARA REGION

The physical geography of the Niagara region is of a relatively simple type, its main topographic features being readily interpreted. Unfortunately no very satisfactory birdseye view of the entire

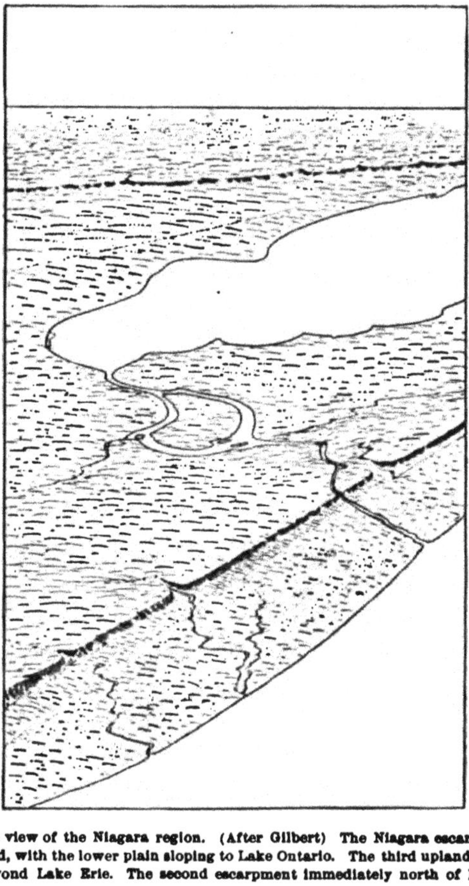

Fig. 2 Birdseye view of the Niagara region. (After Gilbert) The Niagara escarpment is shown in the foreground, with the lower plain sloping to Lake Ontario. The third upland belt is shown in the distance beyond Lake Erie. The second escarpment immediately north of Lake Erie, is not shown.

region can be obtained from any of the elevated points of the district; for the chief features are delineated on a scale too vast to be visible from a single vantage point. The best available spot from which a comprehensive view may be obtained is the summit of

Brock's monument, which commands the hights above Queenston, on the Canadian side of the river. Looking northward from this elevation, the observer sees an almost level plain, cut only by the winding lower Niagara, stretching from the foot of a pronounced and often precipitous escarpment to the shores of Lake Ontario, 7 miles away. Ordinarily the distant northern shore of the lake is not readily recognized by the unaided eye, though on clear days a faint streak of land may be seen between sky and water on the distant horizon. A good field glass however will generally disclose the opposite shore, and the much eroded cliffs of Scarborough. Far beyond these, fully a hundred miles to the north of the observer, the crystalline rocks of the Laurentide mountains rise from beneath their covering of Paleozoic strata, as formerly they rose above the waters of the Paleozoic sea. These ancient Canadian highlands, together with the Adirondack mountains of New York, and the old crystalline regions of the Appalachians, constitute the chief visible remnants of the old pre-Cambric North American continent. The erosion of these ancient lands has furnished much of the material from which beds of later date in this region were derived. Some of these beds may be seen in the sections cut by the rivers through the deposits in comparatively recent times, and no more instructive example than the gorge of the Niagara need be cited.

In the banks of the lower Niagara gorge may be seen the cut edges of the red shales and sandstones of the Medina group, the brilliant color of which is in striking contrast with the greenish blue of the water, and the darker green of the foliage which fringes its borders. The plain above is dotted with farms, orchards and hamlets, and is one of the richest agricultural and fruit districts of the country. In the foreground, on opposite banks of the river, lie the sister towns of Queenston and Lewiston, former rival guardians of the head of navigation of the lower Niagara, but now for the second time joined by bands of steel across the intervening gulf. Farther down the stream Niagara-on-the-Lake and Youngstown crown respectively the left and right bank of the river. These four towns of the lower Niagara, hold daily communication by ferry, steamboat or electric railway; the last and the steam railway keeping them in touch with the cities of the upper Niagara and the world at large. This office is also performed by the well appointed steamboats which ply the lower Niagara, and carry passengers across Lake Ontario, to and from Toronto, the capital and metropolis of the province of Ontario. As these steamers enter or leave the Niagara river, they pass Forts Massassauga and Niagara which stand guard on opposite shores at the mouth of the river. The latter fort was established in 1678, and is rich in historic associations, while the Canadian fort is the modern successor of old Fort George, which was destroyed during the war of 1812.

When the observer on the Brock monument turns to the west or to the east, he sees the escarpment on which he stands, and the plain at its foot stretching in either direction beyond his field of view. The continuity of the escarpment is broken at intervals by ravines or gorges which dissect it, the most pronounced of these being the Niagara gorge in the immediate foreground. Westward from Queenston the escarpment is practically continuous for more than 3 miles, when, at the little town of St Davids, it is seen to recede abruptly, and a gap over a mile in width intervenes, beyond which it continues in force, with only minor interruptions, to Hamilton (Ont.), 40 miles west of the Niagara river. The gap at St Davids marks an ancient valley or gorge cut into the upland plateau which terminates at the escarpment. This old valley is traceable southeastward as far as the whirlpool, in the formation of which it has played a prominent part. It is filled throughout its greater extent by sand and clay, into which modern streams have cut gullies of greater or less magnitude.

Beyond St Davids, the escarpment, though indented by numerous streams, is as stated, continuous to Hamilton (Ont.). Here a larger and more pronounced interruption occurs, the escarpment being breached by a broad and deep channel, locally known as the Dundas valley. This ancient channel, with an average width of 2 miles or more, is traceable westward for a number of miles, when it becomes obliterated by drift deposits. Beyond the breach made by the Dundas valley, the escarpment continues in force, its direction however having changed to west of north, or nearly at right angles to its direction south of Lake Ontario. The eastern face of the Indian peninsula between Georgian bay and Lake Huron and the bold bluff of Cabot's head mark the northward extent of this escarpment, which, after an interruption by a broad transverse channel, is farther traceable in the northern slope of the Manitoulin islands. Eastward the escarpment continues to the vicinity of Lockport, where its continuity is interrupted by two pronounced gulfs, through one of which the Erie canal descends to the lowland of Lake Ontario. Beyond Lockport the escarpment becomes less pronounced; at first it separates into several minor steps or terraces and later it is replaced by a more or less continuous and gentle slope. Beyond the Genesee river it is no longer distinguishable, the surface of the country ascending gently and uniformly from Lake Ontario southward.

Turning now toward the south, the observer sees a second plain extending from the edge of the Niagara escarpment to where its continuity is blended with the horizon. This plain is not as uniform as the Ontario plain, which is fully 200 feet below it, and it is sharply divided by the Niagara gorge, from its northern edge at the escarpment to where, in the distance, a cloud of spray marks the location of the great cataracts. In the walls of the gorge can be seen the cut edges of the strata which enter into the structure of this higher plain, and attentive observation will reveal the fact, that the uppermost of these is a firm-looking limestone bed, which increases perceptibly in thickness toward the north. This thickening of the capping limestone bed, whose upper surface is essentially level, brings out a fact not otherwise readily noticed, namely that the strata all have a gentle inclination or dip to the south. The surface of the upper plain, aside from minor, mainly local irregularities, is essentially level, scarcely rising above the 600 foot contour line. This is the elevation, above the sea, of the base of Brock's monument, and it is the average elevation of the plain in the vicinity of Buffalo, the location of which, 20 miles to the south, is indicated by a perpetual cloud of smoke above the horizon.[1]

[1] A very satisfactory view of the level character of this plain is obtained during a ride by rail from Niagara Falls to Lockport, and thence by train or electric car to Buffalo.

For many miles to the east and west of the Niagara river the plain does not change perceptibly in elevation. Nevertheless, there is a gradual eastward descent, till, on the Genesee river, the surface of the plain, where not modified by superficial deposits, is fully a hundred feet lower than at Niagara. Westward the plain rises gradually, its elevation near Hamilton averaging 500 feet above Lake Ontario, or considerably more than 700 feet above the sea.

Owing to the southward inclination of the strata of this region, the limestone bed which forms the capping rock at the escarpment, eventually passes below the level of the plain, having previously increased in thickness to over 200 feet. The disappearance of the limestone as a surface rock occurs near the northern end of Grand island, as shown by the accompanying geologic map, and from this point southward the surface rock is formed by the soft gypsiferous and salt-bearing shales of the Salina group, which overlie the limestone and in turn pass below the higher strata in Buffalo, where beds of limestone again become the surface rock. Throughout the area where the shales form the surface rock, the plain is deeply excavated on both sides of the Niagara river, a longitudinal east and west valley, now largely filled by surface accumulations of sand and gravel, being revealed by borings. Tonawanda creek occupies this valley on the east, though flowing on drift, considerably above its floor, and Chippewa creek occupies it in part on the west of the Niagara river. This valley, as will be shown later, can be traced westward into Canada and eastward to where it joins the Mohawk valley, with which it forms the great avenue of communication across the state of New York. The northern boundary of the Tonawanda and Chippewa valleys is formed by a limestone cliff similar to, though less pronounced than, the Niagara escarpment. This cliff, generally known as the second limestone terrace of western New York (the Niagara escarpment being the first), is formed by the upper Siluric limestones (Waterlime and Manlius limestone) and the Onondaga limestone of the Devonic series. The latter is a very durable rock and hence it forms a very resistant capping stone. This escarpment is scarcely visible at Black Rock, where it is crossed by the Niagara river, for here it is low, and, in addition, extensive drift accumulations have obliterated its topographic relief. Eastward and westward however it becomes prominent. A drive along Main street from Buffalo to Akron at the Erie county line will reveal the fact that it gradually increases in hight and boldness, till at the latter place it rises nearly a hundred feet above the Tonawanda valley, which itself is drift filled to a not inconsiderable extent. If we trace this escarpment into eastern New York, we find it progressively increasing in hight, owing to the interpolation, between the Manlius and Onondaga limestones, of the thick beds of the Helderbergian series, which, with the other lower Devonic beds, are entirely absent in the Niagara region, where their place is marked by an unconformity. (*See* figs.1 and 21-24)

If the observer changes his position to some elevated point near Buffalo, he may note that the plain which extends southward from the edge of the second escarpment, presents again a scarcely modified and almost level surface, which south of Buffalo gently descends to a third lowland, that of Buffalo creek and Lake Erie. Like the other lowlands, this one is carved out of soft rocks (Marcellus and Hamilton shales) and has subsequently been filled to some extent by drift deposits. This has been proved by borings which show that the bedrock in the valley of Buffalo creek is 83 feet below the surface of Lake Erie.[1] There are other excellent reasons for believing that the western end of this lowland, now occupied by Lake Erie, was once considerably lower than at present.

On the south the Erie lowland is defined by a range of hills, the northern edge of the great Allegany plateau, which forms the highlands of southern New York and northern Pennsylvania. There are no very pronounced declivities in the northern edge of this plateau in the Lake Erie region, owing no doubt to the relatively uniform character of the rocks composing it, there being no resistant capping bed of sufficient magnitude to produce an escarpment. Farther east, however, owing to the increasing thickness of the beds and their more resistant character, a prominent escarpment is developed, which near the Hudson unites with the escarpment of the lower series, and with it constitutes the prominent Hel-

[1] Pohlman. Life history of Niagara. 1888. p. 4.

derberg range, which culminates in southeastern New York in the high plateau of the Catskills. The Allegany plateau is everywhere much dissected by streams whose gorges have made the scenery of southern New York famous.

We have now seen that the topographic features of the Niagara district are arranged in a series of six east and west extending belts of alternating lowlands and terraciform elevations. The lowlands are the Ontario, Tonawanda-Chippewa, and Erie, the uplands are defined by the Niagara escarpment, the Onondaga escarpment and the hills of southern New York which constitute the northern edge of the Allegany plateau. The northern boundary of this belted country is formed by the old Canadian highlands.

We must now briefly consider the various strata of which the area under consideration is constructed, their origin, and the manner in which the topographic features of this region were produced. A brief review of the table of Paleozoic strata, given on pages 20 and 21, will be helpful to an understanding of the succeeding pages.

Development of the Paleozoic coastal plain

The Laurentian old-land is composed of rocks older than the Cambric period of the earth's history. These are largely of igneous origin, and such as were originally sediments have generally suffered much alteration through heat, pressure and other causes, and in most cases have assumed a more or less crystalline character. Though many of these pre-Cambric rocks may show apparent stratification, the present attitude of the beds does not often bear a close relation to their original condition. Indeed, these ancient rocks are generally much disturbed, their beds folded and flexed, and their laminae much contorted. Nor do the layers of the pre-Cambric rocks bear any normal relation to those of later date, the two series being wholly discordant with each other. The older beds are much worn, vast portions of the ancient folds having been swept away by erosion, and on the truncated edges of the remaining portions the newer strata were deposited in an essentially horizontal position. This *unconformity* of relation between the newer and older strata is a marked feature wherever the two series are exhibited in contact with each other. Generally the older rocks have been worn down to an undulating plain (or peneplain), and the succeeding beds made from the fragments which were worn from the old land.

In the area under consideration, the ancient erosion surface of pre-Cambric rocks was overspread by a deposit of sand and occasionally gravel, which commonly possesses characteristics pointing to a very local origin. Thus the pebbles found in the lowest layers of the covering sands, i. e. the Potsdam sandstone, are sometimes of the same lithic character as the crystalline rocks near by. The Potsdam sandstone is a shallow-water rock, and during its accumulation a progressive subsidence of the sea floor took place, thus allowing the deposition of beds of considerable thickness. This subsidence brought with it a northward migration of the shore line of the sea, so that the region of the former coast line gradually became more remote from the shore. As a consequence, land-derived material became less abundant in this off-shore district, being deposited mainly along the new coast line, while farther out to sea calcareous deposits, resulting in part from the shells of organisms, became relatively more abundant. A profile through the strata of this region, such as would be obtained in a well or shaft sunk to the crystalline floor, would show a progressive decrease in the land-derived, or terrigenous material from the Potsdam sandstone upward to the top of the Trenton limestone, and a correspondingly progressive increase in the amount of calcareous matter. This indicates a sustained subsidence of the sea floor, and hence a migration of the shore with its attendant terrigenous deposit. It will also be seen that the lithic character of any particular formation is not the same throughout its extent, but that the local characteristics, or *facies*, show considerable variation. Close to the shore each formation would present a terrigenous character, i. e. would show gravelly, sandy or clayey facies, while away from shore each formation would pass into its calcareous facies, which would increase in purity with the increase in distance from the source of supply of terrigenous sediment. Thus the Potsdam formation has calcareous as well as sandy facies, with facies of intermediate type connecting them.

The Utica shales and the arenaceous Lorraine shales which follow on the Trenton limestones show a return of land-derived deposits due probably to a shoaling of the water. This may have been caused by an upward movement of the sea bottom or by a partial withdrawal of the water into deepening oceanic basins. Some abrupt change is indicated by the sudden transition from limestone to black shale. Another abrupt change occurred at the close of the Ordovicic era, as indicated by the marked contrast between the Lorraine shales and the Oneida and Medina beds which immediately succeed them.

The Siluric deposits of this region began as shallow water accumulations, the lowest bed being the Oswego sandstone, which farther east, is replaced by the conglomerates of Oneida county and the Shawangunk range. The marls and shales of the Medina series succeed these sandstones with an aggregate thickness exceeding 1100 feet. A heavy stratum of gray quartzose sandstone, varying in thickness up to 25 or 30 feet, separates, in the Niagara region, the lower from the upper Medina shales and sandstones, which have an approximate thickness of 100 feet. The Clinton shales and heavy limestones follow on the Medina, with a thickness averaging 30 feet. The Rochester shales, with a thickness of 60 to 70 feet, follow the Clinton limestones and are in turn succeeded by the Lockport limestone, whose average thickness, obtained from well records, approximates 250 feet in this region. The Salina shales succeeding the Niagara beds (Rochester shales and Lockport limestone) have an aggregate thickness of less than 400 feet, and are followed by the Waterlime and the Manlius limestone, the former averaging 50 feet in thickness, the latter from 7 to 8 feet. The lowest Devonic beds are absent in this region, the Onondaga limestone resting directly on the Manlius beds, there being, as before noted, an important though not very pronounced unconformity between the two. A glance at the geologic map of this region will reveal the fact that the lower strata rise from under the covering newer beds on the north, and occupy a belt of country of greater or less width according to the thickness of the beds. Where they come to an end, the next lower beds make their appearance. The discontinuation of the higher beds northward is due to a thinning out of the exposed portion of the strata, as can be readily seen in the Lockport limestone bed, which is less than 30 feet thick at Lewiston, but more than 80 feet at the falls, increasing in thickness southward to 250 feet or more. Where, however, the strata are not exposed on the surface, i. e. where they are only shown in sections under cover of the overlying rock, no such thinning is seen. This may be observed in the case of the Clinton beds and the upper Medina sandstones. In some cases these beds are seen to even thin southward, as proved by borings. The thinning of these strata does not, as is often assumed, mark the original thinning of the beds toward the shore on the north, but is evidently due to erosion. A brief résumé of the origin of the various strata will make this clear.

The Medina sandstone is an ancient shore and shallow water deposit, as will be more fully pointed out in chapter 3. The sands and gravels, which with some finer muds, make up this rock, are all derived from some preexisting land. The only source of supply was the old Laurentian land on the north and the Appalachian old-land on the south. It is true that, owing to the elevation at the beginning of Siluric time, some of the pre-Siluric stratified rocks may have been raised above the sealevel and added to the old-land, and that part of the Medina sands may have been derived from these. Even then the largest amount of detritus was probably derived from the crystalline old-lands, the progressive accumulation of 1200 feet of Medina rock marking a corresponding subsidence and a concomitant encroachment of the seashore of the Medina sea on the old-land. Thus the Medina deposits gradually overlapped the Ordovicic and Cambric deposits and probably eventually came to rest entirely on the crystalline pre-Cambric rocks. Continued subsidence, at least in the Niagara region, produced the purification of the water, so that eventually the limestones of the Clinton epoch could be formed in a region remote from that in which terrigenous material was accumulating. This was likewise true of the Lockport limestone, which was deposited after an interval, during which the calcareous shales separating the two limestone series accumulated. While these deposits, particularly the limestones, point to a considerable distance from the shore line, we are by no means at liberty to assume that no shore formations accumulated during this period. In fact, it would be difficult to understand the non-accumulation of terrigenous material along the shores of any land during any period of the earth's history unless such land was without even moderate relief. As will be shown in chapter 3 there are reasons for supposing that a considerable land barrier existed in the north as well as the east and southeast, and thus we may assume that shore deposits of terrigenous material were formed while the limestones were accumulating in the clearer waters. That the shores of this period did not consist of Medina sandstone is indicated by the absence of any such material in the shales of either the Clinton or Niagara series. It is highly probable that the shore was still formed by the old crystalline highlands, and that the accumulating Clinton and Niagara sediments overlapped and completely buried the Medina beds. The limestones are chiefly fragmental in origin, being composed of calcareous and magnesian sands. These, as will be shown later, were largely derived from the destruction of coral reefs and shells growing in the immediate neighborhood. They indicate shallow water, a conclusion emphasized by the occurrence of well marked cross-bedding structure in some of the beds of limestone. We may assume a gradual passage from pure calcareous beds to beds consisting more and more of terrigenous detritus as we approach the old shore line, where quartz sands probably constituted the chief material of the deposits.

We may obtain an approximate indication of the former extent of these strata if an attempt be made to restore the portions which must have been removed by erosion. We may consider the Clinton and Niagara as a unit, assuming that near the old shore their beds were practicably indistinguishable. The average dip of the strata of this region is 25 feet to the mile (a moderate estimate, as the dip ranges up to 40 feet), and the base of the Clinton-Niagara is approximately 400 feet above sea level. Continuing this dip northward for a hundred miles to where the present borders of the oldland are exposed, the base of this group would have risen 2900 feet above the sea, an elevation sufficient to overtop the highest peak of the present Laurentides; for, according to Logan, "in the country between the Ottawa and Lake Huron the highest summits do not appear to exceed 1500 or 1700 feet, though one . . . probably attains 2300 feet".[1] We assume of course with good reason that the Laurentides at that period were much higher than now, for they must have suffered enormous erosion during the long interval since the close of Siluric time.[2]

Since the deposition of these Siluric strata the region under consideration has suffered an enormous amount of denudation, having been brought to the condition of a low nearly level tract or *peneplain*, but little above sea level, not once, but probably a number of times, separated by periods of elevation and at least one of sub-

[1] Logan. Geol. Canada. 1863. p. 5.

[2] The Niagara beds of Lake Temiscaming, in the great pre-Cambric area of Canada and 150 miles distant from the nearest beds of the same age, are of interest in this connection. They occupy an area about 300 miles due north of Lewiston and on the north side of the present Laurentide chain. According to Logan they do not properly belong to the former extension of the Niagara beds of the region under consideration, but rather to the Hudson bay area on the north. They are of interest however as showing the great former extent of these formations. They lie unconformably on the pre-Cambric rocks, and the basal members are generally sandstones and often conglomerates "containing large pebbles, fragments, and frequently huge boulders of the subjacent rock" (Logan, p. 335). The thickness of the formation here is estimated at between 300 and 500 feet. The Ordovicic and Cambric strata are absent, showing a progressive encroachment of the sea on the old-land, and a consequent overlapping of the strata. Outliers of earlier strata are found in more southern portions of Canada, resting on the pre-Cambric surface, and many of these indicate a progressive overlapping of later over earlier beds. Lawson holds that this indicates, that most of the Canadian old-land was covered by the early Paleozoic strata, and that erosion since Paleozoic time has resulted in simply removing these overlying rocks. (Bul. geol. soc. Am. 1 : 169 et seq.) He holds that comparatively little erosion of the old-land has occurred since Paleozoic time, the present surface being essentially pre-Cambric and only revealed by stripping of the overlying rocks. It is not improbable however that some of these distant outliers may have been preserved during the extensive denudation of the old-land, by having been faulted down previously in a manner well known to have occurred in the Scandinavian old-lands, a solution suggested to me by my friend, A. W. G. Wilson, of Harvard university.

sidence. The present surface of the Niagara plateau is therefore not to be considered as identical with the old surface of deposition, but as due to prolonged peneplanation, or erosion to near sealevel, completed probably toward the close of Mesozoic or the beginning of Cenozoic time. The following diagram (fig. 3) will illustrate the relation between the strata and the surface of the land at 1) the close of Siluric time, 2) late Mesozoic or early Cenozoic time, after the completion of the last cycle of erosion and the reduction of the land to peneplain condition, and 3) the present surface.

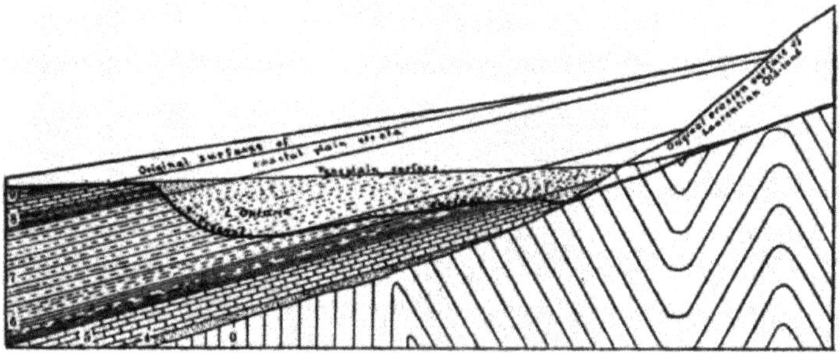

Fig. 3 Diagram of ancient Paleozoic coastal plain, and its relation to the Mesozoic peneplain surface and the present land surface. The numbering of the beds corresponds to that of the table.

Between the close of the Siluric and beginning of Mesozoic time a long period intervened, during which this region was at first a land surface, suffering considerable erosion, but later was resubmerged, and covered with extensive deposits of Devonic limestones, shales and sandstones. The final emergence took place at the close of Paleozoic time, the succeeding Mesozoic time being in this region probably an uninterrupted period of erosion, during which the land suffered the combined attacks of the atmosphere and of running water.

Development of the drainage features

The water which falls as rain or snow on the land either evaporates, runs off on the surface, or sinks into the ground, where it constitutes the ground water. That which evaporates, accomplishes little or no direct geologic work, but both the surface and underground waters are important geologic agents. If the surface on which the water falls is a perfectly smooth but inclined plain, the water will run off in the form of a thin sheet. A perfectly smooth land surface is however unknown, and the run-off of the surface waters is always concentrated along certain lowest lines, thus constituting brooks, streams and rivers. While there may be numerous drainage lines of this type, they generally unite into a few master streams, the direction of whose flow is down the inclination of the surface of the land. Such streams are known as *consequent* streams, their direction of flow being consequent on the original slope of the surface.

When the strata of the Niagara region became a part of the dry land, from the relative lowering of the water level (which may have been due to rise of the land or to drawing off of water by the deepening of the oceanic basins), they formed a broad, essentially monotonous belt of country fringing the old-land on the north, i. e. a marginal coastal plain. The strata of this plain had a gentle southward inclination, a feature shared by the surface of the plain. Consequent streams quickly made their appearance on this plain, a number of them probably coming into existence almost simultaneously and running essentially parallel from the old-land, across the new coastal plain into the sea. These streams soon cut down into the coastal plain, carving channels for themselves and thus establishing definite lines of drainage. As the streams at first consisted entirely of the run-off of the moisture which fell on the plain and in the higher old-land portion, it is evident that, unless the rainfall was continuous, or unless extensive snow fields were present to supply water, these young streams must have fluctuated greatly in volume of water, and at intervals become entirely dry. This condition continued till the valleys, cut by these streams of run-off water, had become sufficiently deep to reach the level of the underground water, when the supply, augmented by springs, became much more constant. Thus in course of time large valleys, supplied with large rivers, came into existence. Meanwhile the sides of the river valleys were attacked by the atmosphere, and degradation of the cliffs cut by the stream resulted.

As long as a river is narrow and vigorously undercuts its banks, the latter will be steep, and the river channel will have the character of a gorge. This generally continues as long as the river is cutting

Plate 4

American and Luna Falls from below, with limestone fragments f: from the cliff above (Copyright by Underwood & Underwood, York)

downward, i. e. till the grade of the river bottom is a very gentle one, when lateral swinging widens the gorge by undercutting the banks, and atmospheric degradation quickly destroys the steep cliffs which the river does not keep perpendicular.

During the process of drainage development, numerous side streams come into existence, which join the main stream as branches. These begin as gullies formed by the rainwater running over the sides of the banks into the main stream. A slight depression in the surface, or a difference in the character of the material composing the banks, may determine the location of such a gully, but, once determined, it will become the cause of its own farther growth. For the existence of this gully will determine the direction of flow of succeeding surface waters, and so in the course of time the gully will become longer and longer by headward gnawing, till finally a channel of considerable magnitude is produced. Streams of this type are known as *subsequent* streams, and they very generally have a direction varying from a moderately acute to nearly a right angle with reference to the main or consequent stream.

As the dissection of the Niagara coastal plain continued, the higher portions of the strata, i. e. those nearer the old-land, were slowly removed, and the beds lying beneath these were thus exposed. The latter strata were generally of a more destructible character than the overlying ones, and on this account great lowlands, parallel to the old shore line, or the line of strike of the strata, were worn in them by subsequent streams. The more resistant beds, meanwhile, favored the formation of more or less prominent cliffs or escarpments which faced the lowlands, and being undermined slowly retreated southward, thus increasing the width of the lowlands. These features are today repeated in the Niagara escarpment which faces the Ontario and Georgian bay lowlands, and the escarpments formed by the outcrops of the Ordovicic limestones farther north. The diagram, fig. 4, illustrates the probable condition during early Mesozoic time. The great master consequent streams indicated are: the Saginaw, the Dundas and the Genesee, flowing from the old-land on the northeast, southward or southwestward into the Mesozoic interior sea. There were probably other consequent rivers, whose location may be in part indicated by some of the valleys now occupied by the Finger lakes of New York. Subsequent streams, flowing along the strike of the beds and capable of accomplishing much erosion by undermining the resistant capping beds of the escarpments, continued to widen the longitudinal (i. e. eastwest) lowland areas, while the transverse valleys of the consequent streams remained relatively narrow.

The topographic relief feature produced by this normal development of drainage on a young coastal plain consisting of alternating harder and softer strata, has been named a " cuesta ",[1] and may be briefly defined as an upland belt of slightly inclined coastal plain strata, with a surface gently sloping toward the newer shore, and a steep escarpment, or *inface,* fronting a low belt, or *inner lowland,* which separates the cuesta from the old-land upon which its strata formerly lapped. The existence of the cuesta form is usually due to a more or less resistant stratum overlying a less resistant one, as, for example, the limestones overlying the upper Medina shales. The inface of the cuesta is continually pushed back by the undermining subsequent streams, aided by atmospheric attack, and thus the belt of low country, lying between the cuesta and the old-land, is continually widened, while during the same time the valley of the transverse consequent stream which carries out the drainage increases comparatively little in width. It must be remembered however that the lowland can never be deepened below the depth of the valley of the consequent stream which carries its waters through the breach in the cuesta.

While the main drainage of this region was undoubtedly southwestward by consequent streams, which flowed through the cuesta in gorges, and by subsequent streams flowing into the former, and occupying the inner lowlands, short streams, flowing toward the old-land, down the inface of the cuesta, were probably not uncommon. These streams began to gnaw gullies back from the inface

[1] Davis, W. M. Science. 1897. New series. 5:362; *also* Textbook of physical geography. 1899. p. 133. Pronounced kwesta, a word of Spanish origin " used in New Mexico for low ridges of steep descent on one side and gentle slope on the other ".

of the cuesta, and ultimately prolonged these gullies into gorges, and carried the drainage into the subsequent streams. Streams of this type, which have their representatives in all coastal plain regions, have been called *obsequent* streams,[1] their direction of flow being opposite to that of the consequent streams. The following diagram (fig. 4) illustrates this type of a stream and its relation to the subsequent and consequent streams. To this type of stream belongs the ancient St Davids gorge, as will be shown more fully in subsequent pages.

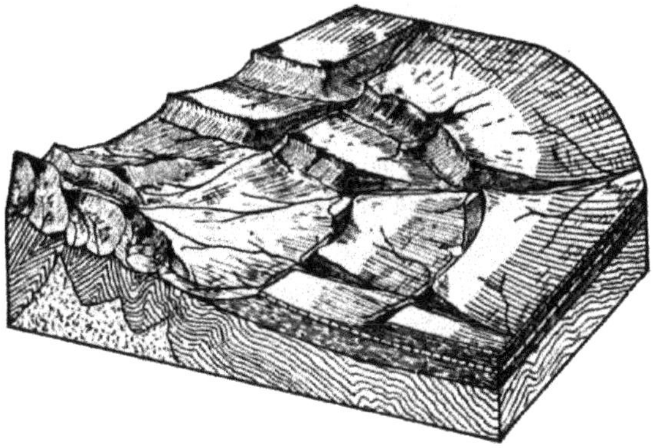

Fig. 4 Diagram of a portion of a dissected coastal plain, showing old-land on the left, and two cuestas with their accompanying inner lowlands. Three consequent streams have breached the cuestas, and subsequent streams from the lowlands join them. An obsequent stream is shown in the center of the outer cuesta.

If we assume that during the greater part of the Mesozoic era, the land in this region remained in a constant relation to the sealevel, it becomes apparent that the southward retreating infaces of the cuestas formed by the resistant members of the Paleozoic rocks, became lower and lower, as the southward inclination of the strata carried the resistant beds nearer and nearer to sealevel. Eventually the escarpment character of the infaces must have become obsolete, from the disappearance, beneath the erosion level, of the weaker lower strata, which permitted the undermining of the capping beds. When this occurred, the capping strata alone continued exposed to the action of the atmosphere, and, from a cliff character, their exposed ends were planed off to a wedge shape, thin-

[1] W. M. Davis

ning northward at a rate proportional to the dip of the beds. The ultimate result of all this erosion was the reduction of the land to a low peneplain, which did not rise much above the sealevel. Portions of this peneplain are today preserved in a scarcely altered condition, in the Niagara upland, the region about Buffalo and other localities. The slight change which these regions have subsequently undergone leads to the supposition that the peneplain was completed in comparatively recent geologic time, possibly at the beginning of the Tertiary era, or even more recently. This is also shown by the comparative narrowness of the valleys cut into the peneplain surface in preglacial times. The present altitude of this peneplain in the vicinity of the Niagara river is approximately 600 feet above sealevel, while southward it rises. There is however good presumptive evidence, some of which will be detailed later, that, during a period preceding the glacial epoch, the land in the north stood much higher than at present, so that the slope of the surface was southward. An accentuation of slope would cause a rejuvenation of the consequent streams, which, in the later stages of peneplanation, had practically ceased their work of erosion on account of the low gradient of the land. As a result of the renewal of erosive activity the early Mesozoic topography was in a large measure restored, but the inface of the Niagara cuesta, the top of which is now found in the Niagara escarpment, occupied in the restored topography a position considerably farther to the south than that characteristic of early Mesozoic time.

We may now examine more in detail the channels of the consequent streams which dissected this ancient coastal plain, and the extent of the inner lowlands drained by the subsequent streams tributary to them.

Dundas valley. The Dundas valley appears to have been the outlet for the master consequent stream of this region, the Dundas river. This valley, as before noted, breaches the escarpment at Hamilton (Ont.), near the extreme western end of Lake Ontario. The valley has been carefully described by Spencer, who considered it the pathway of the preglacial outlet of Lake Erie into Lake Ontario, the drainage of the Erie valley being in his opinion by a river which followed the present course of the Grand river, above Cayuga, past Seneca and Ancaster into the western end of the Ontario valley. It is extremely doubtful that such a stream ever existed, certainly it is highly improbable that the Dundas valley owes its existence to any stream which flowed eastward or toward the old-land, for it is altogether too broad, and continues too uniformly to permit its being regarded as the valley of an obsequent stream. Moreover, its peculiar position at the elbow of the escarpment is most suggestive of a consequent origin, for we would expect the face of the cuesta to make a reentrant where the master stream gathers its converging tributaries and flows out through a great breach in the cuesta.

The Dundas valley is 5 miles wide at Hamilton but rapidly decreases in width to 2 or 2½ miles at the top, where the limestone forms decidedly sharp summit angles (Spencer). Its northern wall has been traced westward for 6 miles to Copetown, and its southern for 3½ miles to Ancaster. Beyond these points the valley is filled with drift which has been much dissected by modern streams. The axis of the gorge is about n 70° e, and the glacial scratches observed on the rock surfaces at its summit, with few exceptions, make angles of 30° or more with it (Spencer).

At Hamilton the bedrock was found to be absent to a depth of 227 feet below the surface of Lake Ontario. The well from which this record was obtained is about 1 mile distant from the southern side of the Dundas valley, which is here 5 miles wide. The total known depth of the canyon is, according to Spencer, 743 feet, but he calculates that it reaches 1000 feet near the center.[1] Along the northern shore of Lake Erie well records have shown the absence of drift to a considerable depth. Thus, according to Spencer, at Vienna, 100 miles due west of Buffalo, the drift is absent to a depth of 200 feet below the surface of Lake Erie, while at Port Stanley, 20 miles farther west, it is absent to a depth of 150 feet below the lake. At Detroit the drift is 130 feet deep. At St Marys on the northwest and Tilsonburg on the southeast of a line connecting

[1] Spencer. Pa. geol. sur. Q 4. p. 384–85.

Port Stanley with Dundas, Devonic limestones occur at a considerable elevation above Lake Erie (Spencer). Hence the southwestward continuation of the Dundas channel must be placed between

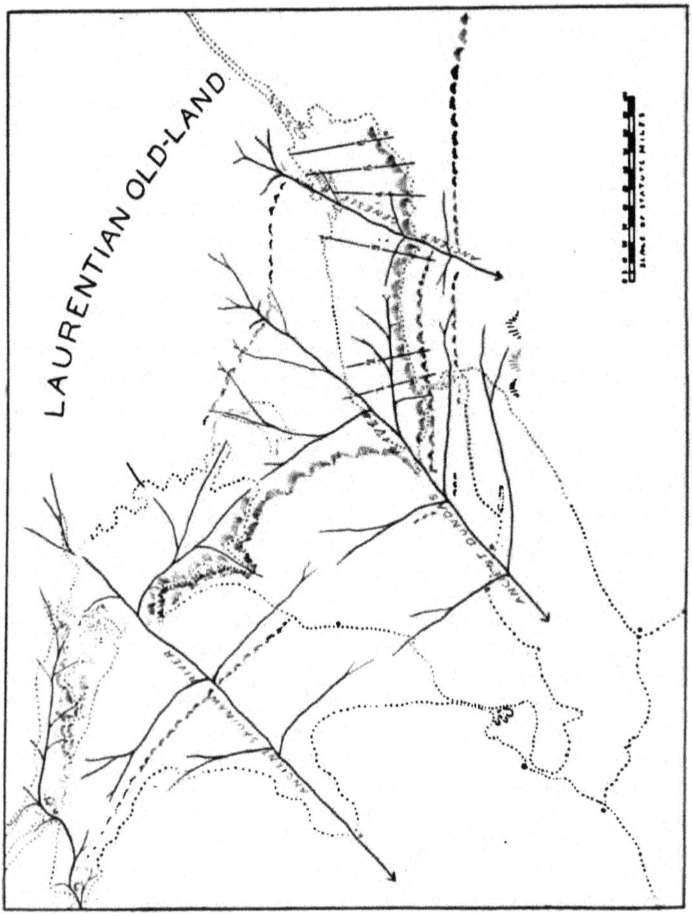

Fig. 5 Diagram showing type of drainage existing in Tertiary times in the Laurentian region.[1]

[1]These maps are intended merely to illustrate the *kind* of drainage, which it is believed existed in preglacial times in the Laurentian region. The ancient consequent streams are probably correctly located; yet it must be stated that the region between Hamilton and Port Stanley has not been sufficiently explored to make the course indicated certain. These consequents may have had a more indirect course, for if the country was worn down to peneplain condition, as appears to have been the case, these streams may have learned

these two points. On the southern shore of Lake Erie borings have revealed numerous deep channels. Thus the bottom of the ancient channel of the Cuyahoga river is reached, according to

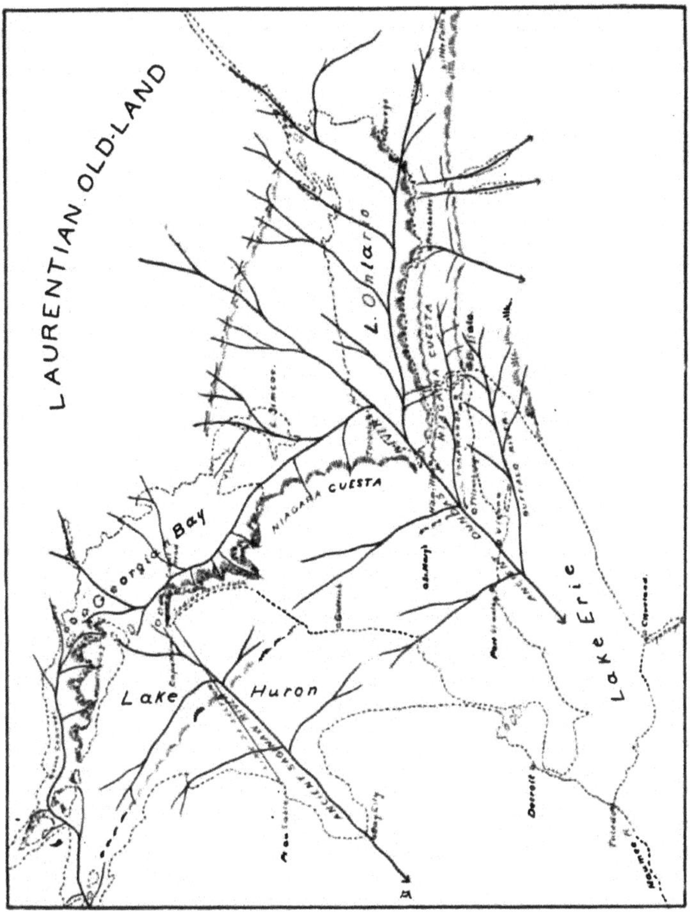

Fig. 6 A hypothetic later stage, showing adjustments which are suggested by existing relief features.

to meander on this surface, the meandering course being retained on re-elevation. The depth of the bed rock at Port Stanley and Vienna, however suggests that a direct channel exists as shown on the map. The principal subsequents are probably located with approximate correctness, but the smaller branches are added without attempt at correctness. They were probably much more numerous than here shown.

Upham,[1] at a depth of more than 400 feet below Lake Erie. Whether this marks the former southward continuation of the preglacial Dundas river or whether that river turned more to the west, following in general the course of the present Maumee, must for the present remain unsettled. The Dundas undoubtedly became eventually tributary to the Mississippi.

Preglacial Saginaw river. The existence of an ancient river, flowing southwestward from the Canadian old-land across the valley of Lake Huron and the lower peninsula of Michigan, and finally becoming tributary to the ancient Mississippi, is indicated by the present character of the topography of that region. The Niagara cuesta is breached by a deep channel which now connects Georgian bay with Lake Huron, and which, north of Cove island, an outlier from the Indian peninsula, has been sounded to a depth of over 300 feet. This channel is in direct line with that of Saginaw bay, and, though this latter is at present very shallow, borings at Bay City show an absence of rock to a depth of at least 200 feet below the surface of the bay. At Alma (Mich.) the rock was shown to be absent to a depth of 350 feet below Lake Huron (Spencer); and, as this locality lies to the southwest of Saginaw bay and in line with the trend of its axis, we may assume that our preglacial Saginaw river was located here. Our limited knowledge of the preglacial topography of this region forbids tracing this channel beyond this point. Dr Spencer many years ago traced out this line of drainage, but he assumed that the river which occupied this channel, and which he has named Huronian, flowed northeastward to join that part of the ancient St Lawrence, or Laurentian river, which he supposed to have occupied Georgian bay.

Preglacial consequent Genesee river. Among the numerous consequent streams which flowed from the old-land southward or southwestward and which eventually became tributary to the preglacial Mississippi, probably through the ancient Ohio,[2] the pre-

[1] Bul. geol. soc. Am. 8: 7.

[2] Westgate, Lewis. Geographical development of the eastern part of the Mississippi drainage system. Am. geol. 1893. 11:245-60. The Ohio, according to Newberry, flows nearly throughout its entire course in a channel, the rock bottom of which is nowhere less than 150 feet below the present river. The rocks at the "falls of the Ohio" show that at that point the river is not following the ancient course.

glacial Genesee river is the only other that can be mentioned here. Though now flowing northward on account of the tilting of the land, we may assume that much of its valley was carved by a southward flowing stream, the bottom of which, as shown by borings, was considerably below the floor of the present river. Whether Irondequoit bay is a part of this ancient channel, or whether it marks the position of an obsequent stream, must remain for the present an open question. Soundings in Irondequoit bay show a depth of 70 feet, though the rock bottom is probably much deeper.

As soon as the consequent streams began cutting down their valleys again after the continental uplift which followed the period of peneplanation, the lateral subsequent streams began once more to open out broad lowlands in the weaker beds which now had become extensively exposed. These lowlands, in part now filled by drift deposits, are the Ontario and Georgian bay valleys, the latter continued in the North Passage, all carved out of the weak Medina and Lorraine shales; the Tonawanda-Chippewa valley, with the deeper portion of the Huron valley farther west, carved out of the soft shales of the Salina group; and the valley of Lake Erie cut out of the softer middle and upper Devonic shales. A few of these may be considered in greater detail.

Ontario valley. It is a well known fact that Lake Ontario is deeper in its eastern than its western part. In the following six cross-sections (fig. 7), constructed from the lake survey charts, the greatest depths from west to east are 456, 528, 570, 738, 684 and 576 feet. The section showing the greatest depth is that from Pultneyville to Point Peter light, in the eastern third of the lake. As the present level of Lake Ontario is 247 feet above the sea, the deepest sounding recorded in these sections is 491 feet below present sealevel. From this point of greatest depth, the floor of the lake rises eastward, at first at the rate of 3 feet in the mile, and later at an average rate of 9 feet a mile. The valley appears to be continued south of the Adirondacks in New York along the present course of the Mohawk river, which flows at present several hundred feet above the rocky floor of the valley.[1] This floor ascends eastward, till at Littlefalls

[1] Carll. Pa. geol. sur. I³:363.

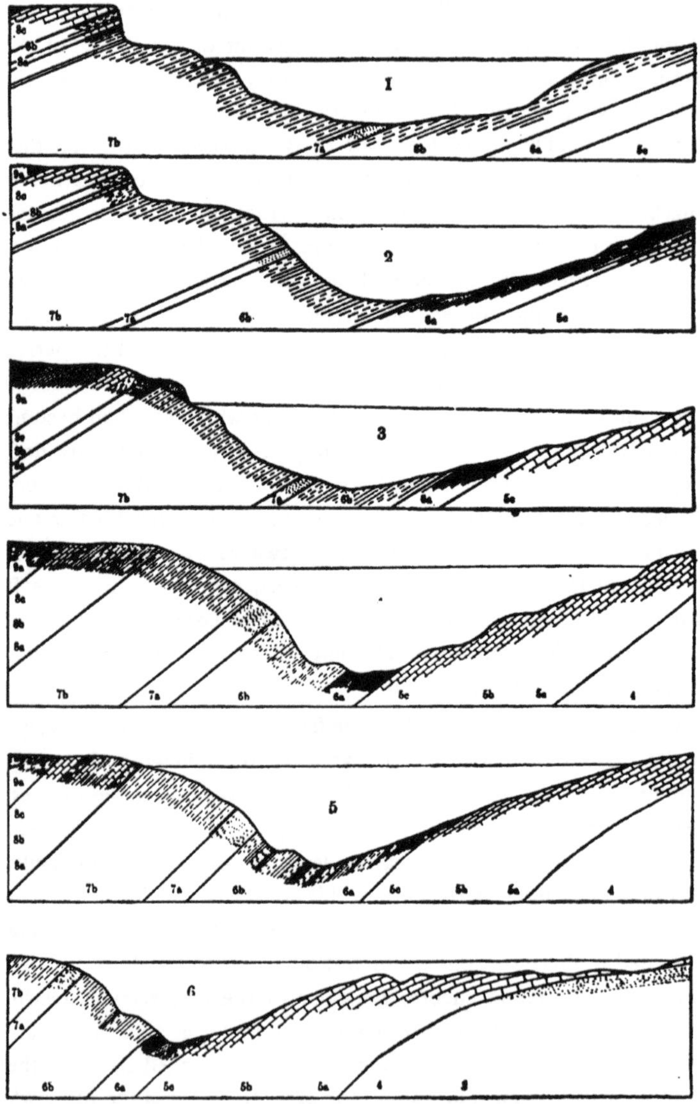

Fig. 7 Six cross-sections of Lake Ontario showing topography and geology. Vertical scale 1 inch = 1280 feet; horizontal scale 1 inch = 15¾ miles. Numbering of beds as in table; location of sections indicated in fig. 5. Section 1) E. of Niagara to E. of Pickering light. 2) Lockport to Darlington light. 3) West of Genesee to Presque Isle light. 4) Pultneyville to Point Peter. 5) West of Fair Haven light to False Duck light. 6) Oswego to Kingston.

the preglacial divide has an elevation of 440 feet above sealevel.[1] The following diagram (fig. 8) shows the present relation of the deepest part of the channel of Lake Ontario to sealevel, and the relation which would result by a tilting of the land back to its probable position in preglacial times. The last profile shows a continuous westward slope of the floor of the valley, steeper in the eastern portion, where the rocks are harder and the valley narrower, and more gentle in the western portion, where the softer rocks have allowed the opening of a broad lowland.

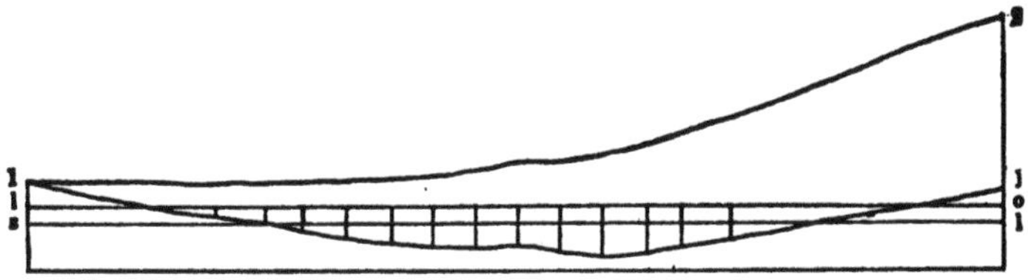

Fig. 8 Diagram showing the present deepest e–w channel of Lake Ontario along line 1–1, and its relation to sealevel s. l. and the level of Lake Ontario l o. At 1, left side of diagram is represented the bottom of the channel at Vienna, 200 ft below level of Lake Erie or 370 ft above sealevel. At 1, right side, is the divide at Littlefalls 440 ft A. T. The line 1–2, is the line 1–1, but elevated on the east (right) so as to give a continuous westward drainage. Horizontal scale 1 inch = 100 miles. Vertical scale 1 inch = 4000 feet.

Numerous theories have been advanced to account for the deep basin of Lake Ontario. Spencer believed it to have been formed by an eastward flowing stream, the ancient Laurentian river, which received the Erian river as a tributary through the Dundas valley. The eastward continuation of this river Spencer believed to have been essentially along the course of the modern St Lawrence, the present great elevation of the rocky bed of this stream, above that of Lake Ontario, being explained by a warping of the land. Upham also believes that the basin is in part due to warping, but he considers it the valley of a westward flowing stream. Russell also holds this latter view; for he says[2] that, "previous to the glacial epoch, the greater part of the Laurentian basin discharged its waters southward to the Mississippi and . . . during the first advance of the ice from the north the drainage was not obstructed so as to form important lakes". Westgate[3], in tracing out the de-

[1] Bigelow. Bul. geol. soc. Am. 9·183.
[2] Lakes of North America, p. 97.
[3] *Loc. cit.* p. 92.

velopment of the Mississippi drainage system, considers that the flow of the Laurentian drainage system was southward into the predecessor of the Ohio river. As has already been shown, Spencer's eastward flowing river system can be originated only by a complete readjustment of the drainage, resulting from a great relative depression of the eastern uplands. Such a system could only come into existence after the valleys had been formed for it, and hence, as far as the history of the lake basins is considered, no such river system is required, and, unless positive proof of its former existence is forthcoming, it may be dismissed as hypothetic. One of the most important theories of the origin of the Ontario and other lake basins, and one which has had, and still has many prominent advocates, is that of glacial erosion, either entire or preceded by river erosion. This explanation was first most strongly urged by Prof. Newberry, and it has found its most recent able supporter in Prof. Tarr. It is impossible to do full justice to this view in the present limited space. Ice erosion is a factor the potency of which has often been overlooked, but of the importance of which there can be no question. We may however question whether a valley which, like that of Ontario, lies transverse to the general direction of ice movement in this region, can owe much of its depth to this agent. The following considerations will be helpful in understanding the influence of glacial erosion on preexisting topography. If a valley like that of Lake Ontario is occupied by a glacier the motion of which is parallel to the trend of the valley, the topographic relief is likely to be accentuated by ice erosion. If the motion of the ice is transverse to the direction of the valley, the erosion tends to obliterate or at least reduce the relief features. If, however, a mass of ice remains stagnant in the valley, the upper strata of ice may override it, and the amount of glacial erosion is reduced to a minimum. The striae in this region, together with the direction of slopes from the oldland, point to a southward movement of the ice, and Gilbert has shown that the amount of erosion on the edge of the escarpment in western New York is comparatively slight.[1] Hence we may assume that the basin of Ontario was mainly *occupied* by ice during the

[1] Bul. geol soc. Am. 11:121.

glacial period, but that comparatively little erosion was accomplished. This is farther borne out by minor relief features, such as the benches shown in sections 4, 5 and 6, in the southern wall of the basin, and which probably consist of harder beds which erosion has left standing out in relief. On the theory of glacial erosion, we might expect these to be absent, or at least much less prominent, since ice would hardly show such selective power as is attributable to running water and atmospheric agents.

With the failures of the theories that an eastward flowing stream or glacial ice produced the Ontario valley, we are forced, with Upham, Russell and others, to look on a westward flowing stream as the most probable agent in the production of this valley. As has before been shown, such a stream would be the normal result of a gradual development of a drainage system on an ancient coastal plain of the type here considered.

Ancient St Davids gorge. Since the time of Lyell, the old buried channel from the whirlpool to St Davids has played a prominent part in the discussion of the life history of Niagara. For a long time it was considered to be the preglacial channel of Niagara, or its predecessor, the Tonawanda. More recently it has been considered of interglacial age, eroded by an interglacial Niagara, during a temporary recession of the ice sheet from this region, and filled with drift during a readvance of the glacier. The most satisfactory interpretation of this channel however makes it independent of the Niagara, and considers it one of many preglacial or interglacial channels which were formed by streams flowing over the edge of the escarpment and which increased in length by headward gnawing of their waters. This type of stream we have learned to call *obsequent*, its direction of flow being contrary to that of the master stream to which its waters eventually become tributary. An illustration of channel-cutting by streams flowing over the edge of a cliff, may be seen today in the chasm near the Devil's hole, on the American side of the gorge below the whirlpool. This gulch was cut by the little stream known as the Bloody run, which during the summer season dries away entirely.

The St Davids gorge has a width of nearly 2 miles at the edge of the escarpment. As will be seen by a glance at the map, it narrows perceptibly southward, till at the whirlpool its width is less than the average width of the Niagara gorge. What the depth of the gorge is has not been determined, though from the depth of the whirlpool, we may assume that its floor is 200 feet or more below the level of Lake Ontario. At, and to the north of the escarpment it probably equals in depth Lake Ontario, opposite to it. The channel is undoubtedly much more irregular than is shown on the map, the sides being probably much diversified by lateral gullies. The great width of the channel at St Davids may perhaps be due in some small degree to widening by glacial erosion; for we know that the channel was occupied by ice, from the glacial scratches which are preserved on its walls, where these are exposed in the present ravine of Bowman's creek near the whirlpool. The influence of this buried channel on the direction and width of the Niagara gorge will be discussed later.

Valley of Georgian bay. Georgian bay is in many respects the analogue of Lake Ontario. Like the latter, it also occupies a valley lying between the Niagara escarpment and the crystalline old-land on the northeast. As has previously been shown, the Niagara escarpment extends northward from Hamilton into the Indian peninsula between Georgian bay and Lake Huron, and, after passing the Cove island channel, it reappears in the northwestern face of Grand Manitoulin island. At Cabot's head, on the Indian peninsula, the escarpment rises to 324 feet above the surface of the water, while just off the promontory soundings show a depth of 510 feet, thus making the total hight of the escarpment at this point 834 feet. In some places the summit of the escarpment rises to an elevation of 1700 feet above tide, or more than 1100 feet above Georgian bay (Spencer). The depth of the transverse channel connecting Georgian bay and Lake Huron has been found to be 306 feet, which is more than 200 feet less than the depth of the channel of Georgian bay. It is possible however that the soundings do not show the absolute depth of the rock bottom in the channel: for there may be a filling of drift which raised the bottom of the channel above that of the bay.

The valley of Georgian bay is continued northwestward in the channel known as North passage, a narrow body of water lying between the Manitoulin islands and the Canadian old-land. The southward continuation of the lowland is blocked by drift; but a number of borings, between the southern end of Georgian bay and Lake Ontario, east of Toronto, have developed the existence of a buried channel, which connects these two valleys. This channel is considered by Spencer to mark the pathway of his former Laurentian river. It is clear however that this valley is merely the buried connecting part of the inner lowland which extends along the base of the entire Niagara escarpment. This portion of the lowland was originally occupied by two streams flowing, the one northwesterly into the ancient Saginaw, the other southeasterly into the Dundas. The divide between the two may have been in the neighborhood of Lake Simcoe. It is however not at all improbable that the tributary of the Dundas may have, owing to favorable conditions, gained an advantage over that of the Saginaw, and pushed the divide northward. Such a migration of the divide might have resulted in the diversion of the upper waters of the Saginaw by capture, so that they eventually became tributary to the Dundas. This would account for the greater depth of the Georgian bay lowland, which, after the capture of the upper Saginaw waters, could be deepened independently of the notch in the cuesta through which its waters were formerly carried out. This of course is merely suppositional, and the truth can be established only by more detailed study of the ground. It is however what we might expect to happen in the normal adjustment of a coastal plain drainage. This hypothetic relation is illustrated in fig. 6.

The Huron lowland and the Chippewa and Tonawanda valleys. On the yielding strata of the Salina group a second lowland was carved out by subsequent streams, leaving an escarpment capped by the Devonic limestones on the south. This, as we have seen, becomes prominent eastward in the Helderberg range, where the third upper Devonic escarpment unites with it. In the Niagara region it faces the Tonawanda and Chippewa lowlands, which were probably opened out by a subsequent stream tributary to the ancient Dundas river. Throughout western Ontario this escarpment is buried by drift, but its presence is indicated by borings, which also prove the continuance of the lowland accompanying it. This escarpment, the inface of the second cuesta, becomes a very prominent feature in Lake Huron, where it is entirely submerged. It is however perfectly traceable from north of Goderich in Canada to the island of Mackinaw. Soundings prove it to have a hight of from 350 to 500 feet or more above the lowland which it faces. This lowland constitutes the deeper portions of Lake Huron, the shallower southwestern area being a part of the upland drowned by the backward setting of the water over the top of the escarpment. The following cross-section (fig. 9) from Point au Sable, north of Saginaw bay, to Cape Hurd, the northern extremity of the Indian peninsula, passes across the highest portion of this escarpment at the 9 fathom ledge and diagonally across the deepest portion of the Huron lowland, where the soundings reach a depth of 750 feet. This apparently marks the location of the preglacial Saginaw river, which probably breached the second cuesta to the south of the 9 fathom ledge, though no channel is indicated by the soundings.

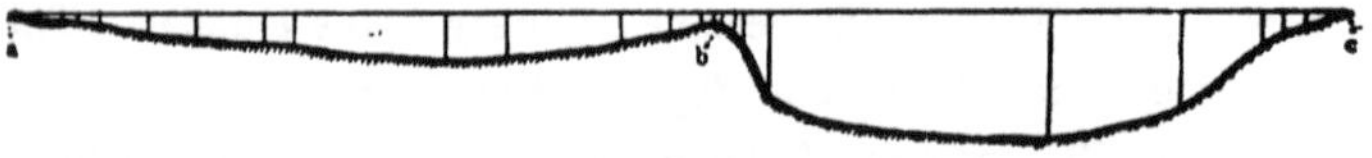

Fig. 9 Section across Lake Huron from Point au Sable, a) across 9 fathom ledge, b) to Cape Hurd, c) (For location of section *see* fig. 6).

We have now traced the development of the topographic features of the Niagara district, and have found this to be in conformity with the laws governing the normal development of drainage systems on an ancient coastal plain. The only abnormal features which need to be considered now are the tilting of the land and the filling of most of the old channels by drift, converting the lowlands into lake basins and reversing the drainage of the unfilled channels. These were the catastrophes which immediately preceded the birth of Niagara and which were directly responsible for its existence. To these and the life history of Niagara, attention will now be invited.

Plate 5

"Rock of Ages," the largest of the fallen limestone fragments at the foot of Luna Falls, on the American side (Copyright by Underwood & Underwood, New York)

Chapter 2

LIFE HISTORY OF NIAGARA FALLS

Glacial period

Two important events immediately preceded the birth of Niagara. The first was the formation of a series of great lowlands and cuestas by stream and atmospheric erosion during a period of time when, according to all indications, the land stood from 2000 to 5000 feet higher than it does now. This was outlined in the preceding chapter. The second event was the accumulation of a great mantle of glacial ice over most of northeastern North America, and the modifications of the previously formed erosion topography, either by the erosive action of the ice or by deposits left on its melting. The time equivalent of the latter event is commonly known as the glacial period of the earth's history, a remote period as time is ordinarily counted, but a very recent one in the chronometry of the geologist. Contemporaneous with this great accumulation of ice was probably the subsidence of the northern part of this region, thus changing the slope of the land surface from a southward to a northward one.

The greatest accumulation of ice during the glacial period appears to have been in the region to the north and northeast of the great lakes, or in general over the area of the Laurentian old-land. The immediate causes which brought about such accumulation, were the extensive refrigeration of the climate and the increased precipitation of moisture, so that a greater amount of snow fell during the winter seasons than could be removed by melting during the succeeding summers. The partial melting and refreezing of the snow, which continued over a long period of time, eventually resulted in producing glacier ice, after the manner of the formation of glaciers at the present time.

The thickness of the great Laurentian glacier, which eventually covered all the land of this region, including even the highest mountains, must be estimated at thousands of feet in its central part with a progressive diminution of thickness toward the margin. The ice of glaciers, as is well known, has a certain amount of plasticity and will flow under the pressure of its own weight, somewhat after the manner of a mass of pitch. The flow of the great Laurentian glacier was outward in all directions from the center of accumulation, local topographic features exerting a deflecting influence only in the more attenuated marginal portions. In its basal portions, the ice was well supplied with rock debris, from the finest rock flour and clay to boulders often of very great size. This material was derived from the surface over which the ice flowed, and it measured in part the amount of erosive work which the ice had accomplished. The rock fragments frozen into the bottom of the moving ice mass, served as efficient tools for grooving and scratching the bedrock over which the ice flowed, while at the same time the finer material smoothed and polished the rock surfaces. The direction of the grooves and striae on the rock surfaces in general indicate the direction of the movement of the ice which produced them, but this may not always represent the direction of general ice movement for the region, since, at the time of making the striae, the ice may have been thin enough to be influenced by the local topographic features of the region. In the Niagara district the striae have a direction extending about 30° west of south (Gilbert) which direction, being inharmonious with the trend of the lowlands, indicates that these striae were formed by the general movement of the ice, rather than by local movements, controlled by topography.[1]

While the surface rocks of this region were everywhere scratched and polished by the ice, these markings are only exhibited where the protecting mantle of loose surface material or drift has been recently removed. For where the polished rock surfaces are exposed for any considerable period of time, weathering usually obliterates these superficial markings. The best place in which the striae of the region about Niagara river may be studied is near the quarries on the edge of the escarpment, a mile or more west of Brock's monument, where the ledges are progressively uncovered previous to quarrying.

[1] For an account of the glacial sculpture in this region, *see* Gilbert. Bul. geol. soc. Am. 1899. 10:121.

Throughout the greater part of the district, the polished rock surfaces are covered by a coating of drift of very varying character and thickness. This was the ground moraine or till of the Laurentian glacier, and represents the rock debris which was frozen into the bottom of the ice, and carried along in its motion, till liberated by the melting of the ice. This ground moraine, either in its original heterogeneous character or modified by the agency of running water, filled most of the old river gorges through which the drainage of preglacial times found its exit. Some of the shallower lowlands, like that of the Tonawanda, were also filled with drift, while the more profound ones, like the Erie and Ontario lowlands, received only a partial drift filling.

The partial obliteration of the old drainage channels, which was thus brought about, together with a depression of the land on the northeast to a depth below that at which it now stands, converted the unfilled lowlands into lake basins, apparently reversed the drainage of many streams, forcing them to cut gorges where their old channels were drift-filled, and finally became the immediate factors in the formation of Niagara.

Lacustrine period[1]

During the slow melting of the glaciers in the Laurentian region, and the resultant northward retreat of the front of the ice, large bodies of water, of varying depth and extent, were held in front of the ice sheet, which formed a dam across the northeastern part of the lowland country, the general slope of which was now toward the ice instead of away from it. The elevations of these glacial lakes were determined by the lowest uncovered passes in the margins of the lake basins across which the discharge took place, and, as during the continued melting of the ice dam, lower passes were progressively uncovered, the outlets were successively transferred to them and the levels of the lakes sank correspondingly.

[1] For a detailed account of the successive stages in the development of the great lakes, the shore lines, outlets and extent of each, the reader is referred to the papers by Gilbert, Spencer, Taylor, Leverett, Fairchild and others, cited in the appendix.

Though of a temporary nature, these bodies of water endured sufficiently long to permit the formation of well marked beaches with their accompaniment of bars, sand-spits and other wave-formed features. These have been carefully studied and mapped by a number of observers, and the general extent and outline of these lakes is today pretty accurately determined.

The largest of these glacial lakes, though not the first to come into existence, was glacial Lake Warren. "At its maximum extent Lake Warren covered the south half of Lake Huron, including Saginaw bay, the whole of Lake Erie and the low ground between it and Lake Huron; extended eastward to within twenty or thirty miles of Syracuse, N. Y. and probably covered some of the western end of Lake Ontario."[1] The retaining ice wall on the east extended in a northwesterly direction, across western New York, Lake Ontario and the northeastern end of Lake Huron. This position of the ice front is in part inferred from the existence of moraines of sand and gravel along a portion of that line. The total area of this ancient lake has been variously estimated as including from one hundred thousand to two hundred thousand square miles of surface but this estimate is based on the assumption that the lake occupied the greater part of the area of the present upper Great lakes, with the intervening land, a supposition which Taylor holds to be incorrect. The area of Lake Warren was probably less than 50,000 square miles, or approximately half that of the state of Kansas. The extent and level of this lake was not constant, there being many oscillations, due chiefly to warpings of the land surface. These oscillations are recorded in the various beaches which have remained to the present time. The chief outlet of Lake Warren was by way of the Grand river valley into the valley of Lake Michigan, the southern end of which was then much expanded and occupied by the waters of "Lake Chicago." The outflow of this lake was to the Mississippi by way of the Illinois river, across the divide near where Chicago now stands, thus temporarily reestablishing the southward drainage of this region.

[1] Taylor. A short history of the Great lakes, p. 101.

As the ice front continued to melt away, retreating northeastward, drainage at a lower level was permitted along the ice front to the Hudson valley, and the sea. As a result, the water level sank, the Chicago outlet was abandoned, and Lake Warren became much contracted and in part cut up and merged into new bodies of water. The largest of these was glacial Lake Algonquin, which occupied the basins of the three upper Great lakes, and seems to have been for a long time independent of Lake Erie, which after the division of Lake Warren was for a time much smaller than it now is. (Fig. 11 and 13)

The critical period in the development of the lakes, with reference to the birth of Niagara, was the uncovering of the divide at Rome (N. Y.) and the consequent diversion of the drainage into the present Mohawk valley. This brought with it a subsidence of the waters north of the Niagara escarpment to the level of this outlet, which was considerably below that to which the other lakes could subside, owing to the rocky barriers which kept them at greater altitudes. As a result Niagara river came into existence, though at first it was only a connecting strait between Lake Erie and the subsiding predecessor of Lake Ontario. The overflow from Lake Erie occurred at the present site of Blackrock, because there happened to be the lowest point in the margin of the lake. It is not improbable that a small preglacial stream had predetermined this point, either flowing southward into the river occupying the Erie basin, or northward as an obsequent stream into the Tonawanda. The course of the river below Blackrock was determined by the directions of steepest descent of the land surface, which was probably predetermined to some extent by preglacial streams. As soon, however, as the level of the waters of the Ontario valley sank below the edge of the Niagara escarpment at Lewiston, a fall came into existence, which daily increased in hight as the level of the northern lake was lowered. From that time to the present, Niagara has worked at its task of gorge-cutting, the present length of the gorge, from Lewiston to the falls, marking the amount of work accomplished.

When the waters north of the escarpment had subsided to the level of the outlet at Rome, a long period of stability ensued, during which extensive and well marked beaches were formed by the waves. This comparatively long-lived body of water has been named Lake Iroquois, and its outline is shown in the accompanying map (fig. 10) reproduced from Gilbert's *History of the Niagara river*. The Iroquois shore lines in this region may be seen in the ridge road which extends eastward from Lewiston, and westward from Queenston, closely skirting the foot of the escarpment.

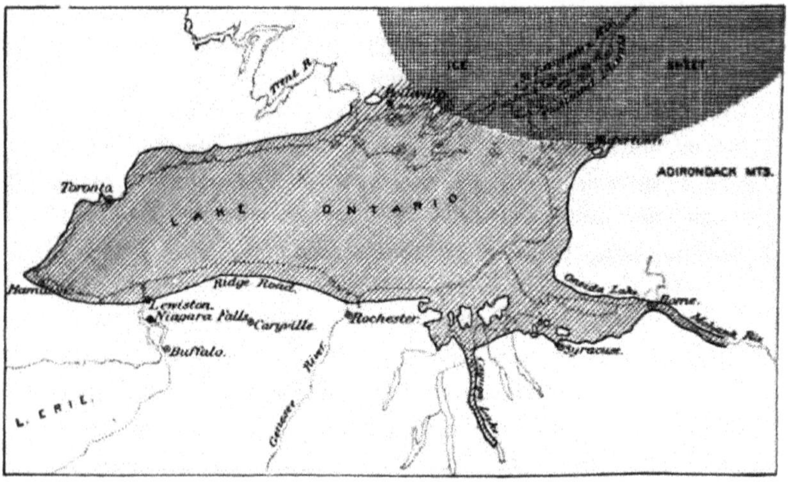

Fig. 10 Map of Lake Iroquois; the modern hydrography shown in dotted lines. (After Gilbert)

A fine section of this old beach is seen just behind the railroad station at Lewiston. Here the layers of sand and gravel slope steeply toward the southeast, and many of them are irregular and wedge-shaped. Some of the beds, a foot or more in thickness, consist entirely of rounded pebbles, with little or no sand between, forming a porous mass of "loose gravel". The prevailing rock of the pebbles is the Medina sandstone, derived from the neighborhood, and the pebbles are always well waterworn, and commonly of the flattened type characteristic of thin bedded rocks. Mingled with the beds of coarse material are layers of fine sand, the structure of which is well brought out by exposure to wind and weather. Not infrequently masses of sand and pebbles are cemented into a conglomerate by calcite or other cementing agents.

The terminal portion of the beach at the Lewiston station is rather exceptional. It has here the character of a sand spit, extending toward the Niagara river. Between this spit and the escarpment there is a low area of irregular outline, something over half a mile in width along the river and extending perhaps three fourths of a mile eastward from it. This area is bounded by steep erosion cliffs of unconsolidated material, and is from 30 to 50 or more feet lower than the level of the ridge road. The suggestion presents itself, that these features may be due to the current of the Niagara at its *embouchure* into Lake Iroquois, at a time when the falls were probably not far distant. (*See* plate 3 and map)

There is evidence that the level of Lake Ontario at one time stood much lower than it does at present; for the bottom of the lower Niagara, from Lewiston to the lake, is from 100 to 200 feet below

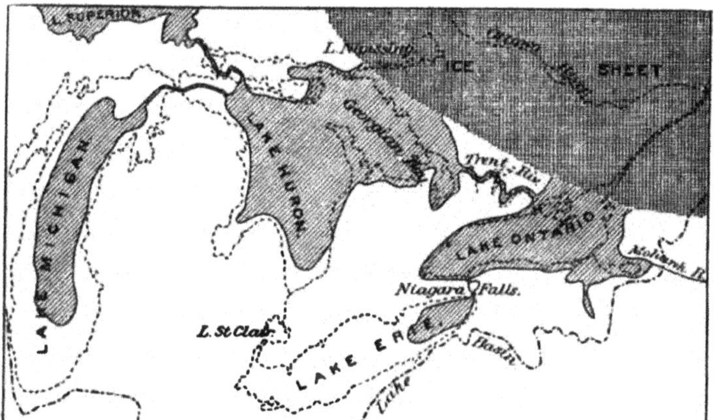

Fig. 11 Gilbert's map of the Great lakes at the time of the Trent river outlet. Modern hydrography dotted.

the present water level. In fact, the old beaches about Lake Ontario indicate a number of oscillations of level, similar to those recorded in the other glacial lakes, and due chiefly to crust warpings.

Lakes Algonquin and Iroquois were probably contemporaneous, and it is believed that for a time the former discharged its waters to the latter by way of Balsam lake and along the course of the Trent river. This discharge by way of the Algonquin river, as this old outlet of Lake Algonquin has been called, robbed the Niagara river of seven eighths of its water supply, which up to then had reached it by the present course through the Detroit river. As a result, the volume and erosive power of the river were for a time enormously diminished. (Fig. 11 and 13)

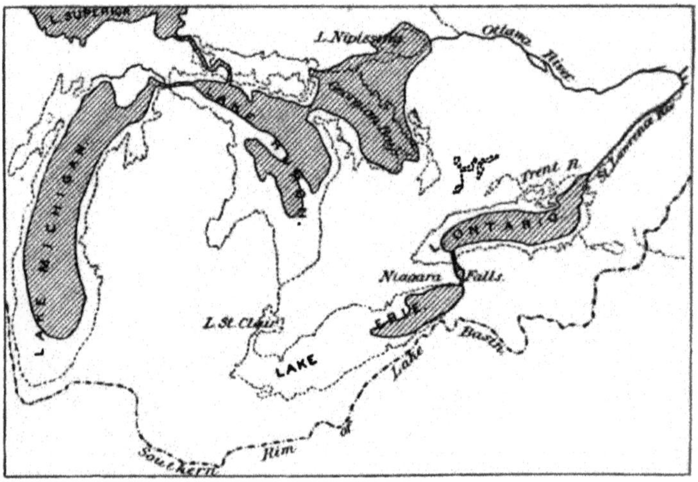

Fig. 12 Gilbert's map of the Great lakes at the time of the Nipissing outlet. Modern hydrography dotted.

During the farther retreat of the ice front, a still lower pass was opened by way of Lake Nipissing and the Mattawa river into the Ottawa. By the time this outlet was opened, the ice had also disappeared from the St Lawrence valley, and the outlet of the waters of the great lakes was transferred from the Rome channel to the one at the Thousand islands, Lake Iroquois at the same time subsiding to Lake Ontario. (Fig. 12 and 14)

The successor of Lake Algonquin, after the change from the Balsam lake to the Nipissing lake outlet, has been named by Taylor, Nipissing great lakes, while the river which carried its discharge to the Ottawa was called by him the Nipissing-Mattawa (fig. 14).

With the gradual melting away of the great ice sheet, the land on the northeast began to recover from its last great depression,

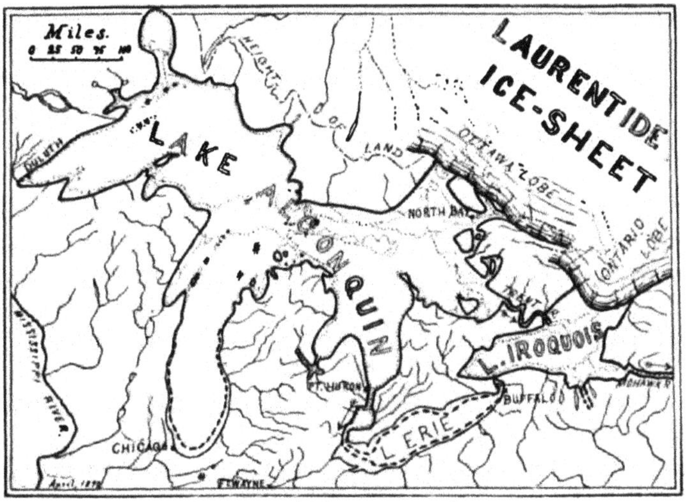

Fig. 13 Taylor's map of Lakes Algonquin and Iroquois.

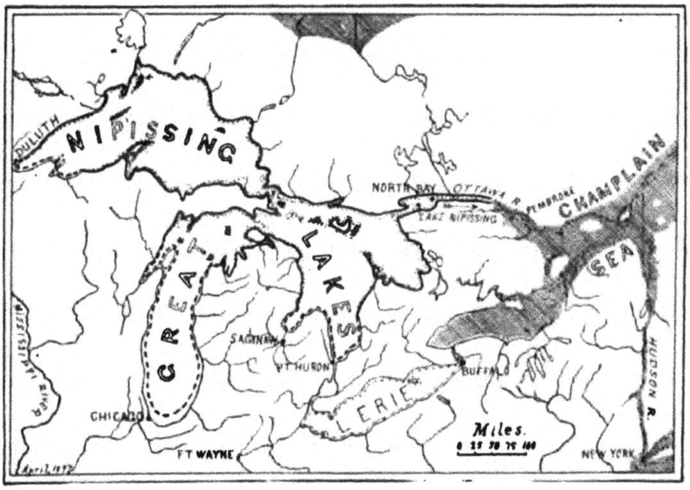

Fig. 14 Taylor's map of Nipissing great lakes and the Champlain submergence.

and, though there had been many oscillations, the balance of change was toward a slow but steady elevation of the Laurentian region. As a consequence the beaches of the old glacial lakes, which of course had a uniform elevation while forming, are no longer of uniform hight above sealevel, but rise progressively toward the northeast. This slow rising of the land caused a gradual canting of the basins, which brought with it a relative fall of the waters along the northeastern shores and a corresponding relative rise of the waters along the southwestern shores. Such a progressive change eventually carried the Nipissing and Balsam lake outlets above the level of the outlet at Port Huron, and the present drainage was reestablished. As the canting affected the Erie basin as well as the others, it caused a progressive elongation of that lake toward the southwest, thus finally giving it its present size and shape. This same canting also resulted in the farther separation of the upper lakes into their present divisions.

While this general outline of the lake history is held by many geologists, others, notably Upham, combat it strongly. Mr Upham holds that the elevation of the land in the northeast had progressed to such an extent by the time the ice had uncovered the northern outlets of Lakes Algonquin and Nipissing, that these passes had been raised above the altitude of the outlet at Port Huron, and that hence these passes never, or but for a brief period of time, served as outlets for the waters of the upper lakes. If this is the case, Niagara always carried the drainage of the upper great lakes as well as Lake Erie, and its volume was approximately uniform throughout its history. The strong erosion features, however, which are found in the Mattawa valley indicate that a large stream discharged here for a considerable period of time; and, if such was the case, it is highly probable that the present Port Huron outlet was not then utilized, and that consequently the Niagara was robbed of the discharge of the upper lake area. The influence on the erosion of the gorge by such a withdrawal of the water must have been a pronounced one, and we shall see later that certain portions of the gorge may well be explained by this hypothesis. During the time of the overflow of the upper waters by way of the Nipissing-Mattawa river it is not improbable that, as held by Taylor and others, the sea had access to the St Lawrence and Ontario basins and possibly to the basins of the upper lakes. This would account for the occurrence of marine types of organisms in the deeper portions of some of the present great lakes as well as for the maritime species of plants found in the lake district. It must however be borne in mind that this marine invasion was not till after the time of Lake Iroquois, for fresh-water fossils have been found in the beaches of this lake.

The tilting of the land, which is recorded in the deformed beaches, has not yet ceased, as recent investigations in the lake regions clearly prove. Mr Gilbert has made an extended study of this problem; and he has been led to the assumption "that the whole lake region is being lifted on one side or depressed on the other, so that its plane is bodily canted toward the southsouthwest, and that the rate of change is such that the two ends of a line 100 miles long and lying in a southsouthwest direction are relatively displaced .4 of a foot in 100 years". From this it follows that "the waters of each lake are gradually rising on the southern and western shores or falling on the northern or eastern shores, or both". This implies of course a drowning of the lower courses of all streams entering these lakes from the southwest and an extension of those entering from the northeast. Assuming that the rate and character of change will be constant in the future, the following interesting results have been predicted by Mr Gilbert. The waters of Lake Michigan at Chicago are rising at the rate of 9 or 10 inches a century; and "eventually, unless a dam is erected to prevent, Lake Michigan will again overflow to the Illinois river, its discharge occupying the channel carved by the outlet of a Pleistocene glacial lake. . . Evidently the first water to overflow will be that of some high stage of the lake and the discharge may at first be intermittent. Such high water discharge will occur in five hundred or six hundred years. For a mean lake stage such a discharge will begin in about one thousand years, and after one thousand five hundred years there will be no interruption. In about two thousand years the Illinois river and the Niagara will carry equal portions of the surplus water of the great lakes. In two thousand five hundred years the discharge of the Niagara will be intermittent, falling at low stages of the lake, and in three thousand five hundred years there will be no Niagara. The basin of Lake Erie will then be tributary to Lake Huron, the current being reversed in the Detroit and St Clair channels."[1]

[1]Gilbert, G. K. Recent earth movements in the great lake region. 18th an. rep't U. S. geol. sur. 1896-97. pt 2.

Fluvial period

Niagara falls came into existence when the waters of Lake Iroquois, the predecessor of Lake Ontario, fell beneath the level of the escarpment at Lewiston. At first it was only a small cataract, but day by day, as the lake subsided, it gained in hight and consequently in force of fall, as well as efficiency in cutting its channel. That the entire gorge from Lewiston to the present falls is the product of river erosion is scarcely questioned by any one today, but there are excellent reasons which lead some to believe that this cutting was not wholly the work of the Niagara. When the falls were at Lewiston, the Niagara was a placid stream from Lake Erie to near the falls, much as it is today from Buffalo to the northern end of Grand island. Its banks consisted chiefly of glacial till, into which terraces were cut by the stream, most of which are visible at the present day. The lower ones are well marked in Prospect park, though there they have been grassed over and modified to a considerable extent. From Niagara falls to the railroad bridges at Suspension Bridge, on the New York side of the river, the old bank runs parallel to the edge of the gorge and at a short distance inland from this. From Suspension Bridge to the whirlpool it makes a curve somewhat more crescentic than that of the margin of the gorge, and a similar curve from the whirlpool to Bloody run at the Devil's hole. On the Canadian side these old river banks can be traced from above the falls almost to Brock's monument, and in some cases two or three successive terraces are recognizable. In Queen Victoria park they constitute the steep slope which bounds the park on the west, and parts of which are still actively eroded. Less than a mile below the carriage bridge, the old banks approach close to the modern one and continue, almost coincident with it, to the railway bridges at Clifton. From here to the whirlpool the old river margin has a nearly straight course, while the modern one is curved, and a similar relation holds below the whirlpool, though here, from the great curvature of the modern channel, the old banks are in places nearly a mile distant.[1] (Plate 6)

[1] These old river banks are indicated on the geologic map by dotted lines; the localities where shells have been found are shown by crosses.

Old banks of the Niagara on the New York side, below the railroad bridges (U. S. geological survey)

Within the old channel thus outlined, which was much broader than the modern channel below the falls, accumulations of stratified sands and gravels were formed in the more protected places, much as such deposits are formed in streams today, where sands are swept into protected areas. With these sands and gravels were swept together the shells of those mollusks which lived in the river water, and many of which were of the species now found living in the upper Niagara.[1] Most of the shells thus swept together were probably of dead individuals, though living ones may also have been carried into these growing deposits. Many excavations have been made in these ancient deposits, fragments of which are preserved in various places between the former and present banks of the river. The most notable of these and the one longest known is on Goat island, perhaps a quarter of a mile inland from the edge of the cliff, at the Biddle stairway. In the section opened here, most of the material is seen to be coarse and rudely stratified. The pebbles are subangular, often quite angular, while some appear to be scarcely worn at all. Blocks a foot or more in diameter are not infrequent, the material being generally limestone from adjoining ledges, though fragments of sandstone and of crystalline rocks are not uncommon. Occasionally a lens of fine sand occurs which shows cross-bedding structure, the laminae pointing in a northwesterly direction. The shells are found on the cross-bedding planes, conforming with them, and indicating that they were spread there by the current which moved the sand grains. Among the coarse material the shells are mixed indiscriminately. In many cases the gravels are of the loose type, with scarcely any sand between them, indicating deposition by a powerful current. Along these zones air and water have most readily penetrated, and a deposition of iron oxid has been formed which stains both pebbles and shells. The shells are generally very fragile, and commonly show signs of wear. Gastropods are most abundant in the Goat island gravels.

In Prospect park several excavations formerly exposed these gravels. The deposit here consists of sand and gravel with the pebbles moderately rounded, though occasionally subangular, and

[1]For descriptions and illustrations of these shells, *see* chapter 5.

varying in diameter up to 6 inches or a foot. The stratification is rude, and shells are abundant. These are mostly fresh-water mussels (Unio, Alasmodonta, etc.) and the valves are generally found in conjunction, a fact which may indicate that these shells lived here. Small gastropod and pelecypod shells are plentifully mingled with the pebbles and sands. Below this are coarser deposits where boulders up to several feet in diameter occur, and below this occurs a bluish clay. In all of these beds shells have been sparingly found.

Several excavations have been made in Queen Victoria park, and here shells are common. The Unionidae appear to be most abundant, though small gastropods are not uncommon. All appear to have been more or less waterworn. The mussel shells are generally decayed, owing no doubt to percolating waters. Below Clifton, the lower of two terraces is of a somewhat sandy character, though many boulders occur in it. Shells of unios occur sparingly in these deposits, and a few small gastropods were found in the lowest terrace. Farther north several excavations in the lower terraces of the old river show loose gravels alternating with a sort of till, a few Goniobasis and other gastropod shells being found here. In some cases the gravels have become cemented into a conglomerate by a deposit of calcite between them, often of considerable thickness. Boulders of similarly cemented gravels are found in the gorge below, at the whirlpool.

It will thus be seen that, throughout the greater part of the young Niagara, deposition was going on as well as erosion. The amount of erosion of the river bed was probably very slight, that of the banks being much more pronounced. The chief part in the cutting of the gorge was enacted by the cataract, which cut *backward* from Lewiston, the amount of *downward* cutting by the river being insignificant. The manner in which the cataract performed its work of cutting may today be observed in both the American and Canadian falls, as well as in waterfalls of other streams falling over strata, the arrangement of which is similar to that obtaining at Niagara. The essentials are a resistant stratum overlying a weak one, the latter being constantly

Plate 7

Cliff on the Canadian side of the gorge, showing the receding base. The giant icicle marks the edge of the overhanging ledge (Copyright by Underwood & Underwood, New York)

worn away by the spray generated by the falling water, thus undermining the resistant layer. Such undermining may be seen in the Cave of the Winds. In course of time this undermining progresses so far that the projecting portion of the capping stratum breaks down for want of support, and the crest line of the fall becomes abruptly altered. The fallen fragments accumulate at the foot of the fall, where they will remain if the force of the water is unable to move them, as illustrated by the rock masses lying at the foot of the American fall. If, however, the force of the falling water is great as at the Horseshoe falls, these blocks will be moved about, perhaps even spun about, and so made to dig a deep channel below the falls. In the soft rocks which lie at the foot of the Horseshoe falls a channel probably not less than 200 feet in depth has been dug in this manner. (Fig. 15)

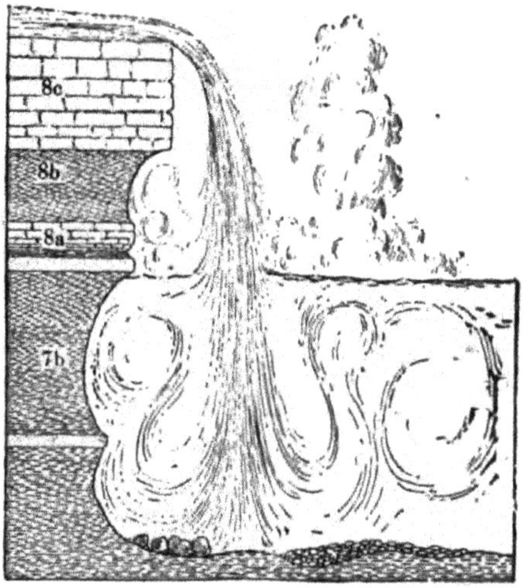

Fig. 15 Sectional view of the Horseshoe falls showing arrangement of strata, and depth of water below falls. (After Gilbert) The numbering of beds corresponds with that of table.

When we consider the Niagara gorge in detail we find it to be much more complex than would at first appear. The first abnormal feature which presents itself in a map view of the entire gorge is the bi-crescentic character of its course, with the rectangular turn at the whirlpool, a course very different from that which we are accustomed to find in large rivers whose direction of flow has been uninfluenced by preexisting relief features. (Fig. 16) Another feature of impórtance is the varying width of different parts of the gorge, and the corresponding increase in velocity of current in the narrower parts. The depth of the channel also varies in different portions of the gorge, being in general greater in the wider and less in the narrower parts. (Fig. 18)

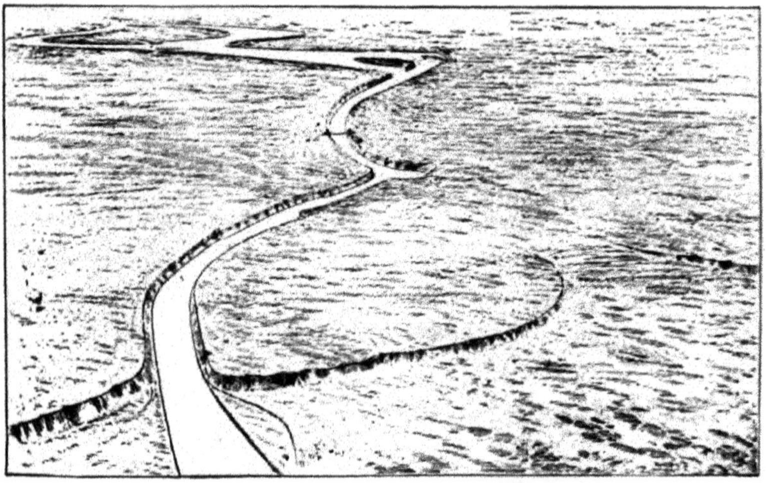

Fig. 16 Birdseye view of Niagara gorge showing the course of the river; the falls, the railroad bridges, whirlpool, location of Fosters flats, escarpment at Queenston and flaring mouth of old St Davids gorge. (After Gilbert)

The first mile and three fourths of the gorge, or that portion marking the retreat from the escarpment to the Devil's hole, extends nearly due south, and is fairly uniform in width, comparatively narrow, and with a current of great velocity. The narrowness of this stretch, when compared with the channel made by the present cataract from the railroad bridges southward, seems to indicate a smaller volume of water during its formation than that now passing over the falls. An alternative hypothesis accounts for the narrowness of this section of the gorge by assuming it to be a preglacial drift-filled channel, made by an obsequent stream flowing northward to the Ontario lowland, similar to that which made the old St Davids channel, but reexcavated by the Niagara. It is highly probable that there was at least a shallow channel which served as a guide to the young Niagara. The southward continuation of this channel beyond the Devil's hole, is found in the valley of Bloody run, a shallow but distinct depression now followed in part by the Lewiston branch of the New York Central railroad and evidently of preglacial origin, as its floor is covered with till.

Next above this lowest section of the gorge is one, in general much broader, and extending in a southwest direction from the Devil's hole to the whirlpool, a distance of a little less than two miles. This section is contracted near its middle by the projection from the Canadian bank, known as Fosters flats, or Niagara glen.

The river is here scarcely 300 feet wide, though the tops of the banks are in places over 1700 feet apart. Above Fosters flats and almost as far as the whirlpool, the river is very calm, and apparently deep, while at the point of contraction at the southern end of Fosters flats, the waters suddenly become tumultuous and rush through the narrow channel with great velocity. This sudden change has been attributed to a sudden decrease in depth of the river at this point, but it is evident that, even if the channel had the same depth as above, the sudden contraction would produce a similar effect, for the waters, spread out over a broad and deep channel, on being suddenly forced to pass through a narrow one, would from mere crowding into a smaller space assume a violent aspect.

Niagara glen, or Foster's flats

PLATE 8

This is one of the most interesting places along the whole Niagara river, though generally little visited by tourists. From the Canadian side a platform of limestone projects, whose surface is a little below that of the general level of the upland plain, from which it is separated by a steep bluff. The platform is known as Wintergreen flat, and, though sparingly wooded, is very deficient in soil. The bluff which bounds it on the west is a part of the old river bank. On the remaining sides this platform is limited by abruptly descending cliffs, at the base of which are extensive talus slopes descending to a lowland of considerable extent. This lowland, which is known as Fosters flats, has its surface well strewn with huge boulders of limestone. The cliff which limits Wintergreen flat on the northern or downstream side is the highest and most precipitous, and from its base a well marked, dry channel leads northward for a third of a mile to the river's edge. This channel is separated from the present river channel on the right by a ridge which appears to consist of huge limestone blocks, though its base is probably formed by undisturbed remnants of the lower strata of the region. The floor of this old channel is strewn with huge limestone boulders, such as are found at the foot of the American falls today, and its left bank is the precipitous west wall of the Niagara gorge. (Fig. 17)

Fig. 17 View of Niagara glen or Foster's flats, looking south. Forests omitted (After Gilbert)

These various features have been well explained by Mr Gilbert,[1] who holds that a narrow island comparable to Goat island, divided the fall in two, when it had receded to the northern end of Fosters flats. The foundations of this island, which has since crumbled away, are seen in the ridge which divides the old dry channel on the left from the main bed of the river. The eastern or American fall at that time was the larger of the two, and it receded more

[1]Nat. geog. monographs. Niagara falls and their history.

Wintergreen Flat looking south; showing the platform which was formerly the river-bed, and the cliff in the center which

rapidly. "When the Canadian fall reached the head of the island, the American had just passed it, and part of the sheet of water on Wintergreen flat was drained eastward into the gorge opened by the American fall. The Canadian fall, through the loss of this water, became less active, and soon fell out of the race."[1] By the final retreat of the American fall beyond the southern end of Wintergreen flat, the latter was left as a dry platform with precipitous sides, over which once poured a portion of Niagara's torrent.

While the occurrence of an island in the position pointed out by Gilbert was undoubtedly the immediate cause of the division of the falls, the more fundamental cause, and the one to which the island itself owed its existence, is to be sought elsewhere. From an inspection of the map the suggestion presents itself that there may be a vital connection between the abandoned falls at Fosters flats and the great bend of the river at the whirlpool. When a great river runs for a mile or more in a straight line, as the Niagara does above the whirlpool, and then abruptly turns to the right, the current is deflected by this sudden change in direction to the right bank of the river below the bend, which it continues to hug till again deflected. It is thus that the greatest amount of water will be carried along the right bank of the river, causing a deeper channeling there. When Niagara falls had receded to the present northern end of Foster's flats, the greatest amount of water was carried over its right side. The resulting deepening of the channel on the right, and the consequent drawing off of the water toward that side, was the cause of the appearance of the island (if such existed, as seems probable from the remaining foundation) above the water and the consequent division of the falls. A precisely analogous feature occurs in the lower falls of the Genesee river below Portage. Here, however, no island was formed, though in other respects the two cases are nearly alike. In the Genesee the change has occurred in comparatively recent times, and records of earlier conditions have been preserved. An abrupt bend of the river to the right, deflected the current to the right bank below the bend, and thus caused the deepening of the river bed on that side, as well as the more rapid

[1]Gilbert. Nat. geog. monographs. Niagara falls and their history.

recession of the right hand portion of the falls. In the course of a comparatively short time the channel became so deep on the right, and the falls receded so fast on that side, that all the water was drawn off from the larger portion of the river bed on the left, which today remains as a triangular platform comparable to Wintergreen flat, with steep sides, and several hundred feet wide, at its downstream end. The river now flows in a channel, in places less than 10 feet wide, and 100 feet below the level of the platform which was its bed less than 100 years ago. The present lower fall, having mostly receded beyond the upstream end of the platform, again extends across the entire bed of the river. The water in the river has not, as far as known, changed in average volume, though above and below the narrow part the gorge is many times as wide. All the water which passes in a thin sheet over a broad fall above the narrow gorge is forced to pass through this contracted portion, and presents a rushing current, though the bed is deeper here than where the gorge is broader. The time required for the recession of the fall over the space of the 2000 feet of narrow gorge, must have been much shorter than that required for the recession through a similar length in the broader portion of the gorge, for the concentration of the waters here enabled it to do much more effective work.

Judging by analogy, we may assume that the narrow channel opposite Foster's flats was cut by a stream of the full power of the present Niagara, but whose main mass of waters was carried over the right side of the fall on account of the bend in the stream above. The present Horseshoe falls is cutting a much narrower gorge than that to the north of it, owing to its peculiar position at the angle of a second great bend. (Fig. 19) From the fact that the cutting was most profound on the eastern or right bank of the river at Foster's flats, this bank has received the precipitous character which it has retained to the present day.

An interesting fact bearing on the interpretation of the history of Foster's flats, is the occurrence in the sands among the huge boulders near the foot of the ancient falls, of shells of the small fresh-water gastropod, Pomatiopsis lapidaria Say,[1]

[1] *See* chapter 5.

which is found living in the Niagara river today, but only on the rocks and boulders lying in the constant spray of the modern cataract.

After passing Foster's flats, the scene of greatest erosive activity seems to have been transferred to the left bank of the river. This is indicated by the verticality of this side of the gorge south of Foster's flats, which suggests active erosion, while the lowland known as Ongiara park opposite to this on the New York side of the river, with its enormous boulders scattered about, recalls the dry channel on Fosters flats or the foot of the present American fall, and suggests an amount of water insufficient to remove them. This may be accounted for by assuming that the nearness of the fall had

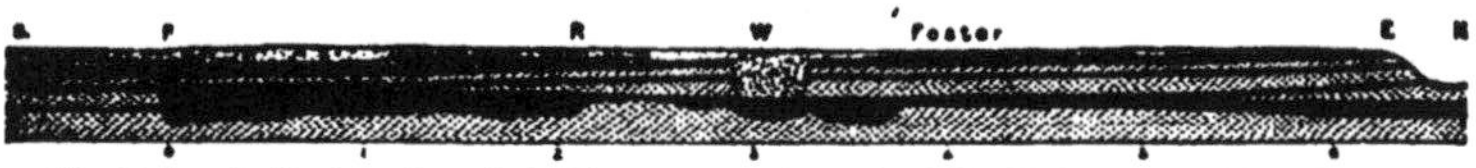

Fig 18 Longitudinal section of the Niagara gorge from the falls F to Queenston hights E, showing strata of west bank and depth of channel. (After Gilbert) R railway bridges. W whirlpool. Foster = Foster's flats. Figures indicate miles.

given the river itself greater momentum above the fall, and that hence it dug deeper into the old drift-filled valley of the St Davids at the whirlpool. As a result, the deflection of the current to the right bank became more abrupt, striking the New York bank immediately south of where Ongiara park now is, and, being again deflected toward the Canadian side, it reached this just at the southern end of Foster's flats, thenceforth for a time causing the most active erosion on that side. The washing out of the drift from the old St David's channel furnished the river with tools with which it was able to cut down into its bed, so that in this portion erosion was probably both by backward cutting of the falls and downward cutting of the river above the falls.

We have so far considered the falls as of simple type, but it is by no means certain that such was the case. If we judged from analogy with other streams which have cut gorges in the same strata as those found at Niagara, we should suppose that, as in the case of these streams, a separate fall was caused at Niagara by each resistant layer. Thus in the lower Genesee river, at Rochester, one fall is caused by the upper hard bed of the Medina formation, another by the limestone of the Clinton group, and a third by the Lockport limestone. In the Niagara river we might suppose that at least three, and possibly four, falls had existed at one time. The lowest of these would have been over a hard bed of sandstone, about 25 feet thick, and about 100 feet below the top of the Medina group. Another might have been caused by the hard capping stratum of Medina sandstone, 10 feet thick. A third over the 30 feet of Clinton limestone; while a fourth would have been formed over the Lockport limestone. The second and third would perhaps unite in one, as the shale bed between the two resistant layers is only 6 feet thick. It may however be objected that in a great cataract the force of the falling water is such as to cause uniform recession of all the layers, and that hence only one great fall existed.

The whirlpool

PLATE 9

Perhaps the most remarkable part of the entire gorge is its great swollen elbow, the whirlpool. Here the current rushing in from the southeast with great velocity, circles around the basin and finally escapes, by passing under the incoming current, through the comparatively narrow outlet, in a northeasterly direction. The waters in the whirlpool have probably a depth of 150 or 200 feet, but both the outlet and the inlet are shallow, for here ledges of the hard quartzose bed of the Medina formation project into the river, extending in the latter case probably across the channel. An examination of the walls of the whirlpool basin shows that rock is absent on its northwestern side, the wall here being formed of unconsolidated material or drift. This is best seen on descending to the edge of the whirlpool on the Canadian side, through the ravine of Bowmans creek. It will be observed that the Niagara has here exposed a cross-section of the ancient drift-filled channel which extends southeastward from St Davids. This channel appears to have been that of a preglacial stream of the obsequent type,[1] which was tributary to the streams of the Ontario lowland. Some geologists however, notably Mr Taylor, believe that this old channel may have

[1] *See* chapter 1.

been formed by a river and cataract similar to the Niagara of today, during interglacial time. That this old channel was once occupied by ice is shown by the glacial scratches on the limestone ledges exposed in the western wall of the old gorge, where this has been cleared of drift by Bowmans creek, and it is apparent that the filling in by drift must have occurred after the ice occupation. An inspection of the map will show that a part of the present Niagara gorge, that containing the whirlpool rapids, is in direct continuation of the old St Davids channel, and that, a little above the railroad bridges, the Niagara makes a pronounced bend, which brings it in conformity with the direction of this channel. This suggests that there was at least a shallow depression, the insignificant southeastward continuation of the St Davids channel, which guided the waters of the Niagara in this direction. Here a question of great importance in the history of the Niagara presents itself. Did the ancient St Davids gorge end where is now the south side of the whirlpool, with only a shallow surface channel extending beyond this point, or was the gorge of the whirlpool rapids a part of the old St Davids channel, which was merely cleared by the Niagara of the drift that filled it? The latter condition was assumed to be the true one by Dr Julius Pohlman of Buffalo, a pioneer in the study of the Niagara gorge and the first to recognize the complexity of the channel and attempt to account for its varying character. The theory is still held by many geologists. On the other hand, Taylor and others think it more likely that the ancient gorge stopped where is now the inlet to the whirlpool, and that the gorge above it is the product of post-glacial erosion. If this view be accepted, the narrowness and shallowness of the gorge of the whirlpool rapids must be accounted for by some change in the volume of water during its formation. Taylor, who has studied this problem, has come to the conclusion that, during the time that the gorge of the whirlpool rapids was being excavated, the upper great lakes (then united into Lake Nipissing) discharged by way of the Nipissing-Mattawa river as already outlined, and that therefore Niagara drained only the shallow Lake Erie, the amount of water in the river being only one eighth its present volume. It is easy to see that such a reduc-

other by the limestone of the Clinton group, and a
Lockport limestone. In the Niagara river we mi[illegible]
at least three, and possibly four, falls had existe[illegible] at
lowest of these would have been over a h[illegible] we
about 25 feet thick, and about 100 feet bel[illegible] upper
group. Another might have been c[illegible] As a
stratum of Medina sandstone, 10 fe[illegible] and the
feet of Clinton limestone; while a [illegible] e railway
over the Lockport limestone. [illegible] size of the
haps unite in one, as the shal[illegible] the water now
is only 6 feet thick. It m[illegible] ation between change
cataract the force of the [illegible] ange in size of the gorge of
recession of all the laye[illegible] estive, and seems admirably to ac-
[illegible] observed in the gorge. For example, it
[illegible] the sudden widening of the gorge just before
[illegible] hirlpool, forming what Taylor has called the Eddy

Perhaps th[illegible] the strong eddy which characterizes this portion of the swollen elb[illegible] this wider part of the gorge Taylor believes was formed by southeas[illegible] large-volume river which cut out the broad channel north escap[illegible] the whirlpool, and he farther thinks, that the sudden change from par[illegible] the broad channel to the narrow one of the whirlpool rapids marks in[illegible] the reduction in volume of water on the opening of the Nipissing-Mattawa channel, which had hitherto been blocked by the remnant of the Laurentian glacier. There are however several features which must be satisfactorily explained before this theory (which Upham rejects on grounds already stated) can be accepted. It is highly probable that the gorge of St Davids was worn back beyond the whirlpool. From the great depth of the whirlpool basin, and the presence of the quartzose sandstone bed at the inlet to it, it seems probable that a fall existed here in the ancient stream which carved the St Davids channel. That channel has probably a depth similar to or greater than that of the part now constituting the whirlpool basin. Now, if, as we have reason to believe, this old channel was formed by an obsequent stream of moderate volume flowing northward to the Ontario lowland, it can hardly be assumed that there was but one continuous fall of from four hundred to five hundred feet in hight, with such a pronounced alternation of hard and soft layers. We must rather assume that a separate fall existed over each hard layer, and that, as in the other streams flowing northward over these same strata, these falls were separated from one another by considerable distances. If then, as is clearly indicated by the quartzose sandstone ledge at the inlet to the whirlpool, the lowest of these falls was at that place, the other two or three must have been at some distances up stream, and in that case it is not too much to assume with Pohlman, that the upper old falls over the Lockport limestone were somewhere near where the gorge is now spanned by the railway bridges. Taylor, however, does not encounter this difficulty, for he assumes that the St Davids gorge was formed by an interglacial Niagara, the great cataract of which, just before its cessation (probably through a southward diversion of the drainage) plunged as a single fall over the cliff into the basin now holding the whirlpool. To this view it may be objected that the old St Davids gorge is not such as would be formed by a single great cataract, since it flares out northward, having a width at St Davids of perhaps two miles. Such a form is more readily accounted for if one assumes that the valley was made by the headward gnawing of an obsequent stream and its various branches. Taylor meets this objection by invoking the action of readvancing ice to broaden the gorge, but, unless the last ice advance was from a very different direction from that indicated by the striae of this region, this hypothesis will scarcely hold. That direction, as already noted, is 30° west of south, while the direction of the old gorge is almost due northwest. Why may we not assume that only a portion, the southern one of the gorge of the whirlpool rapids, was carved by the Niagara during the time that its volume was diminished, and that the greater portion of this gorge was preglacial? This would greatly reduce the length of time during which the upper lakes discharged by way of the Nipissing-Mattawa river, though probably leaving time enough for the waters from these lakes to produce all the erosion features found in this ancient stream channel. This would still leave the Eddy basin to be accounted for, a difficulty which may perhaps be diminished by assuming that the second of the ancient falls was situated at the point where the gorge contracts to the width of the narrower channel of the whirlpool rapids.

It will thus be seen that this interesting problem of the origin of the gorge of the whirlpool rapids, propounded nearly 20 years ago by Dr Pohlman, is by no means wholly solved. We may return to the original solution of the propounder of the question or we may find new evidence which will corroborate Taylor's explanation. And who shall say that still other explanations of these features may not be forthcoming in the future, when those now demanding attention will be no longer regarded as plausible or sufficient?

The upper gorge and the falls

PLATES 1, 2, 4, 5, 11

Whatever may be believed with reference to the narrow gorge of the whirlpool rapids, most observers agree that the broad and deep gorge from Clifton to the present falls was made by a cataract carrying the full supply of water. This, the latest and most readily interpreted part of the gorge, has come to an end at the Horseshoe falls of today, and the character of the channel hereafter to be made can only be conjectured. The river has reached another of its critical points, where a rectangular turn is made, and it is not improbable that, as at the other turns, so here the character of the gorge will change. Already a short channel, considerably narrower than that of the last preceding portion, has been cut by the Horseshoe falls. (Fig. 19) This narrowness of the channel is due to the concentration of the water at the center of the stream. It is easy to see that Goat island and the other islands owe their existence to this concentration of the water; for at one time, as shown by the shell-bearing gravels, these islands were under water. The channel above the Horseshoe falls has been cut more than 50 feet below the summit of Goat island at the falls, while the upper end of the island is still at the level of the water in the river.

Goat island lies on one side of the main mass of forward rushing water, which passes it and strikes the Canadian bank, from which it is deflected toward the center of the cataract, which portion is thus deepened and worn back most rapidly. The directions of the currents may be seen from the upper walks in the Canadian park, and the effectual erosion of the banks may also be observed. In many cases the shores have been ballasted and otherwise protected against the current. During an earlier period, when the falls were situated farther north, and before the central part of the stream had been deepened to its present extent, the water, then at the level of the river above Goat island, flooded what is now the Queen Victoria

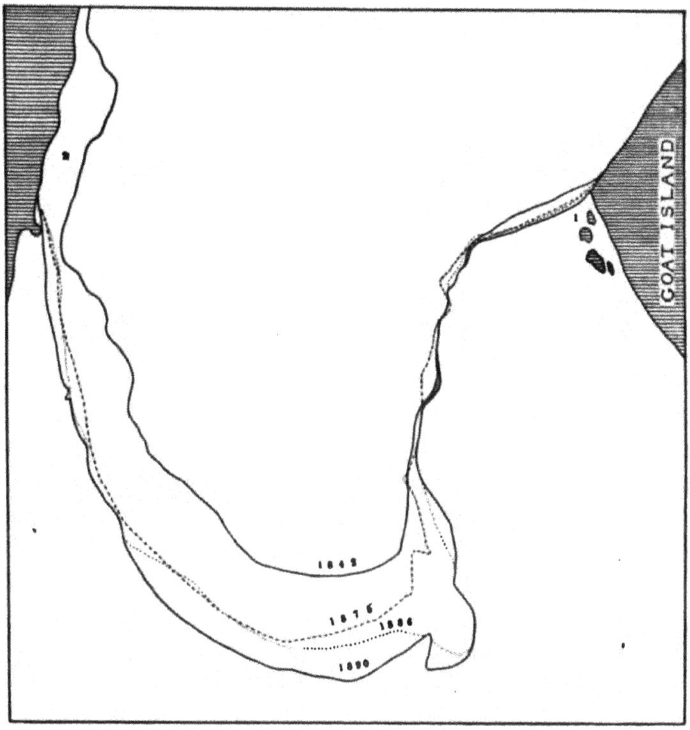

Fig. 19 Crest lines of Horseshoe falls. From the original tracing of the surveys by courtesy of the New York state engineer and surveyor. 1 Terrapin rocks. 2 Former Table rock.

park, and carved from the till the pronounced concave wall which now bounds the park on the west. A local eddy, probably during very recent times, carved the steep and still fresh semicircular cliff which incloses the Dufferin islands.

The fate of Goat island is not difficult to foresee. In a thousand years from now, at the present rate of recession, the Horseshoe falls will have reached the upper end of the island and will draw off all the waters from the American falls, which by that time will have receded only about half way to the Goat island bridge. All the islands will then be joined by a dry channel to the mainland, an event which was anticipated in the year 1848, when, owing to an ice blockade in the Niagara river near Buffalo, the American fall was deprived of all its waters for a day. As already indicated by Gilbert's forecast, in from two to three thousand years from now, or long before the falls have even reached the head of Grand island, the drainage of the great lakes will be reversed, provided the land continues to rise northward as it has in the past, and Niagara will carry only the drainage of the immediate neighborhood. From a majestic cataract it will dwindle to a few threads of water falling over the great precipice, such as may be seen during the summer season in the upper falls of the Genesee at Rochester.

Age of Niagara

Speculations as to the age of Niagara have been indulged in ever since men began to recognize that the river had carved its own channel. The length of time required for the excavation of Niagara gorge is not merely of local interest but serves as a basis for estimating the length of time since the disappearance of the Laurentian glaciers from this region, and incidentally it has served as a chronometer for approximately measuring the age of the human race on this continent. From insufficient data Sir Charles Lyell estimated the age of Niagara at 36,000 years, while others have assumed an age as high as 100,000 years or more.

No reliable basis for estimating the age of the gorge was known till a series of surveys were made to determine the actual recession of the cataracts. From these the following variable rates of recession of the two falls have been obtained.[1]

[1] Report N. Y. state engineer. 1890.

The American falls

	Feet a year
From 1842 to 1875	.74
1875 " 1886	.11
1886 " 1890	1.65
averaging	
From 1842 to 1890	.64

The Horseshoe falls

From 1842 to 1875	2.01
1875 " 1886	1.86
1886 " 1890	5.01
averaging	
From 1842 to 1890	2.18

This shows a most rapid increase in the rate of recession during the four years between the last two surveys. From this we may assume that the mean recession of a cataract combining the volumes of both American and Horseshoe falls, such as existed throughout the greater period of gorge excavation, is at least three feet a year and may be as high as four or even five feet a year.

The first to make use of this known rate of recession in estimating the age of the gorge was Dr Julius Pohlman. He considered that the gorge of the whirlpool rapids and other portions of the present gorge were of preglacial origin, and so reduced the length of post-glacial time to 3500 years. Since that time numerous estimates of the age of the gorge have been made, the results often varying widely, owing to different interpretations given to the narrow portions of the gorge. It is perfectly evident that, if Niagara was deprived of seven eighths of its water supply, for the period of time during which the gorge of the whirlpool rapids was excavated a very slow rate of recession must have obtained, and hence the age of the gorge is greatly increased. Upham, who does not believe in the withdrawal of the waters, makes the age of the gorge between 5000 and 10,000 years. Spencer and Taylor are ardent advocates of the reduction of the volume of water during a prolonged period, when the supply from the upper Great lakes was cut off.

The former makes the age of the gorge in round numbers 32,000 years, the latter places it tentatively at 50,000 years, though recognizing the uncertainty of many of the elements which enter into his calculations. Prof. G. F. Wright has recently applied a most ingenious method to the solution of this question, and one which seems to eliminate the doubtful factors.[1] This method is based on the measured rate of enlargement of the oldest part of the gorge by atmospheric action. The present width of the river at the mouth of the gorge is 770 feet, and Prof. Wright thinks that it was probably not less at the time when the formation of the gorge began. Assuming that the bank at that time was vertical, he finds that since then the stratum of Lockport limestone at the top has retreated 388 feet. Careful measurements show that the total amount of work accomplished here by the atmosphere since the beginning of gorge formation, was the removal from the side of the gorge of a mass of rock constituting in section an inverted triangle 340 feet high and with a base of 388 feet. This would be similar to a mass with a rectangular section of the same hight but with a base 194 feet wide. The rate of waste of the banks was measured by Prof. Wright as accurately as possible and found to be over one fourth of an inch a year, or a total amount of 610 cubic yards of rock from one mile of the gorge wall. From this he finds that 10,000 years is the maximum amount of time required for the entire change which has occurred in the bank since it was left exposed by the recession of the cataract.

The most recent and most detailed estimates of the age of the gorge have been made by Prof. C. H. Hitchcock.[2] He assumes that the present rate of recession is four feet annually, and finds accordingly that the last formed section of the gorge, from the present falls to the point where it suddenly contracts above the railroad bridges, was formed during 2962 years, which closely agrees with Pohlman's estimate. Thus the beginning of the great cataract at the northern end of the upper great gorge "dates back to 1062 B.C., 300 years before the time of Romulus, or

[1] Pop. sci. monthly. 1899. 55:145-55.
[2] Am. antiq. Jan. 1901.

to the reign of King David at Jerusalem." Prof. Hitchcock believes that the gorge of the whirlpool rapids was formed while Niagara drained only the diminished Lake Erie, and he allows a period of 7800 years for the accomplishment of this task. For the erosion of the remaining portion of the Niagara gorge Prof. Hitchcock allows 8156 years. Thus the total length of time required to carve out the Niagara gorge is considered by Hitchcock to be 18,918 years.

The reader should here be reminded that all such estimates are little more than personal opinions, and that they necessarily vary according to the individual predilections as to greater or less power of erosion possessed by the cataract under the given circumstances. The leading questions concerning the extent of the preglacial erosion in this region, and the changes in volume of water during the lifetime of the Niagara, which are of such vital importance in the solution of this problem, are by no means satisfactorily answered. Nor can we assume that we are familiar with all the factors which enter into the equation. There may be still undiscovered causes which may have operated to lengthen or shorten the lifetime of this great river, just as there may be, and probably are, factors which make any estimates of the future history of the river and cataract little more than a mere speculation. We may perhaps say that our present knowledge leads us to believe that the age of the cataract is probably not less than 10,000 nor more than 50,000 years.

Chapter 3

STRATIGRAPHY OF THE NIAGARA REGION

The stratigraphy of the Niagara region, or the succession of fossiliferous beds, their origin, characteristics and fossil contents, has since the time of Hall's investigations barely received cursory attention from American geologists, whose interest has chiefly centered in the problem of the physical development of the gorge and cataract. A careful examination of the strata of this region and of their fossils reveals problems as interesting and profound as those furnished by the gorge and cataract, and many of them are of far more fundamental and far-reaching significance. Profoundly interesting and instructive as is the " Story of Niagara " and of the physical development of the present surface features, it becomes insignificant when placed by the side of that great history of the rise, development and decline of vast mlutitudes of organic beings which inhabited the ancient seas of this region and whose former existence is scarcely dreamt of by the average visitor to the falls. These ancient hosts left their remains embedded in the rocks of this region; and from the record thus preserved the careful student is able to read at least in outline the successive events in the great drama which was enacted here, in an antiquity so remote that it baffles the imagination which would grasp it. But he who would decipher these records must bear in mind the maxim of La Rochefoucauld: "*Pour bien savoir une chose, il faut en savoir les détails.*" A knowledge of details is necessary to an understanding of the stratigraphic and paleontologic history of this region, and there is no better way of obtaining this knowledge than by a close study of the various sections which expose the strata here described.

The strata of the Niagara region belong to the Siluric series of deposits, which accumulated during the Siluric era of the earth's history.[1] Rocks of Devonic age occupy the southern portion of the district, resting on and concealing the Siluric strata which dip beneath them. (*See* fig. 1, p. 19) As has already been noted, all

[1] *See* table in chapter 2.

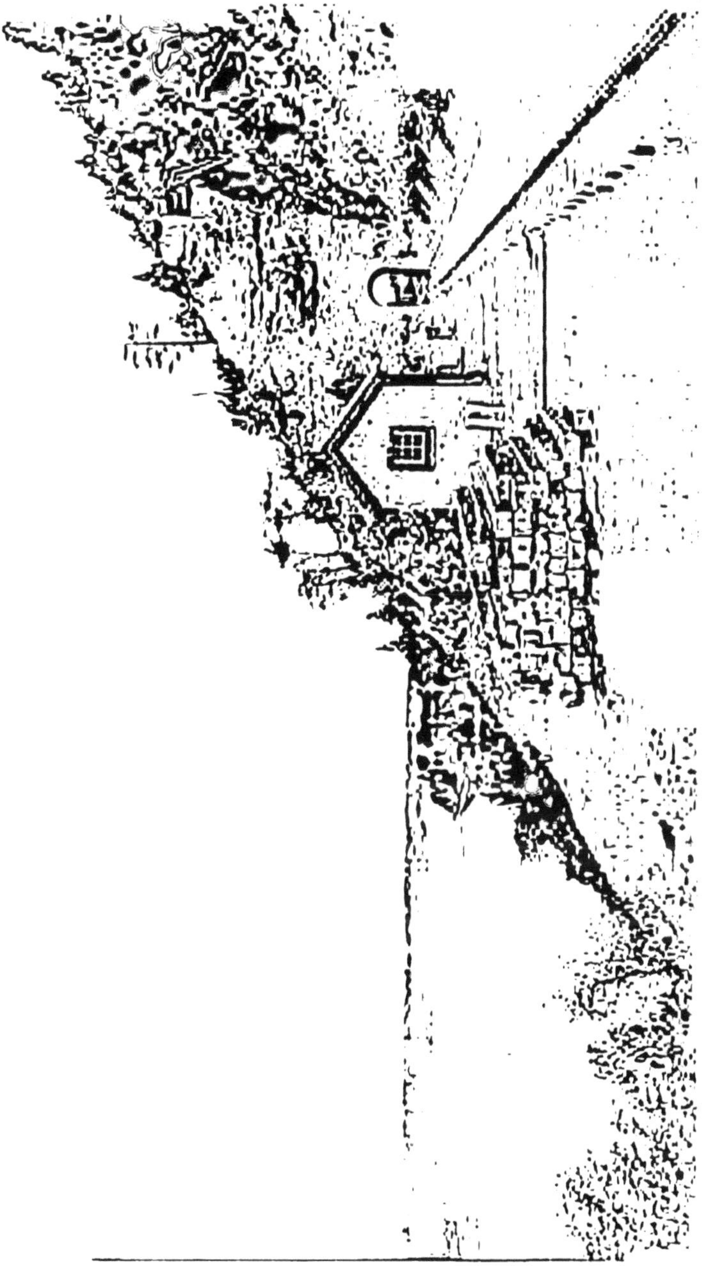

the rocks of this region have a gentle southward dip, which permits the lower members to appear progressively as we proceed northward over the surface of the old erosion plane. We may now proceed to describe the various members of this series in ascending order.

Oswego sandstone

This, the lowest member of the Siluric, is not exposed in the Niagara region, as its point of outcrop is now covered by the waters of Lake Ontario. (*See* sections 1 and 2, fig. 7) From borings, however, we know its character and thickness, which in this region is 75 feet.

Medina sandstones and shales

Only the upper portion of this formation is exposed in the Niagara district, where the total thickness is more than 1200 feet.

Red Medina shales. The upper beds of this division are the lowest exposed beds in this region. They are bright red sandy shales, generally of a very uniform character, though occasionally a bed which might almost be called a sandstone occurs. Wherever this rock is exposed to the atmosphere, it rapidly breaks down into small angular fragments, which quickly form a debris slope or talus at the foot of every cliff. In the faces of the older cliffs this rock is so friable, that it can readily be removed by the hand, the fragments themselves being easily crushed between the fingers. In the course of time these fragments disintegrate into a fine reddish clay soil, which when wet has a rather tenacious character.

As the lower part of the Niagara river from Lewiston to Lake Ontario is wholly excavated in this rock, it may be seen wherever the banks are kept fresh by the river, or where small lateral streams enter the Niagara. Where erosion is not active, the shale bank is soon reduced to a slope of red clayey soil, which generally becomes covered with vegetation.

A good place for the study of this shale is on the New York side of the Lewiston suspension bridge, where a fresh cut reveals about 50 feet of the rock. The bridge is 65 feet above the river, and the total thickness of red shale above the water at this point is therefore 115 feet. The shale here as elsewhere will be found to be seamed by whitish or greenish bands, both parallel with and at right angles to the stratification plane. In the latter case they are seen to lie on both sides of a joint fissure, which indicates that the discoloration of the rock, often extending to an inch on either side of the joint, is due to percolating air and water, the latter probably carrying organic acids in solution. The horizontal bands, often several inches in thickness, are probably similarly discolored portions along lines of greater permeability.

No fossils have been found in these shales.

Gray quartzose sandstone. The red shales terminate abruptly and are succeeded by a stratum of gray quartzose sandstone, which is very resistant, and wherever exposed, produces a prominent shelf. This rock varies somewhat in different portions of its exposure, but it averages perhaps 25 feet in thickness. This bed is exposed along the gorge from its mouth to the whirlpool, where it forms a ledge at the water's edge, beyond which it passes below the water level. It is well shown at Niagara glen, where a spring of cool water issues from beneath it, near the water's edge. In the bank on the opposite side, where a fine section of the rocks of the gorge is shown, this quartzose bed is seen in its full thickness, lying between the red shale below and the shales and sandstones above. The red shale at the water's edge has crumbled away, leaving the quartzose bed projecting from the wall in some cases to a considerable extent.

The quartzose sandstone usually forms beds of considerable thickness in this region, though near the top of the stratum a number of thin beds generally occur. The best exposure for the examination of this rock is in the quarries opened up in the terrace on which the Lewiston tower of the suspension bridge stands. In these quarries the sandstone slabs often show smooth surfaces, which generally bear markings similar to those formed by waves on a surface of fine sand. These wave marks are found in most of the sandstones of the Medina group, but they are nowhere in this region so well developed as in the upper thin bedded layers of the quartzose sandstone. No fossils have as yet been found in the gray sandstone on the Niagara river, though farther east a similar quartzose rock shows shells of the Medina Lingula on the surfaces of the layers, which also show wave marks.

The succeeding beds of the Medina as well as the Clinton, Rochester and Lockport beds, are best exposed along the railroad cut of the Lewiston branch of the New York Central and Hudson River railroad. This cut is reached from the Lewiston end through a short tunnel cut in the Medina sandstone (plate 12). As the beds dip southward, and the roadbed rises in the same direction, we pass rapidly across all the formations from the lowest to the highest exposed.

Upper shales and sandstones. The contact between the quartzose sandstone and the overlying Medina shales is not generally well exposed, except in one place. This is in Evan's gully, the first of the small excavations in the roadbed, made by the streams of water which in the spring time cascade from the banks. The quartzose sandstone forms the bed of the gully below the bridge on which the railroad crosses it, and it also forms the capping rock over which the stream cascades to a lower level.

1 The lowest beds of this division of the Medina are gray shales, 25 feet in thickness and readily splitting into thin layers and generally smooth to the touch, indicating the absence of sand. There are however beds of a more sandy character, even to fair sandstones, interbedded with the shales, and this is particularly the case near the middle of this shale mass. These sandstone beds are similar in character to the quartzose sandstone below the shales, but they occur in thin layers, separated by shaly masses. These same beds are exposed in the cutting which leads to the tunnel on the north, where they are shown near the base of the section. They vary in thickness up to 8 inches, and in some cases contain a few fossils, notably the shells of Lingula cuneata (fig. 81). The shales below the sandstone layers are mostly below the level of the roadbed, the greatest thickness exposed above that, being about 6 feet.

The upper 13 or 14 feet of this shaly series are well shown in the cutting north of the tunnel, where they may be seen above the sandstones just alluded to. These rocks present in places an almost perpendicular wall, where the overlying sandstones have not been removed, while from the rapid weathering of the shale, the capping stone generally projects beyond the face of the shale cliff. The undermining of the upper layers thus results in their ultimate breaking down from non-support, and the resulting fall of rocks may be of a dangerous character. Care is therefore necessary in the examination of these sections, and the warnings of the section guards should always be heeded. These men patrol the tracks continually from early morning till after the last train has passed at night. This is necessary, as the fall of rocks is continuous, and often of such amount as to obstruct traffic for some time. Any one who will watch these cliffs for a time from one of the projecting points where a comprehensive view may be obtained, and note the almost incessant fall of rock particles, will receive an impressive object lesson in the processes by which cliff retreat is effected.

In many cases the shale banks are covered with a coating of red mud carried by rains from the red soil above them. This creates the impression that the color of these lower shales is red like that of the shales higher up in the series, and only after breaking off fresh particles can the true color be seen.

2 These gray shales are succeeded by sandstones and sandy shales, some of the former massive, quartzose and in beds 6 or 7 inches in thickness, separated by shaly layers. The sandstone is gray and often porous, as if it had undergone some internal solution, which suggests that fossils may have been present which were dissolved by percolating waters. Fragments of fossils are occasionally found, but mostly in an unidentifiable condition. Many of the thinner and more clayey beds have raised markings on their under side, which may be indicative of the former presence of seaweeds in the muddy beds of this period. Small black phosphatic pebbles, often very smooth, are not uncommon in some of the layers, and larger masses of black, apparently carbonaceous shale are occasionally found mixed with the sand. In the gray shaly sandstone beds the Medina gastropods and bivalves (pelecypods) occur sparingly, and usually in a poor state of preservation. Some of the thin layers are calcareous, though still containing a large proportion of argillaceous matter. These are generally fossiliferous, the most common organism being a small cylindric bryozoan.[1] Fragments of these

[1] Identified provisionally as Helopora fragilis (fig. 74).

beds with the bryozoan weathered out in relief on their surfaces, may be found at the base of the cliff in the cut north of the tunnel.

3 In the northern end of the section the sandstones and sandy shales have a thickness of about 5 feet, and are in turn succeeded by 6 feet of shale, weathering readily into a clayey earth, which accumulates, as a talus on the underlying sandstone ledges. As in the other shale cliffs, so here weathering causes a more rapid retreat of the shale than of the overlying sandstone, which therefore projects beyond the shale cliff till it breaks down.

These shales are mostly gray, sometimes greenish gray, with occasional sandstone bands. Toward the top they become intercalated with reddish bands, and finally the prevailing color of the shale becomes red.

4 Following these shales is a mass of sandstone from 35 to 40 feet thick and consisting mostly of beds which vary from 4 to 6 inches in thickness. The sandstone is compact and solid, reddish in color or gray mottled with red. The beds are separated by red shaly partings, with occasional beds of red shale 2 to 4 feet thick. About 20 feet above the base of this sandstone mass is a concretionary layer from 1 to 2 feet thick, which appears not unlike a bed of large rounded boulders. These concretions vary in size up to 3 or 4 feet in greatest diameter, and they lie in close juxtaposition, not infrequently piled on each other, thus still more simulating the blocks of a boulder bed.

This sandstone cliff is in general quite perpendicular, and the thin and comparatively uniform layers, which are regularly divided by vertical joint fissures, produce the appearance of a vertical wall of masonry, for which many people, seeing it only from the rapidly moving train, have no doubt mistaken it. The regularity of these successive beds is at times interrupted by a heavier layer, either red or gray and mottled, which may be traced for some distance, after which it thins out and disappears. This thinning out of the layers in one or another direction is a common and characteristic feature of these sandstones, and is a direct result of the irregularities of current action during the deposition of the sands. We may trace a sandstone mass for some distance, and then find that it disappears by thinning, either bringing the layers above and below it in contact or giving way to a bed of shale.

A careful examination of these individual beds will show the presence of ripple marks in many of them. This indicates moderately shallow water during the accumulation of these sands; for ripple marks are found only down to the depth to which wave action penetrates. These ripples vary greatly in size, a bed about 10 feet above the concretionary layer showing examples in which the crests are from one to one and a half or more feet apart.

The fossils found in these sandstones are the characteristic Medina pelecypods, and the common Medina Lingula cuneata.

5 The thin bedded sandstone layers are followed by 12 or 15 feet of massive sandstones in beds from one to several feet in thickness, and varying in color from reddish to grayish. This rock generally shows strongly marked cross-bedding structure on those faces

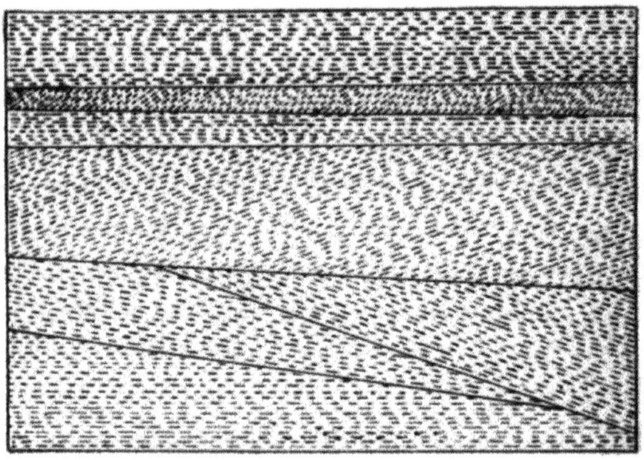

Fig. 20a Cross-bedding in Medina sandstone, Niagara gorge.

which have been exposed for some time. This structure illustrated in figure 20a, copied from a ledge of this rock, indicates diverse current and wave action in the shallow water in which this rock was forming. While the deposition of the strata was essentially horizontal, the minute layers made up of the sand grains were for a time deposited at a high angle, much after the manner of deposition of the layers in a delta. After a while the activity of the current changed to another direction, and the layers already deposited were in part eroded, or beveled across the top, and new layers, inhar-

Section on New York Central railroad cut south of first shanty, looking south. Medina sandstone at base; gray band of upper

monious with the preceding ones, were laid down on the eroded surface. This was repeated a number of times, as is shown by the succession of changes in the sandstone layers.[1] This structure is sometimes shown on a large scale, as in the case of a bed shown about 200 feet north of " Milk cave ravine ", the second of the small ravines met with in coming from the north. Here some of the layers are very gently inclined, and may be traced for some distance. They are obliquely truncated, other horizontal beds resting on the truncated edges (fig. 20b). (*See also* plate 14)

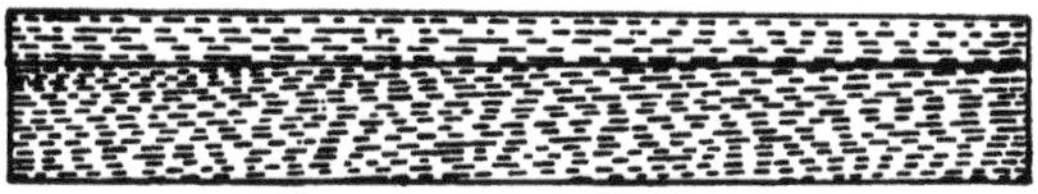

Fig. 20b Contemporaneous erosion and deposition in Medina sandstone, Niagara gorge.

The Medina Lingula (L. c u n e a t a) is found in these sandstones as in the lower ones, but other fossils are rare. Occasionally on the sections the hollows left by the removal of the shells may be seen, while similar cavities, caused by the removal of small black pebbles like those found in the lower layers, also occur. In the upper portions of this mass, on the under side of some thin sandstone lenses resting on and separated by shaly partings, occurs the so-called " jointed seaweed " of the Medina formation, known as A r t h r o p h y c u s h a r l a n i, and illustrated on plate 16. This is a characteristic Medina sandstone fossil, but in this region it has not been found in any of the other sandstone strata. Specimens of this fossil were obtained in digging the great power tunnel at Niagara, but only from the sandstone layers near the bottom of the tunnel, which is about the horizon in which they are found in the gorge section.[2]

6 The highest member of the Medina in this region is a hard, massive bedded and compact quartzose sandstone similar to the

---

[1]Compare with this the cross-bedding structure shown in the unconsolidated sands and gravels in the Goat island gravel pit, and in the section of the old Iroquois beach at Lewiston.

[2]The restriction of this characteristic Medina fossil to these upper layers of sandstone at Niagara was pointed out to me by John MacCormick, the watchman of this part of the road, who collects these specimens and keeps them for sale. As he is continually handling these rocks and has handled them for years, he has become familiar with their characters, and is therefore in a position to obtain knowledge of such facts.

quartzose bed terminating the lower shales. While nearly white when fresh, this rock generally weathers to a grayish yellow color and often exhibits yellow iron stains. On the weathered edges cross-bedding structure is well brought out. When separated from the rocks below by a shaly bed, this rock generally projects from the bank for a sufficient distance to form a shelter for the watchman in case of a sudden shower. Where this sandstone comes down to the level of the roadbed, at a projecting cusp of the cliff, it has been cut through and a portion of it left between the track and the gorge. In the shadow of this rock mass stands the second of the watchmen's shanties which we meet with in approaching from the mouth of the gorge.[1] The upper quartzose bed has here a thickness of 7½ feet. Several hundred feet south of this point, where the top of this sandstone is level with the roadbed, a huge ripple, 15 feet from crest to crest, and nearly 2 feet deep, is shown on the river side of the track. This "giant ripple" was described and illustrated by Gilbert,[2] who found other ripples of similar size in the Medina sandstone at Lockport, as well as in the quartzose sandstone near Lewiston.

On the surfaces of the flagging stones which are derived from the Medina sandstones, ripple marks of small size are not uncommon, and the sidewalks of Buffalo and other cities where this rock is utilized, often exhibit fine examples of such rippled rock surfaces.

In the cliff of Milk cave falls (or St Patrick's falls), which is the second lateral fall below the mouth of the gorge, the upper beds of the Medina formation are well shown. The concretionary layer is near the level of the roadbed, and has a thickness of 3 feet. 29 feet above it is the base of the upper gray quartzose sandstone, before reaching which we find that the red sandstone gradually loses its bright color, at first being mottled, and then at times losing its red color altogether, though the thin partings of shale still retain

[1] This is occupied by John Garlow, on whose beat most of the "Niagara crinoids" (*Caryocrinus ornatus*) are to be found. Specimens may generally be obtained from him at a small price.

[2] Bul. geol. soc. Am. 10:135-40. pl. 13. fig. 2.

it. The quartzose capping rock consists at the base of a white bed, from 1½ to 2 feet thick and showing cross-bedding structure, followed by shale 1 to 1½ feet thick and of a reddish color in places, and finally by a solid bed of white quartzose sandstone 5 feet in thickness, and like the lower bed, showing cross-bedding structure on the weathered sections. A few thin layers of sandstone overlie this bed, having a total thickness of less than half a foot. On these follow the shales of the Clinton formation.

The upper Medina sandstones and shales may be traced in both walls of the gorge nearly to the falls. From the southward dip, the beds progressively pass below the water level, till near the falls only a small portion of the upper beds remains. These may be seen at the river margin in the bottom of the gorge, between the *Maid of the Mist* landing and the carriage bridge on both sides of the river. On the New York side only a few feet of the red sandstones are exposed, the remainder being covered by talus. During high stages of the river these exposed beds are covered by the water. On the Canadian side an extensive ledge of the red Medina sandstone is exposed opposite the inclined railway on the New York side. In the banks behind this ledge the white quartzose sandstone which forms the top of the Medina occurs, its top being at least 25 feet above the water level. It here forms a projecting shelf on which rest huge blocks of limestone broken from the cliff above. From this we may judge that at the foot of the Horseshoe falls the upper layers of the Medina may still be above the water level.

Clinton beds

The Clinton beds at Niagara aggregate about 32 feet in thickness and consist of a stratum of shale at the base and two distinct strata of limestone above this. (*See* Plate 14)

Clinton shale. Resting immediately on the quartzose layers which terminate the Medina formation, is a bed of olive green to grayish or sometimes purplish gray shale, which readily splits into very thin layers with smooth surfaces, and is quite soft enough to be easily crumbled between the fingers. Fossils are rare in it, but occasionally layers are found which have their surfaces covered with crushed valves of small plicated brachiopods, among which Anoplotheca hemispherica and A. plicatula may be mentioned. Other fossils are rarely found except reed-like impressions which are not uncommon. Some impressions have been found which probably belong to Pterinaea emacerata, a pelecypod occurring higher in the Clinton and also in the Rochester shales. The total thickness of these shales is 6 feet.

Clinton lower limestone. On the shale rests a stratum of limestone 14½ or 15 feet in thickness. The lower three or four feet of this rock are compact to granular or finely crystalline, having a sugary texture. Small masses of iron pyrites are not uncommon in this rock, this being the only representative of the ferruginous matter so characteristic of this part of the Clinton beds on the Genesee river and eastward, where a well marked bed of iron ore succeeds the shale. Hall[1] states that "the lower part of the limestone, as it appears on the Niagara river, is highly magnesian, and from the presence of iron pyrites rapidly decomposes, giving rise to the production of sulfate of magnesia, which at favorable points along the overhanging mass upon the river bank, may be collected in quantities of several pounds."

Fossils are not uncommon in this division of the Clinton limestone, though the variety is not very great. The most abundant species are a small brachiopod, Anoplotheca plicatula (fig. 133) with a strongly plicated surface, and a larger flat brachiopod, Stropheodonta profunda, which at times seems scarcely more than an impression on the rock surface. The remaining part of this stratum is a massive dark gray limestone with occasional thin bands of a shaly character separating the individual beds. Recognizable fossils are not very abundant in this rock. Many of the thin bedded portions of the lower Clinton limestone contain numerous shining black phosphatic nodules, very smooth and resembling small black pebbles. These are probably concretionary masses, though some have the aspect of being waterworn organic remains. Where the thin limestone layers are covered with a shaly or sandy coating, impressions of the beautiful, little

---

[1]Rep't 4th dist. 1842, p. 63.

View on the New York Central railroad cut, looking south, just south of the second shanty; the third is shown in the distance. The formations shown are, from below upwards; the top of the upper gray band of the Medina; the Clinton shale; Clinton lower

branching seaweed, *Bythotrephis gracilis*, may be found. This occurs also on some of the shaly partings of the limestones. The impressions vary from the slender variety of great delicacy to a coarse one in which the frond consists of broad irregular lobes.

This stratum generally forms a vertical wall with the next overlying stratum projecting beyond it.

Clinton upper limestone. In the region of the Genesee river the lower limestone is succeeded by a mass of shale which is generally fossiliferous, and on which lies the upper limestone. In the Niagara region this shale is wholly wanting, the upper limestone resting directly on the lower. The line of separation is however well marked, both by the diverse characters of the two rocks and by the different way in which each resists destruction by atmospheric agencies. The upper stratum is a crystalline and highly fossiliferous limestone, often pinkish in color, though chiefly light gray with yellowish or brownish particles where oxidation has occurred. Portions of the beds consist almost wholly of crinoid stems or joints, which give the rock a coarsely crystalline and sometimes porous aspect. Fossils are abundant in this rock, though the variety is generally not large. The most common species is a rotund variety of the brachiopod, *Atrypa reticularis* (fig. 112), which is generally very robust and sometimes almost globular in form. Of the other fossils in this rock several Stropheodontas may be mentioned, among them *Stropheodonta profunda*. A number of rhynchonelloid shells occur, readily recognized by their pointed beaks and strong plications. Among these are some small specimens of *Camarotoechia acinus*, a species characteristic of the Niagara beds of the west. It is readily recognized by its smooth umbonal area, and its single plication in the mesial depression or sinus, corresponding to which, on the opposite valve occur two plications. Among the more abundant fossils of this rock are smooth elongate and rather strongly biconvex brachiopods of the genus *Whitfieldella*. The most common is W. *intermedia*, but other species occur as well. The thickness of this stratum is 11 feet. The upper beds of this series contain species which on the whole are of a strongly marked Niagaran type, such as Spirifer niagarensis and others. A common brachiopod is Strophonella patenta, a flat, thin, subsemicircular shell with a straight hinge line and fine surface striations.

A characteristic feature of this upper limestone stratum is the strong development of stylolite structures. These stylolites are vertically striated columns, from a fragment of an inch to several inches in length, and ranged on either side of a horizontal suture or fissure plane in the limestone bed. Projecting from both upper and lower beds, they interlock with each other and so produce a strongly marked irregular suture. This structure is characteristic of limestone beds of this type, but its origin is still obscure. Pressure of superincumbent layers of rock seems to have been the chief cause of their production, this pressure acting unequally on the rock mass, from the presence of fossils or from other causes. A characteristic feature is the open suture at the ends of the columns, which gives the layers the aspect of having separated by shrinkage along an irregular plane. The vertical striations indicate motion either upward or downward.

The Clinton limestones may be seen in both banks of the river where not covered by vegetation, from the mouth of the gorge to within a short distance of the falls, near which they are covered by talus. They always form a cliff in the profile of the gorge, the 6 feet of shale below them forming a sloping talus-covered bank, below which there is another cliff formed by the hard upper Medina sandstone, the lower members forming one or more talus-covered slopes down to the quartzose bed of the Medina. This latter is again a cliff-maker, and generally projects from the bank, while the soft red shale below invariably produces a sloping talus-covered bank. Above the Clinton limestones is another slope and talus formed by the soft Rochester shale, above which a precipitous cliff is formed by the Lockport limestone.

At the base of the cliffs, fallen rocks of the Clinton limestones are mingled with those from the overlying Lockport limestones, and care must be exercised in discriminating between these when collecting fossils. Halfway between the third and fourth watchman's shanties on the railroad, where the top of the Clinton limestone is on a level with the roadbed, this rock was formerly quarried on the river side, and here a good opportunity is afforded to collect fossils from the limestone fragments. Blocks of the various limestones are also seen by the side of the track between the second and third shanties.

At the whirlpool on the Canadian side the Clinton limestones are seen in both banks of the old St Davids gorge, the section on the west showing glacial striae. Near the foot of the eastern wall of this old gorge and on the talus heaps which flank it, are large masses of calcareous tufa often inclosing leaves, moss or other vegetable structures. These masses appear to come from the horizon of the Clinton limestone, though they have not been seen in place, and it is not improbable that a "petrifying spring" carrying a strong solution of carbonate of lime issues from this rock. Springs issue abundantly from between the two members of the Clinton limestone, and they carry lime in solution, as is indicated by the deposit of soft calcareous ooze on the rocks and other substances over which this water flows. On exposure to the atmosphere this ooze will dry and harden. The joint faces of the Clinton limestone are everywhere veneered over with a thin deposit of calcium carbonate.

Limestone lenses of the Clinton. At intervals in the upper Clinton limestone may be seen large lenticular masses of a compact, hard and apparently structureless limestone, often concretionary and not infrequently showing numerous smooth and striated surfaces of the type known as "slickensides" and which are indicative of shearing movements. One of these masses is visible in the bank opposite the third watchman's hut. Its greatest thickness is about 8 feet, and it lies between the upper limestone and the overlying shale, being partly embedded in both. The rock is often cavernous or geodiferous, the cavities when freshly broken being filled by snowy gypsum or grayish anhydrite. Fossils are abundant in this rock.

Several other lenses of this type are visible in the upper Clinton limestone where it is crossed by the Rome, Watertown and Ogdensburg railroad below Lewiston hights. These masses are however entirely inclosed by the limestone, from which they are differentiated by their structureless character. The lenses exposed on the Rome, Watertown and Ogdensburg road are rich in shells of orthoceratites and shields of trilobites (Illaenus ioxus), while the lens in the gorge yields chiefly brachiopods, the most abundant of which are the smooth Whitfieldellas, the small W. nitida and the larger W. oblata being the most common.

The following species have been obtained from the lens in the gorge:

Brachiopoda

1 Whitfieldella nitida — abundant

2 W. nitida oblata — abundant

3 W. intermedia — common

4 Atrypa reticularis; specimens with strong, rounded bifurcating striae, noded at intervals by strong concentric striae, and apparently intermediate between the typical form of the species as it occurs in the Clinton and upper limestone and A. nodostriata, the most abundant form of the Rochester shales.

5 Atrypa nodostriata; rather common, convex and more elongate than in the shale above, with the plications generally sharper and bifurcating near the front. The pedicle valve has a distinct sinus bordered by strong plications, the corresponding fold being marked merely by strong plications. Anterior margin distinctly sinuate. The nodulations are not well preserved except in specimens from the shaly portions.

6 Atrypa rugosa; several small specimens, both valves very convex, with strongly defined sinus in pedicle valve, in the center of which is a small plication. Plications bifurcate and also increase by intercalation; crossed by strong rugose lines.

7 Rhynchotreta cuneata americana — rare

8 Camarotoechia neglecta — rare

9 Anastrophia interplicata — rare

10 Spirifer niagarensis; common, large and robust, with long hinge line and moderately high area, and strongly incurved beak. The sinus is flanked by two stronger plications and extends to the beak. The plications are flattened on top.

11 Spirifer radiatus; common but generally crushed; with an extended hinge line and form and proportions similar to the preceding species. The striae are fine and flat on top with very narrow interspaces altogether very similar to those covering the plications of S. niagarensis. A scarcely defined plication appears on each side of the sinus in some specimens, and in these the sinus is rather sharply defined and angular at the bottom. In others the sinus is shallow rounded and not definitely outlined by incipient plications. In the more elongated specimens the cardinal angle is well defined, but in the shorter specimens it is rounded.

12 Spirifer crispus — rare

13 Spirifer sulcatus — rare

14 Dalmanella elegantula; rare and with greater convexity than that of the specimens in the overlying shale.

15 Plectambonites transversalis — rare

16 Leptaena rhomboidalis — rare

17 Stropheodonta corrugata — rare

18 Orthothetes subplanus — rare

19 Strophonella patenta — rare

Gastropoda

20 Platyostoma niagarensis — rare

Trilobites

21 Illaenus ioxus; fragments of caudal and cephalic shields crowded together into masses sometimes of considerable size.

22 Calymene blumenbachi — rare

Bryozoa

23 Lichenalia concentrica; common in very irregular and much distorted masses.

Corals

24 Enterolasma caliculus — common

Crinoids

25 Eucalyptocrinus; fragments of root stem and calyx.

In the lenses below Lewiston hights the same species except nos. 2, 3, 9, 10, 14, 15, 17 to 20 and 25 have been found. Rhynchotreta cuneata americana has more the features of the same species from the western Niagara than those of the Rochester shale species.

Spirifer crispus is commonly deficient in plications approaching in this respect and in the character of the sinus, S. eriensis from the Manlius limestone. Atrypa nodostriata is robust, convex, with coarse rounded plications and rather faint concentric striations, characters intermediate between A. reticularis of the Clinton and A. nodostriata of the Rochester shale. Besides these species and some not yet identified, the following occur.

Cephalopoda

26 Orthoceras annulatum

27 O. medullare (?) rare

28 O. sp.

Pelecypoda

29 Modiolopsis cf. subalatus?

The origin of these lenses is still obscure. Many of the fossils found in them are characteristic of the Niagara group of the west, but are rare or wanting in the Niagaran of New York. This is specially the case with the trilobites (Illaenus ioxus) and the cephalopoda. Dr E. N. S. Ringueberg many years ago studied these limestone masses as exposed at Lockport and other more eastern localities, and he termed them the "Niagara transition group".[1] He found in this rock 32 Niagara species, 11 species common to the Clinton and Niagara, two species found otherwise only in the Clinton, and two species not found outside of this rock. The origin and significance of these unique deposits are being carefully studied by the state paleontologist.

Rochester shale

The Rochester (Niagara) shale has a total thickness of about 68 feet in the gorge of the Niagara. It is here divisible into a lower and an upper half. The lower portion is a highly fossiliferous shale with numerous limestone bands, and terminates in a series of thin calcareous beds with shaly partings in all about 4 feet thick, and extremely rich in bryozoa. The upper 34 feet are quite barren and have few limestone layers.

[1]Am. nat. 1882. 6:711-15.

Plate 15

View of the New York bank a short distance north of the Devil's hole. The gorge-road is shown below, and the New York Central railroad cut near the middle. The fourth shanty is shown a short distance to the left (north) of the bridge.

Lower shales. The beds immediately succeeding the Clinton limestone are calcareous shales with frequent thin limestone layers. The latter are the most fossiliferous; being in general entirely made up of organic remains. The calcareous beds of the lower 5 or 10 feet are particularly rich in crinoid remains. Chief among these organisms, on account of its abundance and perfection, is the little triangular Stephanocrinus ornatus, which may be found in most of the calcareous layers. Fragments of Eucalyptocrinus are always common, while the characteristic Niagara cystoid Caryocrinus ornatus is also found, though not so abundantly as in the upper part of the lower division. The most abundant brachiopod of the lower shales is Whitfieldella nitida oblata, similar to the specimens found in the limestone lenses. The little Orthis, Dalmanella elegantula, is also common, ranging throughout the lower division of the shales. Spirifer niagarensis is common above the lowest 3 or 4 feet of the shale. Orthothetes subplanus, a large, subsemicircular and nearly flat brachiopod, is abundant in some of the calcareous layers, which at times seem to be composed of it, so thickly are these shells piled one on the other. Atrypa nodostriata is the commonest representative of the genus, the larger A. reticularis, so abundant in the upper Clinton, being comparatively rare and subordinate in development. In the limestone bands A. nodostriata is usually rotund, but in the shaly beds it is most commonly compressed. Trilobites are comparatively rare in these lower shales, though representatives of all the species found in this region have been obtained from them. Bivalve molluscan shells are also uncommon, but the gastropods, Diaphorostoma niagarense and Platyceras are not infrequent.

Some of the calcareous bands are almost barren of organic remains, but in most cases these beds will be found to constitute the chief repositories of the fossils.

Bryozoa beds. A short distance south of the third watchman's hut, the section comes to an end, being for some distance replaced by a soil-covered and more or less wooded bank. Where the section ends the upper Clinton limestone is only a few feet above the roadbed, and the shale above it is accessible. 29 to 30 feet above the top of the limestone, a group of calcareous beds rich in bryozoa project from the bank, being readily traceable for some distance on account of their compact nature. Their total thickness is about 4 feet, and they consist of numerous thin limestone layers with shale partings of greater or less thickness. On the weathered surfaces of the limestone layers, the bryozoans stand out in relief, and such surfaces will often be found completely covered with these delicate organisms. The cylindric types prevail, but the frondose forms are also common. With them occur brachiopods and other organisms. Slabs of this rock are often found on the talus slopes, and they are among the most attractive objects that meet the collector's eye. The section begins again, after an interruption of perhaps a quarter of a mile, near the old quarry in the Clinton limestone. (Plate 15) Between the river and the railroad are several mounds of shale, which were left in place when the railroad cut was made. These are subject to disintegration, and the fossils in consequence weather out. They may be picked up on these mounds completely weathered out, and often in perfect condition. The best of these mounds is about halfway between the old Clinton limestone quarry and the fourth watchman's hut. Here the top of the mound is on the level of the top of the Bryozoa beds, the whole thickness of which is therefore included in this remaining mass. As these beds are extremely fossiliferous, this mound is a productive hunting ground.[1]

An equally productive locality for weathered-out fossils is the slope of disintegrated shale rising from the Rome, Watertown and Ogdensburg railroad tracks above Lewiston hights. The best hunting ground is in the little gullies made by the rivulets of rain water in the bank. Some glacial till is here mingled with the clay from the decomposed shales, and it requires a little attention to distinguish the two.

---

[1] The fossils here obtained are extremely delicate and brittle. They should be placed at once on layers of cotton batting, in a small box and covered with similar material, the box being completely filled. This is the only way in which many of these delicate fossils can be carried away without breaking.

Upper shales. Above the Bryozoan beds the shale is soft, and more evenly and finely laminated, splitting often into thin slabs of moderate size. Hard calcareous beds are generally absent, though occasionally found near the top. The stratification and lamination is much more strongly marked in this than in any other division of this rock. When freshly broken, the shale has a brownish earthy color, which changes to grayish when the rock decomposes to clay. Fossils are rare, those found being seldom well preserved. In most cases the shells are dissolved away, leaving only the impressions of the fossil, which from compression become faint, and are not readily recognized without careful scrutiny. The most common remains found in these rocks are bivalve mollusks (pelecypods) and trilobites. Among the former Pterinaea emacerata is the most abundant, while Dalmanites limulurus is the chief among the trilobites of these beds. Other trilobites also occur in these shales, notably Homalonotus delphinocephalus, as well as a number of brachiopods.

Toward the top fossils become rarer, and finally are wanting altogether. The shale becomes more heavy bedded, and calcareous layers begin to increase. The last 10 feet or more are quite calcareous and compact, and have an irregular fracture. They grade upward into the basal layers of the Lockport (Niagara) limestone.

Lockport (Niagara) limestone

The limestone which succeeds the Rochester or Niagara shales forms the summit rock of the series from the edge of the Niagara escarpment to south of the falls. It consists of a number of distinct strata, of varying characters, most of them very poor in organic remains. The total thickness exposed in the Niagara region is not over 130 feet, but borings show that the thickness of the limestone lying between the Rochester shale and the Salina shales is from 200 to nearly 250 feet. Some of the upper beds of this limestone mass may represent the Guelph dolomite and others may belong to the base of the Salina beds. Nevertheless we may confidently assume that the thickness of the Lockport limestone in this region, is at least 150 feet.

Hydraulic cement beds. 1) The lowest stratum of the series is a hard, compact, bluish gray silicious limestone, weathering whitish on the exposed faces, and breaking into numerous irregular fragments larger near the bottom of the stratum but becoming small, angular and subcubical near the top, where the weathering is similar to that obtaining in the upper parts of the shales. This stratum varies from 7 to 8 feet in thickness being in places divided into two tiers, the upper one, 4 feet thick, appearing as a distinct bed. This weathers to a creamy gray color, and breaks into small angular fragments with no regularity of fracture, and independent of the plane of stratification. On some of the weathered edges of this rock irregular stratification lines are visible, giving the beds the appearance of a fine grained sandstone. Occasionally small geoditic cavities occur lined with dolomite or gypsum. The line of contact between this stratum and the underlying shale is an irregular one, the shale surface having a wavy character.

2) This rock is succeeded by a 4 foot stratum of arenaceous limestone which shows no well marked stratification lines on the weathered surfaces, though in places a distinct cross-bedding structure appears. It peels off in irregular slabs parallel to the cross-section, i. e. at right angles to the stratification plane. Near the top of this stratum are a few thin beds which show the finer stratification structure on the weathered edges, the character of this structure being such as is found in fine grained sandstones.

Both these strata appear to be wholly destitute of fossils. It is not improbable however that the scattered geodes represent the places where corals or crinoids occurred, which have subsequently been altered or dissolved out. Aside from this, there is no evidence that this rock ever was fossiliferous, and it is most probable that it represents the accumulation of fine calcareous mud or sand.

Crinoidal limestone. 3) The compact hydraulic rock is abruptly succeeded by a stratum of highly crystalline limestone, on the weathered surfaces of which joints of crinoid stems and other organisms stand out in relief, particularly in the lower part of the stratum. The rock is entirely composed of fragments of organisms which were ground up and mingled together in great profusion. Oblique bedding lines may be observed occasionally, indicating that the fragments were subject to wave action. The stratum varies in thickness from 5 to 6 feet, and is occasionally divided by horizontal sutures which show a marked stylolitic structure similar to that found in the crystalline upper Clinton limestone. The contact between this and the underlying stratum is wavy. This rock has been quarried at Lockport under the name of Lockport marble.

Geodiferous limestones. The crinoidal limestone is succeeded by strata all of which are more or less geodiferous, though varying considerably in composition and structure.

4) The rock immediately following on the crinoidal bed is a 4 foot stratum of compact, gray fossiliferous limestone, the fossils being of a fragmentary character. Stratification structure is well marked on the weathered surfaces, specially in some of the lower beds of the stratum. Sometimes there is only one thick bed, at others the stratum consists of a number of thin beds with a heavy one near the center. The thin beds show the stratification structure best, having at the same time a strongly granular character. As the fossils are fragmentary, and only accessible on the weathered surfaces, little is known of the organisms that constitute it. Crinoid joints occur, but they are less characteristic of this than of the lower stratum. Geodes however are not uncommon, the cavities being lined with crystals of pearl spar (dolomite) or filled with masses of snowy gypsum.

5) The fifth stratum of limestone in this series is a finely crystalline magnesian rock, like the others destitute of fossils except in so far as these are represented by geodes. The latter are common and filled with alabaster, or sometimes with massive or crystallized anhydrite. The latter is distinguished from the crystallized gypsum or selenite, which it closely resembles, and which occasionally occurs in the same beds, by the cleavage, which is rectangular and nearly equally perfect in three directions in anhydrite, while it is perfect in one direction only in the selenite.

6) A finely crystalline, somewhat concretionary dolomitic limestone, 3 feet thick, next succeeds, the weathered sectional surfaces of which, buff in color, show the fine stratification structure, which

is of the type of the cross-bedding structure in sandstone. Such structure indicates that the bed possessing it was a fine calcareous sand, subject to shifting movements by waves and deposited in moderately shallow water. We need look for organic remains in such a rock with no more assurance of finding them than we bring to the examination of uniform bedded shales. They may be abundant or they may be rare or absent altogether. Thus a limestone need not be necessarily a fossiliferous rock.

Geodes of the usual type are common, the dolomitic lining predominating.

7) On the preceding thin stratum follows a limestone mass of very uniform character, hardly separable into district strata, though consisting of numerous beds.[1] 27 feet of this stratum are shown at the quarry near the northern end of the section, where the upper exposed bed forms the surface rock of the plateau above. The beds are generally of considerable thickness, but the fine stratification structure is not so well marked as in the strata below. The rock may be considered a compact granular dolomite, in which considerable change has taken place since its original deposition. It is of a grayish color but weathers to a lighter tint. Geodes are plentiful, often quite large, and in these, minerals of great beauty are not infrequently found. The most common are the snowy variety of gypsum or alabaster, the darker gray, massive, fine anhydrite and the uniform, fine, dolomite rhombohedra with curved faces, generally of a pinkish tint and familiarly known as pearl spar. Long slender crystals of calcite, generally in the form known as scalenohedra, or dogtooth spar, are not uncommon. These are commonly of a golden color, and large enough to show well their crystal faces. In the new power tunnel which was excavated in the neighborhood of the falls, large masses of transparent gypsum of the selenite variety were found in cavities in this rock. Some of these pieces were 6 inches in length. Masses of limestone lined with pinkish dolomite crystals and occasional large masses of silvery selenite, and set with

[1]The distinction between stratum and bed is an important one. A stratum is a rock mass having throughout the same lithic character, and may be thick or thin. A bed, on the other hand, is that portion of a stratum limited by horizontal separation planes. *See* Geology and paleontology of Eighteen Mile creek pt 1. Introduction.

amber crystals of calcite, were also found in these cavities, the combination being such as to produce specimens of great beauty. Among the rarer minerals found in this rock is the crystallized and cleavable anhydrite, which like gypsum is a sulfate of calcium, but without the water which is characteristic of that mineral. Anhydrite crystallizes in the orthorhombic system, and its cleavage is in three directions, at right angles to each other (pinacoidal), thus yielding rectangular fragments and enabling one to distinguish it from selenite with little difficulty. It is also a trifle harder than selenite which is easily scratched with the finger nail. This form of anhydrite is rather rare, the principal localities for it being foreign. Masses of considerable size have been found in the limestone of this quarry, and small pieces are not uncommon in the geodes of these strata. Both selenite and the cleavable anhydrite are commonly called "mica" by the uninitiated; that mineral however does not occur at Niagara. Small masses of fibrous gypsum or satin spar have been found, but these are very rare. The satin spar of which the cheap jewelry sold in the curiosity shops is made is not from Niagara.

Among the metallic minerals found in this rock, zinc blende or sphalerite is most common. It is generally of a yellowish or light brownish color and brilliant resinous luster. Large masses however are rare. Galenite or lead sulfid crystals are also occasionally found, but this mineral is comparatively rare. In addition to these, iron pyrite, iron-copper pyrite (chalcopyrite), green copper carbonate (malachite), fluor spar (fluorite), iron carbonate or brown spar (siderite, generally ferruginous dolomite), strontium sulfate (celestite) and native sulfur as well as other minerals are met with.

The total thickness of the limestone exposed in this section is thus somewhat more than 55 feet. At Lewiston hights, on the edge of the escarpment, only about 20 feet are exposed. This includes the two lower strata of hydraulic limestone, the crinoidal limestone and a few feet of the lowest geodiferous beds (stratum 4). Over this lie some two or three feet of glacial till. The distance between the edge of the escarpment and the quarry at the end of the section, is a little over a mile and a half, the increase in thickness of the limestone and the rate of dip (since the surface is about level) is therefore a trifle less than 25 feet in the mile.

The crinoidal limestone is the most prominent stratum on the edge of the escarpment. From its base springs of cold and clear water issue at numerous places along the outcrops, both on the edge of the escarpment and in the gorge. The most prominent of these is at the head of "Milk cave" or St Patrick's falls, and here as almost everywhere at the base of the crinoidal limestone, shallow caverns abound. One of these caverns near the head of the falls, has a depth of 35 or 40 feet and is high enough to permit one to walk upright. No stalactites are found in these caverns, but the walls are much disintegrated and in places covered with a fine residual sand.

In the fields above this cavern are several sink holes of moderate depth, which serve as catchment basins for the waters of the surrounding country, which issue from these caverns during the wet seasons.

The cavern known as the Devil's hole belongs to this category. As in the other caverns, the roof is formed by the crystalline crinoidal limestone (stratum 3), the cavern itself being hollowed out in the hydraulic cement rock. This cavern is deeper than most others, and at the end a spring of deliciously cool water issues from between the two beds, the upper "spring line" of this region. There is no evidence that the cavern extended any deeper than it does at present, nevertheless the spot is worth visiting, as it is the only accessible one of the numerous springs and caverns. The fall of the Bloody run at this place is over a thickness of almost 60 feet of limestone, and the chasm which this stream has worn is interesting both from its historic and scenic points of view.[1]

West of the Niagara river on Queenston hights several quarries have been opened in these limestones, some distance south of the edge of the escarpment. The rock quarried is the crinoidal limestone and overlying beds. The total depth of rock in the quarry is 27 feet, of which the lower 14 or 15 feet are bluish gray and the upper of a lighter gray color. The limestone is here much more uniform, crystalline throughout and more fossiliferous. This may

[1] *See* brief mention of Bloody run massacre in Introduction.

indicate a nearness to the reef of growing organisms which supplied the material for these beds. Geodes lined with dolomite crystals occur in this rock, though not so plentifully as at the quarry in the gorge. Below the crystalline limestone is found the cement rock, which is from 4 to 10 feet thick and is quarried in a tunnel under the limestone quarry.

Owing to the resistant character, the limestone is everywhere exposed in the gorge, forming cliffs which are almost invariably perpendicular. Large blocks of this rock cover the talus everywhere, one of the largest of these being " Giant rock " along the gorge road. This is a block of the upper geodiferous limestone which has fallen from above, and now lies with its stratification planes at an angle of about 45°.

The limestones are well exposed along the gorge road, south of the railroad bridges, but without a special permit no one is allowed to walk on this roadbed. The contact between the limestone and the shale is here very irregular, indications of erosion of the shale prior to the deposition of the limestone occurring. The limestone is also somewhat concretionary, rounded masses projecting down into the shales. The succession of strata is here as follows:

1 Concretionary, irregularly bedded gypsiferous limestone, often earthy and with occasional thin, shaly layers; it splits readily into slabs perpendicular to the stratification. Thickness 6-8 feet.

2 Fine grained limestone with sandy feel, sometimes massive, sometimes in shattered layers with earthy or shaly partings, and separated from the underlying rock by an earthy layer. It weathers to an ashy or sometimes an ochery color, and varies somewhat in thickness. The upper layer is however a solid and fine grained limestone. Thickness 4-4.5 feet.

Strata 1 and 2 are the equivalent of the cement beds.

3 Crystalline and crinoidal limestone abruptly succeeding the lower bed. It is massive though somewhat thin bedded and contains geoditic cavities filled with gypsum. This continues uniform for a thickness of about 19 feet.

4 Compact limestone; concretionary with cavities containing gypsum and other minerals, and with sphalerite embedded in the rock.

The bedding and upper contact lines are irregular. Thickness 14-15 feet.

5 Compact, finely crystalline and homogeneous dolomitic rock, showing traces of fossils and slickensides. Beds showing Stromatopora common. In places the rock has a porous appearance and is rich in geoditic cavities, which are lined with dolomite and calcite crystals. Thickness 19 feet.

This stratum forms the lower portion of the cliff at the first cut on the gorge road, and the basal part of the mass left standing on the river side. Heads of Stromatopora may be seen in this rock, some of the geoditic cavities having replaced this fossil. This is about the summit of the beds exposed in the quarry at the end ot the railroad section.

6 Earthy, compact dolomite in thin layers, which give the cliff the appearance of a stone wall. Toward the top the rock becomes more compact and heavy bedded, this giving the appearance of an overlying stratum. This rock is full of geodes lined with pearl spar or dolomite, the cavities ranging in size up to that of a fist or larger. The beds are generally less than a foot in thickness, the average being from 3 to 6 inches. Toward the top of the cut, the rock becomes more compact and finely crystalline, but otherwise remains similar. Pearl spar geodes remain common to the top. The thickness of this mass, at the beginning of the gorge road, is about 45 feet.

The total thickness of the limestone exposed on the gorge road is in the neighborhood of 110 feet. This is double the thickness found at the quarry, the distance between the two points in a straight line being about three miles or nearly four following the curvature of the river. The rate of increase in thickness, or the amount of dip of the strata is therefore about 20 feet to the mile.

Almost the only recognizable fossils found in these limestones, excepting the crinoid fragments, are the hydro-coralline Stromatopora (concentrica Hall) and the coral Favosites. Both occur in the middle and upper portions of the exposed mass, and may generally be seen in the weathered upper surfaces of the limestone beds. Thus wherever these beds are exposed on the surface, as at the whirlpool on the Canadian side, at the fall of Muddy brook, and near Clifton, these fossils are generally weathered out in relief. They are however not readily separated from the rock. Many of the geodes still show traces of coral structure, which is sometimes shown in the included gypsum.

The limestone is well exposed in the cliff at Goat island, where it has a total thickness of about 110 feet. The contact between the shale and limestone can be seen near the entrance to the Cave of the Winds, where it is about a foot above the top of the stairs. The roof of the Cave of the Winds is formed by the crystalline crinoidal limestone, the same bed which forms the roofs of all the minor caverns along the gorge. The cement beds, about 10 feet thick, together with the 70 feet of Rochester shale, are removed by the spray to a depth of perhaps 30 or 40 feet, the floor of the cave being probably on the upper Clinton limestone, thus making the hight of the cavern 80 feet. Floored and roofed by resisting beds of crystalline limestone, this great cavern is a fit illustration of selective erosion by falling water on rocks of unequal hardness.

The massive limestone which forms the vertical cliff of Goat island is 68 feet thick, its base being on a level with the foot of the Biddle stairway. The top of this cliff marks approximately the level of the falls on either side of Goat island, which therefore have a total thickness of nearly 80 feet of limestone, of which however the lowest 10 feet yield to erosion as does the underlying shale. We may thus say that at the falls there are 70 feet of resistant limestone on top, and 80 feet of yielding shales and limestones below. As the crest of the falls approximates 160 feet above the river below, at least 10 feet of Clinton limestone are found above the water level.

From the top of the vertical cliff at Goat island a sloping bank exposing thin bedded limestones, overlaid by about 10 feet of shell-bearing gravels, rises to a hight of about 40 feet, while on either side of Goat island these thin bedded limestones form the rapids above the two falls. As the total hight of the rapids is about 50 feet, and, as they are formed along the strike of the beds owing to the right-angled turn in the river at this point, the thickness to be added to the known limestone mass is not over 50 feet, giving a total thickness of 130 feet of limestone exposed within this region.

Guelph dolomite

This rock, named from its occurrence at Guelph (Ont.) about 75 miles northwest of Niagara falls, is, so far as known, absent from the Niagara district. As before noted, it may however be represented in the buried hundred feet of limestone (more or less) which lie above the 130 feet of known rock, as shown by the borings in this region.

Salina beds

The basal beds of the Upper Siluric are the saliferous shales and calcareous beds of the Salina stage, so named from the salt-producing village of Salina in Onondaga county. This is the horizon which furnishes all the salt, as well as the gypsum of New York state and the adjoining territory. In the Niagara region this formation is not well exposed, owing to the soft character of the rock which has permitted deep erosion in preglacial times, and to the extensive drift deposits which cover it. The only known exposures on the Niagara are on Grand island and on the Canadian side of the river opposite North Buffalo. On Grand island the Salina rocks may be seen at Edgewater about 200 yards below the boat landing. Here the following section is exposed.[1]

3 Light colored, soft, friable gypseous shales, 5 feet
2 Greenish shales containing nodules of gypsum, 1½ feet
1 Black shale in the river bed

The exposure extends 300 yards down the river bank.

At the extreme northern end of the island, where it divides the river, an impure, thin bedded limestone of this series is exposed. The exposures on the Canadian bank begin a short distance south of this, and extend to the International bridge, the rock here being a more or less gypsiferous shale.

From the numerous borings in this region we have however gained a fair knowledge of the character and thickness of this rock, the latter averaging, according to Bishop, 386 feet. The best available record of the rocks lying between the Waterlime and the Niagara series of limestones is the core of a well drilled on the land of the Buffalo cement co. in North Buffalo. This core, which has a

[1]Bishop. 15th an. rep't N. Y. state geologist. 1895. p. 311.

length of 1305 feet, is now preserved in the museum of the Buffalo society of natural sciences, and from it the following succession of strata can be demonstrated.[1]

		Feet
Rondout waterlime	Waterlime above the mouth of the well, about	7
	Shale and cement rock in thin streaks	25
	Tolerably pure cement rock	5
	Shale and cement rock in thin streaks	13
Salina	Pure white gypsum	4
	Shale	2
	White gypsum	12
	Shale	1
	White gypsum	4
	Shale and gypsum mottled	7
	Drab colored shale with several thin layers of white gypsum	58
	Dark colored limestone	2
	Shale and limestone	4
	Compact shale	3
	Gypsum and shale mottled and in streaks approximating	290±

The gypsum of this formation has never been mined in this district, owing to the strong flow of water through these strata. No salt beds are found in the Salina of this region, though they are characteristic of the formation farther east. Salt water is however obtained. Fossils are very rare throughout these beds; none have been found in the exposures on the Niagara river.

Rondout waterlime

The Salina beds of this region grade upward into a magnesian limestone which contains a considerable amount of aluminium silicate. The upper portion of this series, which in the Niagara region has a thickness of about 50 feet, is very uniform in character and suitable for the manufacture of hydraulic cement. In North Buffalo, extensive quarries have been opened in this rock by the Buffalo

[1] Pohlman. Cement and gypsum deposits in Buffalo. Am. inst. min. eng. Trans. Oct. 1888.

cement co., and here a stratum nearly 6 feet thick is quarried and converted into cement. As the quarries are opened south of the second escarpment (inface of the Onondaga cuesta[1]), the surface rock of Onondaga limestone and the Manlius limestone have to be stripped off before the cement rock is reached.

The characters of the several strata have been briefly enumerated in the section derived from the gas well core. The upper beds, which are alone accessible in this region, may generally be seen in the escarpment, specially where it is crossed by streams, as at Williamsville, or where quarries have been opened. The rock is fine grained, often showing a marked banding or lamination, and breaks with a conchoidal fracture, producing rounded surfaces.

In this rock we find entombed the remains of those remarkable crustacea, the Eurypterids, whose bizarre form, remotely fish-like, has excited more interest than any other fossil found in this region. These Crustacea have made the Waterlime of Buffalo famous, and the Buffalo society of natural sciences, whose collections embrace a magnificent series of these fossils, has fittingly adopted it as chief among its insignia.

Besides these crustacea several other organisms have been found in the Waterlime strata of north Buffalo. Among these are a number of undescribed brachiopods, including at least one species of Lingula.

Manlius limestone

The waterlime of north Buffalo is succeeded by a stratum of impure limestone from 7 to 8 feet in thickness and known locally by the name of "bullhead" rock. The line of demarkation between the two formations is not a very pronounced one, for the inferior rock grades upward into the superior one. The rock is a dolomitic limestone of a very compact semicrystalline character, with a high percent of argillaceous material, and not infrequently a strong petroleum odor. It is mottled, having frequently the appearance of a limestone breccia, and consists of purplish gray, angular or rectangular pieces and similar light colored and more yellowish ones. The latter appear to be more argillaceous than the former. There

---

[1] *See* chapter 1.

is no conclusive evidence that the rock is brecciated, nevertheless the coloration strongly suggests it.

This rock is commonly very porous in its upper portion, the cavities being often lined with crystals of calcite or other minerals. The smaller of the cavities are due to the dissolving out of the small coral, Cyathophyllum hydraulicum, which was exceedingly abundant in the upper part of the stratum. This coral is generally found in a prostrate position, with the mold perfectly preserved in the inclosing rock matrix, so that a perfect cast of the coral can be obtained by the use of gutta percha or dentist's wax. The best exposure of this rock is in the walls of the quarries of the Buffalo cement co. It may also be seen in the face of the Onondaga escarpment at Williamsville and eastward. In many places in the cement quarries, the upper part of this limestone is rich in iron pyrites, which commonly occurs in small cubes, not infrequently oxidized to limonite. Green stains of hydrous carbonate of copper, or malachite, are not uncommon, these resulting probably from the decomposition of chalcopyrite, which is disseminated in minute grains through portions of the rock. Many of the geode cavities contain scalenohedra or acute rhombohedra of calcite, as well as sulfate of strontian.

A remarkable feature of the Manlius limestone of the Niagara region is the nature of the fossil fauna which it contains. This fauna shows an intimate relation to the Coralline limestone fauna of Schoharie county (N. Y.) a rock which is regarded the eastern equivalent of the Lockport (Niagara) limestone of this region. Several of the species found in the Manlius limestone of this region are identical with those of the Coralline limestone, while between other representative species of the two formations there exists a very close relationship. It is difficult to escape the conclusion that the Manlius limestone fauna of the Niagara region is a late return of the Coralline limestone fauna, at the close of the long interval during which the Salina shales were deposited in the Siluric seas of this region.

The Siluro-Devonic contact

The Manlius limestone of the Niagara region is succeeded by the Onondaga limestone of Devonic age. The latter rests unconformably on the former, this unconformity being emphasized by the absence of all Lower Devonic strata in this region, with the exception of thin lenses of sandstone which may be correlated with the Oriskany. The upper surface of the Manlius limestone is knotty and concretionary, producing minor irregularities, but in addition to these there are well marked traces of the erosion of these strata, prior to the deposition of the overlying beds. These traces are of the nature of channels and irregular truncations of the strata, the former in some cases assuming considerable importance. (Fig. 21-23)

Fig. 21 Unconformable contact between Manlius and Onondaga limestones, Buffalo cement quarry.

Fig. 22 Erosion of Manlius limestone prior to deposition of Onondaga limestone, Buffalo cement quarry.

In the east wall of the quarry, not far from the stamp mill, the surface of the Manlius limestone is strongly excavated, the excavation being mainly filled by beds of the Onondaga limestone. Between the two limestones occurs a mass of shale and conglomerate having a total thickness, in the central portion, of something over a foot. The lower 6 or 8 inches are a limestone conglomerate, the pebbles of which are fragments of the underlying limestones. These pebbles are flat, but well rounded on the margins, showing evidence of protracted wear. They are firmly embedded in a matrix of indurated quartz sand, which surrounds them and fills in all the interstices. This bed thins out toward the sides of the channel. On the conglomerate lie about 6 inches of shale and shaly limestone, and these are succeeded by the Onondaga limestone. The width of the channel, which is clearly an erosion channel, is about 18 feet, and its depth is about 3½ feet. (Fig. 23)

From the point where this channel is seen, the contact can be traced continuously for a thousand feet or more eastward, along the quarry wall. It frequently shows a thin shaly bed, often containing quartz grains, lying between the two limestones.

Not very far from the channel just described, a remarkable "sandstone dike" penetrates the Siluric limestones of the quarry wall.

This dike, which can be clearly traced in the wall of the quarry for a distance of perhaps 30 feet in an east and west direction, was

Fig 23 Channel in Manlius limestone with Oriskany sandstone and conglomerate layers, capped by Onondaga limestone, Buffalo cement quarry.

caused by the filling of an ancient fissure in the Siluric strata, by sands forcibly injected from above. The fissure had a total depth of about 10 feet; its walls were very irregular, and at intervals lateral fissures extended in both directions. (*See* Fig. 24) All of these are now filled with pure quartz sand, firmly united into a quartzose sandstone by the deposition of additional silica in the interstices between the sand grains.

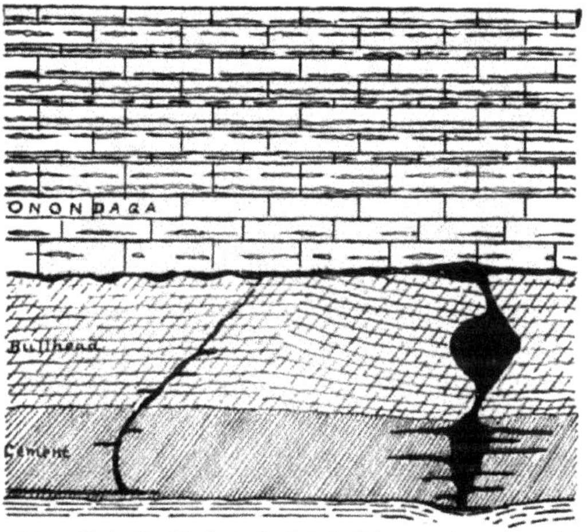

Fig. 24 Sandstone dike in the Siluric strata of the Buffalo cement quarries. (After Clarke)

The dike penetrates the "bullhead" rock and enters the waterlime to a depth of from 2 to 3 feet. It is squarely cut off at the top, where the Onondaga limestone rests on its truncated end and on the limestone flanking it. The Onondaga limestone is entirely unaffected by the dike, being evidently deposited after the formation and truncation of this remarkable mass of sandstone. The width of the filled fissure is scarcely anywhere over 2 feet, but the lateral offshoots extend many feet into the walls of Manlius limestone. These offshoots or rootlets of the dike are irregular, commonly narrow, and often appear as isolated quartz masses in the Manlius or the waterlime rock, the connection with the main dike not being always discernible. Such masses of quartz sandstone have been traced for more than 30 feet from the dike. The irregularity of the walls of the fissure is very pronounced. Angular masses of limestone project into the quartz rock, while narrow tongues of sandstone everywhere enter the limestone. Extensive brecciation of the limestone has occurred along the margin, and the sandstone there is filled with angular fragments of the limestone, which show no traces of solution or wear by running water. These limestone fragments are themselves frequently injected with tongues of the quartz sand. Microscopic examination shows evidence of a certain amount of shearing along the margin of the dike, accompanied by a pulverizing or trituration of the limestone, and followed by reconsolidation. These and other features point to a cataclysmic origin of the fissure which contains the dike and a more or less violent injection of the sand. The fissure must have been formed and filled before the deposition of the Onondaga limestone and while the Manlius limestone was covered by a stratum of unconsolidated sand. The formation of the fissure and the injection of the sand into it from above must have occurred simultaneously; for this appears the only way to account for the inclusion of large fragments, or "horses", of the wall rock in the loose sand, and the injection of the sand into all the cracks and crevices. It seems probable that the fissure records an earthquake shock during the period intervening between the close of the Siluric age and the deposition of the Devonic limestones. This is borne out by the occurrence of numerous small faults or displacements in the underlying strata of waterlime.

Devonic strata

The Lower Devonic is represented in the Niagara region by the thin beds of shale and sandstone before mentioned as occupying erosion hollows in the Manlius limestone. These are perhaps the equivalent of the Oriskany sandstone of eastern New York, though no fossils have been found in them. With the exception of these layers the Lower Devonic strata are wanting in this region.

The Middle Devonic is however well represented in the Niagara region by the Onondaga limestone. This rock, which, as has been shown, rests in most cases directly on the Manlius limestone, consists of a lower crystalline and highly fossiliferous portion, and an upper mass full of layers of hornstone or chert which on weathered surfaces stand out in relief. This part of the formation is generally known as the Corniferous limestone, in reference to the layers of chert which make the rock unfit for other use than rough building. Owing to the presence of the hornstone, this rock effectually resists the attacks of the atmosphere, and hence its line of outcrop is generally marked by a prominent topographic relief feature, the second escarpment of western New York i. e. the inface of the Onondaga cuesta.

The chert-free lower member of this formation varies greatly in thickness even within a limited territory. It is in places extremely rich in corals, and outcrops of this rock show all the characteristics of an ancient coral reef.

History of the Niagara region during Siluric time

We have now gathered data for a brief synopsis of the history of this region during Siluric time. Much still remains to be learned, but from what is known we can trace at least in outline the sequence of geologic events which characterized that ancient era of the earth's history in this vicinity.

When the Siluric era opened, New York, with portions of Pennsylvania and southern Ontario, was covered by the shallow Medina sea. This sea appears to have been of the nature of a mediterranean body of water, which later changed to a bay opening toward the southwest. This "Bay of New York", as we shall call it, came into existence by the orogenic disturbances which marked the transition from the Ordovicic to the Siluric era, and as a result of which the Taconic mountain range, with the Green mountains and the corresponding Canadian ranges, were elevated. This cut off the communication between the open Atlantic and the interior Paleozoic sea which existed during Ordovicic time. This bay was thus surrounded by old-lands on the north, east and southeast, and its waters appear to have been very shallow. We do not know just what the conditions were under which the early Siluric deposits of this region were made; for the lower beds are so barren of organic remains, that we are forced to look for evidences other than that furnished by fossils, of the physical conditions during this period. It is not improbable that the waters of the early Medina sea were cut off from the ocean at large, at least sufficiently to prevent a free communication. This may not have been the case at first; for Arthrophycus harlani flourished in these waters during the deposition of the Oswego beds,[1] and this species characterizes the rocks of late Medina age, during the deposition of which we have reason to suppose that a junction of the Medina sea with the ocean at large had been effected.

Along the eastern and southeastern margin of this interior water body were deposited the thick beds of conglomerate, which now constitute the capping rock of the Shawangunk and other ranges of hills, while farther west, at a distance from the source of supply, the Oswego sandstone was accumulating. Later the character of the deposit changed in this region, from the gray silicious sands to the impalpable muds and fine sands of the lower Medina. Whatever the source of these sands, ferruginous matter was plentiful, as shown by the red color of the deposits, and this leads to the supposition that they were derived from the crystalline rocks of the Adirondacks and the Canadian highlands and not from Ordovicic or Cambric deposits.

It is not improbable that, during the early Medina epoch, the waters of this basin were of a highly saline character. No deposits of salt were formed, or if these existed, they were subsequently leached out. The Medina beds are however rich in saline waters, salt springs being common throughout this region,[2] and this may

[1] This species is found in the eastern part of the district, at the base of the Oneida conglomerate.

[2] In the early part of the century salt was not infrequently manufactured from these springs.

indicate a high degree of salinity of the waters of the early Medina sea. If such was the case, it may have been accompanied by a more or less arid climate, which favored the concentration of the sea water. Thick beds of terrigenous material accumulated in the center of the Medina basin reaching in the Niagara region a thickness of over a thousand feet. These early deposits probably did not extend far west for, though in northern Ohio and Michigan, Medina beds from 50 to 100 feet or more in thickness are known, these are probably to be correlated with the upper Medina of the Niagara region.

Toward the close of the Medina epoch, the Siluric sea had encroached on the lands to such an extent as to effect a junction with the Medina basin, whereupon normal marine conditions were again established. This is indicated by the marine fauna and flora which characterize the upper Medina beds. The first deposit in this region, on the reestablishment of normal marine conditions, was the white quartzose sandstone which caps the red shale of the lower series. Mud and sand now alternated, indicating an oscillation of conditions with numerous changes in the currents which distributed the detrital material. Thin beds of limestones also formed at rare intervals, chiefly from the growth of bryozoans in favorable localities. In the Bay of New York the waters continued moderately shallow, as shown by the well developed cross-bedding structure in the sandstones. At intervals large tracts seem to have been laid bare on the retreat of the tide, as indicated by the wave marks and other shore features which give the surfaces of some Medina sandstone slabs such a remarkable resemblance to a modern sand beach exposed by the ebbing tide. In fact, we may not inaptly compare this stage of the Siluric bay of New York with the upper end of the modern bay of Fundy, where the red sands and muds are laid bare for miles on the retreat of the tide.

After the last sandstone bed of the Medina stage had been deposited, the water probably became purer and deeper, and the 6 feet of Clinton shales were laid down in the Niagara region. In the eastern part of the Bay of New York, sandstones were deposited even during the Clinton epoch, while the conditions favoring the deposition of limestone existed only during the short interval in the Niagara period, when the Coralline limestone of Schoharie was laid down. Westward, however, the adjustment of conditions went on more rapidly, and the Clinton limestones, with the calcareous shales and limestones of the upper Niagaran, became the characteristic deposits. During nearly the entire Niagara period life was abundant in the Siluric sea, and the Bay of New York had its marvelous succession of faunas, which have made these strata the standard for the Siluric beds of this continent.

All the Siluric limestones of the Niagara section show characters pointing to a fragmental origin, and in this respect they contrast strongly with the Devonic limestones in the southern part of the district. The latter, as before mentioned, show the characteristics of an ancient coral reef, and we may therefore assume that they were built up in situ by the polyps and other lime-secreting organisms. Not so with the Siluric limestones. These, to be sure, were derived from similar deposits by lime-secreting organisms, but these deposits were originally made in a different place from that in which we find the limestones today. A sedimentary limestone or lime-sandstone is similar to a quartz sandstone or a shale, in that the material of which it is formed is the product of erosion of preexisting rocks. In the case of the quartz sandstone, this is generally an inorganically formed rock, while the sedimentary limestones are most usually derived from organically formed rocks. In the former case, the source of the material is often a distant one, while in the latter it is generally, though not necessarily always, close at hand. A coral reef growing in moderately shallow water is attacked by the waves, as are all rocks which come within their reach. Erosion results, and the product of this activity is carried away and deposited on the ocean floor as a calcareous sand. Thus stratified deposits of limestones are formed, whereas in the original organic reef, no stratification is to be expected. In the immediate neighborhood of the growing reef, the beds of calcareous sand will slowly envelop the original deposit from which they were derived, and thus the source of supply is chiefly the upper growing portion of the reef. On the lime-sandstone strata which flank the reef, independent masses of coral may at times grow, while other organisms, such as mollusks and brachiopods, will also find this a convenient resting place. Thus the organically formed limestone masses and the fragmental limestones will interlock and overlap each other around the borders of a growing reef. It follows then that in the neighborhood of the growing coral masses the sands derived from their destruction will be coarser, the finer material being carried farther out to sea, and deposited at a distance from the source. Thus an approximate criterion for the determination of the distance of any given bed of calcareous sand from its place of origin is furnished. If deposits of such calcareous sand are made in shallow water, cross-bedding and ripple marks will be found just as in the quartz sands, and, as we have seen, the former structure is characteristic of most of the strata of Lockport limestone exposed in the gorge section at Niagara. It may be added that, as the organic limestone will continue to form as long as the conditions are favorable, the supply of calcareous sand is practically inexhaustible. Hence thick beds of such lime-sandstones may form.

In the Niagaran seas the chief reef-building corals were Favosites, Halysites and Heliolites, together with the hydro-coralline Stromatopora. Bryozoans also added largely to the supply of organically formed limestone of the various reefs. But perhaps the most important contributors in this connection were the crinoids and related organisms, which may at times have constituted reefs of their own. Their abundance is testified to by the frequent thick beds of limestone, which are almost wholly made up of broken and worn crinoid fragments. The crinoids fell an easy prey to the waves, for, on the death of the animal, the calyx, arms and stem would quickly fall apart into their component sections, and hence yield fragments readily transported by the waves. In the case of the corals and the shells, which latter probably formed no unimportant part of the organic contributions to the reefs, the work of grinding the solid limestone masses into a sand probably required the aid of tools, such as large blocks that could be rolled about by the waves, or it may have been aided by the omnipresent reef-destroying organisms.

The infrequency of exposure of the fossil reefs, which furnished the calcareous sand, need not disturb us. We must remember that the actually exposed sections of these limestone strata are very few when compared with the great extent of the beds themselves. It must also be borne in mind, that vast portions of these limestone beds have been removed by erosion during the long post-Siluric time. When we realize that the actual reefs must have been widely scattered in the Niagara sea, and that our sections through these strata are random sections, we need feel no surprise at the unsatisfactory character of these exposures. It must however be added that sections farther east, as at Lockport or other localities, generally show much more of the reef character of the deposit, the corals in these being correspondingly abundant. The upper geodiferous beds of the limestone at Niagara were probably much more fossiliferous than the lower. As before mentioned, the geode cavities most likely are the result of alteration or solution of some fossil body, probably a coral. Though fossils may have been plentiful, none of these beds, so far as examined, show the characteristics of true reefs. They have more the aspect of beds of coral sand, on which isolated heads of corals and other organisms grew rather plentifully.

During the dolomitization of these limestone beds, which was probably brought about by chemical substitution before the consolidation of the coral sand, many of the fossils which were included in these sands probably suffered alteration and more or less complete destruction. Thus it will be seen that even the few organisms which were embedded in these coral sands, did not survive the subsequent changes, and thus the barrenness of these great limestone masses appears to be fully accounted for. The fossiliferous character of the upper Clinton limestone, as well as the coarseness of the calcareous fragments of which it is composed, points to a nearness of this rock to the source of the material; for in the vicinity of the coral and crinoid reefs the food supply for other organisms would be most abundant, and hence these would develop most prolifically in such a neighborhood.

A careful comparative study of the Niagaran deposits of New York and those of the middle states has brought out some important and interesting facts. These may be summed up in the statement, that the New York fauna is more individualized, showing characteristics stamping it in some degree as a provincial fauna. The Niagaran fauna of the central states however is more closely allied to the European Mid-Siluric fauna than to that of New York state, from which we may conclude that the pathway of communication between the American and European Siluric seas was not by way of New York, a conclusion which is in entire harmony with those derived from the physical development of this region and the characteristics of the strata.

Weller[1] has collected data which indicate that the pathway of migration of faunas between the two continents was by way of the arctic region. According to Weller's interpretation of the facts, there existed in North America during Siluric time ". . . a north polar sea with a great tongue stretching southward through Hudson bay to about latitude 33°. There were doubtless islands standing above sealevel within this great epicontinental sea; and at the latitude of New York there was a bay reaching to the eastward, in which the Siluric sediments of the New York system were deposited. Labrador, Greenland and Scandinavia were in a measure joined into one great land area, though perhaps with its continuity broken, with a sea shelf lying to the north of it and another to the south. Another epicontinental tongue of this northern sea extended south into Europe, bending to the west around the southern part of the Scandinavian land and connecting with a Silurian Atlantic ocean. The sea shelf to the north of the Labrador-Scandinavian land was a means of intercommunication between northern Europe and the interior of North America, and the sea shelf to the south of this land was a pathway between England and eastern Canada." That portion of North America lying to the west of a line drawn from the Mississippi to the Mackenzie appears to have been dry land during the Niagara period, and connected with the Appalachian land on the east by the westward trending axis of the latter in the southern United States.

At the close of the Niagara period, there appears to have been an elevation of the continent which converted the Bay of New York

[1] Nat. hist. sur. Chicago acad. sci. bul. 4 and Jour. geol. 4:692-703.

and the greater part of the interior Siluric sea into a vast partially or entirely inclosed basin. This elevation appears to have been accompanied by climatic desiccation which brought about a rapid evaporation of the waters and a consequent increase in salinity. Thus this great interior water body was changed from a richly peopled mediterranean, to a lifeless body of intensely saline water, a veritable Dead sea. As the concentration of the brine continued, deposition of gypsum began, and later on the extensive beds of rock salt of this formation were laid down. Some of these salt beds in Michigan are reported to be a thousand feet thick, but none of the New York beds approach this thickness. The clastic strata of the Salina series were probably derived from the destruction of the sediments which were formed during the early periods of the Siluric and during preceding periods. This would account for the presence of limestone beds in deposits formed in a lifeless sea. All these limestones were more or less mixed with clayey sediments; they may in fact be regarded as consolidated argillo-calcareous muds derived from older limestones and shales. This is the character of the Waterlime and Manlius limestone which succeed the Salina beds, and which, though fossiliferous, could have no other source of origin than preexisting limestone beds.

The Waterlime has been regarded as a fresh-water formation. It is more likely however that it represents a return of marine conditions through the opening of channels between this interior basin and the ocean at large. This is indicated by the fauna, which includes undoubted marine forms. Whether this connection was through the old northern channel, or whether a new channel toward the east was opened is not apparent. The former is indicated by the character of the Manlius limestone which succeeds the Waterlime, and which in the Niagara region has features associating it with the corresponding deposits of Ohio, Michigan and Ontario, rather than its eastern equivalents. Whatever the nature of the transgression of the sea which took place in the late Siluric, it was not of long duration. The epoch of the Manlius limestone and with it the Siluric era were brought to a close with the withdrawal of all the waters from this portion of the continent, which thereafter for a long period of time remained above the sea. During this time, the Helderbergian and other Lower Devonic strata were deposited in the Appalachian region, which by that time had established a southern connection with the open Atlantic.

Finally, toward the middle of the Devonic era, the sea once more transgressed on the abandoned continent, and again all this region was covered by oceanic waters. On the land surface of early Devonic times, now grew corals in great luxuriance; and reefs of great extent, with their accompanying deposits of coral sands, and their wealth of new life, again characterized the interior Paleozoic sea. It was not till long ages after, that this portion of the continent was again raised above the sea. This last elevation, which took place toward the close of Paleozoic time, was a permanent one, with the exception of a possible slight resubmergence of some parts of this region after the close of the glacial period. With the last great emergence of the land were inaugurated those long cycles of erosion outlined in chapter 1, which resulted in the formation of the great topographic features of this region, and which came to a close only with the envelopment of this region in the snow and ice of the great glacial winter.

APPENDIX

Partial bibliography of the geology of Niagara and the Great lakes

Ashburner, C. A. The geology of Buffalo as related to natural gas in New York state. Am. inst. min. eng. Trans. 16:1-54. 1888.

Bakewell, R. Observations on the falls of Niágara, with references to the changes which have taken place, and are now in progress. Am. jour. sci. 2d ser. 23:85-95. 1857.

——— Observations on the whirlpool and the rapids below the falls of Niagara; designed by illustrations to account for the origin of both. Am. jour. sci. 2d ser. 4:25-36. 1847.

Ballou, H. W. Niagara river. Scientific Am. sup. 13:5045-46. 1882.

Bishop, I. P. Salt wells of western New York. 5th an. rep't N. Y. state geologist. p. 12-47. 1886.

——— Structural and economic geology of Erie county [N. Y.]. 15th an. rep't N. Y. state geologist. 1:17-18, 305-92, pl. 1-16, fig. 1-6; 49th an. rep't N. Y. state mus. 2:305-92, pl. 1-16, fig. 1-6. 1898.

Chalmers, Robert. Pleistocene marine shore lines on the south side of the St Lawrence valley. Am. jour. sci. 4th ser. 1:302-8. 1896.

Clarke, J. M. The Oriskany Fauna of Becraft mountain. N. Y. state mus. Mem. 3. p. 95-103.

Claypole, E. W. Pre-glacial formations of the beds of the great American lakes. Can. nat. n. s. 9:213-27. 1881.

——— Evidence from the drift of Ohio, Indiana and Illinois in support of the pre-glacial origin of the basins of Lakes Erie and Ontario. Am. ass'n adv. sci. Proc. 30:147-59. 1882.

——— Eccentricity theory of glacial cold versus the fact. Edinburgh geol. soc. Trans. 5:534-48. 1885.

——— The old gorge at Niagara. Science. 8:236. 1886.

——— Buffalo and Chicago or "What might have been." Am. nat. 20:856-62; 1886. Am. ass'n adv. sci. Proc. 35:224. 1887.

——— On the pre-glacial geography of the region of the great lakes. Can. nat. n. s. 8:187-206. 1887.

——— Falls of rock at Niagara. Nature. 39:367. 1889.

Coleman, Arthur P. Glacial and interglacial deposits at Toronto (Canada). Brit. ass'n adv. sci. Rep't 1897. p. 650, 651.

——— Lake Iroquois and its predecessors at Toronto (Canada). Abstracts, Geol. soc. Am. Bul. 10:165-76; Am. geol. 33:103-4; Science. n. s. 9:143. 144.

Davis, William Morris. Was Lake Iroquois an arm of the sea? Am. geol. 7:139-41. 1891.

——— The ancient outlet of Lake Michigan. Pop. sci. mo. 46:217-29. 1895.

——— [Review of "Origin of the gorges of the whirlpool rapids at Niagara" by F. B. Taylor]. Science n. s. 7:627. 1898.

——— Ancient coastal plain of western New York. Textbook of physical geography. p. 137, 138, fig 86. 1899.

Dawson, George M. Inter-glacial climatic conditions. Am. geol. 16:65, 66. 1895.

Desor, E. Note sur l'existence de coquilles marines des mers actuelles dans le bassin du Lac Ontario (Canada) jusqu' a l'altitude de 310 pieds. Soc. geol. France. Bul. 2d ser. 8:420-23. 1851.

—— On the ridge road from Rochester to Lewiston, and on other similar terraces. Bost. soc. nat. hist. Proc. 3:358-59. 1851.

—— Ueber Niagara falls. Deutsche Geologische Gesellschaft, Zeitschrift. 5:643-44. 1853.

—— The falls of Niagara. Pottsville sci. ass'n. Bul. p. 5-10. 1855.

Fairchild, H. L. Glacial lakes of western New York. Geol. soc. Am. Bul. 6:353-74, pl. 18-23. 1895.

—— Lake Newberry the probable successor of Lake Warren. Abstract, Geol. soc. Am. Bul. 6:462-66. 1895.

—— Glacial Genesee lakes. Geol. soc. Am. Bul. 7:423-52, pl. 19-21. 1896.

—— Glacial geology of western New York. Geol. mag. Dec. 4. 4:529-37; Brit. ass'n adv. sci. Rep't 1897. p. 664.

—— Lake Warren shore lines in western New York and the Geneva beach. Geol. soc. Am. Bul. 8:269-84, pl. 30. 1897.

—— Glacial lakes Newberry, Warren and Dana in central New York. Am. jour. sci. 4th ser. 7:249-63, fig. 1, pl. 6. 1899.

Featherstonhaugh, G. W. On the ancient drainage of North America and the origin of the cataract of Niagara. Am. jour. of geol. and nat. sci. 1:13-21. 1831.

Fleming, Mary A. Pot holes of Foster's flats (now called Niagara glen) on the Niagara river. Abstract, Am. ass'n adv. sci. Proc. 48:226, 227; Science. n. s. 10:489. 1899.

Foerste, Aug. F. On Clinton conglomerates and wave marks in Ohio and Kentucky. With a résumé of our knowledge of similar occurrences in other Silurian strata of these states, and their evidence upon probable land conditions. Jour. of geol. 3:50-60, 169-97. 1895.

—— Account of the Middle Silurian rocks of Ohio and Indiana, including the Niagara and Ohio Clinton, and the bed at the top of the Lower Silurian strata, formerly considered the Medina. Cin. soc. nat. hist. Jour. 18:161-99. 1896.

—— Report on the Niagara limestone quarries of Decatur, Franklin and Fayette counties, with remarks on the geology of the Middle and Upper Silurian of these and neighboring counties of Indiana. Ind. dep't geol. and nat. res. 22d an. rep't. p. 195-255, pl. 14-18. 1898.

Foot, Lyman. Notices of geology and mineralogy (of Niagara falls region). Am. jour. sci. 4:35-37. 1822.

Gebhard, John. Observations on the geological features of the south side of the Ontario valley. Albany institute. Trans. 1:55-59. 1830; Am. jour. sci. 11:213-18. 1826.

Geer, Gerard de. On Pleistocene changes of level in eastern North America. Am. geol. 11:22-44; Bost. soc. nat. hist. Proc. 25:454-77. 1893.

Gibbes, L. R. Remarks on Niagara falls. Am. ass'n adv. sci. Proc. v. 10, pt 2, p. 69-78. 1857.

—— On some points which have been overlooked on the past and present condition of Niagara falls. Elliott soc. nat. hist. Proc. (S. C.) 1:91-100. 1859.

Gilbert, G. K. Old shore lines of Lake Ontario. Science. 6:222. 1885.

—— On a prehistoric hearth under the Quaternary deposits in western New York. Sci. Am. sup. 23:9221-22. 1887.

—— The place of Niagara falls in geological history. Am. ass'n adv. sci. Proc. 35:222-24. 1887.

—— Old shore lines in the Ontario basin. Can. inst. Proc. 3d ser. 6:2-4. 1888.

—— Changes of level of the Great lakes. Forum. 5:417-28. 1888.

—— History of Niagara river. 6th an. rep't. N. Y. state com. res. at Niagara. p. 61-84, 7 plates. 1890.

—— Discussion of the papers "Relationship of the glacial lakes Warren, Algonquin, Iroquois and Hudson-Champlain," and the two papers by J. W. Spencer, "The Iroquois shore north of the Adirondacks" and "Channels over divides not evidence per se of glacial lakes". Geol. soc. Am. Bul. 3:492-94. 1892.

——Itinerary, Chicago to Niagara falls. Int. cong. geol., Compte Rendu, 5th session. p. 453-58. 1894.

—— Old tracks of Erian drainage in western New York. Abstract. Geol. soc. Am. Bul. 8:285-86. 1897.

—— Recent earth movements in the Great lakes region. U. S. geol. sur. 18th an. rep't. pt 2, p. 601-47, pl. 105, fig. 93-101. 1898; Nat. geog. mag. 8:233-47, fig. 1-7. 1897; Abstract, Nature. 57:211-13, fig. 1. 1897; Am. geol. 23:126, 127; Am. jour. sci. 4th ser. 7:239-41; 15th an. rep't com. state res. at Niagara. p. 69-138, fig. 93-101. 1899.

—— Ripple marks and cross bedding. Geol. soc. Am. Bul. 10:135, 140, pl. 13, fig. 5; Abstract, Am. geol. 23:102; Science. n. s. 9:138. 1899.

—— Glacial sculpture in western New York. Geol. soc. Am. Bul. 10:121-30; Abstract, Am. geol. 23:103; Science. n. s. 9:143. 1899.

—— Niagara falls and their history. Nat. geog. Monog. v. 1, no. 7. 1895.

Gordon, C. H. [Review of "Correlation of Erie-Huron beaches with outlets and moraines in southeastern Michigan" by F. B. Taylor]. Jour. geol. 5:313-17. 1897.

Grabau, A. W. Siluro-Devonic contact in Erie county, New York. Geol. soc. Am. Bul. 11:347-76, pl. 21-22. 1900.

Grant, C. C. Geological notes. Hamilton ass'n. Jour. and proc. no. 12, p. 140-45. 1896.

Grote, A. R. & Pitt, W. H. New specimens from the Waterlime group at Buffalo, N. Y. Am. ass'n adv. sci. Proc. 26:300-2. 1878.

Gunning, W. D. Past and future of Niagara. Pop. sci. mo. 1:564-73. 1872.

Hall, James. Niagara falls, their physical changes and the geology and topography of the surrounding country. Bost. jour. nat hist. 4:106-34. 1834.

—— Geology of New York. Pt 4. Fourth geological district. Albany. 1843.

—— Glaciated surfaces of cherty limestone from near Niagara. Am. jour. sci. 45:332. 1843.

—— On the geology of the region of Niagara falls. Bost. soc. nat. hist. Proc. 1:52. 1844.

—— Paleontology of New York, v. 2. Organic remains of lower middle division of the New York system. Albany. 1852.

—— Note (on recession of Niagara falls). Am. ass'n adv. sci. Proc. v. 10, pt 2, p. 76-79. 1857.

—— (Description of Siluric formations of New York). Fossils of the Waterlime group. Pal. N. Y. 3:20-32, 282-424. 1859.

——On the relations of the middle and upper Silurian (Clinton, Niagara and Helderberg) rocks of the United States. Geol. mag. 9:509-13. 1872.

—— On the relations of the Niagara and Lower Helderberg formations, and their geographical distribution in the United States and Canada. Am. ass'n adv. sci. Proc. v. 22, pt 2, p. 321-35. 1874.

—— Descriptions of fossil corals from the Niagara and Upper Helderberg groups. 35th an. rep't N. Y. state mus. nat. hist. p. 407-64, pl. 23-30. 1884.

—— Niagara falls, its past, present and prospective condition. Rep't 4th geol. dist. N. Y. p. 383-404. 1842; 8th an. rep't com. state res. at Niagara. p. 67-89. 1892.

Hallett, P. Notes on Niagara. British ass'n. Rep't of 54th meeting, p. 744-45. 1885.

Hitchcock, C. H. The story of Niagara. Am. antiquarian. 23:1-24, 12 fig. Jan. 1901.

Holley, George W. Niagara, its history, geology, incidents and poetry, New York. 1872.

—— The proximate future of Niagara, in review of Professor Tyndall's lecture thereon. Am. ass'n adv. sci. Proc. v. 22, pt 2, p. 147-55. 1874; Abstract, Can. nat. n. s. 7:164-65. 1875.

Hovey, Horace C. Niagara river, gorge and falls. Sci. Am. sup. v. 22, no. 558, p. 897. 1886.

Hyatt, Alpheus. Rock ruins (Niagara falls). Am. nat. 2:77-85. 1869.

Leverett, Frank. On the correlation of New York moraines with raised beaches of Lake Erie. Am. jour. sci. 3d ser. 1:1-20. 1895.

—— Pre-glacial valleys of the Mississippi and tributaries. Jour. of geol. 3:740-63. 1895.

—— The Pleistocene features and deposits of the Chicago area. Chicago acad. sci. Bul. no. 2, 86p., pl. 1-4, fig. 1-8. 1897.

—— Correlation of moraines with beaches on the border of Lake Erie. Am. geol. 21:195-99. 1898.

Lesley, Joseph P. On the glacial erosion and outlets of the Great lakes. Am. phil. soc. Proc. 20:95-101. 1883.

Lyell, Charles. Niagara falls. Travels in North America, v. 1, ch. 2, p. 22-43. 1845.

Marcou, Jules. Le Niagara quinze ans apres. Soc. géol. France. Bul. 2d ser. 22:290-300, 529-30. 1865.

Maw, George. Geological history of the North American lake region. Geol. mag. n. s. 5:455-56. 1878.

Merrill, F. J. H. A guide to the study of the collections of the New York state museum. N. Y. state mus. Bul. no. 19, p. 107-262. 1898.

Mudge, E. H. Central Michigan and the post-glacial submergence. Am. jour. sci. 3d ser. 1:442-45. 1895.

Newberry, John S. Notes on the surface geology of the basins of the Great lakes. Bost. soc. nat. hist. Proc. 9:42-46. 1865.

—— On the surface geology of the basins of the Great lakes and the valley of the Mississippi. N. Y. lyceum nat. hist. Annals. 9:213-34. 1870; Am. nat. 4:193-218. 1871.

—— On the structures and origin of the Great lakes. N. Y. lyceum nat. hist. Proc. 2d ser. p. 136-38. 1874.

—— On the origin and drainage of the basins of the Great lakes. Am. phil. soc. Proc. 20:91-95.

—— History of the great American lakes. Abstract. Sci. Am. sup. 28:11,505-6; Eng. and min. jour. 48:201-2. 1889.

Nicholson, H. Alleyne. On the Guelph limestones of North America and their organic remains. Geol. mag. n. s. 2:343-48. 1875.

Pohlman, Julius. Cement rock and gypsum deposits in Buffalo. Am. inst. min. eng. Trans. Buffalo meeting 1888.

—— Life history of Niagara. Am. inst. min. eng. Trans. Buffalo meeting p. 1-17.

—— Thickness of the Onondaga salt group at Buffalo. N. Y. Buffalo soc. nat. sci. Bul. v. 5, no. 1, p. 97, 98. 1886.

—— Niagara gorge. Am. ass'n adv. sci. Proc. 35:221. 1887.

—— Life history of the Niagara river. Am. ass'n adv. sci. Proc. 32:202 1884.

—— Fossils from the Waterlime group at Buffalo, N. Y. Buffalo soc. nat. sci. Bul. v. 5, no. 1, p. 97, 98. 1886.

—— On certain fossils of the Waterlime group near Buffalo. Buffalo soc. nat. sci. Bul. v. 4, no. 1, p. 17-22. 1882.

—— Additional notes on the fauna of the Waterlime group near Buffalo. Buffalo soc. nat. sci. Bul. v. 4, no. 2, p. 41-46. 1882.

Quereau, Edmund Chase. Topography and history of Jamesville lake, New York. Geol. soc. Am. Bul. 9:173-82, pl. 12-14. 1898.

Ringueberg, E. N. S. The crinoidea of the lower Niagara limestone at Lockport, N. Y. with new species. N. Y. acad. sci. An. 5:301-6, pl. 3. 1891.

Ringueberg, E. N. S. Evolution of forms from the Clinton to the Niagara group. Am. nat. 16:711-15, fig. a-e. 1887.

—— Niagara shales of western New York; a study of the origin of their subdivisions and their faunae. Am. geol. 1:264-72. 1888.

—— New genera and species of fossils from the Niagara shales. Buffalo soc. nat. sci. Bul. v. 5, no. 2, p. 5-21, pl. 1, 2. 1886.

Rogers, Henry D. On the falls of Niagara, and the reasonings of some authors respecting them. Am. jour. sci. 27:326-35. 1835.

Russell, Israel Cook. Geography of the Laurentian basin. Am. geog. soc. Bul. 30:226-54, 6 fig. 1898.

Schmidt, Friedrich. Eurypterus beds of the Oesel as compared with those of North America. Geol. soc. Am. Bul. 3:59, 60. 1892.

Scovell, J. T. An old channel of the Niagara river. Am. ass'n adv. sci. Proc. 39:245. 1891.

Shaler, N. S. Geology of Niagara. The Niagara book. Buffalo N. Y.

Spencer, J. W. Short study of the features of the region of the lower Great lakes during the Great river age; or, Notes on the origin of the Great lakes of North America. Am. ass'n adv. sci. Proc. 30:131-46. 1882.

—— Discovery of the pre-glacial outlet of the basin of Lake Erie into that of Lake Ontario, with notes on the origin of our lower Great lakes. Am. phil. soc. Proc. 24:409-16, pl. 6, 7. 1882; 2d geol. sur. Pa. Rep't Q 4, p. 357-404.

—— Traces of beaches about Lake Ontario. Am. jour. sci. 24:409-16, pl. 6, 7. 1882.

—— Paleozoic and surface geology of the region about the western end of Lake Ontario. Can. nat. n. s. 10:129-71, 213-36, 265-312. 1883.

—— Terraces and beaches about Lake Ontario. Am. ass'n adv. sci. Proc. 31:359-63. 1883.

—— Age of the Niagara river. Am. nat. 21:269-70. 1887.

—— The Iroquois beach, a chapter in the history of Lake Ontario. Abstract, Science. 11:49. 1888.

—— The St Lawrence basin and the Great lakes. Abstract, Can. rec. of sci. 3:232-35. 1888; Science. 12:99-100. 1888; Sci. Am. sup. 26:10,671-72. 1888; Am. geol. 2:346-48. 1888; Am. ass'n adv. sci. Proc. 37:197-99. 1889; Am. nat. 23:491-94. 1889.

—— Notes on the origin of the Great lakes of North America. Am. ass'n adv. sci. Proc. 37:197, 198. 1889.

—— The Iroquois beach; a chapter in the geological history of Lake Ontario. Roy. soc. Can. Trans. v. 7, §4, p. 121-24. 1890; Abstract, Am. nat. 24:957; Am. geol. 6:311-12. 1890.

—— The deformation of the Iroquois beach and birth of Lake Ontario. Am. jour. sci. 40:443-52. 1890.

—— The northwestern extension of the Iroquois beach in New York. Am. geol. 6:294, 295. 1890.

—— Deformation of the Algonquin beach, and birth of Lake Huron. Am. jour. sci. 3d ser. 41:12-21. 1891.

Spencer, J. W. High-level shores in the region of the Great lakes and their deformation. Am. jour. sci. 41:201-12. 1891.

—— Prof. W. M. Davis on the Iroquois beach. Am. geol. 7:68, 266. 1891.

—— Origin of the basins of the Great lakes of America. Am. geol. 7:86-97. 1891.

—— Channels over divides not evidence per se of glacial lakes. Geol. soc. Am. Bul. 3:491-92; Abstract, Am. geol. 11:58. 1892.

—— Review of the history of the Great lakes. Am. geol. 14:289-301, pl. 8. 1894.

—— Drainage of the Great lakes in the Mississippi by way of Chicago. Abstract, Am. nat. 28:884. 1894.

—— Deformation of the Lundy Beach and birth of Lake Erie. Am. jour. sci. 3d ser. 47:207-12. 1894.

—— Niagara falls as a chronometer of geological time. Abstract, Roy. soc. London. Proc. 1:145-48. 1894.

—— Age of Niagara falls. Am. geol. 14:135, 136. 1894.

—— Duration of Niagara falls. Am. jour. sci. 3d ser. 48:455-72: Abstract, Am. geol. 14:204 (7l.); Abstract, Am. nat. 28:859-62. 1894.

—— [Lake Newberry as the probable successor of Lake Warren]. Geol. soc. Am. Bul. 6:466. 1895.

—— Duration of Niagara falls. Abstract, Am. ass'n adv. sci. 43:244-46. 1895.

—— Geological survey of the Great lakes. Am. ass'n adv. sci. Proc. 43:237-43. 1895.

—— Niagara as a time piece. Pop. sci. mo. 49:1-19, fig. 1-17. 1896.

—— How the Great lakes were built. Pop. sci. mo. 49:157-72, fig. 1-15. 1896.

—— On the continental elevation of the glacial epoch. Brit. ass'n adv. sci. Rep't 1897. p. 661, 662. 1898.

—— On Mr Frank Leverett's "Correlation of moraines with beaches on the border of Lake Erie." Am. geol. 21:393-96, fig. 1. 1898.

—— Another episode in the history of Niagara river. Abstract, Am. ass'n adv. sci. Proc. 47:299; Science. n. s. 8:501, 502; Am. geol. 22:259-60; Am. jour. sci. 4th ser. 6:439-50, 2 fig. 1898.

—— Niagara as a time piece. Can. inst. Proc. n. s. 1:101-3. 1898.

Stose, George W. A specimen of Ceratiocaris acuminata Hall from the Waterlime of Buffalo, N. Y. Bost. soc. nat. hist. Proc. 26:369-71. 1894.

Tarr, Ralph S. Physical geography of New York state. pt 7. The Great lakes and Niagara. Am. geog. soc. Bul. 31:101-17 (4 fig.), 217-35 (10 fig.), 315-43 (21 fig.) 1899.

Taylor, F. B. The highest old shore-line on Mackinac island. Am. jour. sci. 3d ser. 43:210-18; Abstract, Am. ass'n adv. sci. Proc. 40:260-61. 1892.

Taylor, F. B. The limit of post-Glacial submergence in the highlands east of Georgian Bay [Ontario]. Am. geol. 14:273-89, with map. 1894.

—— The ancient strait at Nipissing [Ontario]. Geol. soc. Am. Bul. 5:620-26, pl. 20; Abstract, Am. geol. 13:220-21. 1894.

—— A reconnaissance of the abandoned shore lines of Green bay [Michigan and Wisconsin]. Am. geol. 13:316-27, with a map. 1894.

—— A reconnaissance of the abandoned shore lines of the south coast of Lake Superior. Am. geol. 13:365-83, with map. 1894.

—— The Nippissing beach on the north Superior shore. Am. geol. 15:304-14. 1895.

—— Niagara and the Great lakes. Am. jour. sci. 3d ser. 49:249-70. 1895.

—— The Munuscong islands [Michigan]. Am. geol. 15:24-33. 1895.

—— The second Lake Algonquin. Am. geol. 15:100-20 and 162-79. 1895.

—— Changes of level in the region of the Great lakes in recent geological time. [Letter to J. D. Dana] Am. jour. sci. 3d ser. 49:69-71. 1895.

—— [On the use of the term "Erigan"]. Am. geol. 15:394-95. 1895.

—— Preliminary notes on studies of the Great lakes made in 1895. Am. geol. 17:253-57. 1896.

—— The Algonquin and Nipissing beaches. Am. geol. 17:397-400. 1896.

—— Notes on the Quaternary geology of the Mattawa and Ottawa valleys [Ontario]. Am. geol. 18:108-20. 1896.

—— Correlation of Erie Huron beaches with outlets and moraines in southeastern Michigan. Geol. soc. Am. Bul. 8:31-58, pl. 2. 1897.

—— Scoured boulders of the Mattawa valley [Ontario]. Am. jour. sci. 4th ser. 3:208-18. 1897.

—— The Nipissing-Mattawa river the outlet of the Nipissing great lakes. Am. geol. 20:65-66. 1897.

—— Short history of the Great lakes. Studies in Indiana geography, ed. by C. R. Dryer. 1:90-110, fig. 1-4. 1897.

—— The Champlain submergence and uplift, and their relation to the Great lakes and Niagara falls. Brit. ass'n adv. sci. Rep't 1897. p. 652-53. 1898.

—— Origin of the gorge of the whirlpool rapids at Niagara. Geol. soc. Am. Bul. 9:59-84, fig. 1, 2. 1898.

—— The great ice dams of Lakes Maumee, Whittlesey and Warren. Am. geol. 24:6-38, pl. 2, 3. 1899.

Tyndall, John. Some observations on Niagara. Pop. sci. mo. 3:210-26. 1873.

Upham, Warren. Relationship of the Glacial lakes Warren, Algonquin, Iroquois and Hudson-Champlain. Abstract, Geol. soc. Am. Bul. 3:484-87; Abstract, Am. geol. 11:59. 1892.

—— The Champlain submergence. Abstract, Geol. soc. Am. Bul. 3:508-11; Abstract, Am. geol. 11:119. 1892.

Upham, Warren. The fjords and Great lakes basins of North America considered as evidence of pre-glacial continental elevation and of depression during the glacial period. Geol. soc. Am. Bul. 1:563-67. 1890.

—— Altitude as the cause of the glacial period. Science. 22:75, 76. 1893.

—— Estimates of geologic time. Am. jour. sci. 3d ser. 45:209-20; Sci. Am. sup. 35:14, 403-5. 1893.

—— Epeirogenic movements associated with glaciation. Am. jour. sci. 3d ser. 46:114-21. 1893.

—— Wave-like progress of an epeirogenic uplift. Jour. of geol. 2:383-95. 1894.

—— The Niagara gorge as a measure of the post-glacial period. Am. geol. 14:62-64. 1894.

—— Stages of recession of the North American ice sheet, shown by glacial lakes. Am. geol. 15:396-99. 1895.

—— Departure of the ice sheet from the Laurentian lakes. Abstract, Geol. soc. am. Bul. 6:21-27. 1895.

—— Late glacial or Champlain subsidence and reelevation of the St Lawrence river basin. Am. jour. sci. 3d ser. 49:1-18, with map. 1895; Minn. geol. and nat. hist. sur. 23d an. rep't, p. 156-93. 1895.

—— Beaches of Lakes Warren and Algonquin. Am. geol. 17:400-2. 1896.

—— Origin and age of the Laurentian lakes and of Niagara falls. Am. geol. 18:169-77, fig. 1. 1896.

—— Niagara gorge and Saint Davids channel. Geol. soc. Am. Bul. 9:101-10. 1898.

Walker, A. E. Hamilton sponges [Ontario]. Hamilton ass'n. Jour. and proc. no. 11, p. 85-87. 1895.

Weller, Stuart. The Silurian fauna interpreted on the epicontinental basis. Jour. geol. 6:692-703, 2 fig. 1898.

—— Paleontology of the Niagaran limestone in the Chicago area. The Crinoidea. Chicago acad. of sci. The natural history survey. . . Bul. 4, pt 1. 1900.

Westgate, Lewis G. Geographic development of the eastern part of the Mississippi drainage system. Am. geol. 11:245-60; Abstract, Jour. of geol. 1:420, 421. 1893.

Winchell, Alexander N. Age of the Great lakes of North America. Am. geol. 19:336-39. 1897.

Woodward, R. S. On the rate of recession of the Niagara falls, as shown by the results of a recent survey. Am. ass'n adv. sci. Proc. 35:222. 1886.

Worthen, A. H. Remarks on the relative age of the Niagara and so-called Lower Helderberg groups. Am. ass'n adv. sci. Proc. 19:172-75. 1870.

Wright, G. F. Niagara river and the glacial period. Am. jour. sci. 28:33-35. 1884.

—— The ice age in North America. Appleton. 1889.

Wright, G. F. The supposed post-glacial outlet of the Great lakes through Lake Nipissing and the Mattawa river. Geol. soc. Am. Bul. 4:423-25. 1893.

—— Age of Niagara falls as indicated by the erosion at the mouth of the gorge. Abstract, Am. ass'n adv. sci. Proc. 47:299-300; Abstract, Science. n. s. 8:502; Abstract, Am. geol. 22:260, 261. 1898.

—— New method of estimating the age of Niagara falls. Pop. sci. mo. 55:145-54, 6 fig. 1899.

—— Lateral erosion at the mouth of the Niagara gorge. Abstract, Science. n. s. 10:488. 1899.

Glossary

aberrant—differing from the type
acanthopores—hollow spines occurring between the apertures on the frond of a bryozoan
acinus—a berry
adductor muscles—closing muscles in bivalve shells
agglutinate—firmly united
air-chambers—chambers below the living chamber in the shells of cephalopods
alar—pertaining to wings; the lateral primary septa of the tetracoralla
alate—having wing-like expansions
ambulacral areas—perforated areas in the test of an echinoderm, through which the tubed feet project
anastomosing—uniting to form a net work
angulated—with angles or corners
ankylosed—firmly united; grown together
annulations—rings, or ring-like segments
annulus—a ring; a segment of the thorax of a trilobite
antennae—paired articulated appendages of head of arthropod—trilobite
anterior—front
aperture—opening of shells, cells, etc.
apex—terminal or first-formed portion of gastropod shells
apophysis—a calcareous process (in interior of shells, etc.)
appressed—pressed closely against
arcuate—arched; bent like a bow
articulated—joined by interlocking processes, or by teeth and sockets
asperate—rough
attenuated—tapering; or thinning
auricle—ear, or anterior projection of the hinge of many pelecypods
auriculate—eared
aviculoid—resembling Avicula, winged
axial canal—central canal of crinoid stem
axial furrows—furrows or depressions delimiting the axis in trilobites
axis—central longitudinal division of the body of a trilobite
azygous—unpaired; the azygous side of the calyx of a crinoid has plates differing from those of the regular sides

basals—lowest cycle or cycles (in forms with dicyclic base) of plates in the crinoidea
beak—area of the apex or initial point of a shell
biconvex—both valves convex, as in most brachiopods
bifid—split in two
bifoliate—two-leaved
bifurcating—dividing in two, forking

biserial—with double series or rows
brachial—pertaining to the brachia or arms of brachiopods or crinoids; one of the arm plates of crinoids
brachidium—calcareous support of the arms in brachiopods
branchiae—gills
bryozoum—whole compound colony of the bryozoa
bulbiform—bulb-shaped
byssal notch—notch or opening for the emission of the byssus (supporting-threads spun by the foot) in the pelecypoda

calicinal—pertaining to the calyx or cup
callosity—hardened spot or area
callus—thickened part of the inner lip of gastropods, which usually covers portions of the preceding volutions
calyx—1) cup of corals, limited below by the septa; 2) body, exclusive of the arms, of crinoids, cystoids and blastoids
camerate—chambered; an order of crinoidea
camerae—air-chambers of a cephalopod shell
canaliculate—channeled; having a canal
cancellated—marked by lines crossing each other; lattice-like
carapace—hard shell or shield of crustacea
cardinal—pertaining to the area of the beak in brachiopods and pelecypods
cardinal process—process from under the beak of the brachial valve of brachiopods, to which the diductor (opening) muscles are attached
cardinal quadrants—two quadrants of a Tetracorallum which bound the main, or cardinal, septum
cardinal septum—first or main of the four primary septa of a Tetracorallum; the cardinal septum has the pinnate arrangement of the secondary septa on both sides
cardinal teeth—teeth under the beak in the pelecypods; teeth in the pedicle valve of the brachiopods
carina—projecting ridge running down the center of the branches in some fenestelloid and other bryozoa; the projecting ridges on the septa of Heliophyllum and other corals
carinated—having a ridge or keel
cartilage—compressible, elastic substance between the hinge-margins of the valves of pelecypods. The cartilage is the internal, as the ligament is the external medium for opening the valves
cast—the impression taken from a mold
caudal—pertaining to the tail
celluliferous—cell bearing (bryozoa commonly have a celluliferous and a non-celluliferous side)
cephalic limb—anterior border of the cephalon of a trilobite
cephalon—head-shield of trilobites
cephalothorax—combined head and thorax of crustacea
cercopods—lateral tail spines in the ceratiocarida

cespitose—matted, tangled or growing in low tufts

cheeks—lateral portions of the cephalon, divided into fixed and free cheeks, of a trilobite

chelae—pincer-like claw terminating some of the legs of crustacea

chilidium—covering for the chilyrium

chilyrium—triangular opening under the beak of the brachial valve in those brachiopods in which that valve is furnished with a hinge area

chitinous—composed of chitin, the substance forming the horny wings or elytra of beetles, and the carapaces of crustacea

cicatrix—a scar

cincture—depression anterior to the beak in the shell of some pelecypods

cirri—root-like appendages to the stem of crinoids

clastic—consisting of fragments, i. e. rocks made of fragments of older rocks

clavate—club-shaped

clavicle—heavy internal ridge running downward from the beak in some pelecypods

columella—central or axillary rod

compound corallum—made up of corallites, either separate or closely joined by their walls (ex. Favosites)

composite corallum—compound corallum with coenenchyma or extrathecal calcareous tissue connecting the corallites (ex. Galaxia and many other recent forms)

concavo-convex—shells of brachiopods are normally concavo-convex, when the brachial valve is concave, and the pedicle valve convex; reversed or resupinate, when the reverse condition obtains

confluent—blended so that the line of demarcation is not visible

coniferae—order of arborescent plants to which the pines, firs, etc. belong

consequent stream—type of stream which flows down the original constructional slope of the land

corallites—individual tubes of a compound corallum

corallum—calcareous skeleton of a single, or of a colonial, coral stock

corneous—horny

coronal—crown-like

costae—extrathecal extensions of the septa of the corals

costals—first brachial or arm-plates of the crinoids lying between the radials and the first bifurcation of the arms

counter quadrants—quadrants bounding the counter septum of a Tetracorallum

counter septum—front primary septum of the Tetracoralla, opposite the cardinal septum; the secondary septa are parallel to it

crenulated—notched to produce series of teeth

crura—apophyses to which the brachidium of the brachiopods is attached

cuesta—topographic relief element, resulting from the normal dissection of a coastal plain composed of alternating harder and softer strata (see p. 40)

cuneate—wedge-shaped
cuneiform—wedge-shaped
cyathophylloid—in form like Cyathophyllum; one of the Tetracoralla
cyst—a closed cavity
cystoid—most primitive class of Pelmatozoa or stemmed echinoderms

delthyrium—triangular fissure under the beak of the pedicle valve of the brachiopoda
deltidium—single covering plate of the delthyrium (also called pedicle plate)
deltidial plates—two plates which close the delthyrium in the higher brachiopoda (Telotremata)
dendroid—branching after the manner of a tree
dental plates—internal plates below the teeth in pedicle valve of the brachiopoda
denticles—small teeth, or tooth-like ridges
denticulate—toothed
denticulation—set of denticles or small teeth
depressed—on a level with, or below the general surface
dextral (right handed)—the normal method of coiling in the gastropoda
diaphragm—transverse partitioning plate
dicyclic—with two cycles of basals; applied to crinoids
diductor muscles—opening muscles of the brachiopoda
discinoid—resembling Discina
discoid—disk-like
dissepiments—partitions; the intrathecal connecting plates between the septa of the corals; the connecting bars between the branches of a fenestelloid bryozoum
distal—situated away from the center of the body
distichals—second series of arm plates or brachials of crinoids, situated above the axillary costals
divaricators—opening muscles of brachiopoda; also called diductors
dorsal—pertaining to the back
doublure—infolded margin of a trilobite

ear—anterior cardinal expansion of the pelecypod shell, usually smaller and more distinctly defined than the posterior expansion or wing
echinate—spinous
endoderm—inner cellular body layer
emarginate—with a notched margin
endoderm—inner cellular body layer
endothecal—within the theca; intrathecal; used for corals
epicontinental—encroaching on the continent
epidermal—pertaining to the skin
epitheca—outer calcareous covering of a corallum or bryozoan
equilateral—with similar sides
equivalve—with similar valves
escharoides—like Eschara (a bryozoan)

escutcheon—depression behind the beak of the pelecypod shell
exfoliate—peeling off
exothecal—outside of the theca of corals
explanate—spread out in a flat surface
extrathecal—outside of the theca of corals
extroverted—turned base to base; applied to spirals of brachiopods

facetted—having facets or numerous faces as the eye of an insect, etc.
facial sutures—sutures in the cephalon of trilobites which separate the free from the fixed cheeks
facies—local characteristics
falcate—curved like a scythe or sickle
fasciculate—clustered
fathom—a measure of length equaling 6 feet used chiefly for depths of the sea
fenestrule—open spaces between the branches and dissepiments of a fenestella frond
filament—a fine thread or fiber
fimbriae—a fringe
fixed cheek—that part of the cephalon of a trilobite which lies between the glabella and the facial suture
fission—the act of splitting or cleaving into parts
flabellate—fan-shaped
flange—a projecting rim
flexibilia—an order of crinoids characterized by the loose jointing of the plates of the calyx
fold—the central elevation of the valve, usually the brachial of a brachiopod
foliate—leaf-like; in the form of a thin leaf-like expansion
foramen—an opening or pore; specifically the opening for the pedicle in the pedicle valve of the brachiopoda
fossula—groove in the calyx of a coral, usually due to the abortion of a septum
free cheeks—lateral portions of the cephalon of trilobites separated off by the facial sutures
frond—foliaceous or leaf-like expansion of the skeleton of bryozoa and other organisms
fruticulose—resembling a small shrub
fucoid—a seaweed, particularly of the type similar to the modern Fucus, or rockweed

galeate—with a helmet-like covering
gastric—pertaining to the stomach
genal angles—posterior lateral angles of the free cheeks of trilobites
genal spines—posterior prolongations, or spines, of the free cheeks of trilobites
geode—a hollow concretion usually lined with crystals, but also filled completely with foreign mineral matter

geodiferous—containing or abounding in geodes
geodetic—geode-bearing, pertaining to geodes
gibbous—swollen or humped
glabella—central, most prominent portion of the trilobite cephalon, bounded by the fixed cheeks
glomerate—growing in dense heads or clusters, generally of an irregular character
gonopolyp—reproductive polyp of Hydrozoa
granulated—having small and even elevations resembling grains
granulose—bearing or resembling grains or granules

hexacoralla—class of corals built on the plan of six
hinge area—flat area bordering the hinge line of many brachiopods
hinge line—line of articulation
hydrocoralline—order of Hydrozoa which build calcareous skeletal structures
hydroid—animal belonging to the class of Hydrozoa
hydrotheca—cup inclosing the nutritive polyp in thecaphore Hydrozoa
hyponome—water tube, or squirting organ, of squids, cuttlefish, and other cephalopods
hypostoma—underlip of the trilobites, usually found detached

imbricate—overlapping serially
implantation—planting between, as a new plication suddenly appearing between two older ones
inarticulate—not articulating by teeth and sockets; of brachiopoda
incised—cut into
incrusting—covering as with a crust
inequilateral—having unequal sides
inface—steep face or escarpment of a cuesta, facing toward the old-land
inferior—lower in position
inflated—distended in every direction and hollow within
inflected—bent or turned inward or downward
infrabasals—lower cycle of basal plates in the crinoids with dicyclic base
infundibuliform—funnel-shaped
inosculating—connecting, so as to have intercommunication
interambulacral—between the ambulacra
interapertural—between the apertures
interbrachials—plates in the calyx of a crinoid, lying between the brachials
intercalation—irregular interposition
intercellular—between the cells or meshes
interdistichals—plates in the calyx of a crinoid, lying between the distichals
interradials—plates in the calyx of a crinoid, lying between the radials
interstitial—pertaining to an intervening space; between lines, plications, etc.
intervestibular—between the vestibules or circumscribed areas
interzooecial—between the zooecial tubes in bryozoa, etc.

intrathecal—within the theca; endothecal
introverted—turned apex to apex; applied to the spirals of brachiopods
involute—rolled up, as a Nautilus shell

joints—component segments of the stem of a crinoid
jugum—yoke-like connection between the two parts of the brachidium of a brachiopod

keel—strong central carina or ridge (Taeniopora)

lacrymiform—tear-form; drop shaped—pear shaped, but without the lateral contractions
lamellar—disposed in lamellae or layers
lamellibranch—leaf-gilled, the class of molluska with bivalved shell, to which the oyster and clam belong; pelecypod
lamelliform—having the form of a leaf or lamella
lamellose—made up of lamellae
lamina—a thin plate or scale
lateral gemmation—a budding from the sides, as in some corals
lateral teeth—ridge-like projections on either side of the beak, in the interior of lamellibranch shells
laviformia—primitive order of crinoids
ligament—external structure for opening the valves in the pelecypoda
limb—lateral area or marginal band of the cephalon of trilobites on either side of the glabella, corresponding to a pleuron of the thoracic region
lines of growth—lines marking the periodic increase in size, in shells
linguiform—tongue-shaped
linguloid—tongue-shaped; like Lingula
lip—margins of the aperture of univalve shell
listrium—depressed area surrounding the pedicle opening in the pedicle valve of Orbiculoidea and other discinoid brachiopods
lithic—pertaining to stone
living chamber—the last chamber in the shell of a cephalopod, which is occupied by the animal
lobes—backward bending portions of the suture of cephalopod shells
lophophore—ciliated or tentaculated, oral disk of bryozoa; the oral disk and brachia of brachiopods
lunarium—more or less thickened portion of the posterior wall of the cell in many paleozoic bryozoa, which is lunate or curved to a shorter radius, and usually projects above the plane of the cell aperture
lunule—depression in front of the beak of pelecypod shells

macerate—softening and disintegrating by immersion in water
macrocorallites—the larger corallites in a compound corallum
maculae—irregular, usually depressed, areas on the celluliferous face of a bryozoan frond, which are free from cells, or otherwise differentiated
mandibles—first upper or outer pair of jaws of crustacea and insects

mantle—fleshy membrane infolding the soft parts of mollusks and brachiopods and building the shell
medullary rays—the "silver grain" or radiating vertical bands or plates of parenchyma in the stems of exogenous plants
medusa—a jelly fish
membranaceous—pertaining to a membrane
mesial—central
mesogloea—central, non-cellular layer in the body of coelenterates
meso-pores—irregular meshes or cysts on the intercellular spaces of certain bryozoa
mesotheca—median wall separating opposed cells in certain bryozoan fronds
metastonia—underlip of crustacea, composed of small pieces immediately below and behind the mouth
microcorallites—smaller corallites of a compound corallum
mold—any impression of a fossil, in rock matrix, external or internal
moniliform—resembling a necklace or string of beads
monocyclic—of a single cycle
monticuliporoids—corals belonging to the order Monticuliporidae having many points of resemblance with the bryozoa
monticules—elevated areas on the surface of certain coral and bryozoan colonies, commonly carrying larger apertures
mucronate—produced into a long pointed extension
mural pores—pores in the walls of the corallites of the Favositidae
muscle scar—scar in a shell marking the former attachment of a muscle

nacreous—pearly; the nacreous layer of shells is the inner smooth pearly layer
nariform—shaped like a nostril
nasute—projecting, nose-like
nettlecell—one of the nematocysts or stinging cells found covering the tentacles and other body parts of most Coelenterata
node—knob; usually considered as ornamental
nodose—bearing nodes or tubercles
nodulose—knotty, or having nodes

obconical—inversely conical
oblate—flattened at the poles
obovate—inversely ovate or egg-shaped
obsequent stream—a stream flowing down the inface of a cuesta, or toward the old-land, tributary to the subsequent stream which in turn flows into the consequent
occipital—applied to the posterior part of the cephalon of a trilobite
occipital furrow—transverse groove on the cephalon of trilobites, which separates the last or occipital ring from the rest of the cephalon
occipital ring—posterior division of the glabella of a trilobite cephalon
operculiform—resembling an operculum
operculum—lid or cover

paddles—large or last pair of thoracic legs of the eurypterids
pallial line—line on the interior of the shell of mollusks marking the attachment of the mantle
pallial sinus—reentrant angle in the pallial line usually at the posterior end of the shell of pelecypods; it marks the attachment of the siphon muscles
palmars—third series of brachial plates of the Crinoidea, lying above the axillary distichals
palmate—palm-shaped
palpebral lobes—supra-orbital extensions from the fixed cheeks of trilobites
papilla—a small nipple-shaped protuberance
papillose—covered with papillae or fine projections
parabasals—second cycle of basal plates in crinoids
pectinated rhombs—paired pore clusters in the calyx of certain cystoids (Callocystites)
pedicle—fleshy peduncle or stem used for attachment in the brachiopoda
pedicle valve—valve which gives emission to the pedicle in the brachiopoda. Ventral of most authors. Usually the larger valve
pentameroid—five chambered, similar to Pentamerus
pentapetalous—resembling a five-petaled flower
penultimate—next to the last
periderm—outer chitinous covering of Hydrozoa
periostracum—epidermis or outer organic coating of shells
peripheral—pertaining to the circumference
peristome—margin of an aperture, i. e. the mouth of a univalve molluscan shell, the mouth of a bryozoan cell, etc.
peritheca—epithecal covering which surrounds a colony of corallites, i. e. a compound corallum
petaloid—resembling a leaf or petal
pinnate—shaped like a feather
pinnulate—provided with pinnules
pinnules—finest divisions of the arms of crinoids
plano-convex—normally in brachiopods, with the pedicle valve convex and the brachial valve flat
pleura—lateral portions of the thoracic rings of trilobites
plicate—plaited or folded
plications—folds or rib-like plaits of a brachiopod shell
polyp—animal of a simple coelenterate or bryozoan
polypite—individual polyp of a colony
pore-rhombs—pore clusters, arranged in rhombic manner in the calyx of cystoids
poriferous—pore-bearing, corals which like Favosites are furnished with several pores
posterior—situated behind

post-palmars—all the plates, superior to the axillary palmars in the arms of crinoids
prehensile—adapted for seizing
preoral—situated in front of the mouth
produced—drawn out, elongated
proliferous—reproducing buds from the calyx
protoconch—embryonic shell of a cephalous molluscan
proximal—nearest or basal portion
pseudocolumella—false columella in corals, formed by a twisting of the septa
pseudodeltidium—false deltidium (Spirifer), formed by union of the two deltidial plates
pseudosepta—septa-like ridges of Chaetetes, etc., the projecting ends of the lunaria in the cells of certain bryozoa
pseudotheca—false wall or theca in some corals, formed by the expansion of the outer margins of the septa
punctate—dotted, with scattered dots or pits
pustule—small blister-like elevation
pustulose—bearing pustules or projections
pygidium—posterior or tail portion of the carapace of trilobites
pyramidal—having the form of a pyramid
pyriform—pear-shaped
pyriformis—pear-shaped

quadrangular—four angled
quadrate—with four equal and parallel sides
quadrifid—cut into four points
quadrilobate—bearing four lobes
quadriplicate—with four folds
quincunx—five objects arranged in a square with one in the middle

rachis—central stem of a frond in bryozoa. etc.
radials—main plates of the calyx of a crinoid, resting on the parabasals, and alternating with them
radii—ribs or striations diverging from the beak of a shell
ramose—branching
ramus—branch of a skeletal structure
reniform—kidney form
resilium—internal cartilage or compressible substance in the hinge of pelecypods
reticulated—like a network
retractile—capable of being withdrawn
retral—backward
rhynchonelloid—resembling Rhynchonella
root—expanded basal portion of a crinoid stem, used for fixation
rostrum—a beak or snout
rugosa—an old name for the Tetracoralla

saddles—forward bending portions of the suture in the shells of cephalopods
salient—standing out prominently
scabrous—rough or harsh with little projecting points
scalae—small transverse plates in the genus Unitrypa of the bryozoa
scalariform—stair or ladder-shaped
sclerenchyma—calcareous tissue deposited by the coral polyps
scorpioid—scorpion-like, coiled like the tail of a scorpion
semilunar—crescentic, or resembling a half moon
semiovate—half egg-shaped
senile—pertaining to old age
septal radii—radiating ridges taking the place of septa in certain corals
septate—with partitions or septa
septum—partition; in corals, the radiating calcareous plates; in cephalopods, the transverse partitions between the chambers
serrate—notched like a saw
setiferous—bristle-bearing
sigmoid—curved like the Greek letter Σ (sigma)
sinistral—left handed, reversed coiling of some gastropod shells
sinuate—wavy, winding
sinuosity—notch or incision forming a wavy outline
sinus—impression in the surface or margin of a shell
siphonal funnel—siphonal projection from the septum of a cephalopod shell
siphonal lobe—lobe in the suture of an ammonoid shell, corresponding in position to the siphuncle
siphuncle—tubular canal passing through the air chambers in the shells of cephalopods
slickensides—polished or striated surfaces on rock due to motion under great pressure
sockets—hollows in the brachial valve of brachiopods for the reception of the teeth of the opposite valve
spatulate—shaped like a spatula; spoon-shaped
spheroidal—globose, of the form of a spheroid
spiniform—spine-like
spinulose—spine bearing
spondylium—spoon-shaped cavity under the beak of pentameroid brachiopods
squamous—scaly, covered with scales
stalk—stem of crinoids
stellate—star-shaped; arranged in star-like manner
stipe—stalk or stem in plants
stock—main stem or trunk
striae—fine radiating surface lines of shells
stylolites—peculiar columnar and striated rock form seen in limestones at the junction of two layers
sub—in composition indicates a low degree: sub-angular—rather angular; sub-carinate—somewhat toothed, etc.

subfusiform—more or less spindle-shaped
subglobose—more or less globose
sublunate—approaching the form of a crescent
suborbicular—nearly circular
subpentahedral—irregularly five-sided
subpyramidal—approximately pyramid-shaped
subquadrangular—between quadrangular and oval
subquadrate—nearly but not quite square
subspheroidal—imperfectly spheroidal
subtruncate—irregularly cut off
subturbinate—approaching top shape
sulcation—a furrow or channel
sulcus—a furrow
superior—higher in position
suture—in cephalopods, the line of junction between shell and septum, seen on breaking away the former; in gastropods, the external line of junction between the several whorls; in trilobites, the dividing line between fixed and free cheks, commonly called *facial suture;* in crinoids, the line of junction between adjacent plates

tabulae—transverse, continuous partitions or floors in corals, etc.
tabulate corals—group of corals in which the tabulae cross plates are prominent, while the septa are faintly or not at all developed e. g. Favosites, Aulopora, etc.
talus—the mass of rocky debris which lies at the base of a cliff, having fallen from the face of the cliff above
teeth—articulating projections on the margins of the valves of bivalve shells
tegmen—vault or cover of the calyx in crinoids
terebratuloid—like the recent genus Terebratula
terete—cylindric or slightly tapering terrigenous—derived from the land test-shell
tetracoralla—the old group of rugose corals, built on the plan of four
tetrameral—on the plan of four
theca—the proper wall of the individual corals
thoracic—pertaining to the thorax
thorax—central part of the body of the trilobites
trabeculae—projecting bars
trigonal—three-angled
trihedral—with three equal faces
tripartite—divided into three parts
tripetalous—three leaved or petaled
trochiform—in form like a Trochus or top shell
tubercle—small swollen projection
tuberculiform—in form like a tubercle
tuberculous—having or resembling tubercles
tubicola—an order of marine worms which build calcareous or other tubes
tumid—swollen, inflated
turbinate—top-shaped

vestibular area area surrounding the cell apertures of some bryozoans
depressed
viscera the internal organs of the body

whorl single volution of a coiled shell
wing [illegible] larger expansion along the hinge-line of a pelecyp

zoarium aggregation of the polypides of a bryozoan colony
zooecium the bryozoan cell
zooid one of the "persons" or individuals of a zoarium